THE OCEAN BASINS AND MARGINS

Volume 4 A
The Eastern
Mediterranean

THE OCEAN BASINS AND MARGINS

THE OCEAN BASINS AND MARGINS

Edited by
Alan E. M. Nairn and William H. Kanes

Department of Geology
University of South Carolina
Columbia, South Carolina

and

Francis G. Stehli

Department of Geology
Case Western Reserve University
Cleveland, Ohio

Volume 4A
The Eastern
Mediterranean

PLENUM PRESS · NEW YORK AND LONDON

Library of Congress Cataloging in Publication Data

Nairn, A   E   M
   The ocean basins and margins.

   Includes bibliographies.
   CONTENTS: v. 1. The South Atlantic.–v. 2. The North Atlantic. v. 3. The Gulf
of Mexico and the Caribbean.–v. 4A. The Eastern Mediterranean.–v. 4B. The
Western Mediterranean.
   1. Submarine geology. 2. Continental margins. I. Stehli, Francis Greenough,
joint author. II. Title. QE39.N27   551.4'608   72-83046
ISBN 0-306-37774-8 (v. 4A)

© 1977 Plenum Press, New York
A Division of Plenum Publishing Corporation
227 West 17th Street, New York, N.Y. 10011

Printed in the United States of America

CONTRIBUTORS TO THIS VOLUME

Zvi Ben-Avraham
Israel Oceanographic and Limnological
    Research
Haifa, Israel

Daniel Bernoulli
Geological Institute of the University
Basel, Switzerland

Ziad R. Beydoun
Department of Geology
American University of Beirut
Beirut, Lebanon

Paul Celet
Département de Géologie Dynamique
U. E. R. des Sciences de la Terre
Université de Lille
Villeneuve d'Ascq, France

E. M. El Shazly
Egyptian Atomic Energy Establishment
Academy of Scientific Research and
    Technology
Cairo, Egypt

Kenneth J. Hsü
Geological Institute
ETH Zurich, Switzerland

Emin Ilhan
Konur Sok 44/11
Yenisehir
Ankara, Turkey

Hans Laubscher
Geological Institute of the University
Basel, Switzerland

Jennifer M. Lort
Deepwater Studies Group
British Petroleum Co., Ltd.
London, England

J. K. Melentis
Department of Geology and
    Paleontology
University of Thessaloniki
Thessaloniki, Greece

David Neev
Israel Geological Survey
Jerusalem, Israel

Nuriye Pinar-Erdem
Department of Geology
State Academy of Engineering and
    Architecture
Yildiz-Istanbul, Turkey

David A. Ross
Woods Hole Oceanographic Institution
Woods Hole, Massachusetts

Daniel Jean Stanley
Division of Sedimentology
Smithsonian Institution
Washington, D. C.

CONTENTS OF VOLUME 4A

Chapter 3. Post-Miocene Depositional Patterns and Structural Displacement in the Mediterranean

Daniel Jean Stanley

Chapter 4. Geophysics of the Mediterranean Sea Basins

Jennifer M. Lort

Chapter 5A.   **The Dinaric and Aegean Arcs: The Geology of the Adriatic**

Paul Celet

Chapter 5B.   **The Dinaric and Aegean Arcs: Greece and the Aegean Sea**

J. K. Melentis

Chapter 6. **Outlines of the Stratigraphy and Tectonics of Turkey, with Notes on the Geology of Cyprus**

Nuriye Pinar-Erdem and Emin Ilhan

Chapter 7A. **The Levantine Countries: The Geology of Syria and Lebanon (Maritime Regions)**

Ziad R. Beydoun

Chapter 7B. The Levantine Countries: The Israeli Coastal Region

David Neev and Zvi Ben-Avraham

Chapter 8. The Geology of the Egyptian Region

E. M. El Shazly

Chapter 9. **The Black Sea and the Sea of Azov**

David A. Ross

CONTENTS OF VOLUME 4B

xv

Chapter 1

MEDITERRANEAN AND TETHYS*

Hans Laubscher and Daniel Bernoulli
Geological Institute of the University
Basel, Switzerland

I. INTRODUCTION

The boundary between the Eurasian and the African plates, formerly the suture between Eurasia and Gondwana, has been the locus of violent tectonic diastrophism and rapidly changing geography since the Triassic. The Mesozoic seas, and sometimes the Paleozoic seas, of this zone and its extension into the Himalayan region are known as the Tethys (Neumayr, 1883; Bittner, 1896; Suess, 1893, 1901; cf. e.g., Kamen-Kaye, 1972), while Tertiary seas are usually called the Mediterranean. From the viewpoint of plate tectonics, it would appear appropriate to talk in general of the African–Eurasian boundary seas. We can try to trace the history of the Tethyan or Mediterranean seas from the breakup of Pangea at the end of the Triassic through the Mesozoic and Cenozoic. The most ambitious attempt to do this has been by Dewey *et al.* (1973) as a sequel to a model for the opening of the Atlantic Ocean proposed by Pitman and Talwani (1972). However, although the general postulates are valid and, within the framework of plate tectonics, even obvious, the actual implementation of this kinematic jigsaw puzzle is very difficult. It is also ambiguous because of large gaps in information and in the differences in language and interpretation by the various investigators. Indeed, at present, there is no

---

* Contribution No. 5 of the International Geological Correlation Programme, Project 105, Continental Margins in the Alps.

1

model that would not be seriously questioned by one part or another of the earth science community.

The authors, who are land geologists, but also have experience in geophysics and marine geology, will attempt to restate some of the more important rules that, in their opinion, should guide the construction of a model. We will attempt to sketch a general outline of the history of the Africa–Europe boundary region, using these rules.

The rationale of our reconstruction is based on the following restrictions: (1) The initial (Triassic) fit between Europe, North America, and Africa is hereby accepted as proposed by Dewey *et al.* (1973), except for the location of microplates in the belt presently deformed by the Alpine mountain building. On geological grounds, we believe that these were arranged quite differently. (2) Obviously, all the past movements have resulted in the present relation of the three continents, and in the present structure of the intervening fault and nappe systems. The movements within these nappe systems must have produced the present arrangement of ophiolite belts (interpreted as the remnants of oceanic crust and mantle) and of the nappes containing continental margin sediments from the European and African sides of the ophiolite seas. Moreover, the direction and amount of the latest motions that took place in the Alpine systems, those of Tertiary age, must be acknowledged. (3) There is a belt of mainly Mesozoic carbonates which may be followed from Morocco through Sicily, the Apennines, the southern Alps, the Austroalpine nappes, the internal western Carpathians, the external Dinarides and Hellenides, and into the Lycian Taurus of western Turkey. This carbonate belt invariably lies south of the ophiolite belt, except for the ophiolites of the Antalya nappes in southwestern Turkey and those of the Cyprus–Hatay belt (Fig. 1). There is, as yet, no certainty as to the original position of these last two allochthonous ophiolites (Brunn, 1974; Monod *et al.*, 1974) although many geologists now seem to think that they originated from north of the carbonate belt (Ricou *et al.*, 1974). So far as we can see, this carbonate belt could have been interrupted by deep sea with oceanic crust in only three places; there its continuity is presently questionable: (a) Along the Malta escarpment, with a connection below the Peloritan nappes into the Pennine–Ligurian juncture. (b) Through the Antalya nappes into the Isparta corner, in a maze of superposed nappes, and well hidden below Tertiary and Quaternary rocks. (c) Along the Levantine–Van system.

A direct connection of the Vardar zone with the eastern Mediterranean, as proposed by Dewey *et al.* (1973), seems inconceivable to us.

We call the Mesozoic carbonate belt the "southern continental margin of Tethys," and use its continuity as a guide in constructing our model. To the north was the main Tethyan Ocean, which may have had an intricate morphology. The ocean had a "northern continental margin" which, though complexly block-faulted and dissected by minor zones of movement, may be recognized as a continuous element. The smoothing out of nappes and folds that comprise the continental margin rocks, with their essentially continental basement, leads to a minimum figure on the order of 500 km for the original width of the belt of continental crust now compressed in the Alps (cf. Trümpy, 1960; Laubscher, 1970). Its width may have been considerably greater farther east, and narrower farther west. The evidence in the Alps is clear, and there is no physical law limiting the amount of continental crust that may have been subducted, though on the whole, the assumption that oceanic crust and lithosphere are most easily subducted is reasonable and supported by world-wide evidence. However, for a narrow sea such as the Tethyan, a few hundred kilometers of subducted continental crust are essential, even though they might be quite negligible for the Pacific Ocean.

Beside these main pillars of our reconstruction, there are a few additional considerations. They concern the independent mobility of microplates and the origin of isolated basins. Although Dewey *et al.* (1973) devote some effort to subordinating relative movements of microplates to the overall movements of the superplates, they occasionally make exceptions. One of these is the independent Tertiary rotation of Corsica and Sardinia, which has absolutely no

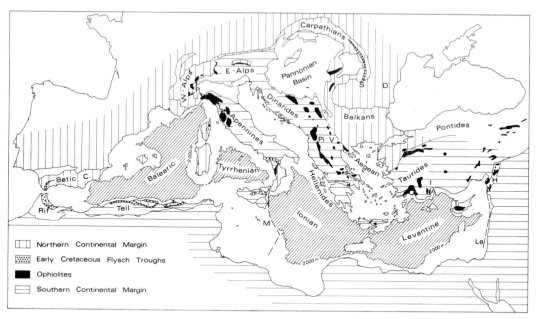

Fig. 1. Occurrences of Tethyan ophiolites with associated oceanic sediments and of the Mauretanian–Massylian–South Ligurian Flysch. Also shown is the interpreted extent of the northern and southern continental margins of the Tethys Ocean. Bathymetry of the Mediterranean basins after *The Mediterranean Sea* (Defense Mapping Agency, Hydrographic Center, Washington, D.C., 1972). Localities mentioned in the text: (A) Antalya Nappes; (D) Dobrodgea; (H) Hatay; (I) Isparta; (L) Ligurian ophiolites; (Le) Levantine Escarpment; (M) Malta Escarpment; (P) Pennine (Piedmont) ophiolites; (Pi) Pindos; (S) Sinaia Flysch; (T) Troodos; (V) Vardar zone.

relation to the overall movements (Laubscher, 1975), but is required to produce the Balearic Basin by sea-floor spreading. Similarly, attempts are made to show that the Po Basin and the Pannonian Basin were produced by spreading, but also in these cases there is much direct geological evidence against it. Thus, many parts of the Pannonian Basin, and particularly of its margins, are well known, e.g., the internal Vienna Basin and the Styria Basin, and they show the foundering of the continental crust along large, normal faults, without significant horizontal spreading. We believe that unprejudiced acceptance of this information, though it is incomplete, leads to the proposition that basins may form by foundering of the crust, and that as geophysical data show the continental crust to be thin below such basins, there must exist a process such as crustal stoping (Stegena *et al.*, 1974). Another example is found in the Aegean Sea, which is in the process of foundering, yet is far from being oceanic. Still another example is the Tyrrhenian Sea, the marginal foundered basins of which are typical for Tuscany. Although we do not propose foundering to be a major process competing with spreading in the major oceans, unbiased weighing of present information would seem to require it for a number of small Mediterranean and marginal basins. We will not introduce arbitrary spreading in order to produce such basins. However, although rational tectonics require subordination of individual movements to those of the superplates, there are some intriguing paleomagnetic data, particularly those concerning a rotation of Italy and the southern and eastern Alps (Soffel, 1975; Lowrie and Alvarez, 1975; Premoli Silva *et al.*, 1974) which, at this time, defy such an integration.

II. THE OVERALL SITUATION

The development of the Tethys Ocean and its margins is only one aspect of the general breakup of Pangea and Panthalassa in the Mesozoic. In particular, Tethyan development is intimately linked with the opening of the Atlantic Ocean in the west and with the independent motion of plates in the Himalaya–India region, which eventually resulted in the replacement of Panthalassa by such Mesozoic and Tertiary Oceans as the Indian. Of these oceans, only the Atlantic has a comparatively simple history, decipherable with some confidence from its magnetic lineations, although even here terrestrial geological data require one to keep an open mind. The dominance of Atlantic influences seems to be restricted to that segment of the Tethys west of the Vardar zone. To account for structural developments farther east, one is forced to call on the poorly-defined but independent movements of Himalaya–Indian plates. Consequently, the restrictions imposed on Mediterranean kinematics by global tectonics are in reality still quite weak. In addition, the nearly complete disappearance of the wedge-shaped space between Arabia and Russia (see Fig. 2), whatever its initial nature, is still hard to understand (Argyriadis, 1975).

In this article we accept the disappearance of Panthalassa and its continental margin, whatever its width, and particularly of that part occupying the wedge between Arabia and Russia, but we do not wish to enter into any arguments about the exact location of the required sutures (Adamia, 1975), even though they must be present somewhere in the belt composing the Caucasus and the Pontides. There is also a nomenclatural question: The area that has been called "Tethys I" by Dewey *et al.* (1973), and for which there is no direct evidence, was an embayment of Panthalassa according to Fig. 2. It is unfortunate that in the literature the faunas and sediments from this embayment are often called "Tethyan" (see below), for it seems advisable to restrict the term "Tethys" to the Mesozoic breakup of Pangea and refer to the Paleozoic embayment as "Paleotethys." This term ought to be sufficiently different, and at the same time keeps to time-honored usage.

III. TETHYAN FAUNAS AND PALEOTECTONICS

The Mesozoic seaways between the Eurasian and African plates were situated along an east–west-trending equatorial belt, as were their Paleozoic forerunners north of Gondwana (Smith *et al.*, 1973; cf. Fig. 2). It is to this tropical belt that the term "Tethyan," as used in paleontological literature, usually refers, often regardless of the paleotectonic or chronostratigraphic context. The so-called Tethyan faunas, as opposed to Boreal or Pacific faunas, may thus simply imply tropical or subtropical Paleozoic seas including the Permo-Triassic embayment of Panthalassa, here called "Paleotethys," or they may refer to the Mesozoic Tethys and its branches into the Himalayan region, or to its Tertiary successor basins (Adams and Ager, 1967). We propose that in paleobiogeography the term "Tethys" also be restricted to faunas related to the Mesozoic Tethys.

In addition to such questions of definition, there are other difficulties in the correlation of faunal character and paleotectonics. One part of these difficulties is due to uncertainty in the palinspastic location of paleontological sampling localities. Another stems from the fact that ecology is related to paleotectonics in a very indirect way, e.g., only by complex current patterns for which geomorphology merely serves as a geometric boundary. As an example of the first kind of difficulty, many publications up to the mid-1960's show sample localities on small-scale maps too vague for tectonic and palinspastic assignment. Since then, a sounder series, kinematically and palinspastically, of faunal distribution maps, based on plate tectonic concepts, have been published. However, although the different continental margins of the Tethys are distinguished (Enay, 1972, 1973; Hallam, 1971, 1975), the causal correlation with plate tectonics is not unique for the reasons stated above. Consequently, there is no point in making more than a few very general remarks.

During the Permian and the Triassic, faunal evidence merely suggests the existence of extensive epeiric seas bordering the Paleotethys (Gobbett, 1967, 1973). During the Jurassic there are characteristic associations of bivalves (*Lithiotis*) and benthic foraminifera (*Orbitopsella*) along the southern margin of the Tethys (Hallam, 1975). According to Ager (1973) some early Jurassic assemblages of brachiopods are restricted to certain areas of the southern Tethyan margin. However, from the bulk of the data, one gets the impression that during the early (Jurassic) stages of Tethyan evolution the control of faunal distributions was exerted largely by local ecological factors rather than by geomorphological (tectonic) barriers, such as deep basins. Thus, in the Early Jurassic, ammonites typically found along the southern Tethyan margin dispersed over large areas of the northern epicontinental seas (Donovan, 1967; Howarth, 1973), and mixed faunas of "Tethyan" and "European" forms occurred throughout the Jurassic. In the Late Jurassic, the Tethys certainly was no physical barrier to the migration of Tethyan ammonites (Enay 1972, 1973). On the other hand, the opening to the central Atlantic Ocean allowed many Tethyan forms to migrate to the Caribbean and Central America (Enay, 1973; Geyer, 1973). Thus Ager's statement (1967, p. 139) that "the Pygopidae (Tethyan brachiopods) were largely restricted to the Mediterranean part of the Tethys simply because it provided them with the right habitat" can be extended to other Jurassic forms, e.g., the phylloceratids and lytoceratids, which most probably inhabited more open and deeper seas, or to many planktonic forms such as calpionellids and radiolarians, which are sometimes mentioned as typical of the Tethys (Hallam, 1975). It therefore seems that the terms "Tethyan," "Alpine," or "Mediterranean" facies in the Jurassic often merely imply pelagic deposits as they are found along the opening ocean. Geographic separation seems to become more important with the expansion and the deepening of the central Atlantic and the opening of the southern Atlantic during the Cretaceous. This, and the isolation of a number of shallow-water domains in the southern Tethyan margin during the Jurassic, according to some authors, resulted in increasing endemism and diversification of the Tethyan shallow-water faunas, particularly from the Albian onward (Dilley, 1973; Kauffman, 1973). However, the data are fragmentary and of doubtful value for paleotectonic reconstructions.

IV. PALEOTECTONIC AND PALEOGEOGRAPHIC EVOLUTION

A. Permian and Triassic: Paleotethys and Early Rifting

Figure 2 shows the reassembly of the principal continental masses in the Triassic (Smith *et al.*, 1973). This global situation has many implications in the development of the Tethys–Mediterranean belt. The first is that Panthalassa

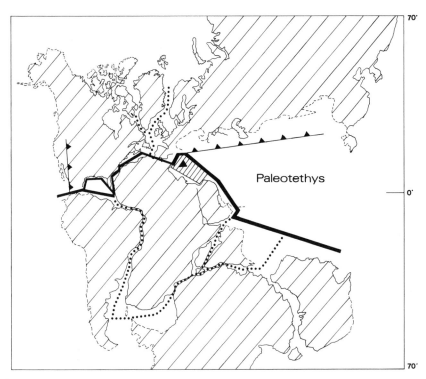

Fig. 2. Paleogeography of the principal continental masses in the Triassic, about 220 ± 20 m.y.b.p., after Smith *et al.* (1973). The heavy black line indicates the later east–west-oriented Mesozoic breakup of Pangea; the barbed lines the approximate locations of Mesozoic sub-duction zones; and the dotted lines the Cretaceous–Tertiary sea-floor spreading centers. The closely cross-hatched area indicates the eastern Mediterranean as a possible relic of Paleotethys, and the triangle shows the location of the Mesozoic triple point between the Eurasian, African, and Paleotethys plates. Discussion in the text.

had a wide embayment into the space between Asia and Arabia, with extensive marginal seas that reached into the western Mediterranean area. This entire ocean was subsequently consumed, along with continental margins of unknown width, in connection with the opening of the Indian Ocean and its forerunners in the Paleotethys domain, which, in turn, have been partly obliterated by subsequent plate movements. Viewed in this global perspective, the ophiolite belts from the Vardar zone to Oman are vestiges of such forerunners of the Indian Ocean. It is evident that the Mediterranean area has been influenced as much by little-known developments in the Paleotethys–Indian domain as by the better-known ones of the Atlantic. In the same way that spreading move-ments which shifted from the Caribbean–central Atlantic–western Tethys belt to a north–south-trending modern Atlantic profoundly influenced movements in the western Tethys, the little-known wanderings of the centers of spreading in Asia must have influenced those movements in the eastern Tethys.

Figure 3 shows the Mediterranean part of the reassembly shown in Fig. 2. Except for the microplates the relative positions of the continental fragments have been drawn according to Dewey *et al.* (1973). During the Permian and Triassic, marginal seas reached as far west as Sicily and Tunisia (Argyriadis, 1975). During the Early Triassic, the rapid transgression of a shallow epeiric sea from the east (Assereto *et al.*, 1973) initiated marine conditions over most of the Mediterranean area as far as the internal zones of the Rif and the Betic Cordilleras. Evaporites and thick neritic limestone deposits record an increasing marine influence which extended into the otherwise clastic Germanic facies

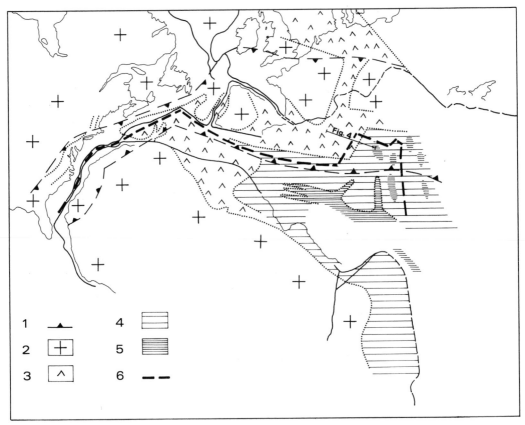

Fig. 3. Palinspastic restoration of the Late Triassic paleogeography of the Mediterranean and central Atlantic areas. The positions of the North American, Eurasian, and African continents are drawn after Dewey *et al.* (1973). Based upon data by Choubert and Faure-Muret (1962), Desmaison (1965), Druckman and Gvirtzman (1975), Jansa and Wade (1974), Wurster (1968), Ziegler (1975), and others. (1) Front of Variscan orogen; (2) continental areas with erosion or local continental sediments; (3) basins occasionally flooded by marine waters including important evaporite deposits; (4) strongly subsiding basins with thick carbonate platform (and evaporite) deposits; (5) major basins with deep-water facies; (6) location of Jurassic breakup and sea-floro spreading.

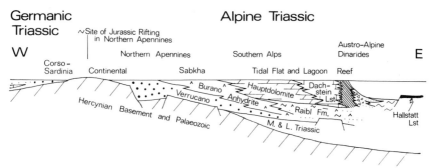

Fig. 4. Paleogeographic profile through the Late Triassic deposits of the central Mediterranean area (for location, see Fig. 3), modified after Bosellini (1973). The section combines depositional areas now occurring in different tectonic units into one palinspastic profile in order to show the general arrangement of facies types (not to scale). The main feature is a wedge of late Triassic sediment transgressing westward onto Hercynian basement.

(Muschelkalk salt and limestones) from time to time and fed the important evaporite deposits of the Canadian and Moroccan Basins (Jansa and Wade, 1974). The neritic deposits of the "Alpine" Triassic may reach a thickness of several thousand meters. These shallow carbonate platforms are interspersed with deeper-water pelagic limestones, cherts, carbonate turbidites, and volcanic sandstones and submarine volcanics (Bosellini and Rossi, 1974). Some of these basins were quite small and short-lived, but others were more extensive and permanent, often persisting throughout the Mesozoic. [The basins are: Baër–Bassit (Lapierre and Parrot, 1972), Mamonia (Lapierre and Rocci, 1969), Antalya (Marcoux, 1970), Pindos (Aubouin, 1959), Lagonegro (Scandone, 1967), and Sicily (Broquet et al., 1966).] Figure 4 shows the typical arrangement of facies belts during the Late Triassic (for general location, see Fig. 3). The later (Jurassic) rifting is slightly discordant with respect to the distribution of Triassic facies: in the transection between the central Apennines and Corso-Sardinia, the rifting coincides with the boundary between the Germanic Muschelkalk province and the "Alpine" Triassic, whereas in the central Alps the later suture lies within the "Alpine" Triassic limestone belt (cf. Figs. 3 and 6) (Laubscher, 1974).

The original width of these marginal belts is unknown, but it must have been considerable to judge from their extensive occurrence in the nappes of the Alpine system. The expanse of the oceanic part of Panthalassa on Fig. 2 is not known, but the pelagic Triassic troughs with their associated submarine volcanics and sedimentary material may indicate its proximity. In particular, there is a possibility that the eastern Mediterranean Basin is, in part, a relic of Panthalassa. Although direct evidence is lacking, the several thousand meters of marine Permian in Libya (Bishop, 1975) suggest this. Also, the basements of some of the decollement nappes with pelagic Triassic deposits are unknown (e.g., Pindos), and there is a remote chance that they have been oceanic. Thus, although there is no compelling evidence for Permo-Triassic oceanic domains in the Mediterranean region, the possibility cannot be ruled out.

It is instructive to compare the pattern of Mesozoic evolution with the outlines of the Late Paleozoic (Variscan) mountain belt, the internal parts of which show metamorphic and igneous activity (Argyriadis, 1975). As the creation of the belt supposedly led to the assembly of Pangea, it lies mostly in the interior of the landmass. The Mesozoic breakup roughly followed this belt in the central Atlantic and in the western Mediterranean, but deviated in the eastern Mediterranean. Similarly, the splitting up of the landmass shown in Fig. 3 by sea-floor spreading in the Jurassic did not follow exactly the Triassic paleotectonic trends (cf. Figs. 3–7). It left most of the thick marine Triassic of the "Alpine" facies on the southern (African) plate, though in some areas (internal units of Rif and Betic Cordillera, central Pennines) a part remained with the northern (European) plate (Laubscher, 1974). In Fig. 3, the future main plate boundary (central Tethys) and those belts of the pre-Jurassic arrangement attributable to the northern and southern continental margin complexes, respectively, are indicated. Attention is also drawn to the fact that the European as well as the African plates have been segmented by Triassic furrows and fault structures similar to those recognizable in Appalachian North America. Such features, often accompanied by intense volcanic activity, include the Lusitanian, Pyrenean–Aquitaine, and the Argana (Morocco) Troughs, and in Europe the Germanic Trough and the Tornquist–Dobrodgea Lineament. It seems that some of them are forerunners of the approximate location of future plate or sub-plate boundaries.

The formation of grabens, with accompanying volcanic activity, along the eastern margin of North America, in Morocco, and in southwestern Europe has generally been interpreted as early rifting, preceding sea-floor spreading in the central Atlantic (Ballard and Uchupi, 1975) and in the western Tethys. Similarly, the occurrence of down-faulted basins with associated alkaline volcanic deposits in the eastern Mediterranean seems to be related to early rifting in the Paleotethys embayment (Juteau *et al.*, 1973).

B. Jurassic: The Oceanic Tethys

During the Jurassic period the Atlantic–Tethyan system developed from the stage represented in Fig. 3 to the one shown in Fig. 5. Magnetic anomalies and deep-sea drilling results suggest initiation of sea-floor spreading and formation of a young oceanic crust in the central Atlantic during the Early Jurassic, about 180 million years ago (Pitman and Talwani, 1972). The opening of this part of the Atlantic obviously implies relative movements between

Africa and Eurasia (Dietz and Holden, 1970) and is roughly contemporary with the formation of oceanic crust and lithosphere in the Ligurian–Piedmont ophiolite zone, as indicated by radiometric dating (Bertrand, 1970; Bigazzi *et al.*, 1972) and the approximate age of the oldest overlying sediments (Decandia and Elter, 1972). A Jurassic age has also been established for many of the ophiolites in the eastern Mediterranean area (Moores, 1969; Hynes *et al.*, 1972). Tectonic movements were then essentially transcurrent, or by a transform fault, with minor rifting from Iberia to the Apennines. The fault seems to be located in the Mauretanian–Massylian Flysch Trough (Didon *et al.*, 1973). In places, the marginal areas bordering this trough have been pelagic from the

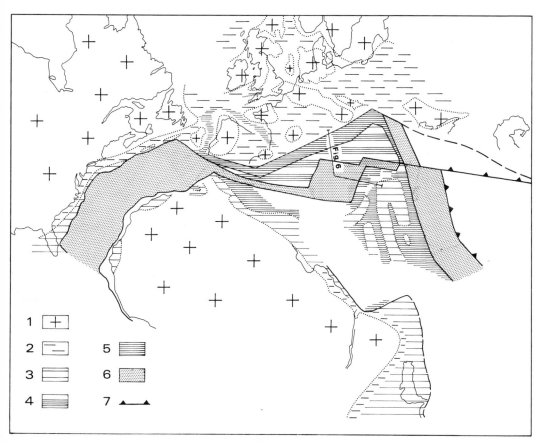

Fig. 5. Palinspastic restoration of the Late Jurassic (Kimmeridgian) paleogeography of the Mediterranean and central Atlantic area, based upon data by Bernoulli (1972), Bismuth *et al.* (1967), Bonnefous (1967), Choubert and Faure-Muret (1962), Enay (1972), Jansa and Wade (1974), Ziegler (1975), and others. The positions of the North American, Eurasian and African continents drawn after Dewey *et al.* (1973). (1) Continental areas with erosion or local continental sediments; (2) generally thin, shallow marine deposits (marls, limestones and subordinate sandstones); (3) thick carbonate platform deposits; (4) unstable, block-faulted areas with thin, pelagic or locally restricted reefal limestones; (5) basinal pelagic deposits; (6) Tethys Ocean; (7) Jurassic subduction zones.

Sinemurian onwards (Dorsale externe, Prédorsale) while flysch sedimentation was initiated south of this zone as early as the Late Jurassic (Andrieux, 1971). Although no ophiolites are known from the central flysch trough (only minor spilitic volcanics are associated with the pelagic sediments at the base of the Massylian sequence), the close similarity of its Early Cretaceous flysch sequence with the South Ligurian sequences of Sicily (Durand Delga, 1960) and the Apennines (Vezzani, 1973, and references therein) and with the Early Cretaceous flysch of the northeastern central Atlantic (Lancelot *et al.*, 1975) and of the Canary Islands (Bernoulli *et al.*, in prep.) supports this interpretation. In contrast to Dewey *et al.* (1973), we find no evidence for oceanic Jurassic basins in the Middle and High Atlas Mountains. Although there was some Early Cretaceous volcanic activity, it was part of a continental episode apparently associated with the formation of half-grabens similar to those of the Appalachian Triassic. There were minor sinistral movements with the creation of small *en échelon* extensional troughs throughout the Mesozoic. The main dextral movements in the Atlas system occurred approximately, in the Late Tertiary–Quaternary.

East of Corso-Sardinia, which we believe was part of the Iberian block, the plate boundary swung to the north and parallel to a true ridge segment where ophiolites formed in the Ligurian–Pennine Trough. The margins of this trough, particularly in the southeast display a paleotectonic and sedimentary evolution closely parallel to that of passive Atlantic-type continental margins (Fig. 6). In the Early Jurassic, synsedimentary normal faulting was widespread in the northern and southern continental margins (Bernoulli and Jenkyns, 1974), and was followed by submergence of many of the former shallow-water platforms, and the inception of pelagic conditions over large areas (for an undeformed counterpart compare Fig. 6 in Bernoulli, 1972, and our Fig. 6 with Veevers, 1974, Fig. 8). Pelagic deposits are particularly frequent along the southern continental margin of the Tethys, and this may be related to arid conditions along the African margin, where evaporites and shallow-water carbonates predominate. However, pelagic "alpine" facies occur on submarine highs and in sheltered furrows along the northern margin, where terrigenous material was trapped in intervening troughs (pelagic deposits in the Brianconnais, some Carpathian zones, the Subbetic Trough; Fig. 6; and Bernoulli and Jenkyns, 1974).

In some areas carbonate platforms persisted throughout the Mesozoic and their spatial arrangement, subsidence rate and facies distribution closely resemble those of the Bahamian platform with its intervening troughs of pelagic and turbiditic carbonate sediments (Bernoulli, 1972; d'Argenio, 1970; d'Argenio *et al.*, 1975). There are also parallels in the evolution of the oceanic central Atlantic and the Ligurian ophiolite zone, at least from the Late Jurassic onward. After pronounced depression of the calcite compensation depth caused by the bloom of coccolithophorids in the latest Jurassic (Garrison and Fischer, 1969), a parallel sedimentary and bathymetric evolution of the central Atlantic and the Ligurian zone is suggested by comparing Sclater curves and the solution facies in both regions (Fig. 7; for the Tethyan part see also Bosellini and Winterer, 1975), although some discrepancies should be expected because of the difference in boundary conditions.

Ophiolites become less important in the eastern Alps (Glockner facies in the Tauern window, Tollmann, 1975), but reappear as impressive masses in the Dinarides and Hellenides. In these mountains, several authors (Dimitrijevic and Dimitrijevic, 1973; Smith, 1971) have argued that there were at least two ocean basins separated by intervening microcontinents. However, these arguments and their possible consequences have not yet been worked out in detail. For this reason, we show the simple version, with a central Tethys occupying what is now known as the Almopias subzone of the Vardar zone (Mercier,

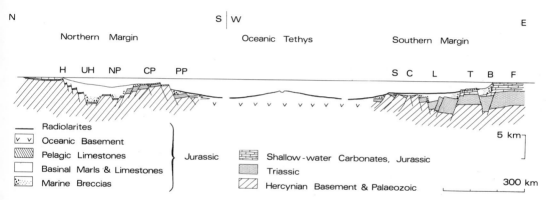

Fig. 6. Palinspastic section through the Liguria–Piedmont Ocean in the Late Jurassic (Kimmeridgian; for location, see Fig. 5). The main features are: (1) Thick "alpine" shallow-water sediments in the Triassic of the central Pennine area (CP) on the northern continental margin (see page 9); (2) the early Jurassic phase of block-faulting in the embryonic continental margins, well documented by marine breccia formations in the Ultrahelvetic–North Pennine (UH and NP) Trough and the southern scarp bordering the central Pennine platform (CP) (Trümpy, 1960; Lemoine, 1975), in the Austro-Alpine nappes (Trümpy, 1975) and in the southern Alps (Bernoulli, 1964; Castellarin, 1972; Kälin and Trümpy, 1977); (3) the onlap of pelagic deep-sea sediments over the deeply submerged continental margins following the initial phase of block-faulting. These sediments are characterized by a sequence of solution facies indicating prolonged subsidence of the distal continental margins (cf. Bosellini and Winterer, 1975); (4) the development of Bahamian-type carbonate platforms along the southern continental margin with adjacent troughs with pelagic and turbiditic carbonate sedimentation; (5) the development of the Tethys Ocean, documented by the oceanic ophiolite suite. Its width and shape are somewhat arbitrary, but 1000 km as a rough figure is consonant with the history of sea-floor spreading in the Atlantic as derived by Pitman and Talwani (1972) (cf. Dewey et al., 1973). It could be reduced by a transform fault into the eastern Mediterranean, or increased by changes in the position of the poles of rotation or influences of Paleotethys plate movements (cf. Fig. 8). Jurassic magmatism in the continental margin of the southern Alps (Casati et al., 1976) and in the North Pennine Trough (Dietrich and Oberhänsli, 1975; see also Laubscher, 1970) has been omitted. (H) Helvetic; (UH) Ultrahelvetic; (NP) North Pennine; (CP) Central Pennine; (PP) pre-Piedmont zone; (S) Sesia zone—Dent–Blanche nappe; (C) Canavese zone; (L) Lombardia zone; (T) Trento zone; (B) Belluno zone; (F) Friuli zone.

1966*b*), from which have come the great ophiolite nappes now lying far to the southwest (Dercourt, 1972; Bernoulli and Laubscher, 1972). This part of the central Tethys trended southeast and was a compressive boundary for the Africa–Europe rotation. For extension to have occurred, there must have been

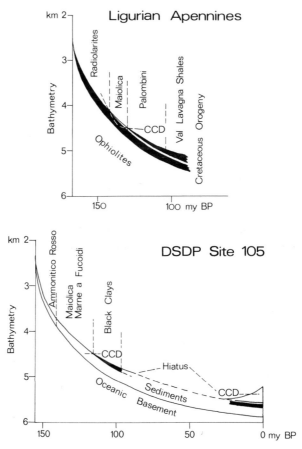

Fig. 7. Time/depth diagrams ("Sclater curves") for the Bracco zone of the Ligurian Apennines and for Site 105 (central Atlantic) of the Deep-Sea Drilling Project. The stratigraphy and age of the oceanic basement in the Ligurian Apennines is based on Decandia and Elter (1972) and Bigazzi *et al.* (1972), and that of Site 105 on Hollister *et al.* (1972). During the early phases (up to Late Tithonian) the central Atlantic and Ligurian–Piedmont basins show a different facies: red, slightly nodular limestones in the Atlantic, and carbonate-free radiolarites in Liguria. This may be explained by a higher calcite compensation depth (CCD) in the Tethys (Bosellini and Winterer, 1975); this interpretation seems plausible to us as the radiolarites encroach on the deeply submerged margins, which can hardly have subsided more than 2500 m below sea level during the early phases of foundering. From the Late Tithonian on, both basins show a remarkably similar development with the deposition of white coccolith oozes (Maiolica; Bernoulli, 1972) and the gradual sinking below the CCD during the Cretaceous (then at approximately 4.5 km).

a special movement of a Paleotethys plate to the northeast, separating the Paleotethys from the African plate, which in turn requires increased subduction somewhere between the Paleotethys and the Eurasian plates (Dewey *et al.*, 1973). In the Jurassic, the Vardar zone must have contained an oceanic central Tethys. There also is ample evidence for compression or subduction and obduction at approximately the same time. This is indicated by Late Jurassic to Early Cretaceous nappes, distal (Flyschs Bosniaque et Béotiens; Blanchet *et al.*, 1969; Celet and Clément, 1971; Aubouin, 1973) and proximal flysch and wildflysch deposits with ophiolite detritus (Diabas–Hornstein Formation; Celet, 1976; Baumgartner and Bernoulli, 1976) in the internal Dinarides and Hellenides. Late Jurassic granites (Borsi *et al.*, 1966; Mercier, 1966*a*) and postorogenic conglomerates (Kockel *et al.*, 1971) occur in the eastern Vardar zone. The simplest explanation for this juxtaposition would seem to be that in this important boundary region between Paleotethys, African and Eurasian plates, the triple point was somewhat unstable and it shifted position with a consequent shifting of convergent and divergent boundaries. These segments do not appear to have been displaced relative to each other by significant amounts during subsequent deformations.

Figure 8 illustrates that there was a possible additional Triassic–Jurassic–Cretaceous spreading center in the eastern Mediterranean (Monod *et al.*, 1974). This would require gaps in the southern continental margin belt, their only plausible locations being along the Malta Escarpment in the west and the Levantine–Van fault zone in the east. These may have served as transform faults, in which case the Paleotethys influence would have extended as far west as the Ligurian–Pennine Trough. The supporting information consists mainly of the Hatay–Cyprus–Antalya ophiolites, interpreted as having a southern origin; a certain polarity in the stratigraphy of Libyan (Sirte Basin) wells, and the Jurassic down-faulting along the coast of Israel (Ginzburg *et al.*, 1975; Freund *et al.*, 1975). In the north (external border of Dinarides and Hellenides), however, there is nothing to suggest that there was an oceanic area to the south.

The northern marginal complex, like its southern counterpart, was dissected by a number of subordinate boundaries of which the most important was the Ultrahelvetic–North Pennine Trough. Its segments may be identified from the Golfe du Lion through the western Alps and into the external western, eastern, and southern Carpathians.

In this belt minor amounts of mafic and ultramafic rocks are found, some of which in the Alps may represent covered frontal parts of the main ophiolite nappe. In comparison with the main ophiolite belt, there are few oceanic rocks here. This may be interpreted as suggesting that there was a minor boundary, perhaps comparable to the Subbetic, which it may have joined across the present Balearic Basin. Exactly how the Ultrahelvetic–North Pennine Trough, which in the eastern Carpathians centered around the Sinaia Trough, rejoined the central Tethys in the Vardar zone, as it probably did, is not clear because of the later contortions and disruptions in the Carpathian–Balkan bends and the widespread submergence below postdeformational deposits. It is possible that

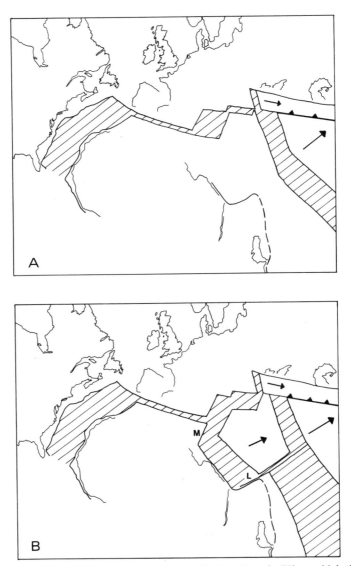

Fig. 8. Two different palinspastic restorations for the Late Jurassic (Kimmeridgian) Tethys: (A) assuming a northern origin of the Antalya–Cyprus–Hatay ophiolites and a young age of the eastern Mediterranean basin; (B) assuming a southern origin of the Antalya–Cyprus– Hatay ophiolites and a Mesozoic ocean in the eastern Mediterranean. (M) Malta Escarpment; (L) Levantine Escarpment. Discussion in the text.

the western boundary of the Paleotethys plate at times coincided with the Sinaia Trough and is responsible for the occasional occurrence of ophiolitic rocks (Burchfiel *et al.*, 1976), particularly as the Balkan triple point seems to have been unstable. On the other hand, the shuffling of microplates seems unwarranted in this area; first inversion of Tertiary deformations would appear necessary.

The opening of the Tethys by movement of the Paleotethys plate and its subduction due to Africa–Europe convergence is made complex in Turkey because tectonic complications are poorly understood there. In particular, there are three zones of Mesozoic ophiolites which may or may not originally have formed one central Tethys: Northwest Anatolian or Izmir–Eskisehir ophiolite belt (Kaya, 1975); the Lykian ophiolites (Brunn *et al.*, 1970); and the Antalya–Cyprus–Hatay ophiolites (Juteau *et al.*, 1973; Ricou 1971). Decisive evidence (Ricou *et al.*, 1974; Monod *et al.*, 1974; Argyriadis *et al.*, 1975) is still lacking, but opinion now seems to favor a former single oceanic basin broken up by a series of discordant compressional belts and complex nappes (Dürr, 1977). Jurassic compression in the Vardar zone is juxtaposed to the Tethys Ocean and diverges into the Pontides. Subsequent deformation, erosional destruction or burial has resulted in a large number of additional possible small-scale zones of movement. They may be assumed to accommodate considerable adjustments for movements which would seem incompatible for rigid blocks.

C. Cretaceous to Recent: Alpine Orogeny and Mediterranean

During the Cretaceous rifting in the South Atlantic began. The Iberian microplate rotated with minor compression far into the western Tethys (Laubscher, 1975). Between the Late Cretaceous and the present, the essential features of relative movement between Africa and Europe may be summarized as follows: Up to the time of the opening of the North Atlantic about 80 million years ago (Santonian), Africa had moved east (sinistral motion) by about 2000 km with respect to Europe. From that time to the present, about 1000 km of dextral movement, associated with several hundred kilometers of north–south displacement, occurred (Pitman and Talwani, 1972). There is excellent evidence for dextral and closing movement in the Alpine systems. The Betic Orocline, taken at face value, involves about 600 km of dextral movement since the Middle Eocene, of which perhaps half is post-Aquitanian. The Atlas system, with perhaps 200 to 300 km dextral movement, has been active mostly in the Neogene, and much of this activity has been in the Pliocene–Quaternary.

Although the amount of dextral movement agrees approximately with that postulated by Dewey *et al.* (1973), the timing does not. According to these authors, most of the dextral movement took place before the Middle Eocene.

However, we are inclined to believe surface observations more readily than the correlation of magnetic lineations. The dextral movements are oblique to the general direction of the Africa–Europe boundary belt, and sometimes the strike-slip and normal compressive movements are distributed unevenly along subordinate mobile zones.

Prior to the Aptian–Albian, compressive movements were restricted to the eastern Tethys and to the boundaries of the Paleotethys plate. From that time on, they also became important in the western Tethys. Compression and the injection of Aptian–Albian peridotites in the Pyrenees are related to trans-current movements along the North Pyrenean Fault (Choukroune *et al.*, 1973). Compression is particularly well documented in the internal zones of the Mediterranean mountain belts, from the Apennines to the Carpathians and to Turkey and beyond. By late Early Cretaceous time, chaotic complexes containing large olistostromes and olistoliths of ophiolites formed in the Apennines and in the Alps (Elter, 1973; Caron, 1972); they are capped by Upper Cretaceous flysch wedges (Helminthoid Flysch). Moreover, ophiolitic detritus in Cretaceous flysch sediments is widespread throughout the area (Oberhauser, 1968). The importance of these Cretaceous movements in the Alps and in the Apennines is uncertain as the resulting structures have been severely deformed by the Tertiary movements.

In the time span from Late Cretaceous to latest Eocene, the central Tethyan Ocean (except perhaps for the eastern Mediterranean) completely disappeared. Flysch sedimentation became more widespread as orogenic movements proceeded, and finally culminated in continental collision all along the plate boundary at the end of the Eocene. We have not attempted to show the Late Cretaceous and Early Tertiary situations on maps because of the many uncertainties in palinspastic arrangements and correlations. However, as the Tethyan Ocean and its margins were largely consumed by subduction, a series of new basins made their appearance, discordantly superposed on the orogenic zones. Those basins, which are still under water, constitute much of the present Mediterranean.

Figure 1 shows the extent of those basins that are more than 2000 m deep in the present Mediterranean, and Fig. 9 shows a rough reconstruction of the major basins that existed from late Middle to early Late Miocene times (i.e., after the Early to Middle Miocene nappe movements but before the onset of the Messinian salinity crisis and the Late Miocene to Pliocene deformations). The basins of the western Mediterranean and the Aegean Sea are clearly postorogenic and discordant with respect to the Alpine chains, and, although they differ greatly in size, shape, age, crustal structure, sedimentary fill, and bathymetry, they might represent different evolutionary stages due to similar subduction instabilities. Their sedimentary evolution depended largely on terrigenous input. The Pannonian Basin, for instance, was filled with thick Tertiary

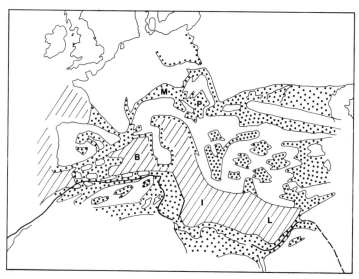

Fig. 9. Paleogeographic restoration for the late Middle to early Upper Miocene. Mainly based on data compiled by Institut Français du Pétrole and Centre National pour l'Exploitation des Océans (1974). The situation shown is that which followed the Middle Miocene nappe movements and preceded the foundering of the Tyrrhenian Basin and the Messinian salinity crisis. The dotted areas are areas of continental, shallow-marine, or proximal flysch sedimentation; the hatched areas are deep basins with hemipelagic or distal flysch sedimentation. (B) Balearic; (I) Ionian; (L) Levantine; (M) extra-Alpine Molasse; (P) Pannonian Basin.

sediments probably supplied by drainage of the northern continent, whereas the Tyrrhenian Basin received only limited amounts of detritus.

The largest and deepest of the basins is the Balearic, which is superposed on the Paleogene orogenic belt, and its foundering is pre-Burdigalian as Burdigalian hemipelagic sediments encroach on its rifted margin along the Minorca Rise (Hsü *et al.*, 1975). In contrast, the Tyrrhenian Basin is superposed on the Late Tertiary orogenic belt in the Apennines, and its foundering postdates Tortonian movements (Toscan phase, Elter, 1973). The Aegean Sea seems to be the youngest of these basins, for although postorogenic hemipelagic and turbiditic sediments date the general nappe structure of Crete as pre-Tortonian (Meulenkamp, 1969), Quaternary movements along the Cretan Arc and Hellenic Trench suggest a young age for the inter-arc and back-arc basins.

The Adriatic and eastern Mediterranean basins lie largely to the south of the Alpine chains and thus differ from the general pattern of successor basins. The northern Adriatic is part of the foredeep of the Neogene Apennines. It lies on sialic crust, and during the Late Tertiary was characterized by an eastward facies change from flysch to hemipelagic sediments with gradual outward migration of the flysch trough (Bortolotti *et al.*, 1970; Sestini, 1970). This foredeep continues west of the Apulian Platform onland through the Fossa

Bradanica into the Gulf of Taranto and along the Calabrian margin (Finetti and Morelli, 1972; Biju-Duval *et al.*, 1974).

A front of Neogene, mainly Pliocene–Quaternary, compression is present all along the northern margin, but not the eastern and southern margins, of the eastern Mediterranean (Cretan Arc and Hellenic Trench—Ryan *et al.*, 1973; Florence Rise–Cyprus Arc—Biju-Duval *et al.*, 1974). This front, together with the Jurassic down-faulting along the Levantine Escarpment and the essentially pelagic Late Cretaceous to Miocene sequence in Cyprus (Baroz and Bizon, 1974; Robertson and Hudson 1974) is another indication that the eastern Mediterranean may actually be a remnant of Tethys.

The genesis of the Mediterranean, particularly that of the postorogenic basins behind the Alpine arcs, is highly controversial, as has been noted in the introduction. For advocates of microspreading, there are many strong objections to overcome. Certainly the basins cannot be a part of the Cenozoic world rift system as they are inserted on a compressive plate boundary. Presumably they would have to result from subduction dynamics which created local lithospheric instabilities of their own. For reasons stated in the introduction, we prefer an explanation involving crustal stoping initiated by subduction. This, however, is not of primary interest in the context of this article. What is quite clear is that while the Mesozoic Tethyan Ocean seems to have been part of a world-wide ridge system associated with breakup of Pangea, the Tertiary successor basins forming much, if not all, of the present Mediterranean are not. This is the essential difference between the Tethys and the Mediterranean.

It is instructive to view the controversy about depositional conditions of the thick Messinian evaporites in this perspective (Drooger, 1973). When they formed, the old Tethys had already been destroyed. The Mauretanian and Subbetic seaways, originally connecting with the Atlantic, now formed part of the Rif–Gibraltar arc–Betic mountain belt. However, within this mountain belt a number of "postorogenic" Late Tertiary basins similar to those of Tuscany and the eastern Alps were formed. Some of them, though unconformable on Middle Miocene nappes, were violently folded, and it is evident because of such movements that the connection between successor basins was severed from time to time, facilitating evaporite formation. New and often different connections between these basins appeared from time to time. The graben of Gibraltar seems to be younger, and at any rate would have been only one of a number of contributing collapse features. The main connection between the Atlantic and the larger Mediterranean basins may have passed through the southern Rif and the Melilla Basin.

One may question whether the Mediterranean successor basins were only the largest of the collapse basins or something special. As there are transitional stages between the extremes, we favor the first view. Also, we think that the well-documented subsidence in post-Messinian times of such basins as the Balearic is a developmental stage of collapse. The larger basins of the Mediterranean (Balearic, Ionian) certainly were in existence prior to the Messinian salinity crisis, as is shown by restriction of halite deposits to their central parts. They must have been deep, as is shown by the pelagic to hemipelagic and turbiditic sediments underlying the evaporites in the Balearic Basin in the Mediterranean Ridge and on the Florence Rise (Hsü et al., 1975). The Messinian salinity crisis is thus a pan-Mediterranean event documented in a series of individual basins which greatly differed among each other as to both their pre- and post-Messinian evolution.

V. CONCLUSIONS

In our considerations we have accepted a Permo-Triassic reassembly of the major continental masses as proposed by Dietz and Holden (1970) or Smith et al. (1973). After the Variscan orogeny, the central Atlantic and western Mediterranean areas were part of one large continental mass. No remnants of pre-Jurassic oceanic crust are known from this area and Permo-Triassic deposits are predominantly continental, with marine incursions from time to time. However, in the east the pre-drift configuration shows a wedge-shaped space between Africa and Eurasia. This area, probably once oceanic, and called Paleotethys here, fell victim to subduction in Mesozoic–Tertiary times and only the extensive deposits of its shallow continental margins are conserved in the nappes of continental origin in the Alpine–Himalayan system.

The separation of North America and Africa in the Early Jurassic implies relative movements between Africa and Eurasia and the generation of oceanic crust and lithosphere in the central Atlantic and in the Liguria–Piedmont and Vardar ophiolite zones of about the same age. We call this generally east–west-trending, Jurassic–Cretaceous oceanic seaway the Tethys. From kinematic considerations, it seems plausible that Tethyan oceanic areas belonged to the Jurassic–Early Cretaceous world rift system and were associated with a spreading ridge rather than occurring in small basins behind island arcs. Evidence is seen in the close affinities in the sedimentary and most probably also bathymetric evolution of the central Atlantic and the Liguria–Piedmont Ocean basin and in the evolution of typical, subsiding continental margins of Atlantic type. In the area of the triple point between the African, Eurasian and Paleotethys plates, extensional and compressional movements were closely juxtaposed in space and time.

When, in the Cretaceous, rifting began in the South Atlantic, the Iberian microplate rotated, probably with some compression, far into the western Tethys. With the opening of the North Atlantic in the Late Cretaceous, sinistral and extensional movements between Africa and Eurasia were replaced by dextral and compressional ones. By Late Eocene times, the entire central Tethys north of the southern Tethyan carbonate belt had disappeared, and a potential relic of the Paleotethys or a separate Jurassic–Cretaceous ocean basin is preserved only in the eastern Mediterranean, south of that belt.

The western Mediterranean and the Aegean are postorogenic basins which are not linked to the world rift system but are related to instabilities associated with Tertiary subduction. They occur as discordant successor basins behind and across the Alpine arcs. The Late Miocene evaporites have to be seen in this context; they represent a pan-Mediterranean event, but occur in a series of individual basins of varying origin whose evolution was not completed until after Late Miocene.

ACKNOWLEDGMENTS

Both authors are indebted to the Swiss National Science Foundation, which has continuously supported study of the Mesozoic evolution of the Tethyan margins at the Geological Institute of Basel University. They extend their acknowledgments to the many colleagues and friends who have helped them by generously communicating their knowledge and ideas, by field demonstrations and stimulating discussions, and who are too numerous to be named individually. D. Bernoulli is very grateful to the Deep-Sea Drilling Project for the opportunity to work on "Tethyan" sediments of the central Atlantic and to participate in the second cruise of *Glomar Challenger* in the Mediterranean.

REFERENCES

Adamia, S. A., 1975, Plate tectonics and the evolution of the Alpine system: Discussion, *Geol. Soc. Amer. Bull.*, v. 86, p. 719–720.

Adams, C. G. and Ager, D. V., eds., 1967, *Aspects of Tethyan Biogeography*, London: The Systematics Association.

Ager, D. V., 1967, Some Mesozoic brachiopods in the Tethys region, in: *Aspects of Tethyan Biogeography*, Adams, C. G. and Ager, D. V., eds., London: The Systematics Association p. 135–151.

Ager, D. V., 1971, Space and time in brachiopod history, in: *Faunal Provinces in Space and Time*, Middlemiss, F. A., Rawson, P. F., and Newall, G., eds., Liverpool: Seel House Press, p. 95–108.

Ager, D. V., 1973, Mesozoic brachiopoda, in: *Atlas of Palaeobiogeography*, Hallam, A., ed., Amsterdam: Elsevier Scientific Publishing Company, p. 431–436.

Andrieux, J. 1971, La structure du Rif central. Etude des relations entre la tectonique de compression et les nappes de glissement dans un tronçon de la chaîne alpine, Serv. Géol. Maroc, Notes Mém., 235, p. 1–155.

Argyriadis, I. 1975, Mésogée permienne, chaîne hercynienne et cassure téthysienne, *Soc. Géol. France Bull.*, v. (7) 17, p. 56–67.

Argyriadis, I., Brunn, J. H., Ricou, L. E., de Graciansky, P. C., Poisson, A., and Marcoux, J., 1975, Elements majeurs de liaison entre Taurides et Hellénides [abstract], Orsay, France: Vème Colloque sur la Géologie des Régions Égéennes.

Assereto, R., Bosellini, A., Fantini Sestini, N., and Sweet, W. C., 1973, The Permian–Triassic boundary in the Southern Alps (Italy), in: *Permian and Triassic Systems and Their Mutual Boundary*, Logan, A., and Hills, L. V., eds., Calgary: Can. Soc. Petr. Geol., p. 176–199.

Aubouin, J., 1959, Contribution à l'étude géologique de la Grèce septentrionale: les confins de l'Epire et de la Thessalie, *Ann. Géol. Pays Hellén.*, v. 10, p. 1–403.

Aubouin, J., 1973, Des tectoniques superposées et de leur signification par rapport aux modèles géophysiques: l'exemple des Dinarides; paléotectonique, tectonique, tarditectonique, néotectonique, *Soc. Géol. France Bull.*, v. (7) 15, p. 426–460.

Ballard, R. D., and Uchupi, E., 1975, Triassic rift structure in Gulf of Maine, *Amer. Assoc. Petr. Geol. Bull.*, v. 59, p. 1041–1072.

Baroz, F., and Bizon, G., 1974, Le Néogène de la chaîne du Pentadaktylos et de la partie nord de la Mésaoria (Chypre). Etude stratigraphique et micropaléontologique, *Inst. Franç. Pétrole Rev.*, v. 29, p. 327–358.

Baumgartner, O., and Bernoulli, D., 1976, Stratigraphy and radiolarian fauna in a Late Jurassic–Early Cretaceous section near Achladi (Evvoia, Eastern Greece), *Eclogae Geol. Helv.*, v. 69, p. 601–626.

Bernoulli, D., 1964, Zur Geologie des Monte Generoso (Lombardische Alpen), *Beitr. Geol. Karte Schweiz.*, n. s., v. 118, p. 1–134.

Bernoulli, D., 1972, North Atlantic and Mediterranean Mesozoic facies: a comparison, in: *Initial Reports Deep-Sea Drilling Project*, Hollister, C. D., Ewing, J. I., *et al.*, eds., Washington, D. C.: U. S. Govt. Printing Office, Bull. no. 11, p. 801–871.

Bernoulli, D., and Jenkyns, H. C., 1974, Alpine, Mediterranean, and Central Atlantic Mesozoic facies in relation to the early evolution of the Tethys, in: *Modern and Ancient Geosynclinal Sedimentation*, Dott., R. H., Jr., and Shaver, R. H., eds., Soc. Econ. Paleont. Mineral. Spec. Publ. 19, p. 129–160.

Bernoulli, D., and Laubscher, H., 1972, The palinspastic problem of the Hellenides, *Eclogae Geol. Helv.*, v. 65, p. 107–118.

Bertrand, J., 1970, Etude pétrographique des ophiolites et des granites du flysch des Gêts (Haute-Savoie, France), *Arch. Sci. (Genève)*, v. 23, p. 279–542.

Bigazzi, G., Ferrara, G., and Innocenti, F., 1972, Fission track ages of gabbro from northern Apennines ophiolites, *Earth Planet. Sci. Lett.*, v. 14, p. 242–244.

Biju-Duval, B., Letouzey, J., Montadert, L., Courrier, P., Mugniot, J. F., and Sancho, J., 1974, Geology of the Mediterranean sea basins, in: *The Geology of Continental Margins* Burk, C. A., and Drake, C. L., eds., New York: Springer-Verlag, p. 695–721.

Bishop, W. F., 1975, Geology of Tunisia and adjacent parts of Algeria and Libya, *Amer. Assoc. Petr. Geol. Bull.*, v. 59, p. 413–450.

Bismuth, H., Bonnefous, J., and Dufaure, Ph., 1967, Mesozoic microfacies of Tunisia, in: *Guidebook to the Geology and History of Tunisia*, Martin, L., ed., Tripoli: Petr. Explor. Soc. Libya, p. 159–214.

Bittner, A., 1896, *Bemerkungen zur neuesten Nomenclatur der alpinen Trias*, Vienna: Selbst-Verlag des Verfassers.

Blanchet, R., Cadet, J.-P., Charvet, J., and Rampnoux, J.-P., 1969, Sur l'éxistence d'un important domaine de flysch tithonique–crétacé inférieur en Yougoslavie: l'unité du flysch bosniaque, *Soc. Géol. France Bull.*, v. (7) 11, p. 871–880.

Bonnefous, J., 1967, Jurassic stratigraphy of Tunisia: a tentative synthesis (northern and central Tunisia, Sahel and Chotts areas), in: *Guidebook to the Geology and History of Tunisia*, Martin, L., ed., Tripoli: Petr. Explor. Soc. Libya, p. 109–130.

Borsi, S., Ferrara, G., Mercier J., and Tongiorgi, E., 1966, Age stratigraphique et radio-métrique jurassique supérieur d'un granite des zones internes des Hellénides (Granite de Fanos, Macédoine, Grèce), *Géogr. Phys. Géol. Dyn. Rev.*, v. 8, p. 279–287.

Bortolotti, V., Passerini, M., Sagri, M., 1970, The miogeo-synclinal sequences, *Sédiment. Geol.*, v. 4, p. 341–444.

Bosellini, A., 1973, Modello geodinamico e paleotettonico delle Alpi Meridionali durante il Giurassico–Cretacico. Sue possibili applicazioni agli Appennini, in: *Moderne Vedute sulla Geologia dell'Appennino*, Roma: Accad. naz. Lincei, p. 163–213.

Bosellini, A., and Rossi, D., 1974, Triassic carbonate buildups of the Dolomites, northern Italy, in: *Reefs in Time and Space*, Laporte, L., ed., Soc. Econ. Paleont. Mineral., Spec. Publ. no. 18, p. 209–233.

Bosellini, A., and Winterer, E. L., 1975, Pelagic limestone and radiolarite of the Tethyan Mesozoic: A genetic model, *Geology*, v. 3, p. 279–282.

Broquet, P., Caire, A., and Mascle, G., 1966, Structure et évolution de la Sicile occidentale (Madonies et Sicani), *Soc. Géol. France Bull.*, v. (7) 8, p. 994–1013.

Brunn, J. H., 1974, Le problème de l'origine des nappes et de leurs translations dans les Taurides occidentales, *Soc. Géol. France Bull.*, v. (7) 16, p. 101–106.

Brunn, J. H., de Graciansky, P. C., Gutnic, M., Juteau, T., Lefèvre, R., Marcoux, J., Monod, O., and Poisson, A., 1970, Structures majeures et corrélations stratigraphiques dans les Taurides occidentales, *Soc. Géol. France Bull.*, v. (7) 12, p. 515–556.

Burchfiel, B. C., Bleahu, M., Borcos, M., Patrulius, D., and Sandulescu, M., 1976, Geology of Romania, Geol. Soc. Amer. Spec. Pap. (in press).

Caron, C., 1972, La nappe supérieure des Préalpes: subdivisions et principaux caractères du sommet de l'édifice préalpin, *Eclogae Geol. Helv.*, v. 65, p. 57–73.

Casati, P., Nicoletti, M., and Petrucciani, C., 1976, Età (K/Ar) di intrusioni porfiritiche e leucogabbriche nelle prealpi Bergamasche (Alpi meridionali), *Soc. Ital. Mineral. Petrol. Rend.*, v. 32, p. 215–226.

Castellarin, A., 1972, Evoluzione paleotettonica sinsedimentaria del limite tra "piattaforma veneta" e "bacino lombardo" a nord di Riva del Garda, *Giorn. Geol.*, v. (2) 38, p. 11–212.

Celet, P., 1976, A propos de mélange de type "volcano-sédimentaire" de l'Iti (Grèce méri-dionale), *Soc. Géol. France Bull.*, v. (7) 18, p. 299–307.

Celet, P., and Clément, B., 1971, Sur la présence d'une nouvelle unité paléogéographique et structurale en Grèce continentale du sud: l'unité du flysch béotien, *Soc. Géol. France C. R.*, p. 43–44.

Choubert, G., and Faure-Muret, A., 1962, Evolution du domaine atlasique marocain depuis les temps paléozoiques, in: *Livre à la Mémoire du Professeur Paul Fallot*, Paris: Soc. Géol. France, p. 447–527.

Choukroune, P., Seguret, M., and Galdeano, A., 1973, Caractéristiques et évolution struc-turale des Pyrénées: un modèle de relations entre zone orogénique et mouvement des plaques, *Soc. Géol. France Bull.*, v. (7) 15, p. 601–611.

D'Argenio, B., 1970, Evoluzione geotettonica comparata tra alcune piattaforme carbonatiche dei Mediterranei Europeo ed Americano, *Accad. Pontaniana Atti*, n. s. v. 20, p. 3–34.

D'Argenio, B., De Castro, P., Emiliani, C., and Simone, L., 1975, Bahamian and Apenninic limestones of identical lithofacies and age, *Amer. Assoc. Petr. Geol. Bull.*, v. 59, p. 524–530.

Decandia, F. A., and Elter, P., 1972, La "zona" ofiolitifera del Bracco nel settore compreso fra Levanto e la Val Graveglia (Appennino ligure), *Soc. Geol. Ital. Mem.*, v. 11, p. 503–530.

Dercourt, J., 1972, The Canadian Cordillera; the Hellenides and the sea floor spreading theory, *Canad. J. Earth Sci.*, v. 9, p. 709–743.

Desmaison, G. J., 1965, The Triassic salt in the Algerian Sahara, in: *Salt Basins around Africa*, London: Inst. Petr., p. 91–111.

Dewey, J. F., Pitman, W. C., Ryan, W. B. F., and Bonnin, J., 1973, Plate tectonics and the evolution of the Alpine system, *Geol. Soc. Amer. Bull.*, v. 84, p. 3137–3180.

Didon, J., Durand-Delga, M., and Kornprobst, J., 1973, Homologies géologiques entre les deux rives du détroit de Gibraltar, *Soc. Géol. France Bull.*, v. (7) 15, p. 77–105.

Dietrich, V., and Oberhänsli, R., 1975, Die Pillow-Laven des Vispertales, *Schweiz. Min. Petrogr. Mitt.*, v. 55, p. 79–87.

Dietz, R. S., and Holden, J. C., 1970, Reconstruction of Pangaea: breakup and dispersion of continents, Permian to present, *J. Geophys. Res.*, v. 75, p. 4939–4956.

Dilley, F. C., 1973, Cretaceous larger foraminifera, in: *Atlas of Palaeobiogeography*, Hallam, A., ed., Amsterdam: Elsevier Scientific Publishing Company, p. 403–419.

Dimitrijevic, M. D., and Dimitrijevic, M. N., 1973, Olistostrome mélange in the Yugoslavian Dinarides and late Mesozoic plate tectonics, *J. Geol.*, v. 81, p. 328–340.

Donovan, D. T., 1967, The geographical distribution of Lower Jurassic ammonites in Europe and adjacent areas, in: *Aspects of Tethyan Biogeography*, Adams, C. G., and Ager, D. V., eds., London: The Systematics Association, p. 111–134.

Drooger, C. W., ed., 1973, *Messinian Events in the Mediterranean*, Amsterdam: North-Holland Publishing Co.

Druckman, Y., and Gvirtzman, G., 1975, Distribution and environments of deposition of Upper Triassic on the northern margins of the Arabian shield and around the Mediterranean sea, IXème Congrès Internat. Séd., Thème 5, p. 183–189.

Dürr, S., 1977, Über Alter und geotektonische Stellung des Menderes-Kristallins/SW-Anatolien und seine Aequivalente in der mittleren Aegaeis, *Geol. Jb.* (in press).

Durand Delga, M., 1960, Le sillon géosynclinal du flysch tithonique-néocomien en Méditerranée occidentale, *Accad. Lincei Atti, Rend. Cl. Sci. Fis. Mat. Nat.*, v. 29, p. 579–585.

Elter, P., 1973, Lineamenti tettonici ed evolutivi dell'Appennino settentrionale, in: *Moderne Vedute sulla Geologia dell'Appennino*, Roma: Accad. Naz. Lincei, p. 97–118.

Enay, R., 1972, Paléobiogéographie des ammonites du Jurassique terminal (Tithonique/Volgien/Portlandien *s.l.*) et mobilité continentale, *Geobios*, v. 5, p. 355–407.

Enay, R., 1973, Upper Jurassic (Tithonian) ammonites, in: *Atlas of Palaeobiogeography*, Hallam, A., ed., Amsterdam: Elsevier Scientific Publishing Company, p. 297–307.

Finetti, I., and Morelli, C., 1972, Wide scale digital seismic exploration of the Mediterranean Sea, *Geofis. Teor. Appl.*, v. 14, p. 291-342.

Freund, R., Goldberg, M., Weissbrod, T., Druckman, Y., and Baruch, D., 1975, The Triassic-Jurassic structure of Israel and its relation to the origin of the eastern Mediterranean, *Bull. Geol. Surv. Israel*, v. 65, p. 1–26.

Garrison, R. E., and Fischer, A. G., 1969, Deep-water limestones and radiolarites of the Alpine Jurassic, in: *Depositional Environments in Carbonate Rocks: A Symposium*, Friedman, G. M., ed., Soc. Econ. Paleont. Mineral., Spec. Publ. 14, p. 20–56.

Geyer, O. F., 1973, Das präkretazische Mesozoikum von Kolumbien, *Geol. Jb.* (Reihe B), v. 5, p. 1–155.

Ginzburg, A., Cohen, S. S., Hay-Roe, H., and Rosenzweig, A., 1975, Geology of Mediterranean shelf of Israel, *Amer. Assoc. Petr. Geol. Bull.*, v. 59, p. 2142–2160.

Gobbett, D. J., 1967, Palaeozoogeography of the Verbeekinidae (Permian foraminifera), in: *Aspects of Tethyan Biogeography*, Adams, C. G., and Ager, D. V., eds., London: The Systematics Association, p. 77–91.

Gobbett, D. J., 1973, Permian Fusulinacea, in: *Atlas of Palaeobiogeography*, Hallam, A., ed., Amsterdam: Elsevier Scientific Publishing Company, p. 151–158.

Hallam, A., 1971, Provinciality in Jurassic faunas in relation to facies and palaeogeography, in: *Faunal Provinces in Space and Time*, Middlemiss, F. A., Rawson, P. F., and Newall, G., eds., Liverpool: Seel House Press, p. 129–152.

Hallam, A., 1975, *Jurassic Environments*, Cambridge: Cambridge University Press.

Hollister, C. D., Ewing, J. I., Habib, D., Hathaway, J. C., Lancelot, Y., Luterbacher, H., Paulus, F. J., Poag, W. C., Wilcoxon, J. A., and Worstell, P., eds., 1972, *Initial Reports of the Deep-Sea Drilling Project*, Washington, D. C.: U. S. Govt. Printing Office, Bull. no. 11.

Howarth, M. K., 1973, Lower Jurassic (Pliensbachian and Toarcian) ammonites, in: *Atlas of Palaeobiogeography*, Hallam, A., ed., Amsterdam: Elsevier Scientific Publishing Company, p. 275–282.

Hsü, K. J., Montadert, L., Garrison, R. E., Fabricius, F. H., Bernoulli, D., Melières, F., Kidd, R. B., Müller, C., Cita, M. B., Bizon, G., and Erickson, A., 1975, *Glomar Challenger* returns to the Mediterranean Sea, *Geotimes*, v. 20 (8), p. 16–19.

Hynes, A. J., Nisbet, E. G., Smith, A. G., Welland, M. J. P., and Rex, D. C., 1972, Spreading and emplacement ages of some ophiolites in the Othris region, eastern central Greece, *Zeitschr. Deutsch. Geol. Ges.*, v. 123, p. 445–468.

Institut Français du Pétrole and Centre National pour l'Exploitation des Océans, 1974, Carte géologique et structurale des bassins tertiaires du domaine méditerranéen, First edition, Paris: Editions Technip.

Jansa, L. F., 1975, The central north Atlantic basin—its birth and disappearance, IXème Congrès Internat. Séd., Thème 5, p. 231–235.

Jansa, L. F., and Wade, J. A., 1974, Geology of the continental margin of Nova Scotia and Newfoundland, *Geol. Surv. Canada Pap.*, v. 2, p. 51–105.

Juteau, T., Lapierre, H., Nicolas, A., Parrot, J.-F., Ricou, L.-E., Rocci, G., and Rollet, M., 1973, Idées actuelles sur la constitution, l'origine et l'évolution des assemblages ophiolitiques mésogéens, *Soc. Géol. France Bull.*, v. (7) 15, p. 476–493.

Kälin, O. and Trümpy D. M., 1977, Sedimentation und Paläotektonik in den Westlichen Südalpen: zur triadisch–jurassischen Geschichte des M. Nudo-Beckens, *Eclogae Geol. Helv.*, v. 70.

Kamen-Kaye, M., 1972, Permian Tethys and Indian Ocean, *Amer. Assoc. Petr. Geol. Bull.*, v. 56, p. 1984–1999.

Kauffman, E. G., 1973, Cretaceous Bivalvia, in: *Atlas of Palaeobiogeography*, Hallam, A., ed., Amsterdam: Elsevier Scientific Publishing Company, p. 353–383.

Kaya, O., 1975, Northwest Anatolian ophiolite zone (Nonmetamorphic and Mesozoic in age) [Abstract], Orsay, France: Vème Colloque sur la Géologie des Régions Égéennes.

Kockel, F., Mollat, H., and Walther, H. W., 1971, Geologie des Serbo-Mazedonischen Massivs und seines mesozoischen Rahmens (Nordgriechenland), *Geol. Jb.*, v. 89, p. 529–551.

Lancelot, Y., Seibold, E., Cepek, P., Dean, W. E., Eremeev, V., Gardner, J. V., Jansa, L. F., Johnson, D., Krasheninnikov, V., Pflaumann, U., Rankin, J. G., and Trabant, P., 1975, The eastern North Atlantic, *Geotimes*, v. 20 (7), p. 18–21.

Lapierre, H., and Parrot, J.-F., 1972, Identité géologique des régions de Paphos (Chypre) et de Baër–Bassit (Syrie), *C. R. Acad. Sci. (Paris)*, v. 274, p. 1999–2002.

Lapierre, H., and Rocci, G., 1969, Un bel exemple d'association cogénétique laves–radiolarites–calcaires, *C. R. Acad. Sci. (Paris)*, v. 268, p. 2637–2640.

Laubscher, H. P., 1970, Bewegung und Wärme in der alpinen Orogenese, *Schweiz. Min. Petrogr. Mitt.*, v. 50, p. 503–534.

Laubscher, H. P., 1974, Evoluzione e struttura delle Alpi, *Le Scienze*, v. 72, p. 48–59.

Laubscher, H. P., 1975, Plate boundaries and microplates in Alpine history, *Amer. J. Sci.*, v. 275, p. 865–876.

Lemoine, M., 1975, Mesozoic sedimentation and tectonic evolution of the Briançonnais Zone in the Western Alps—Possible evidence for an Atlantic-type margin between the European craton and the Tethys, IXème Congrès Internat. Séd., Thème 4, p. 211–215.

Lowrie, W., and Alvarez, W., 1975, Paleomagnetic evidence for rotation of the Italian Peninsula, *J. Geophys. Res.*, v. 80, p. 1579–1592.

Marcoux, J., 1970, Âge carnien de termes effusifs du cortège ophiolitique des nappes d'Antalya (Taurus lycien oriental, Turquie), *C. R. Acad. Sci. (Paris)*, v. 271, p. 285–287.

Mercier, J., 1966a, Mouvements orogéniques et magmatisme d'âge jurassique supérieur-éocrétacé dans les zones internes des Hellénides (Macédoine, Grèce), *Rev. Géogr. Phys. Géol. Dyn.*, v. 8, p. 265–278.

Mercier, J., 1966b, Paléogéographie, orogenèse, métamorphisme et magmatisme des zones internes des Hellénides en Macédoine (Grèce): Vue d'ensemble, *Soc. Géol. France Bull.*, v. (7) 8, p. 1014–1017.

Meulenkamp, J. E., 1969, Stratigraphy of Neogene deposits in the Rethymnon Province, Crete, with special reference to the phylogeny of uniserial *Uvigerina* from the Mediterranean region, *Utrecht Micropal. Bull.*, v. 2, p. 1–108.

Monod, O., Marcoux, J., Poisson, A., and Dumont, J.-F., 1974, Le domaine d'Antalya, témoin de la fracturation de la plateforme africaine au cours du Trias, *Soc. Géol. France Bull.*, v. (7) 16, p. 116–127.

Moores, E. M., 1969, Petrology and structure of the Vourinos ophiolite complex, northern Greece, *Geol. Soc. Amer. Spec. Pap.*, no. 118, p. 1–73.

Neumayr, M., 1883, Klimatische Zonen während der Jura- und Kreidezeit, *Denkschr. Kais. Akad. Wiss. Wien*, v. 47, p. 1–34.

Oberhauser, R., 1968, Beiträge zur Kenntnis der Tektonik und der Paläogeographie während der Oberkreide und dem Paläogen im Ostalpenraum, *Jb. geol. Bundesanst.*, v. 3, p. 115–146.

Pitman, W. C., and Talwani, M., 1972, Sea-floor spreading in the North Atlantic, *Geol. Soc. Amer. Bull.*, v. 83, p. 619–646.

Premoli Silva, I., Napoleone, G., and Fischer, A. G., 1974, Risultati preliminari sulla stratigrafia paleomagnetica della Scaglia cretaceo-paleocenica della sezione di Gubbio (Appennino centrale), *Soc. Geol. Ital. Boll.*, v. 93, p. 647–659.

Ricou, L. E., 1971, Le croissant ophiolitique péri-arabe, une ceinture de nappes mises en place au Crétacé supérieur, *Rev. Géogr. Phys. Géol. Dyn.*, v. 13, p. 327–349.

Ricou, L.-E., Argyriadis, I., and Lefèvre, R., 1974, Proposition d'une origine interne pour les nappes d'Antalya et le massif d'Alanya (Taurides occidentales, Turquie), *Soc. Géol. France Bull.*, v. (7) 16, p. 107–111.

Robertson, A. H. F., and Hudson, J. D., 1974, Pelagic sediments in the Cretaceous and Tertiary history of the Troodos Massif, Cyprus, Int. Assoc. Sediment., Spec. Publ. no. 1, p. 403–436.

Ryan, W. B. F., Hsü, K. J., Cita M. B., Dumitrica, P., Lort, J. M., Maync, W., Nesteroff, W. D., Pautot, G., Stradner, H., and Wezel, F. C., 1973, *Initial Reports of the Deep-Sea Drilling Project*, Washington, D. C.: U. S. Govt. Printing Office, Bull. no. 13.

Scandone, P., 1967, Studi di geologia lucana: la serie calcareo–silico–marnosa e i suoi rapporti con l'Appennino calcareo, *Soc. Nat. Napoli Boll.*, v. 76, p. 3–175.

Sclater, J. G., and Detrick, R., 1973, Elevation of mid-ocean ridges and the basement age of JOIDES Deep-Sea Drilling sites, *Geol. Soc. Amer. Bull.*, v. 84, p. 1547–1554.

Sestini, G., 1970, Sedimentation of the late geosynclinal stage, *Sediment. Geol.*, v. 4, p. 445–479.

Smith, A. G., 1971, Alpine deformation and the oceanic areas of the Tethys, Mediterranean and Atlantic, *Geol. Soc. Amer. Bull.*, v. 82, p. 2039–2070.

Smith, A. G., Briden, J. C., and Drewry, G. E., 1973, Phanerozoic world maps, *Palaeontology*, Spec. Pap. no. 12, p. 1–42.

Soffel, H., 1975, The palaeomagnetism of age-dated Tertiary volcanites of the Monti Lessini (northern Italy) and its implication to the rotation of northern Italy, *J. Geophys.*, v. 41, p. 385–400.

Stegena, L., Géczy, B. and Horvath, F., 1974. Late Cenozoic evolution of the Pannonian basin. *Tectonophysics*, v. 26, p. 71–90.

Sturani, C., 1973, Considerazioni sui rapporti tra Appennino settentrionale ed Alpi occidentali, in: *Moderne Vedute sulla Geologia dell'Appennino*, Roma: Accad. naz. Lincei, p. 119–145.

Suess, E., 1893, Are great oceans depth permanent?, *Nat. Sci.*, v. 2, p. 180–187.

Suess, E., 1901, *Das Antlitz der Erde*, v. 3, Vienna: F. Temsky.

Tollmann, A., 1975, Ozeanische Kruste im Pennin des Tauernfensters und die Neugliederung des Deckenbaues der Hohen Tauern, *N. Jb. Geol. Paläont. Abh.*, v. 148, p. 286–319.

Trümpy, R., 1960, Paleotectonic evolution of the Central and Western Alps, *Geol. Soc. Amer. Bull.*, v. 71, p. 843–908.

Trümpy R., 1975, Penninic–Austroalpine boundary in the Swiss Alps: a presumed former continental margin and its problems. *Amer. J. Sci.*, v. 275A, p. 209–238.

Veevers, J. J., 1974, Western continental margin of Australia, in: *The Geology of Continental Margins*, Burk, C. A., and Drake, C. L., eds., New York: Springer-Verlag, p. 605–616.

Vezzani, L., 1973, L'Appennino siculo–calabro–lucano, in: *Moderne Vedute sulla Geologia dell'Appennino*, Roma: Accad. naz. Lincei, p. 15–48.

Wurster, P., 1968, Paläogeographie der deutschen Trias und die paläogeographische Orientierung der Lettenkohle in Südwest-deutschland, *Eclogae Geol. Helv.*, v. 61, p. 157–166.

Ziegler, P. A., 1975, Geologic evolution of North Sea and its tectonics, *Amer Assoc. Petr. Geol. Bull.*, v. 59, p. 1073–1097.

Chapter 2

TECTONIC EVOLUTION OF THE MEDITERRANEAN BASINS*

Kenneth J. Hsü

Geological Institute
ETH Zurich, Switzerland

I. INTRODUCTION

A. Statement of Three Hypotheses

Compared to the Atlantic Ocean, the opposite shores of which fit together almost perfectly, the Mediterranean is more complex, being bounded on the north by three peninsulas which run at right angles to the trend of the sea. To make matters worse, the Mediterranean Ridge has proved to be a compressional feature quite distinct from a normally extensional mid-ocean ridge. One can hardly postulate a simple splitting apart of the adjacent continents to account for the origin of this inland sea. Unlike the Pacific Ocean, with its circumoceanic orogenic system, the Mediterranean is not rimmed on all sides by coastal mountains. Thus, we cannot invoke circum-Mediterranean subduction to explain the genesis of the Mediterranean.

Conventional interpretations have related the origin of the Mediterranean and the adjacent Alpine chain to the interaction between Europe and Africa. Three schools of thought have evolved, namely: (1) The present Mediterranean is a relic of an ancient ocean, the Tethys; (2) the present Mediterranean is a new

* Contribution No. 81, Laboratory of Experimental Geology, ETH Zurich.

creation after the climax of Paleogene Alpine folding; (3) the present Mediterranean is a combination of a relic Tethys and a neo-Mediterranean.

Tethys is the child of the Viennese school of geology, which dominated thinking on the continent during the last two decades of the 19th century. The idea of an ancient Mediterranean was brought forth by Neumayr (1883). Based upon the distribution of Jurassic fauna, he postulated the existence of an equatorial ocean which extended from India in the east to Central America in the west, separating the Neoarctic Continent on the north and the Brazilian–Ethiopian Continent on the south. This central Mediterranean was later baptized Tethys by Suess (1893). Mojsisovics (1896) postulated a Triassic Tethys through comparison of the pelagic formations of the Austrian Alps and the Himalayas. Later Suess (1901) defined Tethys as being represented by the Mesozoic marine sequence north of Gondwanaland. As this mythical continent was the home of the Late Paleozoic *Glossopteris* flora, its inclusion in the definition of Tethys seemed to imply the existence of this Mediterranean during the Paleozoic. In fact, Du Toit (1937, p. 40) referred to Tethys as an east–west-trending, intercontinental ocean, separating two supercontinents — Laurasia in the north and Gondwana in the south — "from at least the mid-Paleozoic onwards." Carey (1958) entertained the same idea in his portrait of Tethys (Fig. 1).

Concerning the genesis of the Mediterranean, Suess (1901, p. 25) wrote as a matter of fact: "Das heutige europäische Mittelmeer ist einer Reste der Tethys."

This idea persists today, although there has never been a satisfactory explanation of why the Mediterranean should have remained while the lion's share of the Tethys was consumed by intercontinental collisions.

Another school of geology draws attention to the fact that the mountain chains around the Mediterranean do not always run parallel to the coastline. The abrupt termination of the Alpine chain provided motivation for the postulate that the present Mediterranean is a new creation following the destruction of the Tethys by Alpine folding. Argand (1924) was the first to invoke continental drift and rotation of microcontinents to account for the puzzling geography of the western Mediterranean. He postulated that Africa was pushed over Eurasia in the Early Tertiary to form the Alpine and Carpathian Mountains. After an orogenic climax in the Oligocene, the Mediterranean was reduced to a network of epicontinental seas; Italy, Corsica, Sardinia, the Balearic Islands, and Spain were all crowded to one corner in southwestern Europe and joined to France and the stable European block (Fig. 2a). He then postulated the start of an important phase of extension during the Late Oligocene or Early Miocene as Africa moved southeastward relative to Europe (Fig. 2b). Corso-Sardinia and Italy drifted counterclockwise and left behind the Balearic and Tyrrhenian Basins. Egypt and Libya were torn away from Greece and Turkey, thereby creating the eastern Mediterranean.

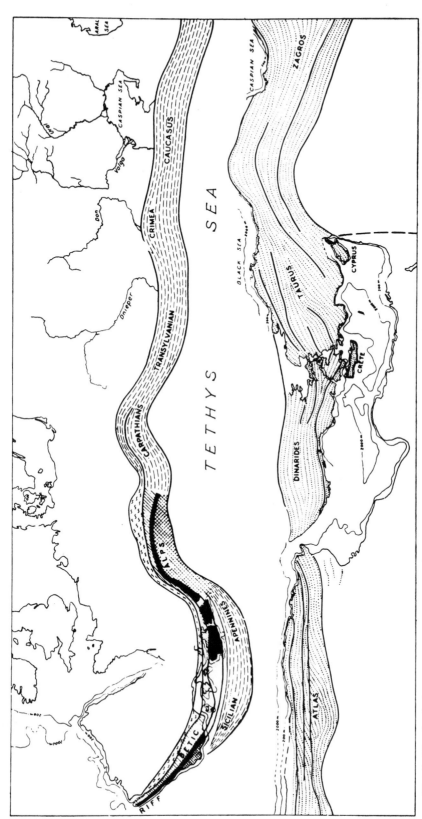

Fig. 1. The classical concept of Tethys: The Tethys was conceived as an east–west-trending ocean bounding Gondwanaland on the north. (From Carey, 1958.)

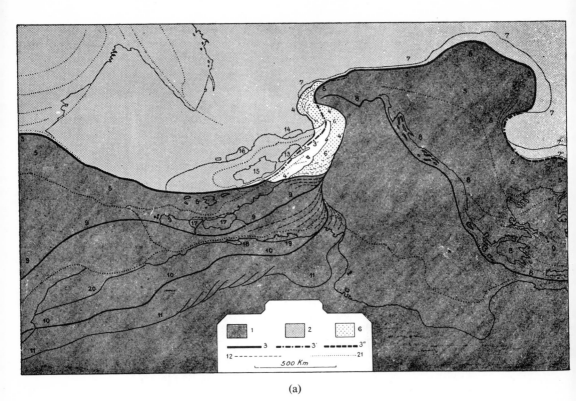

(a)

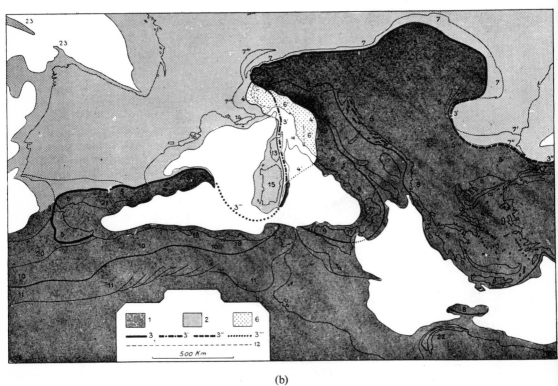

(b)

Fig. 2. Argand's postulate of the neo-Mediterranean: (a) Continental collision eliminated Tethys at the end of Alpine orogenic paroxysm; (b) rotation of microcontinents since Late Oligocene created a neo-Mediterranean. (From Argand, 1924.)

Argand's idea for a Neogene Mediterranean in the west received considerable support from later investigations (e.g., Anonymous, 1969). However, the record of continuous pelagic sedimentation (on Cyprus, for example) suggests that the eastern Mediterranean was already in existence during the Mesozoic (e.g., Robertson and Hudson, 1974). Geophysical and geological investigations have revealed a great tectonic contrast between the western and eastern Mediterranean basins. Whereas the western basins appear to have been created by extensional rifting, the eastern Mediterranean has been under compression (Hsü and Ryan, 1973). The current thinking tends to favor the third idea, that the present Mediterranean is a composite tectonic unit, combining a relic Tethys with a neo-Tethys (e.g., Smith, 1971; Hsü, 1971; Dewey et al., 1973). The purpose of this article is to discuss the tectonics of the Mediterranean basins and to present data concerning the various hypotheses.

B. Physiographic Provinces of the Mediterranean

A quick glance at the physiographic diagram of the Mediterranean floor (Fig. 3) suggests a threefold subdivision of a composite basin: (1) The Balearic Basin, characterized by an abyssal plain almost devoid of relief; (2) the back-arc basins of the Tyrrhenian and Aegean Seas, dotted with numerous seamounts and active volcanoes; (3) the eastern Mediterranean province, dominated by the presence of an arcuate submarine mountain range (the Mediterranean Ridge).

The differences in topography are a manifestation of the distinct tectonic frameworks. It is convenient to discuss the tectonic evolution of the various basins separately, starting from west to east.

II. BALEARIC BASIN

A. Crustal Structure and Age

The Balearic Basin includes the part of the Mediterranean west of Corsica, southeast of the Balearic Islands, and north of the African coast. The westernmost part, between Spain and Morocco, is commonly referred to as the Alboran Basin. Physiographically, the Balearic Basin is dominated by an abyssal plain at a depth of about 3000 m.

Both the eastern and western margins of the Balearic Abyssal Plain are fringed by "non-magnetic basement-highs," as shown by seismic profiles (Figs. 4a, 4c). Deep-sea drilling in 1970 revealed that those "highs" are tilted sialic basement on the upthrown blocks of graben faults (Figs. 4b, 4d). The facies distribution of the Late Miocene sediments indicate that topographic relief

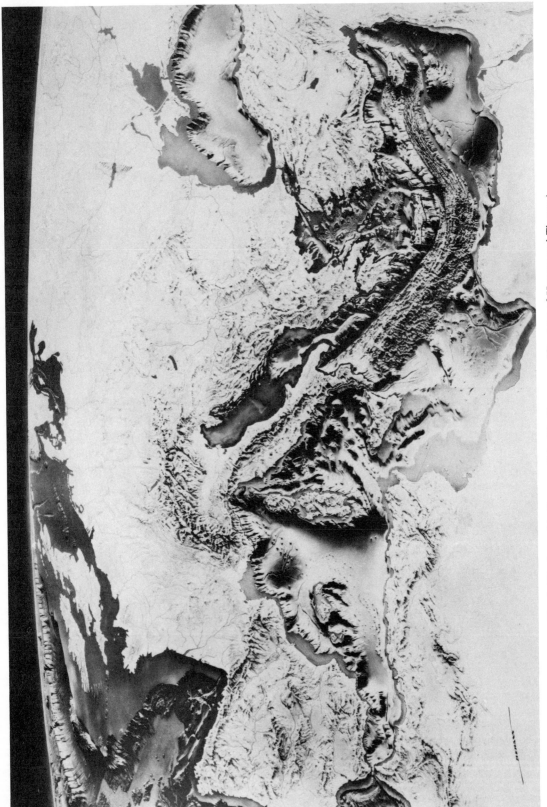

Fig. 3. Physiographic diagram of the Mediterranean. (Courtesy of Heezen and Thorpe.)

was in existence prior to the last evaporite deposition (Fig. 5); the age of faulting was largely pre-Messinian (Ryan *et al.*, 1972, p. 491).

Detailed seismic surveys on the northwestern Balearic margin off Minorca showed that the basement underlying the continental slope was also cut by normal faults, giving rise to the horst-and-graben structure typical of passive continental margins (Fig. 6). Drilling in 1975 proved the pre-Burdigalian (earliest Miocene or Late Oligocene) age of this phase of rifting (Hsü *et al.*, 1975). Drilling in the Gulf of Lyon south of the French coast gave about the same age for the intitial subsidence there (Burollet and Dufaure, 1971).

The Balearic Abyssal Plain is underlain by a thin crust and a thick sedimentary sequence. Seismic refraction (Falhquist and Hersey, 1969) and aeromagnetic surveys (Le Borgne *et al.*, 1971) suggest that the sediments are 5–6 km thick. A deep seismic refraction profile across the south Balearic Basin (= Algerian Basin) between Mallorca and Algeria, *Anna* III (Hinz, 1972), calibrated on the basis of drilling (Hsü and Ryan, 1972), indicates the presence of: (1) Quaternary and Pliocene (up to 3 km/sec V_p), 0.5 km thick; (2) Miocene evaporites and older sediments (3.5–4.5 km/sec V_p), 4 km thick; and (3) basic crust (6–7.4 km/sec V_p), 5 km thick. The French seismic surveys in the central Balearic Basin gave similar results (Mauffret *et al.*, 1973), the Quaternary and Pliocene sediments are up to 1–1.5 km thick. The Messinian evaporites include a 0.6 km upper evaporite unit and 1.2 km of deformed salt. The pre-evaporite formation should be about 3–4 km thick, above a non-sedimentary crust. Finetti and Morelli (1973, p. 53) carried out seismic surveys with wide-angle reflection and obtained comparable results in the central Balearic Basin (e.g., Fig. 7), except that the pre-evaporite formation was estimated to be almost 4.8 km thick locally (WA-6).

Drilling on the edge of the abyssal plain west of Sardinia (Site 134) in 1970 penetrated only the top of the Mediterranean evaporite. Later, at Site 372 south of Minorca, the drill passed through the evaporite and penetrated hemipelagic formations of Miocene age below. The evaporite deposits were largely deposited in shallow water or even subaerially(Hsü *et al.*, 1972*a*). Some considered this shallow-water origin as evidence of the very recent origin of the Mediterranean basins, and postulated thousands of meters of post-Miocene subsidence to account for present depths (Nesteroff, 1972; Stanley *et al.*, 1974; Sonnenfeld, 1975). However, all available evidence points to the existence of deep Mediterranean Basins prior to the Messinian (Late Miocene) salinity crisis. Paleoecological and sedimentological studies of the Quaternary and Pliocene from the Balearic Basin indicate that these sediments were deposited in a deep basin not much different in setting from that of the present. Cores of pre-Messinian sediments from the 1975 drilling campaign yielded further proof that the Balearic Basin was about as deep during the Early and Middle Miocene as it is today (Hsü *et al.*, 1975). The prior existence of a deep basin,

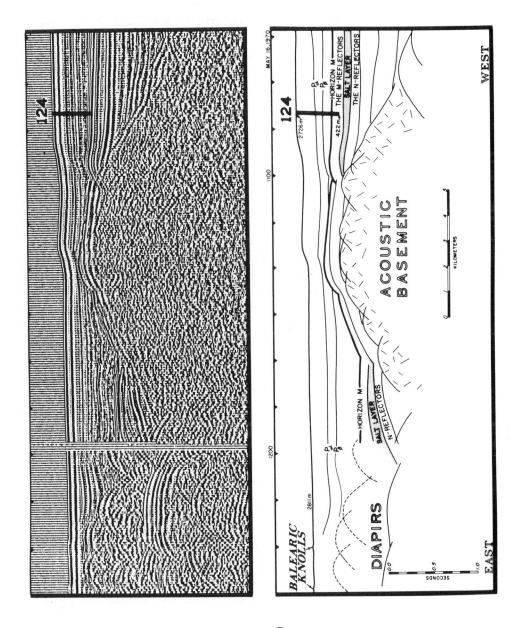

(a)

(b)

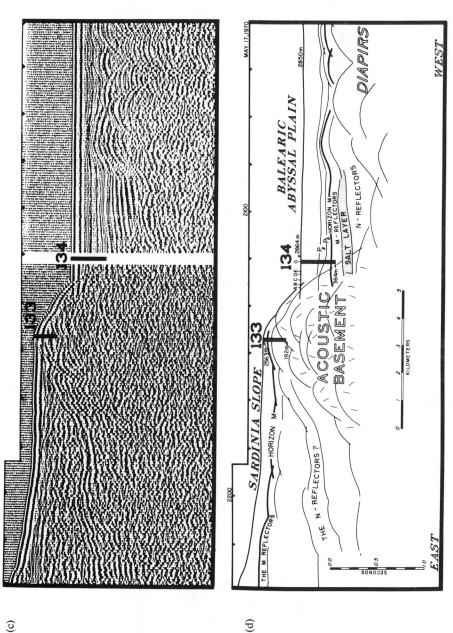

Fig. 4. Structures of Balearic Margins: (a) Seismic reflection profile by *R/V Jean Charcot* across a non-magnetic basement ridge on western Balearic margin (Balearic Rise); (b) interpretation of the same after DSDP Drilling Site 124, the acoustic basement believed to be an upthrown block of sialic crust (vertical exaggeration about 3 : 1); (c) seismic reflection profile by *R/V Charcot* across a buried non-magnetic basement ridge on eastern Balearic margin (Sardinia Slope); (d) interpretation of the same after DSDP Drilling Sites 133 and 134, the acoustic basement being an upthrown block of sialic crust (vertical exaggeration about 4 : 1).

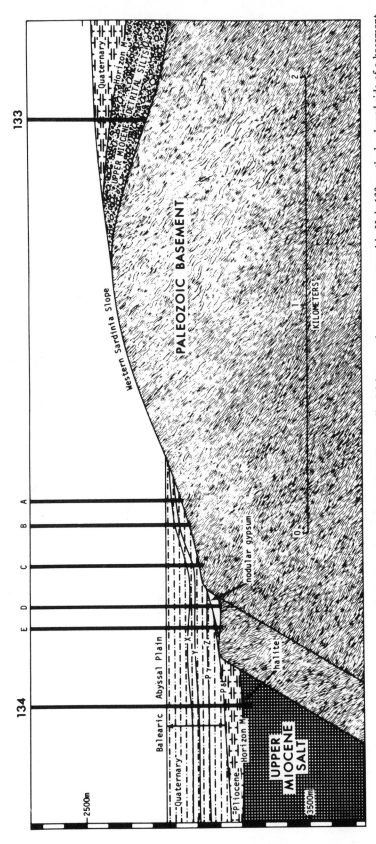

Fig. 5. Cross section of basement ridge beneath the foot of western Sardinia Slope: Alluvial fan gravels were encountered in Hole 133 on the landward side of a basement ridge at 2623 m subsea; subkha-type gypsum was penetrated at 3056 m subsea in Hole 134 D; playa salts were reached beneath the abyssal plain in Hole 134 at 3218 m subsea. The observed facies distribution is related to a Late Miocene topographic relief of more than 500 m elevation difference. (No vertical exaggeration.)

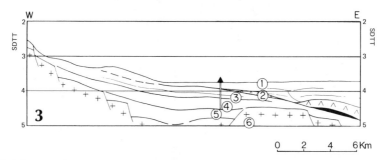

Fig. 6. Minorca Rise, DSDP Site 372: (1) Quaternary and Pliocene, (2) Messinian evaporites, (3) Tortonian to Upper Burdigalian, (4) Upper Burdigalian to Aquitanian, (5) Oligocene (?), (6) Sialic basement. SDTT = Seconds Double Travel Time. (Vertical exaggeration about 4:1.)

combined with the unmistakable indication of shallow-water evaporite deposition, can only be explained by a desiccated deep-basin model for the origin of these evaporites (Hsü et al., 1972a; Ryan, 1973; Cita, 1973).

A consideration of the crustal structure revealed by seismic surveys also suggests the existence of a deep Balearic Basin long before the beginning of the Pliocene inundation. A thin, 5-km oceanic crust would be expected to underlie an isostatically adjusted depression 5.5 km deep (Hsü, 1958). A rough test can be made because the consequence of sedimentary infill and of isostatic adjustment under the load of a sequence T_s thick and averaging 2.4 g/cm³ density can be computed as follows:

$$5.5 \times 1.03 + (T_s + D - 5.5) \times 3.3 = D \times 1.03 + T_s \times 2.4$$

$$D = \text{Basin depth} = 3.5 \text{ km} \qquad \text{for } T_s = 5 \text{ km}$$

$$D = \text{Basin depth} = 3.0 \text{ km} \qquad \text{for } T_s = 6 \text{ km}$$

Since the present sediment thickness is between 5 and 7 km, the computation suggests that the present 3-km-deep basin is isostatically adjusted. Given that the earliest Pliocene basin had 0.5 km less sediments, the Balearic should have been slightly deeper then.

Of course, we should also consider the isostatic load of water. The floor of a desiccated Balearic Basin at the end of the Miocene should have been about 2500 m (see Hsü et al., 1972a, p. 1128). Isostatic response to desiccation (i.e., basinal uplift) is indicated by the landward tilting of the Messinian basinal sediments and their truncation prior to the Pliocene submergence (Fig. 8). Since the end of evaporite deposition, the weight of the water and accumulating sediments should have induced a more than 1000-m subsidence. Evidence of post-Miocene faulting, as shown by seismic profiling records (e.g., Morelli, 1975), attests to the extent of such isostatic adjustments.

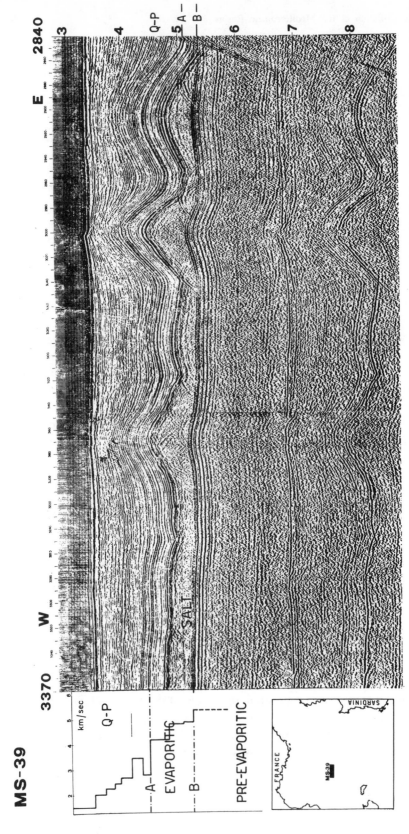

Fig. 7. Seismic reflection profile MS-39, central Balearic Abyssal Plain: (Q–P) Quaternary and Pliocene (0.8–1.2 km); (A–B) Messinian evaporite (2.0–2.8 km). Total sediment thickness about 8 km at about 6.5 SDTT. (From Morelli, 1975.)

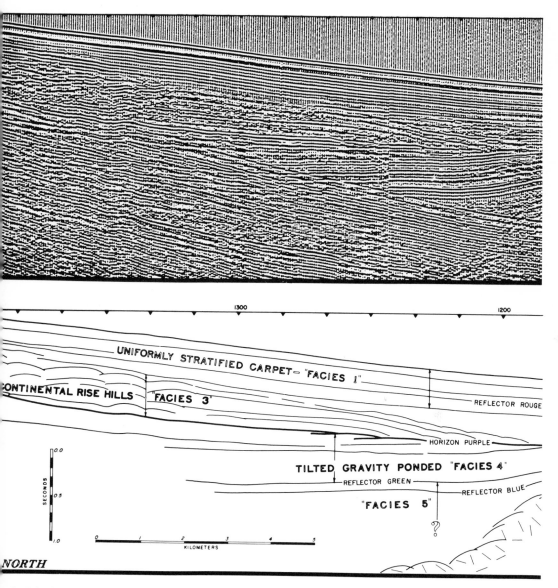

8. Seismic profile of *R/V Jean Charcot* and interpretation after drilling DSDP Site 121 at the base of continental
pe south of Malaga in Alboran Sea. Note the prominent unconformity (Horizon Purple) between the Pliocene ("Facies
and the Upper Miocene (Tilted Gravity Ponded "Facies 4"). We interpret the pre-Pliocene tilting (sloping toward
continent) as having been caused by isostatic uplift after the Alboran Basin was desiccated during the evaporite
osition. The tilted beds were bevelled and the unconformity surface was tilted toward the ocean by isostatic subsidence
r the Alboran Basin was flooded by sea water at the beginning of the Pliocene.

B. Origin

If the term "oceanization" is used in its broadest sense, e.g., that a region once land is now a part of an ocean, then there seems to be a concensus that the Balearic Basin has been "oceanized." The much quoted evidence for "oceanization, *sensu lato*" has been summarized by Pannekoek (1969): (1) Abrupt termination of land tectonic trends; (2) sedimentological evidence indicating transport of sediment from what is now the deep sea toward what is now land; (3) indications of drainage directed inland from the present coast; (4) movement of nappes or olistostromes directed from the present sea toward circum-Mediterranean lands.

However, "oceanization, *sensu stricto*," implies that a sialic crust was somehow transformed *in situ* into an ocean, and that an existing ocean basin was formed after the foundering of a continent. The so-called Atlantic type of oceanization (Van Bemmelen, 1969) is better known as continental drift, sea-floor spreading, or rotation of microcontinents. Foundering through oceanization, *s.s.*, and rotation of Corso-Sardinia microcontinents are thus two very different models that have been formulated to explain the origin of the Balearic Basin.

Many of the arguments for oceanization are equivocal. For example, it seems that a landmass was once present in the position now occupied by the Balearic Basin. This fact might mean that the Balearic originated by the foundering of a sialic crust. Or, alternatively, that the landmass was the Corso-Sardinia continent which has drifted away since Late Paleogene time. The extent of subsidence of the continental margins is also not unequivocal evidence. Drilling proved that the sialic basement west of Sardinia has subsided more than 3000 m since pre-Messinian time (Hsü *et al.*, 1972*b*). One might speculate that the Balearic Abyssal Plain was underlain by a similarly foundered sialic crust. Or, alternatively, one can explain such subsidence by assuming a crustal thinning of a rifted margin as Corso-Sardinia drifted away. For a critical appraisal of the two alternative models, we must have a knowledge of the crust and of the sediments under the central abyssal plain.

The strongest proponents of the foundering hypothesis were two French masters, Bourcart and Glangeaud. Bourcart (1950) first proposed the foundering of the Balearic Basin by continental flexure in order to explain the origin of submarine canyons cut into granite. Those canyons, filled with gravels, must have been cut subaerially, yet they are now submerged below 2000 m. The age of the canyons was found from dredging and submarine geology to be Late Miocene. Bourcart (1962) cited this evidence to postulate a Pliocene revolution in the Mediterranean. He also mapped faults around the Balearic Sea, and referred to a revolutionary change from Miocene compression to Pliocene tension. Bourcart's associate and successor at the Sorbonne, Glangeaud (1967), uncovered further evidence of Late Miocene canyon cutting to support the idea of continental flexure.

The foundering hypothesis received support when salt domes were found under the Balearic Abyssal Plain (Hersey, 1965). In 1966, an extensive series of French surveys mapped the distribution of the Balearic salts. Glangeaud *et al.* (1966) believed that the salt was Triassic and Liassic and considered the central Balearic Basin (Zone A) as the submarine extension of the Briançonnais and Subbriançonnais facies belt of the Maritime Alps. Criticizing Argand's postulate of a neo-Mediterranean, Glangeaud *et al.* (1966) postulated subsidence during the Mesozoic. They recognized a phase of deformation during the Tertiary, and held that Plio-Quaternary foundering was responsible for the subsidence of the present Balearic Basin.

The synthesis by Glangeaud *et al.* (1966) was warmly received by the Geological Society of France. Only a few skeptics counselled caution against the hasty assignment of a Triassic age to the salts (Sarrot-Reynauld, 1966; Cornet, 1968). The Bourcart–Glangeaud scheme became obsolete following drilling by the *Glomar Challenger*, which established the Miocene age of the salt in 1970 (DSDP Leg 13). Drilling gave credence to the idea first proposed by Denizot (1951) of deep erosion during a Messinian eustatic lowering of the Mediterranean base level when the evaporites were being deposited. We no longer need postulate a Pliocene revolution, or catastrophic foundering, or mysterious oceanization to account for the Pontian canyons on the margins of the Balearic Basin.

During the brief reign of the "Pliocene revolution" concept, facets of Mediterranean geology were overlooked. Ruggieri's (1967) account of a pan-Mediterranean salinity crisis and Cornet's (1968) prophecy of Messinian salt under the Balearic Basin were all but ignored. A still more critical hindrance to reaching a correct interpretation was the general failure to acknowledge the Late Oligocene or Earliest Miocene age of the Balearic Basin, the central theme of Argand's analysis.

Argand represented the best of the traditional "cylindrist school" of Alpine geology, which was founded on the premise of lateral continuity of Alpine facies belts (see Fig. 2a). The most prominent tectonic lineament in the Alps is the suture between the Eurasian and Gondwana continents (heavy lines 3, 3', 3'', 3''' in Fig. 2b). Noting the anomalous "detour" of the suture south of Liguria and the apparent disjunction between Sardinia and Minorca, Argand (1924) proposed the creation of a Neogene Balearic Basin by the rifting of Corsica and Sardinia.

During the last half century, several lines of reasoning have yielded arguments in favor of Argand's rifting postulate (with or without rotation). They are: (1) Morphological — matching and fitting of bathymetric contours (Nairn and Westphal, 1968; Smith, 1971; Hsü, 1971; Ryan *et al.*, 1972; West-

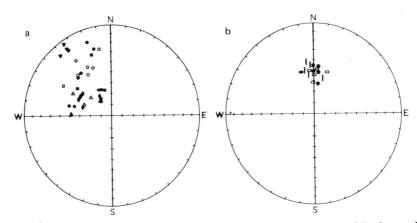

Fig. 9. Equal area projections of the site-mean directions of magnetization of Tertiary rocks from Sardinia. All symbols are projections on the lower hemisphere: Open symbols indicate paleomagnetic directions which had to be inverted to impinge on the lower hemisphere (reversed polarity), full symbols indicate directions with normal polarity. Note the north-westerly declinations of the Oligocene and Lower Miocene rocks (a), and the northerly declinations of the Middle Miocene and younger rocks (b). (From De Jong et al., 1973a.)

phal et al., 1973; Alvarez, 1973); (2) aeromagnetic surveys (Bayer et al., 1973; Auzende et al., 1973) — magnetic anomalies interpreted as magnetic lineations related to the sea-floor spreading; (3) seismological surveys (Fahlquist, 1963; Hinz, 1972; Finetti and Morelli, 1973), indicating the presence of ocean crust under the Balearic Abyssal Plain; (4) structural — matching of geological structures (Hsü et al., 1972b; Alvarez et al., 1974); (5) sedimentological — Corso-Sardinia located as the source of Oligocene and older detrital sediments in southern France and Italy (Stanley and Mutti, 1968); (6) plate tectonics — relating rotation to deformation and volcanism on land, particularly to the pre-Tortonian deformation and Late Neogene volcanism of the Apennines (Hsü, 1971; Boccaletti et al., 1971; Alvarez, 1973); (7) paleomagnetic.

The most critical evidence favoring the Corso-Sardinia rotation comes from paleomagnetic investigations, though the original idea did not come from those studies. Paleomagnetic studies of Corsica and Sardinia revealed a northwesterly dipole position for most pre-Tortonian rocks, exactly as Argand might have predicted decades ago (e.g., Nairn and Westphal, 1968; Zijderveld et al., 1973; De Jong et al., 1969, 1973a; Bobier, 1970; Coulon et al., 1974b; Bobier and Coulon, 1974; Manzoni, 1974). Summarizing the results up to 1972 (see Fig. 9), De Jong et al. (1973a, p. 282) commented on the magnetic declinations of pre-Tortonian rocks of Sardinia:

> The change in declination could be the result of a long lasting Late Oligocene–Early Miocene excursion of the geomagnetic dipole field, but such an event has not been demonstrated for this period. We conclude, therefore, that it is more

likely that the change in declination represents a counterclockwise rotation of Sardinia over an angle of approximately 50°. Because the declination of the tuffites of Mlt are intermediate, it may be argued that the rotation of Sardinia actually occurred during the sedimentation of the Aquitanian–Langhian (Early Miocene) rock sequence of Castelsardo.*

The rotation of Sardinia must have ended prior to the Tortonian. The Messinian evaporites are traceable across the whole basin, and post-Tortonian volcanics show a northerly paleomagnetic direction identical to that of today (De Jong et al., 1973a; Coulon et al., 1974b; Alvarez et al., 1973).

Deep-sea drilling south of Minorca in 1975 showed that the structural evolution of the Balearic margin was similar to that established for margins of the Atlantic type (Hsü et al., 1975). The deposition of the first sediments was controlled by horst-and-graben tectonics, indicating rifting of continental crust, probably in Late Oligocene. If we adopt the Atlantic spreading model

---

* Unfortunately, there is considerable confusion concerning the paleomagnetic data of Sardinia. The conclusion of the Dutch–Italian group is being challenged by a French team, who claim that there has been no rotation of Sardinia since the Late Oligocene (Coulon et al., 1973, 1974b). The former group worked in the Alghero District, the latter on Logudoro and Bosano volcanics nearby. The graphical summaries of dipole positions obtained by both groups are similar with declinations, generally west of north, for the Oligocene–Miocene volcanics (cf. Manzoni, 1974). Yet the Dutch–Italian group concluded that the data indicated counterclockwise rotation, whereas the French insisted that there had been no rotation! Later, the published results of the French team were included in the article written by De Jong et al. (1973b) to contradict the French claim (the average declinations are 332° for Alghero and 335°30′ for Logudoro; see also Alvarez et al., 1973). The north-west declination is particularly shown by the Ignimbrite Series, which has been dated as Early Miocene (17.5 m.y., in Coulon et al., 1974a; see Fig. 9). I might add that the French support the rifting model by sea-floor spreading without rotation (e.g., models of Le Pichon et al., 1971; Auzende et al., 1973; Bayer et al., 1973). Even if we accept their insistence that "La rotation aurait eu lieu bien avant l'Oligocène" (Coulon, 1973, p. 169), their interpretation cannot be cited as evidence against the postulate that the Balearic was created by an Oligocene–Miocene rifting of microcontinents.
A second confusion in interpreting paleomagnetism resulted from a review article by Zijderveld and Van der Voo (1970). They suggested that the angles of rotation of the micro-continents are comparable to that of Africa, and postulated that those blocks were rotated with Africa during the Alpine orogenesis. They admitted (p. 256) that the critical test is the "timing of rotations. If all the landmasses south of the Pyrenean–Alpine mountain belt belong to the African plate, rotations should be synchronous." On this test, their postulate did not hold up. As they stated, Africa rotated counterclockwise continuously between Early Cretaceous and Eocene. Spain, however, was rotated between Late Jurassic and Late Cretaceous (Stauffer and Tarling, 1971), terminating at about 73 m.y. when the Bay of Biscay was completely opened (Williams, 1973), long before the termination of African rotation. Much of the rotation of Italy also took place prior to Late Cretaceous (see discussion in Section IIIB). Finally, Corso-Sardinia was rotated during the mid-Tertiary. Those paleomagnetic ages of rotation are in agreement with those deduced on the basis of inter-preting stratigraphy (see Figs. 13a–f).

for the Balearic Basin, we might speculate that the oceanic crust under the abyssal plain was created mainly during the Late Oligocene or Early Miocene when Corsica and Sardinia were rifted apart and rotated to their present positions. The timing would be in accord with the conclusions of De Jong *et al.* (1973*a*) from their paleomagnetic studies.

The large sediment thickness found in the Balearic Basin has been considered evidence against a Miocene age (Le Pichon *et al.*, 1971; Auzende *et al.*, 1973). In fact, however, the basin has been the accumulation site for turbidites and has filled rapidly. Drilling west of Cyprus (Site 375) indicated that Tortonian turbidites in the Antalya Basin are 2 or 3 km thick, and were deposited at a rate of about one-half meter per thousand years. Thus, we cannot rule out a latest Oligocene or earliest Miocene age (> 25 m.y.) for the Balearic Basin, even though the pre-evaporite sediment may be 5 km thick as reported by Finetti and Morelli (1973). Actually, if we take as an average thickness for the pre-evaporite series the 3.5 km given by the French survey, and assume it was deposited during the 17.5 m.y. (24 m.y. to 6.5 m.y.) of the pre-Messinian Miocene, we obtain an average sedimentation rate of 20 cm/t.y. This rate is about that for the post-Messinian sediments of the Alboran Basin (DSDP Site 121, Ryan *et al.*, 1972, p. 56).

Oligocene–Miocene rifting of the Balearic Basin was certainly related to extensional stress. Argand referred to "les plus grandes distensions et disjunctions" after "les plus grandes paroxysms" of intercontinental collision. Extension started in Late Oligocene and continued until Middle Miocene, when Corso-Sardinia and Italy reached approximately their present positions (see Fig. 13f). Ocean grabens produced at the trailing edges of the rifted microcontinents now house the Balearic and Tyrrhenian Seas. Meanwhile, the leading edge of this microplate was thrust over the Adriatic foreland to induce the formation of the Apennines.

Argand did not neglect the evidence of Miocene deformations in the Subbetics or in the Rif and Atlas. He related these compressional structures to interaction between Spain, Africa, and a laterally drifting block (see Argand, 1924, Figs. 22–25). This idea was developed further by modern students of plate tectonics. The drifting block, encompassing the internal zones of the Betic, Rif, Kabylies, and possibly also Sicily and Calabria, was named the Alboran plate (Andrieux *et al.*, 1971; Hsü, 1972). The Alboran Sea was formed during the Early and Middle Miocene when the small plate was rifted apart (cf. Figs. 13e–f). Movement of the split halves led to collision with Iberia on one side and Africa on another side, and gave rise to the Betic, the Rif, and the Tellian Atlas (see Wezel, 1970; Auzende *et al.*, 1973).

C. Summary of Geologic History

To summarize, we might follow Argand and conclude that the Balearic Basin was part of a neo-Mediterranean, created in the Late Oligocene and Early Miocene by the rifting and by rotation of microcontinents. The ensuing marine transgression reached beyond the present confines of the Balearic Basin during the Burdigalian (Colom and Escandel, 1962). Pelagic sedimentation continued until earliest Messinian, when the salinity crisis started. A thick evaporite deposit, including halite, was formed as the basin was subjected to desiccation. The Balearic Abyssal Plain was covered at times by a shallow hypersaline sea, at times by a brackish lake, and at times by normal marine waters from the Atlantic (Hsü et al., 1972a). At the beginning of the Pliocene, a deep Gibraltar Strait came into existence. The sea rushed in, filling the basin. The weight of the water and of the accumulating sediments induced faulting and subsidence (up to 1000 m) to restore isostatic equilibrium. Hemipelagic sedimentation continued from that time to the present.

III. TYRRHENIAN BASIN

A. Crustal Structure and Age

The Tyrrhenian Basin is characterized by a small abyssal plain about 3500 m deep, which is dotted by a number of elongated, north–south-trending seamounts (Morelli, 1970; see also Fig. 10). Dredging on the edge of the abyssal plain has indicated that the crust around this small central basin is sialic (Heezen et al., 1971). The crust is thin, and may be characterized as oceanic under the central depression. Refraction seismic studies of the central basin indicate the presence of a layer with 6.9–7.5 km/sec velocity below 2.6 km of sediments (Fahlquist and Hersey, 1969); this layer might represent abnormally light mantle such as that in the mid-ocean ridge region (Morelli, 1975). Sonobuoy profiles across another part of the central abyssal plain (Wa-1A) indicate the presence of a thin crust above such a layer (Morelli, 1975, p. 63); the sequence there consists of: (1) Quaternary and Pliocene sediments, 0.2 km; (2) Messinian evaporites, 0.7 km; (3) pre-evaporite sediments, 1.4 km; (4) non-sedimentary crust, 4 km; (5) 7.1 km/sec layer, lower crust?, upper mantle?, 1 km; (6) upper mantle, 8.2 km/sec. The crust between the shelf edge and the coast is continental, and the Moho depth there is 20–25 km (Colombi et al., 1973; Morelli, 1975).

In contrast, in the Balearic Basin the sedimentary infill is thickest under the abyssal plain; the Tyrrhenian Basin sediments are thickest on the periphery. Between Corsica and Elba, the sedimentary sequence is 8 km thick, and is believed to be mainly Miocene (Finetti et al., 1970; Morelli, 1975). The section

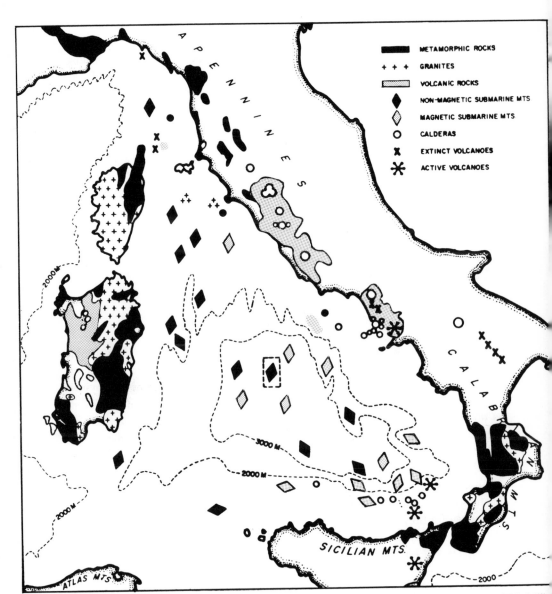

Fig. 10. Distribution of submarine mountains (seamounts) in the Tyrrhenian Basin. The dredging and DSDP drillir
results suggest that non-magnetic seamounts are underlain by sialic basement blocks, the magnetic seamounts of th
abyssal plain province are underlain by tholeiitic basalts, and the magnetic seamounts in the southeastern Tyrrhenia
by calcalkaline volcanics. The seamount framed by the dashed line yielded sialic basement rocks to dredging. (Aft
Heezen *et al.*, 1971.)

east of Sardinia is also mainly Miocene and relatively thick. Seismic data indicate the existence of the Messinian salts in diapiric structures (Finetti and Morelli, 1973). Seismic profiles from the Tyrrhenian margin of Italy and Sicily show very thick Pliocene and Quaternary sediments, where the top of the evaporites (M-reflector) plunges to a depth of about 5 km (Fig. 11). It seems that land-derived detritus was trapped in subsiding areas on the margins of the Tyrrhenian Sea, so that the central abyssal plain was relatively "starved."

The age of the present Tyrrhenian Basin is commonly believed to be Neogene. However, there is considerable controversy concerning the precise date of initial subsidence.

Selli and Fabbri (1971) noted the diachronous nature of the earliest Pliocene sediments in the Tyrrhenian and suggested that subsidence began in the Middle Pliocene after the Messinian salinity crisis. This idea was also entertained by their associate Wezel (1974) at Bologna.

The Bolognese school was opposed by Cita, Ryan, and Hsü, who discussed the fallacy of the arguments for a Pliocene Tyrrhenian Basin (Ryan et al., 1972, p. 428–429). Deep-sea drilling at Site 132 on the Tyrrhenian Rise did not find transgressive deposits indicative of a passage from shallow to deep water during the Pliocene. Shallow-water Messinian evaporites were found, but they were abruptly overlain by deep-water Pliocene and Quaternary hemipelagic sediments. The deep-water character of the latter resulted from abrupt flooding at the beginning of the Pliocene, as a formerly desiccated deep basin was filled, rather than from catastrophic subsidence during the Middle Pliocene. Drilling results from the second Mediterranean DSDP cruise also argue against Pliocene foundering. In Hole 373 under the Tyrrhenian Abyssal Plain, Lower Pliocene hemipelagic ooze unconformably overlies pre-evaporitic sumbarine basalt on the flank of a seamount (Hsü et al., 1975). Partly brecciated Miocene basalts, cemented by calcareous oozes, were apparently extruded onto a deep-sea floor.

While the Tyrrhenian Basin was essentially created prior to the latest Miocene (6 m.y.), we know very little of its beginning. Boccaletti and Guazzoni (1972) reasoned that it could not have started subsiding until after the early Tortonian orogeny (12 m.y.). However, rifting in a central abyssal plain could be synchronous with folding of the periphery as suggested by the genesis of the Alboran Basin. Thus, we can not rule out the postulate of Argand (1924, Fig. 23), who dated the initial rifting as Burdigalian. In fact, the presence of hemipelagic sediments of this age on Pianosa Island (Dallan, 1964) seems to support this idea. Nevertheless, a preponderance of the evidence indicates that the Tyrrhenian is a younger basin than the Balearic. The "oceanization" (*sensu lato*) processes under the Tyrrhenian Basin may have continued to operate long after they had stopped under the Balearic Basin. The Tyrrhenian Basin has, therefore, a basin-and-range type of physiography, in strong contrast to the flat-bottomed Balearic Abyssal Plain.

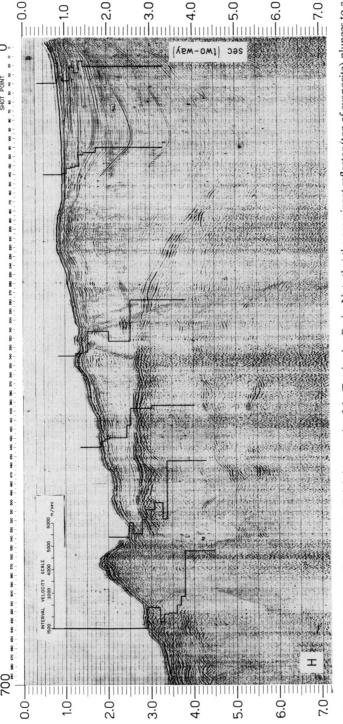

Fig. 11. Seismic reflection profile across the Calabrian margin of the Tyrrhenian Basin. Note that the prominent reflector (top of evaporite) plunges to a depth of more than 5 SDTT under the Calabrian margin. (From Morelli, 1975.)

B. Origin

Hypotheses similar to those postulated for the Balearic Basin have been advanced to account for the origin of the Tyrrhenian Basin. There are the "founderers" who invoke vertical tectonics, and there are the "rifters" who picture lateral movements.

Selli and his associates are among the most vocal "founderers." Their idea received support when sialic materials were dredged from non-magnetic basement highs in the Tyrrhenian Abyssal Plain (Heezen *et al.*, 1971; see also Fig. 10). We might thus conclude that large parts of the Tyrrhenian Basin, underlain by sialic crust, have subsided. Yet the existence of an oceanic structure under the central abyssal plain seem to favor the "rifters."

Argand was the first and the most radical of the "rifters." He placed Italy (prior to rifting) just to the southeast of Corsica and Sardinia. This scheme required a counterclock wise rotation of about 90° since the Oligocene (see Argand, 1924, Figs. 22–25). Italy has rotated that much since the Permian (Zijderveld *et al.*, 1970; Smith, 1971, p. 2049), but the latest paleomagnetic evidence permits only a 25° rotation of Umbria since the Eocene (Lowrie and Alvarez, 1975). It should be pointed out, however, that the Umbrian Apennines are autochthonous, and that only the allochthonous formations in the Apennines were involved in rotation during the Neogene. The idea was developed in some detail by Boccaletti and Guazzoni (1972) and by Boccaletti *et al.* (1971). According to their postulate, the Mesozoic Paleo-Tyrrhenian Oceanic Basin was completely consumed at the beginning of the Tortonian and the Tyrrhenian plate was subducted along a westerly dipping Benioff Zone east of Sardinia (Fig. 12). The present Tyrrhenian Basin was created as a marginal basin behind a late Neogene Periadriatic–Calabrian arc. This idea of viewing the Tyrrhenian as a back-arc basin was based in part upon studies of seismic foci (Ritsema, 1970; Caputo *et al.*, 1970). Investigations of the Tyrrhenian volcanics lent further support to the model. Barbieri *et al.* (1973) noted that the calcalkaline volcanism of the Eolian Islands (Fig. 10) is genetically related to a west-north-west-dipping subduction zone. In contrast, the oceanic tholeiites from the north–south-trending seamounts in the abyssal plain (Fig. 10) could be interpreted as products of sea-floor spreading in a back-arc basin. Drilling on 42-Seamount (Site 373) in the Tyrrhenian Sea penetrated more than 200 meters of olivine tholeiite very similar to that on the Mid-Atlantic Ridge. The drilling results fulfilled the prediction of "rifters" that the magnetic basement highs and seamounts of the Tyrrhenian Abyssal Plain are underlain by basaltic volcanics produced by sea-floor spreading.

The actual difference between "oceanization" and "marginal basins," or that between "founderers" and "rifters," may not be so irreconcilable as the semantic contrast seems to demand. While opting for foundering, Heezen *et al.*

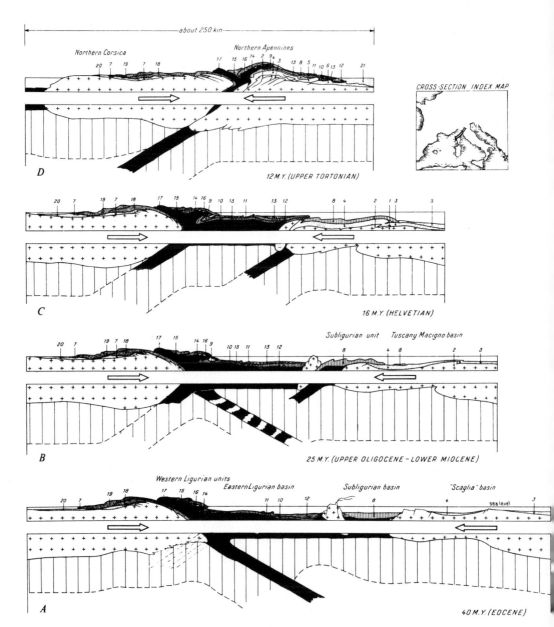

Fig. 12. Scheme by Boccaletti *et al.* (1971) for the Tertiary development of the Alps–Apennine structures. Note the presence of flysch deposits (10, 11, 12, 13) on the northern margin of the Paleo-Tyrrhenian Sea. During the early Tortonian the flysch formations were thrust on land to form the Apennines by collision of Corsica and the Italo-Dinaridian block. Similar continental-margin-type flysch sediments are found in the mountains of Sicily and the Tellian Atlas.

(1971, p. 328) wrote:

> We come to the conclusion that a continental terrain formerly existed in the
> present location of the Tyrrhenian Sea and that it started to subside during the
> Neogene time. The subsidence of Tyrrhenian appears to have been accompanied
> and perhaps caused, by extension of the crust which broke the former conti-
> nental crust into a series of crustal slivers, which are now represented by narrow
> ridges. Volcanic activity along these same faults created the linear volcanic
> piles.

It may be recalled that the intrusion of basalt dikes into a preexisting
crust is a part of the sea-floor spreading process postulated by Moores and Vine
(1971) on the basis of their observations of the Sheeted Dike Complex in the
Troodos Massif of Cyprus. In some regions, the pre-existing crust is not an
ophiolite, and dike swarms intruding into continental crust are well known.
On the Canary Islands, for example, dikes emplaced into preexisting sialic
materials have a spacing comparable to that of the Troodos Sheeted Dike
Complex (Gansser, personal communication). If we substitute "dike swarms"
in place of "linear volcanic piles" in the text of Heezen *et al.* (1971), we would
be describing a model put forward by the "rifters," who wanted to rip continental crust apart to create a back-arc basin. Perhaps, the two schools envision
the same thing. Their differences may be traced to their choice of words or
their placement of emphasis while describing the same phenomenon.

Accepting the Neogene origin of the Tyrrhenian Basin does not necessarily
refute the postulate of a Paleo-Tyrrhenian precursor during the Paleogene.
In North Africa, in Sicily, and in the Apennines there are thick sequences of
Oligocene and Early Miocene flysch sediments (Numidian–Reitano and their
equivalents). Sedimentological studies, mainly by Wezel (1970), suggest that
they represent sediments deposited on the African continental margin, which
then extended from Morocco to central Italy. North and west of this margin
there should have been a Paleogene Tyrrhenian Basin. This small ocean basin
was consumed before early Tortonian (Boccaletti and Guazzoni, 1972; Wezel
and Ryan, 1971), prior to the extension phase which created the present Tyrrhenian Basin. One may question if the Paleo-Tyrrhenian Basin completely
disappeared prior to the birth of the new one. De Roever (1969, p. 10) suggested:

> Part of the landmass, formerly existing in the western Mediterranean was formed
> by upheaval of the ocean floor above sea level (cf. Cyprus, ...). This possibility
> implies later isostatic subsidence in the region under consideration; the upheaval
> ... could have been a purely mechanical process caused by long-lasting influence
> of compressive tectonic forces, and could have ceased and changed into sub-
> sidence of isostatic character after these forces ceased to operate.

Is it possible then that a Cyprus-type island was present in the Paleo-
Tyrrhenian? Would such a model explain the local presence of 7.1 km/sec

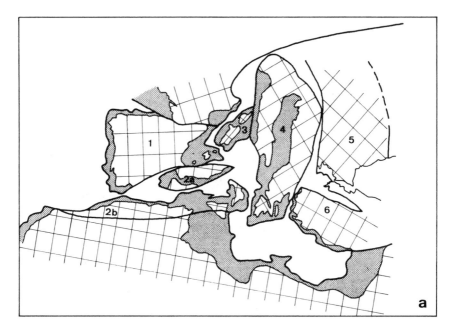

Fig. 13a.

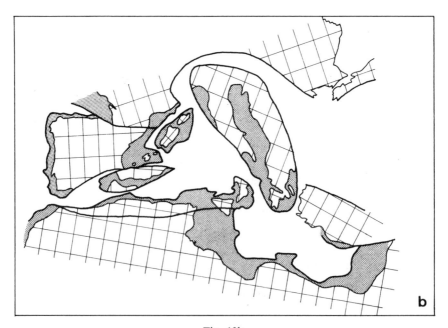

Fig. 13b.

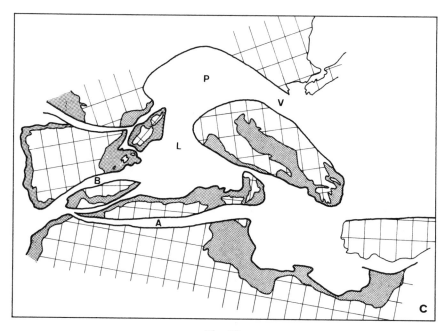

Fig. 13c.

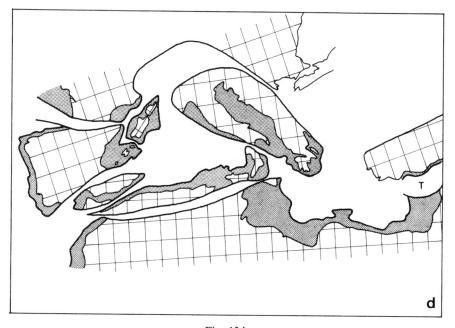

Fig. 13d.

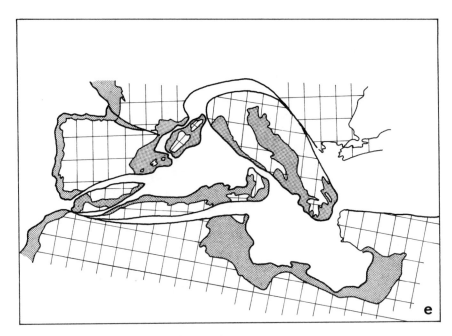

Fig. 13e.

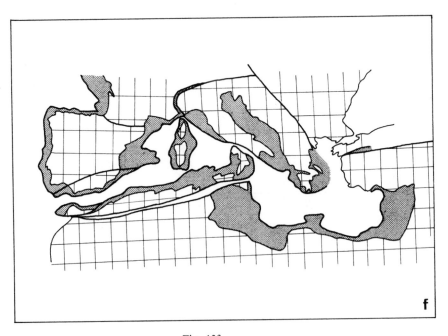

Fig. 13f.

material directly beneath a sedimentary column under the Tyrrhenian Abyssal Plain?

C. A Speculative Summary

Paleogeographic reconstructions (Hsü, 1972) suggested that a small ocean basin was open during the Middle Jurassic between the Italo-Dinaride block and Africa on the one side and the stable European block on the other. I have chosen to name this the Meso-Mediterranean Basin (Fig. 13c). Vestiges of this former ocean are found in the ophiolite melanges of the Ligurian Apennines, the Pennine Alps, and the Vardar Dinarides. True ophiolites are not known from the Betics or from the Tellian Atlas where Mesozoic and Paleogene deep-water hemipelagic and flysch sediments are present. Probably those geosynclinal basins were not rifted widely enough apart to permit the emplacement of a simatic floor.

By the end of the Eocene the oceanic crust of the Vardar and Pennine Troughs was all but consumed; however the configuration of the continents may have prevented a complete suture. A triangular oceanic region may have existed between Italy, North Africa, and Corso-Sardinia (Fig. 13e). The great distension started in the Late Oligocene when Corso-Sardinia started to rotate and the Alboran block was rifted into halves (cf. Figs. 13e–13f). At the same time southern Spain, North Africa, Sicily, the Italian Peninsula, and probably also the eastern edge of Corso-Sardinia was thrown into compression by the

---

Fig. 13. Tectonic evolution of the Mediterranean through a "metamorphosis" of Tethys: (a) Reconstruction of microcontinents in the Late Triassic (modified after Bosellini and Hsü, 1973) — (1) Iberia, (2a) Alboran (Rif–Betic), (2b) Alboran (Kabylia), (3) Corso-Sardinia, (4) Italo-Dinaridian, (5) Bulgarian, (6) Anatolia. (b) Alpine Tethys in the Late Jurassic. (c) Alpine Tethys in the Early Cretaceous — (A) Atlas Geosyncline, (B) Betic Geosyncline, (L) Liguria Eugeosyncline, (P) Pennine Eugeosyncline, (V) Vardar Eugeosyncline. The Alpine geosyncline has reached its widest extent (see Colom and Escandell, 1962). Whether the Anatolia block started to be ripped away from Africa during the Cretaceous or in an earlier episode (Triassic? Jurassic?) depends upon the eventual determination of the ophiolites of the Taurides. (d) Alpine Tethys in the Late Cretaceous. Note the elimination of the Vardar, and the narrowing of the Pennine by plate consumption during the Cretaceous ("pre-Gosau orogeny"). The Tauride Eugeosyncline (T) was fully developed and sea-floor spreading created the Troodos Massif of Cyprus. (e) Alpine Tethys in the Late Eocene. The collision of the Italo-Dinaridian block and Eurasia led to the Late Eocene deformation of the western Alps, when the subduction zone flipped from north-dipping to south-dipping (see Hsü and Schlanger, 1971). Movement of the Alboran block led to deformation of the Betic, Rif, and Tellian Atlas. (f) Mediterranean in the Middle Miocene after the rotation of Corsica and Sardinia. The gaps between Eurasia and microcontinents were all but eliminated by widespread Early and Middle Miocene orogenies. The shape of the Mediterranean began to assume a geometry similar to that of the present sea. Only the movement of the Alboran block during the Tortonian was still to take place, to enlarge the Tyrrhenian Basin.

displacement of the "drifting" microcontinents, resulting in subduction of the crust of the Paleo-Tyrrhenian Basin. The widespread orogenic deformation culminated in the early Tortonian; the processes leading to the creation of the present Tyrrhenian Basin might have started at that time as the crust was "oceanized" by the emplacement of dike swarms (analogous to the Sheeted Dike Complex of Cyprus) while the Ionian Basin was subducted under the Calabrian Arc. Subsidence of the Tyrrhenian Basin might have progressed to such a great extent when the Messinian salinity crisis started that the evaporites were deposited in a basin not much different physiographically from that of today (Fig. 13f). Further distension and subsidence during the Pliocene and Quaternary resulted in continuing volcanism and an enlargement of the basin.

IV. EASTERN MEDITERRANEAN

A. Crustal Structure and Age

The physiography of the eastern Mediterranean is dominated by an arcuate submarine mountain range, the Mediterranean Ridge. Southwest of the ridge is the Ionian Abyssal Plain. On the southeast, between the Mediterranean Ridge and the onlapping Nile Fan, is the narrow Herodotus Abyssal Plain. The Mediterranean Ridge itself consists of a number of parallel chains, and is separated from the Cretan Arc on the north by the Hellenic Trench (Fig. 3). The eastern and southern margins of the Mediterranean are stable or passive continental margins. The northern margin, by contrast, is an active one where the plate moved under the Cretan and the Calabrian Arcs (McKenzie, 1972; Papazachos and Comninakis, 1971). Deep-sea drilling through the inner wall of the Hellenic Trench (Site 127) penetrated a tectonic melange, which probably was produced by this subduction (Hsü and Ryan, 1973). The western boundary of the eastern Mediterranean is complicated. I have mentioned the hypothesis of subduction under the Calabrian Arc. However, there is a curious absence of earthquake foci between 50 to 200 km along the postulated subduction zone (McKenzie, 1972). Also the Calabrian margin is not bounded by a trench, but by a foredeep filled with sediments. Overthrusting took place along this margin during Plio-Quaternary time (Ogniben, 1969). Compression tectonics are also evident on seismic profiles east and south of Calabria (Finetti and Morelli, 1973; MS-20; MS-25). In contrast, the profile (MS-26) between Sicily and Ionia indicates a passive continental margin within the African plate.

Seismotectonic studies by McKenzie (1972) and by Papazachos (1973) suggest that the Adriatic Basin (the Italo-Dinaridian block) is now a part of the African Plate. While the Tyrrhenian block appears to be moving eastward, the Adriatic is moving north or northeast relative to Europe and to the Aegean

(Fig. 14). The complex deformational history of Italy during the Late Pliocene and Quaternary seems to have resulted from such plate motions.

Crustal structure in the eastern Mediterranean is less well known than that in the western part because the very thick sedimentary column commonly prevented penetration to Moho. Morelli (1975, p. 95) attempted to contruct a crustal profile for the Levantine Basin on the basis of poor sonobuoy and good reflection data (WA-3) (Fig. 15). The Moho is placed at 25 km, and is overlain by a simatic crust 7 km thick. Whether the overlying "lower sialic crust" consists of Mesozoic carbonates or of granite cannot be determined on the basis of its 6 km/sec velocity. If this crust is sedimentary, the total sedimentary column would be approximately 16 km thick, comparable to that reported for

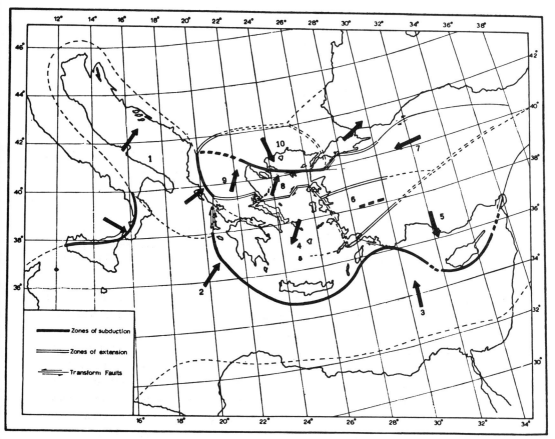

Fig. 14. Lithospheric blocks in the eastern Mediterranean region. The present boundary between Eurasia and Africa consists of numerous lithospheric blocks reminiscent of the situation during the geologic past. Arrows indicate directions of local movement. Note that the Tyrrhenian block (a part of Eurasia) moved east-southeast while the Adriatic (a part of Africa) moved northeast. Note also that the Aegean block is at present under extension. (From Morelli, 1975.)

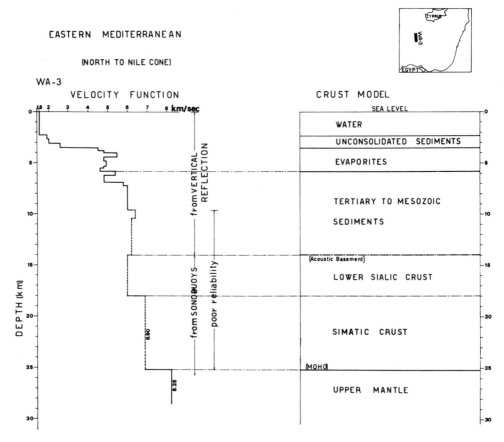

Fig. 15. Morelli's crustal model for the central eastern Mediterranean. The lower sialic crust may not be sialic basement but sedimentary rocks. If so, the model suggests the presence of an old oceanic crust overlain by a very thick sedimentary sequence. (From Morelli, 1975.)

the Black Sea. To the southeast of WA-3 profile, Lort's (1973) R-7 profile near Nile Cone shows the Moho at 27 km, while the 6.4? km/sec crust was interpreted to be 10 km thick. The latter profile is located close to the African continental margin and is probably not representative of the central basin.

Gravity anomalies have permitted several interpretations of crustal structure under the Mediterranean Ridge (see review by Morelli, 1975, p. 88–89). The model presented by Rabinowitz and Ryan (1970) assumes tectonic thickening under the Mediterranean Ridge and might explain the unusually low heat flow data obtained by Erickson (1970; see also Hsü et al., 1975).

The age of the eastern Mediterranean is largely a matter of interpretation as deep-sea drilling has penetrated only the uppermost horizons of the thick sedimentary column. Hemipelagic sediments and distal turbidites of Early or Middle Miocene age were recovered at several drilling sites (Sites 126/377;

129/375), indicating that the Levantine was deep sea long before the onset of the Late Miocene salinity crisis. Sancho et al., (1973) and Morelli (1975, p. 77) thought that the Ionian Basin might have originated by Miocene and/or post-Miocene foundering. However, there is little evidence to suggest it is so young, and deep-sea drilling in the Ionian Abyssal Plain (Site 373) indicated that the basin was deep by the end of the salinity crisis (Hsü et al., 1975).

The best direct evidence for the age of the eastern Mediterranean is afforded by the outcrops at Cyprus where the Troodos Massif is believed to be an uplifted fragment of ancient Mediterranean floor (Gass, 1968; Moores and Vine, 1971). A correlation of sea-floor seismic data with land geology was established by drilling at Sites 375 and 376 west of Cyprus (Hsü et al., 1975). The Troodos Massif is Cretaceous or older (Moores and Vine, 1971) and the main ophiolite complex may well be older (Freund et al., 1975). Drilling on the Mediterranean margins also has suggested that the Ionian and Levantine Basins are Mesozoic. Deep-sea sediments of Jurassic age were encountered in the Delta 1 borehole offshore from Israel (Friedman et al., 1971). This Jurassic formation can be traced by seismic reflection studies from coastal regions far into the deep sea (Neev et al., 1973). This evidence led Freund et al. (1975) to postulate a Late Triassic age for the eastern Mediterranean. On the western passive margin in southeastern Sicily, Cretaceous and Jurassic hemipelagic sediments overlie Triassic shelf carbonates in the oil wells of the Ragusa districts (Sander, 1970), again suggesting an initial subsidence during the earliest Jurassic.

A Mesozoic age could explain the tremendous thickness of sediments in the eastern basins. For the 16 km of sediment deposited there in 160 m.y., we obtain an average rate of 10 cm/t.y., quite reasonable for deposition of hemi-pelagic (plus some pelagic and turbiditic) sediments.

B. Tethys and Mediterranean

The geology of the circum-Mediterranean lands records a history of compression tectonics during the Tertiary as Africa moved ever closer to Eurasia. Geophysical data and deep-sea drilling evidence indicate recurrence of orogenic deformation within the basin during the Neogene (Ryan, 1969; Hsü and Ryan, 1973).

The idea of "relic Tethys" as a working hypothesis for the origin of the eastern Mediterranean seems reasonable. The age of this relic Tethys is probably Late Triassic or Jurassic. What was the configuration of the ocean and continents in the Triassic before the spreading of the Atlantic created a revolutionary reorganization of land and sea?

There is general consensus that the Triassic supercontinent Pangea probably assumed the shape as shown by Fig. 16 (Bullard et al., 1965; Smith, 1971). The Tethys was then a wedge-shaped ocean, not an equatorial Mediterranean.

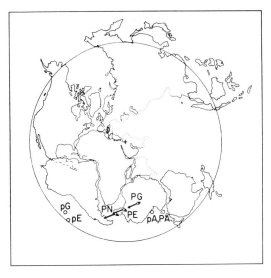

Fig. 16. A possible fit of the principal continental areas during Permo-Triassic time. Average Permian pole positions have been rotated with the continent concerned and plotted on the reassembly relative to Africa: (PA) Africa, (PN) North America, (PG) Greenland, (PE) Eurasia. Unrotated poles are pA, pG, pE. Note the presence of a large oceanic area southeast of Eurasia and northeast of Gondwanaland, which might be designated as the Permo-Triassic Tethys. This ocean is believed to have been consumed along the Dobrogea–Crimea–Caucasus–Kopet Dagh trend during the Late Triassic and Jurassic, while the Mesozoic Alpine Tethys was being created by sea-floor spreading. (From Smith, 1971.)

However, several attempts have been made to reconstruct the arrangement of Mediterranean microcontinents. Smith (1971) used a least-square fit of the 1000-m contour. He recognized four major microcontinents in the Mediterranean region: (1) Iberia, (2) Corso-Sardinia, (3) Italy, and (4) Greece–Yugoslavia–Turkey. At the end of the Triassic, Smith believes all these blocks were linked together to fill in a wedge-shaped space between Europe and Africa and west of the Triassic Tethys. Dewey *et al.* (1973) disapproved of this fitting method because the margins of microcontinents are rarely rifted margins. They claimed (p. 3155) to have achieved reassembly on the basis of "multiple arguments." I studied their opus and could find little factual basis for their reconstruction.*

---

* The geological evidence presented by Dewey *et al.* (1973) is sketched in Figs. 17, 18, and 19. The figures are useful, graphic summaries of pertinent information, but they are inadequate to support their identification of plate boundaries or to warrant their kinematic descriptions of the complicated motions of some 20 ill-defined blocks. In fact, several key postulates by Dewey *et al.* (1973) are contradicted by geological data. For example, they recognized two sialic fragments, Rhodope and Moesia on the south and north shores of the Triassic Tethys, respectively. While Moesia was tucked safely under the belly of the southeastern European platform, with minor exceptions, the Rhodope wandered all over the Mediterranean and was finally welded to Moesia during the Early Miocene (see Dewey *et al.*, 1973, Figs. 8b and 18). One would thus expect the Balkanides to be a major suture between those two sialic fragments. In fact, those mountains have a style of deformation comparable to the folded Jura; nowhere could one find a trace of a suture–melange. All the postulated motions of the Rhodope Block are thus demonstrably false. The separation by Dewey *et al.* (1973) of Apulia across the Triassic Tethys from the Carnics is equally unacceptable.

Like Smith, I believe that the present outline of the various sialic blocks could give us useful information, but his computer-fitting was too rigid a methodology. I attempted a visual fit of the 2000-m contour of the Mediterranean microcontinents (Hsü, 1971). The credibility of this geometrical reconstruction was later checked with an analysis of the Triassic sedimentary facies (Bosellini and Hsü, 1973). In my first attempt I placed a Triassic Tethys between the Italo-Dinaridian Block and Bulgaria and envisioned the Vardar Trough as a relic of a Permo-Triassic Tethys (Hsü, 1971). However, Smith (1973) reviewed the evidence for dating Tethyan ophiolites and concluded that the Alpine ophiolites are mainly the product of Jurassic and Cretaceous sea-floor spreading. His evidence is persuasive, but then where is the Triassic Tethys?

Pelagic sediments of Late Triassic (Hallstatt Facies) extend from the Austrian Alps to the Carpathians and they occur also as exotic blocks in the Balkanides. If those deep-water deposits were laid down on a subsiding continental margin, then truly oceanic Tethys should not be far away. Yet, nowhere can we identify vestiges of Permo-Triassic ophiolites.

A consuming plate margin may be defined by delineating andesite volcanism caused by plate subduction. A belt of Triassic and Jurassic volcanism can be traced from Dobrogea to the Crimea and the Caucasus (Fig. 17). In addition, Triassic flysch and submarine volcanics, deformed during the "Cimmerian" (Triassic–Jurassic) orogeny (Nalivkin, 1973; see also Fig. 18), have been recognized in these regions. We might thus place the northern margin of the Permo-Triassic Tethys along the Dobrogea–Crimea–Caucasus trend. Such a reconstruction implies that Bulgaria was still more or less attached to the Italo-Dinaride block [and to Euro-Africa during the Triassic (see Fig. 13a)].

The southern margin of the Triassic Tethys (or the northern edge of the Africa–Arabian continents) should be sought in the Middle East. A zone of ophiolite melange extends from Cyprus to southern Turkey, to Zagros in Iran, to Oman, and thence to Afghanistan (Gansser, 1974; see also Fig. 19). Yet, like the Alpine ophiolites, the melanges include no known Permo-Triassic ocean floor fragments. It seems that the northern margin of the Triassic African continent must be present near or under the Kopet Dagh Range of northeastern Iran. During the autumn of 1975, I was shown some outcrops of ultramafic rocks occurring as lenticular bodies in Carboniferous (?) slates near Mashad. Those rocks were interpreted as pre-Jurassic picrite dikes (Davoudzadeh, 1975). My own impression is that they could be melange blocks of fragmented sea floor. Perhaps, we have finally found the vanished Triassic Tethys.

Fragmentation of Pangea began in the Late Triassic. Subsidence created a number of "aborted oceans" within the Alpine realm, where submarine basalts, pelagic limestones, and flyschlike sediments accumulated, but most of them were stillborn and never developed into an ocean basin. Major reorganization of the continents and oceans started in Early Jurassic when Pangea was split

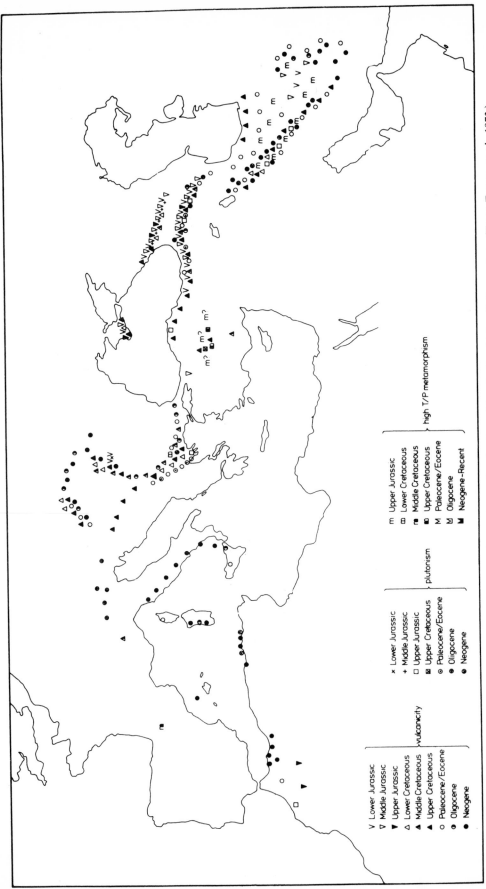

Fig. 17. A graphical summary of the distribution and age of volcanic activites in the Alpine Tethys system. (From Dewey *et al.*, 1973.)

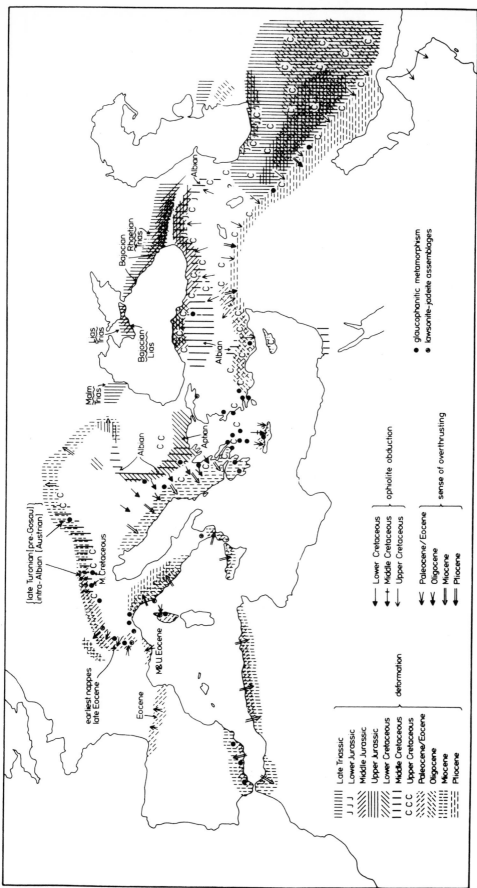

Fig. 18. A graphical summary of the distribution and age of orogenic deformation in the Alpine Tethys system. The belt of Triassic deformation (Cimmerian orogenic belt) shown on the map may be adjacent to the suture zone along which the Permo-Triassic Tethys was consumed. (From Dewey et al., 1973.)

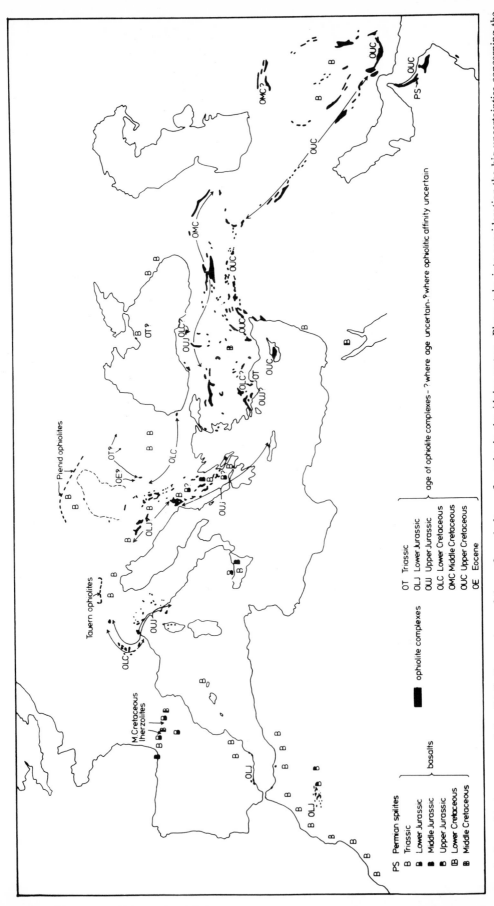

Fig. 19. A graphical summary of the distribution and age of the mafic and ultramafic rocks in the Alpine system. Please take into consideration the big uncertainties concerning the age of those rocks. Also, some of the purported occurrences of ophiolite (e.g., Betic, Rif) may be ultramafic intrusions and cannot be interpreted as fragmented ocean floors. (From Dewey et al., 1973.)

into a northern and a southern continent. The central Atlantic was born where Africa was torn from North America. At the same time, three microcontinents developed between Europe and Africa, namely: (1) the Alboran, (2) the Italo-Dinaride, and (3) the Bulgarian (see Figs. 13a–b).

The southern boundary of the Alboran microcontinent was apparently a transform fault now buried beneath of the flysch of the Tellian Atlas. On the northern margin was the Subbetic Zone. The movement and rotation of the Italo-Dinaride block away from Europe created the meso-Mediterranean geosynclines (Pennine and Ligurian). Meanwhile, a counterclockwise rotation of the Bulgarian block proceeded at an even faster pace, leaving the Vardar Trough between the two microcontinents (Fig. 13b).

The creation of Jurassic and Cretaceous oceanic basins in the west was accompanied by the destruction of the Triassic Tethys in the east along a consuming plate margin. The evidence of this early Mesozoic consumption has been referred to in geological literature as the Cimmerian Orogeny, which extended from Dobrogea, to the Crimea, to the northern Caucasus, to the Middle East, to the Kopet Dagh of northeastern Iran (Fig. 18), and thence through the interior of Southwest Asia to surface again in Indochina and Indonesia.

The eastward movement of Africa during the Jurassic and Early Cretaceous is reminiscent of a military maneuver of "pincer encirclement." Microcontinents, like sentries, were detached forward to meet Eurasia along the Cimmerian front, leaving in their wake, the Vardar, the Tauride, the Zagros, and the Oman Ophiolite Troughs, as well as the eastern Mediterranean. The main mass followed and was jointed, or sutured, with the "advanced guards" during the Late Cretaceous along the ophiolite belt (Figs. 8–9). Trapped by this encirclement was the ancestral eastern Mediterranean (cf. Figs. 13d–e). As Africa moved relentlessly northward throughout the Tertiary, the trapped basin was compressed, and now faces annihilation should the Mediterranean Ridge be raised above sea level.

V. AEGEAN SEA

The Aegean is a marginal sea behind an island arc. The sea floor has a block-faulted topography, and a number of interconnected basins all less than 2000 m deep can be recognized. The main basins are: Cretan Basin, in a back-arc setting; Central Aegean Basin; North Aegean Basin, apparently undergoing rifting, strikes west-southwest–east-northeast with relatively thick sediments (Fig. 14).

Geological and geophysical data indicate that the Aegean is underlain by sialic crust. The Moho-depth is about 25 km under the Cretan Basin, and is

30 km or more in the Central and North Aegean Basins (Makris, 1973). The Aegean was emergent in early Miocene time and was connected to Europe. The subsidence leading to the formation of the Aegean Sea is dated as early Late Miocene (Drooger and Meulenkamp, 1973).

Deep-sea drilling in the Cretan Basin penetrated Plio-Quaternary hemipelagic sediments and cored Messinian gypsum (Hsü *et al.*, 1975). This fact suggests that the basin was already in existence in Late Miocene, though perhaps not as deep as it is now. The stratigraphy indicates that the Aegean was part of the Mediterranean and was separated from the Black Sea Basin during much of the Pliocene and Quaternary.

The origin of the Aegean may be explained by appealing to "foundering" or to "rifting." A new variant on the "foundering" theme assumes crustal thinning by subaerial erosion. In an attempt to explain the subsidence of passive continental margins, I reasoned that a sialic crust should be uplifted and subjected to deep erosion when the underlying mantle was hot and was abnormally light, and that the thinned crust should eventually subside in response to a subsequent cooling and density increase in the mantle (Hsü 1965). Schuiling (1973) applied the principle to the subsidence of Mediterranean basins, citing as evidence the impressive extent of erosion of the Cyclades in the Aegean. He further contended that the Aegean is undergoing only incipient oceanization. With further cooling of the underlying mantle, the Aegean Basin might be turned in to a deep basin like the Tyrrhenian. However, back-arc basins are being heated up, as shown by the abnormal heat flow there. We obtained, for example, a heat flow value of 1.8 HFU for the Cretan Basin (Hsü *et al.*, 1975) and could hardly relate the present depth (2 km) of this basin to mantle cooling. Secondly, some means other than crustal thinning by subaerial erosion is necessary before the relatively thick sialic crust under the Aegean can be converted into a thin ocean crust such as that which underlies the Tyrrhenian Abyssal Plain. I prefer, therefore, to invoke a crustal extension model to interpret the genesis of the Aegean.

VI. SUMMARY AND CONCLUSIONS

I have presented in this paper the major schools of thought on the evolution of the Mediterranean. I have presented facts and arguments which led me to the following conclusions:

1. The Balearic Basin is mainly a Neogene Basin in which rifting started in Late Oligocene, after the paroxysm of the Alpine Orogeny, and ended in Middle Miocene.

2. The Alboran Basin originated with the Miocene (Early to Middle) rifting of the Alboran block; the split halves of the block were welded to Europe

on the north and to Africa–Sicily–Calabria on the south, giving rise to the Betic Cordillera, the Rif, the Tellian Atlas, and to the mountains of Sicily and Calabria.

3. The Tyrrhenian Basin originated with Middle to Late Miocene rifting, which permitted emplacement of dike swarms into a pre-existing crust. A Paleo-Tyrrhenian Basin might have existed in this region before the Early Tortonian deformation of North Africa, Sicily, and Calabria.

4. The eastern Mediterranean is a relic of the Mesozoic Tethys, and originated from Jurassic and Cretaceous sea-floor spreading. Genesis of this and other Tethyan geosynclinal basins was related to sea-floor spreading in the Atlantic, which displaced Africa eastward relative to Europe. The Late Cretaceous and Cenozoic shrinkage of the eastern Mediterranean and the elevation of the Mediterranean Ridge were caused by movements which brought Africa and Europe together again.

5. The Aegean Basin is a back-arc basin underlain by sialic basement, and it started to subside in Middle or Late Miocene. The Aegean Basin might eventually be converted into a Tyrrhenian-type deep basin by processes connected with island-arc tectonics.

ACKNOWLEDGMENTS

In a synthesis paper like this, one must draw greatly from the data and ideas of others. I would like to single out the 1969 Dutch Symposium (Anonymous, 1969) and the 1975 Morelli review as having been particularly helpful to me. I am indebted to the United States, the Bulgarian, and the Romanian Academies of Sciences for their Scientific Exchange Program, which enabled me to repeatedly visit eastern Europe for clarification of problems on Tethyan geology. I am very grateful to many of my friends, shipmates, and colleagues, especially Bill Ryan, Alan Smith, Eldridge Moores, Dan Bernoulli, Rudy Trumpy, Lucien Montadert, Radu Sandulescu, and many others who went with me to sea or to the field and who discussed and enlightened me on many aspects of Mediterranean geology. I reserve my last and very special thanks to the JOIDES organization and to the Deep Sea Drilling Project who entrusted me with the *Glomar Challenger* for her two epic voyages to the Mediterranean.

REFERENCES

Alvarez, W., 1973, The application of plate tectonics to the Mediterranean region, in: *Implications of Continental Drift to the Earth Sciences*, Tarling D. S., and Runcorn, S. K., eds., New York: Academic Press, p. 893–308.

Alvarez, W., Franks, S. G., and Nairn, A. E. M., 1973, Palaeomagnetism of Plio-Pleistocene basalts from Northwest Sardinia, *Nature Phys. Sci.*, v. 243 (123), p. 10–11.

Alvarez, W., Cocozza, T., and Wezel, F. C., 1974, Fragmentation of the Alpine orogenic belt by microplate dispersal, *Nature*, v. 248, p. 309–314.

Andrieux, J., Fontbote, J.-M., and Mattauer, M., 1971, Sur un modèle explicatif de l'arc de Gibraltar, *Earth Planet. Sci. Lett.*, v. 12, p. 191–198.

Anonymous, 1969, Symposium on the problem of oceanization in the Western Mediterranean, *Verhandl. Kon. Ned. Geol. Mijnbouwk. Gen.*, v. 26, 165 pp.

Argand, E., 1924, La tectonique de l'Asie, XIIIe Congr. Géol. Intern., Brussels, Compt. Rend., p. 171–372.

Auzende, J. M., Bonnin, J., and Olivet, J. L., 1973, The origin of the western Mediterranean Basin, *J. Geol. Soc. Lond.*, v. 129, p. 607–620.

Barbieri, F., Gasparini, P., Innocenti, F., and Villari, L., 1973, Volcanism of the Southern Tyrrhenian Sea and its geodynamic implications, *J. Geophys. Res.*, v. 78, p. 5221–5232.

Bayer, R., Le Mouel, J. L., and Le Pichon, X., 1973, Magnetic anomaly pattern in the Western Mediterranean, *Earth Planet. Sci. Lett.*, v. 19, p. 168–176.

Bobier, C., 1970, La signification de l'aimantation remanente des laves de la série "Des Ignimbrites inférieures" — conséquence pour l'étude de la rotation du Bloc Corso-Sarde durant le Tertiaire, *Rend. Sem. Sci. Univ. Cagliari*, Suppl. 43, p. 35–49.

Bobier, C., and Coulon, C., 1974, Résultats préliminaires d'une étude paléomagnétique des formations volcaniques tertiaires et quaternaires du Logudoro (Sardaigne Septentrionale), *C. R. Acad. Sci. Paris.*, v. 270, Sér. D, p. 1434–1437.

Boccaletti, M., and Guazzoni, G., 1972, Gli archi appenninici, il mar Ligure e il Tirreno nel quadro della tettonica dei bacini marginali retro-arco, *Mem. Soc. Geol. Ital.*, v. 11, p. 201–216.

Boccaletti, M., Elter, P., and Guazzoni, G., 1971, Plate tectonic models for the development of the western Alps and northern Apennines, *Nature Phys. Sci.*, v. 49, p. 108–111.

Bosellini, A., and Hsü, K. J., 1973, Mediterranean plate tectonics and Triassic palaeogeography, *Nature*, v. 244, p. 144–146.

Bourcart, J., 1950, Les déformations récentes et leur influence sur le modèle actuel, Congr. Geogr. Intern., Lisbon, Compt. Rend., 1949, v. 2, p. 168–190.

Bourcart, J., 1962, La Méditerranée et la révolution du Pliocène, *Soc. Géol. France, Livre P. Fallot*, v. 1, p. 103–116.

Bullard, E. C., Everett, J. E., and Smith, A. G., 1965, The fit of the continents around the Atlantic, *Roy. Soc. London Phil. Trans.* (A), v. 258, p. 41–51.

Burollet, P. F., and Dufaure, P., 1971, The Neogene Series drilled by the Mistral no. 1 Well in the Gulf of Lion, in: *The Mediterranean Sea, A Natural Sedimentation Laboratory*, Stanley, D. J., ed., Stroudsburg, Pa.: Dowden, Hutchinson, and Ross, p. 91–98.

Caputo, M., Panza, G. F., and Postpischl, D., 1970, Deep structure of the Mediterranean Basin, *J. Geophys. Res.*, v. 75, p. 4919–4923.

Carey, S. W., 1958, A tectonic approach to continental drift — A symposium, Carey, S. W., convener, Hobart, Australia, Univ. Tasmania, p. 177–355.

Cita, M. B., 1973, Mediterranean evaporites: paleontological arguments for a deep-basin desiccation model, in: *Messinian Events in the Mediterranean*, Drooger, C. W., ed., Amsterdam: North Holland Publ. Co., p. 206–228.

Colom, G., and Escandell, B., 1962, L'évolution du Géosynclinal Baléare, *Soc. Géol. France, Livre P. Fallot*, v. 1, p. 125–136.

Colombi, B., Giese, P., Luongo, G., Morelli, C., Riuscetti, M., Scarascia, S., Schütte, K. G., Strowald, L., and De Visintini, G., 1973, Preliminary report on the seismic refraction profile Gargano–Salerno–Palermo–Pantelleria, 1971, *Boll. Geofis. Teor. Appl.*, v. 15 (59), 225 pp.

Cornet, C., 1968, Le graben médian (Zone A) de la Méditerranée occidentale pourrait être pontien, *C. R. Soc. Géol. France*, v. 5, p. 149–150.

Coulon, C., 1973, Données géochronologiques, géochimiques et paléomagnétiques sur le volcanisme Cenozoique calco-alcalin de la Sardaigne nord-occidentale, *Rend. Sem. Sci. Univ. Cagliari*, Suppl. 43, p. 163–169.

Coulon, C., Bobier, C., and Demant, A., 1973, Contribution du paléomagnétisme à l'étude des séries volcaniques cénozoiques et quaternaires du Logudoro et du Bosano (Sardaigne Nord-occidentale). Le problème de la dérivé de la Sardaigne, Monaco: CIESM, Rapport Scientifique No. 6, p. 59–61.

Coulon, C., Demant, A., and Bellon, H., 1974a, Premières datations par la méthode K/Ar de quelques laves cénozoiques et quaternaires de Sardaigne, Nord-occidentale, *Tectonophysics*, v. 22, p. 41–57.

Coulon, C., Demant, A., and Bobier, C., 1974b, Contribution du paléomagnetisme à l'étude des séries volcaniques cénozoiques et quaternaires de Sardaigne Nord-occidentale, *Tectonophysics*, v. 22, p. 59–82.

Dallan, L., 1964, I foraminifera miocenici dell'Isola di Pianosa', *Boll. Soc. Geol. Ital.*, v. 83, p. 167–182.

Davoudzadeh, M., 1975, Excursion guide to Kopet Dagh, ophiolites of Robat-Sefid and metamorphics ophiolites, granites of Navhad, Teheran: Iranian Geol. Survey, Unpubl. Rep., 10 pp.

De Jong, K. A., Manzoni, M., and Zijderveld, J. D. A., 1969, Paleomagnetism of the Alpine Trachyandesites, *Nature*, v. 224, p. 67–69.

De Jong, K. A., Manzoni, M., Stavenga, T., Van Dijk, F., Van der Voo, R., and Zijderveld, J. D. A., 1973a, Palaeomagnetic evidence for rotation of Sardinia during the Early Miocene, *Nature*, v. 243, p. 281–282.

De Jong, K. A., Manzoni, M., Stavenga, T., Van der Voo, R., Van Dijk, F., and Zijderveld, J. D. A., 1973b, Early Miocene Age of rotation of Sardinia: Paleomagnetic evidence, Monaco: CIESM, Rapport Scientifique No. 6, p. 57–58.

Denizot, G., 1951, Les anciens rivages de la Méditerranée française, *Bull. Inst. Océanogr. Monaco*, No. 992, 56 pp.

De Roever, W. P., 1969, Genesis of the western Mediterranean Sea, *Verhandl. Kon. Ned. Geol. Mijnbouwk. Gen.*, v. 26, p. 9–11.

Dewey, J. F., Pittman, W. C., Ryan, W. B. F., and Bonin, J., 1973, Plate tectonics and the evolution of the Alpine system, *Geol. Soc. Amer. Bull.*, v. 84, p. 3137–3180.

Drooger, C. W., and Meulenkamp, J. E., 1973, Stratigraphic contributions to geodynamics in the Mediterranean area: Crete as a case history, *Bull. Geol. Soc. Greece*, v. X(1), p. 193–200.

Du Toit, A. L., 1937, *Our Wandering Continents*, Edinburgh: Oliver and Boyd, 366 pp.

Erickson, A. J., 1970, The measurements and interpretation of heat flow in the Mediterranean and Black Seas, Ph. D. Thesis, Cambridge, Mass.: Mass. Inst. Technology, 272 pp.

Fahlquist, D. A., 1963, Seismic refraction measurements in the western Mediterranean Sea, *Rapp. Comm. Int. Mer Médit.*, v. 17, p. 963–964.

Fahlquist, D. A., and Hersey, J. B., 1969, Seismic refraction measurements in the western Mediterranean Sea, *Bull. Inst. Océanogr. Monaco*, v. 67 (1386), 52 pp.

Finetti, I., and Morelli, C., 1973, Geophysical exploration of the Mediterranean Sea, *Boll. Geofis. Teor. Appl.* v. 15, p. 263–341.

Finetti, I., Morelli, C., and Zarudski, E., 1970, Reflexion seismic study of the Tyrrhenian Sea, *Boll. Geofis. Teor. Appl.*, v. 12, p. 311–346.

Freund, R., Goldberg, M., Weissbrod, T., Druckman, Y., and Derin, B., 1975, The Triassic–Jurassic structure of Israel and its relation to the origin of the eastern Mediterranean, *Geol. Survey Israel Bull.*, v. 65, p. 1–26.

Friedmann, G. M., Barzel, A., and Derin, B., 1971, Paleoenvironments of the Jurassic in the coastal belt of northern and central Israel and their significance in the search for petroleum reservoirs, Jerusalem: Geol. Surv. Israel, Rep. OD/1/71.

Gansser, A., 1974, The ophiolitic mélange, a world-wide problem on Tethyan examples, *Eclogae Geol. Helv.*, v. 67, p. 479–507.

Gass, I. G., 1968, Is the Troodos massif of Cyprus a fragment of Mesozoic ocean floor? *Nature*, v. 220, p. 39–42.

Glangeaud, L., 1967, Epirogenèse ponto-plio-quaternaire de la marge continentale franco-italienne du Rhône à Gênes, *Bull. Soc. Géol. France*, 7ᵉ sér., v. 9, p. 426–449.

Glangeaud, L., Alinat, J., Polvèche, J., Guillaume, A., and Leenhardt, O., 1966, Grandes structures de la Mer Ligure, leur évolution et leurs relations avec les chaînes continentales, *Bull. Soc. Géol. France*, 7ᵉ sér., v. 8, p. 921–937.

Heezen, B., Gray, C., Segre, A. G., and Zarudski, E. F. K., 1971, Evidence of foundered continental crust beneath the central Tyrrhenian Sea, *Nature*, v. 229, p. 327–329.

Hersey, J. B., 1965, Sedimentary basins of the Mediterranean Sea, in: *Submarine Geology and Geophysics*, Whittard, W. F., and Bradshaw, R., eds., London: Butterworths, Colston Paper 17, p. 75–91.

Hinz, K., 1972, Results of seismic refraction investigations (Project Anna) in the western Mediterranean Sea, South and North of the Island of Mallorca, *Centre Rech. Pau, SNPA, Bull.*, v. 6 (2), p. 405–426.

Hsü, K. J., 1958, Isostasy and a theory for the origin of geosynclines, *Am. J. Sci.*, v. 246, p. 305–327.

Hsü, K. J., 1965, Isostasy, crustal thinning, mantle changes, and the disappearance of ancient land masses, *Am. J. Sci.*, v. 263, p. 97–109.

Hsü, K. J., 1971, Origin of the Alps and western Mediterranean, *Nature*, v. 233, 44–48.

Hsü, K. J., 1972, The concept of geosyncline, yesterday and today, *Leicester Lit. Phil. Soc. Trans.*, v. 66, p. 26–48.

Hsü, K. J., and Ryan, W. B. F., 1972, Comments on the crustal structure of the Balearic Basin in the light of deep sea drilling in the Mediterranean, *Centre Rech. Pau, SNPA, Bull.*, v. 6, p. 427–430.

Hsü, K. J., and Ryan, W. B. F., 1973, Deep-sea drilling in the Hellenic Trench, *Geol. Soc. Greece Bull.*, v. 10 (1), p. 81–80.

Hsü, K. J., and Schlanger, S. O., 1971, Ultrahelvetic flysch sedimentation and deformation related to plate tectonics, *Geol. Soc. Am. Bull.*, v. 82, p. 1203–1218.

Hsü, K. J., Cita, M. B., and Ryan, W. B. F., 1972a, The origin of the Mediterranean Evaporites, in: Ryan, W. B. F., Hsü, K. J., et al., *Initial Reports of the Deep Sea Drilling Project*, Washington D. C.: U.S. Government Printing Office, p. 1023–1231.

Hsü, K. J., Ryan, W. B. F., Cocozza, T., and Magnia, P., 1972b, Comparative petrography of three suites of basement rocks from the Western Mediterranean, in: Ryan, W. B. F., Hsü, K. J., et al., *Initial Reports of the Deep Sea Drilling Project*, Washington D. C.: U.S. Government Printing Office, p. 775–780.

Hsü, K. J., Montadert, L., Garrison, R. E., Fabricius, F. H., Bernoulli, D., Melieres, F., Kidd, R. B., Müller, C., Cita, M. B., Bizon, G., and Erickson, A., 1975, *Glomar Challenger* returns to the Mediterranean Sea, *Geotimes*, v. 20 (8), p. 16–19.

Le Borgne, E., Le Mouel, J. L. and Le Pichon, W., 1971, Aeromagnetic survey of southwestern Europe, *Earth Planet. Sci. Lett.*, v. 12, p. 287–299.

Le Pichon, X., Pautot, G., Auzende, J. M., and Olivet, J. L., 1971, Le Méditerranée occidentale depuis l'Oligocène. Schéma d'évolution, *Earth Planet. Sci. Lett.*, v. 13, p. 145–152.

Lort, J. M., 1973, Summary of seismic studies in the eastern Mediterranean, *Bull. Geol. Soc. Greece*, v. 10, p. 99–108.

Lowrie, W., and Alvarez, W., 1975, Paleomagnetic evidence for rotation of the Italian Peninsula, *J. Geophys. Res.*, v. 80, p. 1579–1592.

Makris, J., 1973, Some geophysical aspects of the evolution of the Hellenides. *Bull. Geol. Soc. Greece*, v. 10 (1), p. 206–213.

Manzoni, M., 1974, Discussion of "La significance de l'aimantation remanente des laves de la série 'Des Ignimbrites inférieures'" conséquence pour l'étude de la rotation du Bloc Corso-Sarde durant le Tertiaire, *Rend. Semin. Fac. Sci. Univ. Cagliari*, Suppl. 43, p. 50–56.

Mauffret, A., Fail, J. P., Montadert, L., Sancho, J., Winnock, E., 1973, Northwestern Mediterranean sedimentary basin from seismic reflection profile, *Am. Assoc. Petr. Geol. Bull.*, v. 57, p. 2245–2262.

McKenzie, D. P., 1972, Active tectonics of the Mediterranean region, *Geophys. J.*, v. 29, p. 300–377.

Mojsisovics, E., 1896, Beiträge zur Kenntniss der obertriadischen Cephalopoden-Faunen des Himalaya, *Denkschr. k.k. Akad. wiss. math.-naturw. kl.*, v. 63, 575–701.

Moores, E. M., and Vine, F. J., 1971, The Troodos Massif, Cyprus and other ophiolites as ocean crusts: evaluation and implications, *Roy. Soc. London Phil. Trans.* (A), v. 268, p. 443–446.

Morelli, C., 1970, Physiography, gravity and magnetism of the Tyrrhenian Sea, *Boll. Geofis. Teor. Appl.*, v. 12, p. 274–308.

Morelli, C., 1975, Geophysics of the Mediterranean, Monaco: CIESM, Bulletin de l'étude en commun de la Méditerranée, Special Issue No. 7, p. 29–111.

Nairn, A. E. M., and Westphal, M., 1968, Possible implications of the Paleomagnetic study of Late Paleozoic igneous rocks of northwestern Corsica, *Palaeogeogr. Palaeoecol.*, v. 5, p. 179–204.

Nalivkin, D. V., 1973, *Geology of the U.S.S.R.*, Toronto: Univ. Toronto Press, 855 pp.

Neev, D., Almagor, D., Arad, A., Ginzburg, A., and Hall, K. J., 1973, The geology of the southeastern Mediterranean Sea, Jerusalem: Geol. Surv. Israel, Unpub. Rep. (No. MG/73/5).

Nesteroff, W. D., 1972, Mineralogy, petrography, distribution, and origin of the Messinian Mediterranean Evaporites, in: Ryan, W. B. F., Hsü, K. J., *et al.*, *Initial Reports of the Deep Sea Drilling Project*, Washington, D. C.: U. S. Government Printing Office, v. 13, p. 673–694.

Neumayr, M., 1883, Klimatische Zonen während der Jura-and Kreidezeit, *Denkschr. math.-naturw. kl.*, v. 47, p. 277–210.

Ogniben, L., 1969, Schema introduttivo alla geologia del confine Calabro-Lucano, *Mem. Soc. Geol. Ital.*, v. 8, p. 453–763.

Pannekoek, A. J., 1969, Uplift and subsidence in and around the western Mediterranean since the Oligocene, *Verhandl. Kon. Ned. Geol. Mijnbouwk. Gen.*, v. 26, p. 53–77.

Papazachos, B. C., 1973, *Seismotectonics of the Eastern Mediterranean Sea Area*, Izmir, Turkey: NATO Advanced Study Institute in Modern Developments in Engineering Seismology and Earthquake Engineering.

Papazachos, B. C., and Comninakis, P. E., 1971, Geophysical and tectonic features of the Aegean Arc, *J. Geophys. Res.*, v. 76, p. 8517–8533.

Rabinowitz, P. D., and Ryan, W. B. F., 1970, Gravity anomalies and crustal shortening in the eastern Mediterranean, *Tectonophysics*, v. 10, p. 585–608.

Ritsema, A. R., 1970, On the origin of western Mediterranean Sea basins, *Tectonophysics*, v. 10, p. 609–623.

Robertson, A. H. F., and Hudson J. D., 1974, Pelagic sediments in the Cretaceous and Tertiary history of the Troodos Massif, Cyprus, in: *Pelagic Sediments on Land and under the Sea*, Hsü, K. J., and Jenkyns, H. C., eds., Oxford: Blackwell, Spec. Publ. Inst. Assoc. Sed., No. 1.

Ruggieri, G, 1967, The Miocene and later evolution of the Mediterranean Sea, in: *Aspects of Tethyan Biogeography*, Adams, C. G., and Ager, D. V., eds., London: Syst. Assoc. Publ. No. 7, p. 283–290.

Ryan, W. B. F., 1969, The floor of the Mediterranean Sea, Ph.D. Thesis, Columbia University, 196 pp.

Ryan, W. B. F., 1973, Geodynamic implications of the Messinian crisis of salinity, in: *Messinian Events in the Mediterranean*, Drooger, C. W., ed., Amsterdam: North Holland Publ. Co., p. 26–38.

Ryan, W. B. F., Hsü, K. J., *et al.*, 1972, *Initial Reports of the Deep Sea Drilling Project*, Washington, D.C.: U.S. Government Printing Office, vol. 13, 1448 pp.

Sancho, J., Letouzey, J., Biju-Duval, B., Courrier, P., Montadert, L., and Winnock, E., 1973, New data on the structure of the eastern Mediterrean Basin from seismic reflection, *Earth Planet. Sci. Lett.*, v. 18, p. 189–204.

Sander, N. J., 1970, Structural evolution of the Mediterranean Region during the Mesozoic Era, in: *Geology and History of Lybia*, Alvarez W., and Gohrbandt, K., eds., Tripoli: Petr. Exploration Soc. Libya, p. 43–132.

Sarrot-Reynauld, M. J., 1966, Grandes structures de la Mer Ligure (a discussion), *Bull. Soc. Géol. France*, 7ᵉ sér., v. 8, p. 937.

Schuiling, R. D., 1973, The Cyclads: an early stage of oceanization? *Bull. Geol. Soc. Greece*, v. 10, p. 174–176.

Selli, R., and Fabbri, M., 1971, Tyrrhenian: a Pliocene deep sea, *R.C. Accad. Lincei*, v. 8, p. 104–116.

Smith, A. G., 1971, Alpine deformation and the oceanic areas of the Tethys, Mediterranean and the Atlantic, *Geol. Soc. Am. Bull.*, v. 82, p. 2039–2070.

Smith, A. G., 1973, The so-called Tethyan ophiolites, in: *Implications of Continental Drift to the Earth Sciences*, Tarling D., and Runcorn, S., eds., London: Academic Press, p. 977–986.

Sonnenfeld, P., 1975, The significance of Upper Miocene (Messinian) Evaporites in the Mediterranean Sea, *J. Geology*, v. 83, p. 287–312.

Stanley, D. J., and Mutti, E., 1968, Sedimentological evidence for an emerged-land mass in the Ligurian sea during the Palaeogene, *Nature*, v. 218, p. 32–36.

Stanley, D. J., Got, H., Leenhardt, O., and Weiler, Y., 1974, Subsidence of the western Mediterranean Basin in Pliocene–Quaternary time: Further evidence, *Geology*, v. 2, p. 345–350.

Stauffer, K., and Tarling, D. H., 1971, Age of Bay of Biscay: New paleomagnetic evidence, in: *Histoire Structurale du Golfe de Gascogne*, II 2, p. 1–18, Paris: Editions Technip.

Suess, E., 1893, Are great oceans depth permanent?, *Nat. Sci.*, v. 2, p. 180–187.

Suess, E., 1901, *Das Antlitz der Erde*, v. 3, Vienna: F. Temsky, 508 pp.

Van Bemmelen, R. W., 1969, Origin of the western Mediterranean Sea, *Verhandl. Kon. Ned. Geol. Mijnbouwk. Gen.*, v. 26, p. 13–52.

Westphal, M., Bardon, C., Bossert, A., and Hamzeh, R., 1973, A computer fit of Corsica and Sardinia against Southern France, *Earth Planet. Sci. Lett.*, v. 18, p. 137–140.

Wezel, F. C., 1970, Numidian Flysch: an Oligocene–early Miocene continental rise deposit off the African Platform, *Nature*, v. 228, p. 275–276.

Wezel, F. C., 1974, Primo Riesame delle Carote Raccolte nel Leg 13 del DSDP (Mediterraneo), *Giorn. Geol.*, v. 39, p. 447–468.

Wezel, F. C., and Ryan, W. B. F., 1971, Flysch margini continentali e zolle litosferiche, *Boll. Soc. Geol. Ital.*, v. 90 p. 249–270.

Williams, C. A., 1973, A fossil trip junction in the NE Atlantic west of Biscay, *Nature*, v. 244, p. 86–88.

Zijderveld, J. D. A., and Van der Voo, R., 1970, Paleomagnetism in the Mediterranean area in: *Implications of Continental Drift to the Earth Sciences*, Tarling D. S., and Runcorn, S., eds., New York: Academic Press.

Zijderveld, J. D. A., de Jong, K. A., and Van der Voo, R., 1973, Rotation of Sardinia, *Nature*, v. 226, p. 933–934.

Chapter 3

POST-MIOCENE DEPOSITIONAL PATTERNS AND STRUCTURAL DISPLACEMENT IN THE MEDITERRANEAN

Daniel Jean Stanley

Division of Sedimentology
Smithsonian Institution
Washington, D.C.

I. INTRODUCTION

Unconsolidated Pliocene and Quaternary sediments consisting largely of terrigenous clay, silt, and sand, and to a lesser extent organic-rich ooze, sapropel, volcanic ash (tephra), and wind-blown material, blanket Miocene evaporites, Tertiary infrasalt series, and older deposits throughout most of the Mediterranean Sea. This chapter focuses on the remarkable regional variability of depositional thickness and sediment types as well as on structural displacement of Lower Pliocene to Recent sequences in this physiographically and structurally complex land-bound sea. The geologic changes in time and space recorded are attributed to (1) specific dispersal patterns and transport processes, (2) climatic evolution affecting the flow of water masses and eustatic sea level oscillations, and (3) tectonic activity of variable style and intensity which has disrupted sedimentation and displaced these series during and following their accumulation. This investigation of depositional patterns considers both sedimentological and tectonic factors, and reviews major trends in the various basins based on continuous seismic profile (CSP) records and supplemental drill core and sea-floor sample data.

It has long been recognized that the geological history of the Mediterranean is related to large-scale events which molded the complex Alpine and related mobile belts surrounding it. Until the advent of subbottom seismic surveys, most workers concerned with the origin of this narrow 4000-km-long ocean attempted to interpret its origin and configuration in terms of structure and stratigraphy recorded on land. On the basis of intensified geophysical exploration during the 1960s and of the first *Glomar Challenger* deep-sea drilling cruise (Leg 13, 1970), some workers concluded that the present configuration and depth of the Mediterranean was delineated essentially by the Miocene (Hsü, 1972*b*, and this volume; Ryan *et al.*, 1973, 1976). Others have opted for a much shallower ($<$ 500 m) sea floor at that time (Nesteroff, 1973*a*, *b*). The majority of geologists, however, have stressed the importance of structural mobility, which has continued to modify the Mediterranean physiography from the Miocene to the present (cf. Bourcart, 1950, 1960*b*; Glangeaud, 1962; Mauffret *et al.*, 1973; Mulder, 1973; Storetvedt, 1973; Biju-Duval *et al.*, 1974; Stanley *et al.*, 1974*b*, 1976; Morelli, 1975).

The end of the Miocene (Messinian) was a period of marked structural and physiographic changes as well as of evaporite formation throughout much of the Mediterranean realm; the reader is directed to discussions of this phase, termed the salinity crisis, summarized in Drooger (1973). Some workers adhere to the concept of evaporites forming in an already deep ($>$ 2500 m) Mediterranean desert during the Messinian (Cita, 1973; Hsü, 1973; Ryan *et al.*, 1973). This hypothesis remains a minority viewpoint. It should also be noted that the origin of the Miocene evaporites (shallow marine versus intertidal versus desert or sabkha) on land and in JOIDES cores remains a subject of controversy (Ryan *et al.*, 1973; Selli, 1973; Sonnenfeld, 1974, 1975; Kidd, 1976). In sum, seismic profiles record considerable offset of the sea floor throughout the Mediterranean in recent geologic time; this displacement has affected the Pliocene and Quaternary sedimentary cover. The present configuration of large areas of Mediterranean basins and contiguous lands is an overprint on the ancient Tethys, or Mesogée, and in many sectors is related to pre-Pliocene structural trends (cf. discussions in Aubouin, 1973; Dewey *et al.*, 1973; Boccaletti, 1975; Lort, Chapter 4, and Hsü, Chapter 2, this volume).

The causes, amount, and time of tectonic displacement during and following the Messinian salinity crisis remain points of contention. In the author's view, marked Neogene to Recent vertical movements or lateral translation, or both, have prevailed along many of the basin margins, and the present configuration of this ocean does not define strictly relict paleogeographic attributes. Sedimentological considerations suggest that the major basin plains, such as that of the western Mediterranean, were neither necessarily truly "deep" ($>$ 2500 m) or "shallow" (200–500 m) but probably occupied various depths, for the most part intermediate (1000–1500 m, or *bathyal*), at the end of the Miocene (Stanley *et al.*, 1976).

II. NEOCENE TO RECENT STRATIGRAPHY

A. Continuous Seismic Profiles

The early geophysical studies by Hersey (1965b) and Moskalenko (1965) provided a preliminary view of the stratigraphic and structural configuration of the sedimentary sequences underlying Mediterranean basins. Interpretation of the different sectors of this complex region is now possible as a result of the marked increase of seismic surveys in recent years (cf. review by Morelli, 1975). Geophysical coverage of the relatively deep basins and margins is uneven, but continuous seismic profile (CSP) lines are sufficiently dense in many regions to depict the distribution of unconsolidated sediment (Fig. 1). Invaluable stratigraphic control of the Upper Miocene to Recent sediment sequences has been provided by core recovered at Drill Sites 121 to 134 during the 1970 Mediterranean JOIDES *Glomar Challenger* Cruise 13 (Ryan *et al.*, 1973) and at Sites 371 to 378 during the 1975 IPOD *Glomar Challenger* Cruise 42A (Hsü *et al.*, 1975). Comprehensive reviews of regional stratigraphy based on selected intermediate- (sparker, air-gun) and deep-penetration (Flexotir, sleeve exploder) seismic profiles are available (Ryan, 1969; Ryan *et al.*, 1970, 1973; Finetti and Morelli, 1973; Mauffret *et al.*, 1973; Biju-Duval *et al.*, 1974; Morelli, 1975; Stanley *et al.*, 1976); references also are listed in Table I. Identification of regionally important acoustic reflectors and estimates of their velocities are summarized by various workers (cf. Fahlquist and Hersey, 1969; Alla, 1970; Leenhardt, 1970; Hinz, 1972a; Lort, 1972, and this volume; Finetti and Morelli, 1973; Mauffret *et al.*, 1973; Ryan *et al.*, 1973).

The basic stratigraphic succession determined from deep-penetration CSP systems can be summarized most simply from the water–sediment interface downward, using letter designations as follows (see Mauffret *et al.*, 1973; Biju-Duval *et al.*, 1974; Morelli, 1975; also Fig. 12): A, an upper group of acoustic reflectors (the sequence is generally stratified at the top and acoustically transparent at the base); this upper group is interpreted as the unconsolidated sediment cover which, depending upon the setting, ranges in age from the Lower Pliocene to the Recent; B, an uninterrupted sequence, the top of which has been called "H" by some French authors and which is equivalent to the M horizon or M reflectors of Ryan *et al.* (1970, 1973); group B is identified as interbedded marine sediments and evaporites of a Late Miocene age ($\cong 3.5$ km/sec); C, identified as salt, the top of which is termed "K" and the base termed "L" by French geophysicists; group C includes the Upper Miocene salt (4.3–4.4 km/sec); and D, which includes several reflectors of which the most distinctive is designated the N reflector (Fig. 14); group D is believed to include Upper Miocene to Oligocene infrasalt series (some with acoustic velocities lower than 4.0 km/sec and others ranging from 4.5–5.0 km/sec).

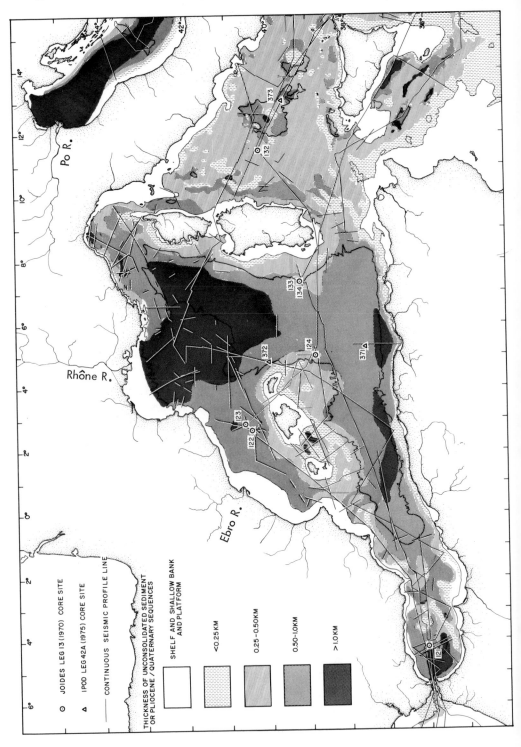

JOIDES LEG 13 (1970) CORE SITE

IPOD LEG 42A (1975) CORE SITE

CONTINUOUS SEISMIC PROFILE LINE

THICKNESS OF UNCONSOLIDATED SEDIMENT
OR PLIOCENE / QUATERNARY SEQUENCES

SHELF AND SHALLOW BANK
AND PLATFORM

<0.25 KM

0.25 - 0.50 KM

0.50 - 1.0 KM

>1.0 KM

Po R.

Rhône R.

Ebro R.

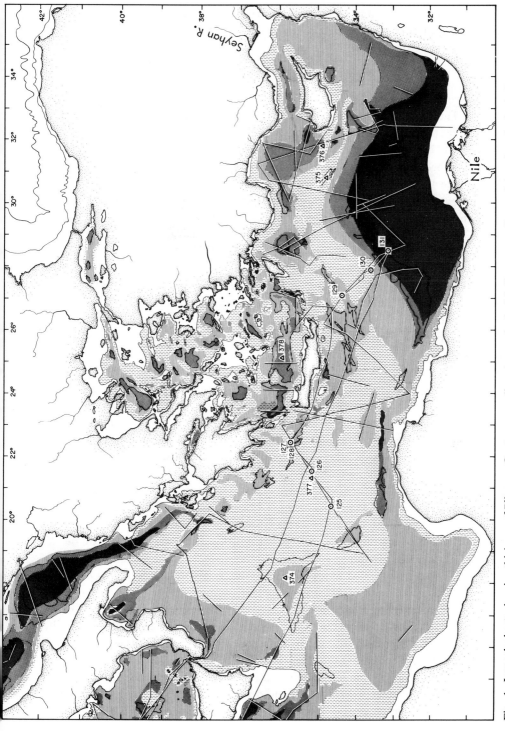

Fig. 1. Isopach charts showing thickness of Pliocene–Quaternary sediment cover in the Mediterranean. Some of the selected seismic lines and deep-sea drill sites used to compile charts are shown. (Published sources used are listed in Table I.)

This lower series, often incomplete, overlies the acoustic basement termed S (Fig. 12) or B (Fig. 13). The above seismic velocities (in two-way travel time), applicable to the western Mediterranean, are based on values quoted by Mauffret *et al.* (1973) and others; analysis of subbottom acoustic properties and stratigraphy in the eastern Mediterranean is provided by Lort (1972; Chapter 4, this volume).

The band of M reflectors is widespread and recorded in many physiographic provinces throughout most areas of the Mediterranean (Ryan *et al.*, 1970, 1973; Biscaye *et al.*, 1972) and serves as a key stratigraphic marker. It has been recovered in some *Glomar Challenger* Legs 13 and 42A cores, and the lateral extent of this sequence, which caps evaporite sequences and underlies the unconsolidated largely terrigenous deposits, has been delineated in both eastern and western basin plains (Ryan *et al.*, 1970, 1973; Morelli, 1975). The configuration of the Upper Miocene evaporites mapped by Finetti and Morelli (1973), Mauffret *et al.* (1973), Biju-Duval (1974), and others shows that thickest evaporite sequences underlie basin plains and usually thin or pinch out along basin margins. In some detailed coring and seismic surveys, the M reflectors have been traced upslope on margins such as off Provence (Glaçon and Rehault, 1973), the eastern Betic chain (Fig. 13), and on the Balearic Platform (Fig. 14) to depths as shallow as 500 m (Stanley *et al.*, 1976).

The upper unconsolidated sediment section has an acoustic velocity ranging from 1.7 to as high as 2.8 km/sec, but a more generalized velocity of about 2.0 km/sec is commonly used when considering the unconsolidated sequence as a whole (Biju-Duval, personal communication). Moderate to high-resolution sparker and air-gun records reveal unconformities, intraformational channeling, small-scale offsets, pinch-outs, allochthonous slump deposits, etc., in these Pliocene–Quaternary units. Faunal and seismic evidence indicates that the age of the base of the unconsolidated sediment series (lying above the Messinian units or Pontian erosional unconformity) is not everywhere the same; determination of the age of the basal units is not easily resolved as in the case of the series off the Rhône in the northwest Balearic–Provençal Basin (cf. Burollet and Byramjee, 1974).

The upper stratified series of the A group is generally interpreted as alternating mud and sand of Pleistocene to Recent age, and the lower acoustically transparent layer as Pliocene mud with diminished proportions of sand (Figs. 13, 14, 19). This conforms with the hypothesis that enhanced amounts of silt and sand transported with mud by markedly increased fluvial flow would have bypassed the subaerially exposed shelves and been carried more directly to the submerged margins and basin plains during Pleistocene low stands of sea level. Detailing of piston cores shows enhanced proportions of sand layers and higher sedimentation rates during the last glacial epoch (Huang and Stanley, 1972;

Stanley *et al.*, 1975*b*). However, there are insufficient data to conclude that the upper acoustically layered units recorded on CSP profiles are necessarily different texturally than those forming the underlying acoustically transparent sequence. It may be noted, for example, that deep-sea drill cores retrieved at several *Glomar Challenger* Leg 13 sites in the western Mediterranean recovered sand strata within Lower Pliocene sequences which appear acoustically transparent on CSP records as well as in the upper (Quaternary) sections, which are acoustically layered (Ryan *et al.*, 1973). The marked vertical transparent versus layered differentiation on seismic records may be in part an acoustic artifact related to (a) the presence of considerably denser evaporites or other consolidated units directly underlying the unconsolidated sediment, and (b) to depth (cf. discussion in Stanley *et al.*, 1976).

B. Correlation between CSP and Sea-Floor Samples

Surficial and Quaternary sediment distribution patterns (U. S. Naval Oceanographic Office, 1965; Frazer *et al.*, 1970; Horn *et al.*, 1972) and interpretations of transport processes (Stanley, 1972) are based on analysis of numerous piston cores, surface grabs, and dredge samples (references listed in Table I). Particularly important stratigraphic data have been obtained on *D/V Glomar Challenger* Cruises 13 (1970, Sites 121 to 134) and 42A (1975, Sites 371 to 378). The drill site locations are shown in Fig. 1 and lithic, faunal, and other pertinent data are summarized in Ryan *et al.* (1973) and Hsü *et al.* (1975, and in prep.).

The *Glomar Challenger* and petroleum drill site surveys in various sectors such as the Strait of Sicily, margins off Italy, Spain, and Greece, and in the Gulf of Lion (Burollet and Dufaure, 1972; Cravatte *et al.*, 1974) have identified some of the more regionally prominent seismic reflectors and established firmer bio- and lithostratigraphic correlation between land and offshore sections. Intrabasin and basin-to-basin correlation of the Late Miocene to Recent sequences now can be reasonably well outlined. Less is known of the Lower Tertiary and older series, although several *Glomar Challenger* drill sites have penetrated deeper Tertiary sections (Sites 372, 375) and the acoustic basement (Sites 121, 373A). Deep-sea drilling has revealed some of the pitfalls inherent to evaluating thicknesses of post-Miocene deposits from seismic profiles. For example, seismic reflectors initially attributed a Miocene age to deposits which subsequently have been identified in drill cores as lithified strata interbedded within Quaternary sections. Thus, it has been necessary to considerably modify earlier interpretations of the Nile Cone (Wong and Zarudzki, 1969; Ryan *et al.*, 1970) and Alboran Sea stratigraphy as a result of recent drilling (Ryan *et al.*, 1973).

In some instances, selective dredging has proved a productive method of recovering older series, including volcanics and Paleogene and pre-Tertiary basement units; most dredging to date has been made in the western basin (cf. Bellaiche and Mauffret, 1971; Bellaiche *et al.*, 1974). Some direct visual observations, particularly in areas of relief (canyons, slopes, salt domes), by submersible (Alla and Leenhardt, 1971; Le Pichon *et al.*, 1975) and underwater camera (Kelling and Stanley, 1972*a*, *b*; Milliman *et al.*, 1972), supplemented by piston cores, also provide a means to identify older units such as Lower Pliocene and Miocene deposits (Bellaiche, 1972; Biscaye *et al.*, 1972; Glaçon and Rehault, 1973; Bellaiche *et al.*, 1974).

Not only is the base of the unconsolidated sediment section not isochronous, even within the same basin, but the benthic microfaunal evidence from drill and piston cores apparently does not provide unequivocal evidence as to the paleodepth ("deep" or "shallow" water) of the muds either lying above or below the evaporites (cf. discussion in Drooger, 1973). Faunas interpreted as deep (Benson, 1973; Cita, 1973), and benthic organisms identified as shallow (Gennesseaux and Glaçon, 1972), have been recovered in the units above the M reflectors. The depth indicated by faunas in the Miocene infrasalt units recently cored east of Minorca at Site 372 in the Balearic Basin remains imprecisely determined at present (a range from 500–2500 m according to Hsü *et al.*, 1975). Clearly, more specific information and continued caution are needed in the application of faunal paleodepth data to larger-scale problems of Mediterranean paleodepth, paleogeography, and evolution in Late Miocene to Early Pliocene time (cf. Drooger, 1973; Stanley *et al.*, 1976).

III. THICKNESS OF POST-MIOCENE SEQUENCES

A. General Regional Depositional Patterns

Two charts (Fig. 1) depict the thickness variations of unconsolidated deposits on the Mediterranean margins and in basins based on published CSP records and on *Glomar Challenger* cores. The position of selected seismic lines and core sites are shown on the two charts (see Table I for more complete list of sources), and isopach contour lines have been simplified. The thickness values are based on a geometric scale (< 0.25, 0.25–0.5, 0.5–1.0, and > 1.0 km) to emphasize important regional variations and depositional trends. Somewhat more generalized thickness maps include those of Wong and Zarudski (1969) in the eastern Mediterranean and of Biju-Duval *et al.* (1974).

The regionally most extensive and thickest (from about 1000 to perhaps as much as 3000 m) Pliocene–Quaternary deposits are those related to delta point-sources. The three most important are the Rhône Cone in the Provençal–Balearic Basin, the thick terrigenous Adriatic Sea sequences in large part supplied by the Po, and the Nile Cone in the eastern Levantine Basin. Smaller delta-related deposits include the Ebro submarine fan in the Valencia Trough, and the thick sequences off the Ceyhan and Seyhan Rivers in the Cilicia Basin and sector north of Cyprus in the Levantine Basin.*

Thick deposits in non-deltaic regions include those resulting from circulating (or channelized) suspensate-rich water masses. Such terrigenous sequences occur in the western Alboran Basin east of the Strait of Gibraltar, the deep basins in the Sea of Marmara between the Dardanelles and Bosporus, the Strait of Otranto between the Adriatic Sea and the Apulian Plateau, the channelized narrows SSE of Sardinia, and the narrow, elongate, fault-bounded basins in the Strait of Sicily.

Topography is a primary consideration: depressions trap (pond) available sediment, and ridges (structural, volcanic, or salt) act as barriers which dam sediment or deviate dispersal paths. The Mediterranean Ridge is probably the largest feature modifying sediment dispersal in the Mediterranean. The steep, narrow margins and topographic highs such as seamounts and shallow platforms are often thinly veneered or, in some cases, almost devoid of unconsolidated sediment cover (the Alboran Ridge, the Emile Baudot and Minorca escarpments off the Balearic Platform and the Malta Escarpment east of the Sicily Strait, seamounts in the Tyrrhenian and Ionian Seas, and structural ridges in the Hellenic Arc and Aegean Sea). Thick, ponded accumulations are recorded in the southern Adriatic Basin, Tyrrhenian Abyssal Plain, the elongate Strait of Sicily troughs, large basins southwest of Rhodes and north and northwest of Cyprus (Rhodes and Antalya Basins), and numerous small depressions in the Aegean Sea. Relatively thick but aerially restricted deposits occur in some major submarine valleys such as the Taranto Canyon, which trends southeast toward the Ionian Basin, the southwest-trending canyons off Genoa in the Ligurian Sea, Spanish canyons on the eastern extension of the Betic Chain, and canyons south of Crete and west of the Peloponnesus which extend into Hellenic Trough basins and trenches.

Thick depositional sequences may define sectors structurally active or formerly active during sedimentation. The elongate (more than several hundred kilometers in length), depressed areas filled with thick ($>$ 1000 m) sequences off western North Africa (the Algerian Trench), the region between eastern Corsica and the island of Elba, and the peri-Tyrrhenian basins are readily identifiable examples.

The patterns in Fig. 1 show that unconsolidated Pliocene–Quaternary deposits are generally thickest (0.5 km and more) in the Alboran Sea, Balearic Basin, and Adriatic Sea, and thinnest (large sectors covered by less than 0.25

---

* The location of the major geographic zones discussed in the text, numerically coded in Table I, is shown in Fig. 31 (p. 128).

km) in the long arcuate region comprising the Mediterranean Ridge in the
Ionian Sea and Levantine Basin north of the Nile Cone and Herodotus Abyssal
Plain. The Tyrrhenian Sea, sectors of the Strait of Sicily, and the Aegean Sea
are covered by post-Miocene terrigenous and bioclastic sediments of highly
variable but usually intermediate thicknesses (0.25–0.50 km). Sediments tend
to be thinner on the basin margins (particularly steep ones) than in basin aprons
and plains (the Balearic Basin is an example). Exceptions include margins close
to deltas and the peri-Tyrrhenian basins where structurally controlled ridges
have served as tectonic dams which have restricted sediment dispersal and
localized ponding.

The thickness of Pliocene–Quaternary deposits in the Ionian Sea is reduced
relative to that in the Tyrrhenian, and deposits in both regions are thinner than
those in the Balearic Basin. This is surprising inasmuch as the Tyrrhenian and
Ionian Seas, and the Balearic Basin as well, are supplied by myriad fluvial
sources actively draining coastal regions on a seasonal basis, and in all three
instances sediment can be transported to basin plains in direct fashion by way
of submarine canyons and/or relatively steep slopes. Inasmuch as the Tyrrhenian Sea is much smaller than the western basin, one might except equal or
relatively larger volumes of sediment per sea-floor surface than in the Balearic
Basin. While depositional thickness is, at least locally, a function of fluvial
input, water mass flow, and topography, it is apparent that another critical
factor has to be considered: structure. That depositional thicknesses in the
Ionian and Tyrrhenian Basins are much reduced relative to those of the Balearic Basin reflect not so much differences in amounts and accessibility to
sediment as differences in age and style of deformation affecting the basins;
i.e., the Tyrrhenian Sea is interpreted as a younger basin according to Selli
and Fabbri (1971). This is borne out by the marked difference of basin plain
depths throughout the Mediterranean: 1500 m in the Alboran Sea, 2800 m in
the Balearic Basin, 3400 m in the Tyrrhenian Sea, 1200 m in the southern
Adriatic Sea, 4000 m in the Ionian Sea, and to as much as 4500–5000 m in the
small Hellenic Trench depressions. These regional variations of topography
and sediment thickness reflect surficial structural responses to large-scale,
deep-seated crustal processes and not sedimentation.

B. Rates of Sedimentation

Carbon-14 dated Mediterranean piston cores show consistently high
sedimentation rates for Late Pleistocene to Recent deposits. Rates ranging
from 10 to over 100 cm per 1000 years have been calculated in different parts
of the western Mediterranean (Gennesseaux and Thommeret, 1968; Huang
and Stanley, 1972; Leclaire, 1972b; Rupke et al., 1974); rates of 15–125 cm/
1000 years are measured in the Strait of Sicily (Stanley et al., 1975b); and rates

of about 10 to over 100 cm/1000 years are recorded in the eastern basins (Mellis, 1954; Pettersson, 1957; Opdyke *et al.*, 1972; Pastouret, 1970; Ryan *et al.*, 1970). A general range of 20–30 cm/1000 years has prevailed during the past 30,000 years. These rates, several orders of magnitude higher than those recorded in the major ocean basins (Heezen and Hollister, 1971), reflect important seasonal fluvial input into relatively small catchment basins.

It is possible to distinguish various mud types in piston cores, and of these hemipelagic mud accounts for about a quarter to half of the Late Quaternary section in some basin plains, and well over that amount in sectors topographically isolated from turbidite and current-influenced turbulent flows. In the Algéro-Balearic Basin, for instance, hemipelagites accumulate at a rate of about 10 cm/1000 years (Fig. 2); in marked contrast, the mud turbidites which

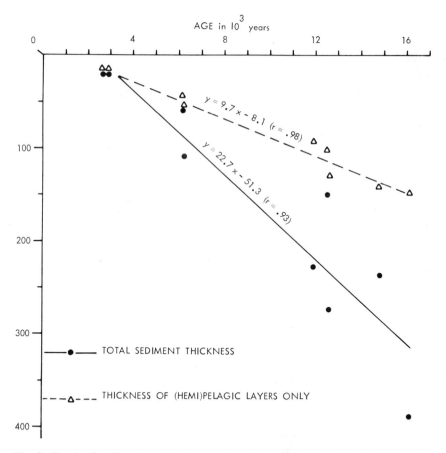

Fig. 2. Graph of radiocarbon dates from carbonate sand fraction of hemipelagic mud *vs.* total depth in core (rate ≅ 23 cm/1000 years; solid line); also combined thicknesses of hemipelagic mud layers only, omitting turbiditic sections of core (rate about 10 cm/1000 years; dashed line). Samples are from cores shown in Fig. 3. From Rupke and Stanley (1974).

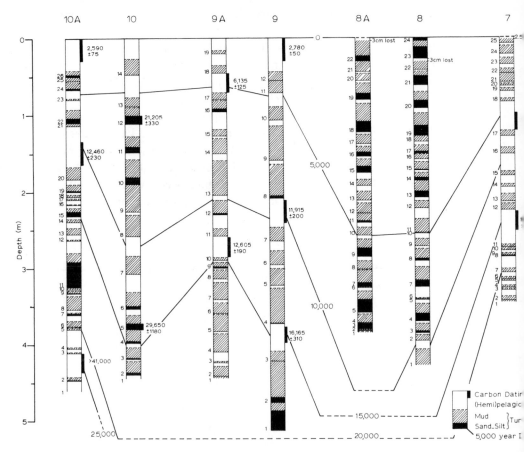

Fig. 3. Cores collected in the Algéro-Balearic Basin showing alternating hemipelagic and turbiditic mud and turbiditic sand-and-silt layers in the Late Quaternary basin plain sequences. Correlation based on radiocarbon dates. Core locations shown in Fig. 8. From Rupke and Stanley (1974).

represent well over 50% of core sections in this area (Fig. 3), are deposited at considerably accelerated rates (Rupke *et al.*, 1974).

Rates are often higher on slopes than in basins (about 50 cm *vs.* 30 cm/1000 years in the sector off the French Côte d'Azur, according to Gennesseaux and Thommeret, 1968). The fact that the Pliocene–Quaternary sediment cover is usually thicker in the basins reflects the transient and temporary nature of slope accumulation and the importance of downslope displacement of sediment by gravity on margins. The highest sedimentation rates are off deltas; a value of about 1 cm/year has been recorded off the mouth of the Po in the northern Adriatic (Schreiber *et al.*, 1968).

Carbon-14 dates of piston cores in the Alboran Sea, Balearic Basin, and

Strait of Sicily indicate generally higher sedimentation rates in the Late Pleistocene than during the Holocene, a phenomenon attributed largely to Quaternary climatic changes. These climatic changes (cf. Mars, 1963; Farrand, 1971; Rossignol and Pastouret, 1971; Rotschy *et al.*, 1972; Ryan, 1972; Huang and Stanley, 1972; Letolle and Vergnaud-Grazzini, 1973; Stanley *et al.*, 1975*b*) produced eustatic oscillations and modified water mass circulation patterns. These in turn affected oxygen levels, the amount and texture of the fluvial contribution, composition of neritic deposits (terrigenous versus bioclastic components), and the biomass.

Initial estimates of pre-Pleistocene sedimentation rates based on CSP records in the different basins in many instances have not proven reliable. Certain marked reflectors (such as those recovered in Alboran Sea and Nile Cone deep-sea drill cores and now identified as Pleistocene) were originally attributed a Miocene or older age; thus values based on assumed ages are unrealistically low. [One case in point are rates of 0.5–2.5 cm/1000 years in the eastern Mediterranean calculated by Wong and Zarudski (1969)]. The *Glomar Challenger* deep-sea cores now permit a more accurate extrapolation of sedimentation rates from the Pleistocene back to the end of the Miocene. In the Alboran Sea (Site 121), for example, the 3 cm/1000 years rate prevalent from the Miocene to Early Pliocene increased to over 22 cm/1000 years from the Early Pliocene to the Quaternary. Other Pliocene to Quaternary rates calculated by Ryan *et al.* (1973) are: on the Valencia Basement Ridge (Sites 122, 123), from about 5 to 8 cm/1000 years; just east of the Balearic Rise (Site 124), 9 cm/1000 years; in two cores in clefts on the Mediterranean Ridge (Sites 125, 126), about 2–4 cm/1000 years; in the Hellenic Trench (Sites 127, 128), 26 cm/1000 years; on a sector of the Mediterranean Ridge north of the Nile Cone (Site 130), a decrease in time from 40 cm to 7 cm/1000 years as the ridge formed and the uplifted sea floor was no longer accessible to Nile sediments; on the lower part of the Nile Cone (Site 131), > 30 cm/1000 years; on the Tyrrhenian Rise (Site 132), 3–4 cm/1000 years; on the boundary of the Sardinia Slope with the Balearic Basin plain (Sites 133, 134), 4–13 cm/1000 years. Thus, it appears that extrapolated Pliocene to Quaternary rates in some of these sectors are not markedly different from those measured in the Late Quaternary.

The regional differences of Pliocene and Quaternary rates cited above reflect relative accessibility to hemipelagic and sand and mud turbidite deposits as well as structural changes which have modified the sea floor and adjacent land during sedimentation; the evolution of the Mediterranean Ridge at JOIDES Site 130 is one such example. The importance of concurrent structural displacement and deposition in the post-Miocene is discussed at greater length in later sections.

IV. PLIO-QUATERNARY SEDIMENTATION

Depositional patterns depend largely upon the relative supply of sediment, its availability to the sea floor, and a surface, preferably a topographically isolated depression, likely to retain the deposit. River-borne mud, sand, and pebbles are redistributed along the coast (Nesteroff, 1960), and on the narrow shelves (Stanley et al., 1975a) and platforms (Poizat, 1970; Maldonado and Stanley, 1976a) by nearshore currents and surface water circulation. Mediterranean flow patterns in this tideless sea are interpreted by Lacombe and Tchernia (1960, 1972), Wüst (1961), and Miller et al. (1970). River input, nearshore circulation, and storm patterns, all of which influence sedimentation, vary seasonally.

Lithofacies maps (U.S. Naval Oceanographic Office, 1965; Frazer et al., 1970) show that the sediment facies covering the shallow (< 200 m) shelves and broad platforms are considerably more diverse in composition, texture and distribution patterns than those in deeper environments. Bioclastic sand and uniform (bioturbated) mud (Figs. 4C–D) form the prevalent deposits on shallow banks in the Alboran Sea (Milliman et al., 1972), on the margins of the Algéro-Balearic Basin including Provence (Blanc, 1972) and Algeria (Caulet, 1972a, b), in the Strait of Sicily (Blanc, 1958; Fabricius and Schmidt-Thomé, 1972; Maldonado and Stanley, 1976a), and off eastern North Africa and shallower sectors of the Hellenic Arc and the Aegean Sea (Blanc, 1972; Schwartz and Tziavos, 1975). Detailed petrographic studies reveal aerially distinct patterns of Recent and of relict (Quaternary) biogenic sediment as well as zones covered by mixes of both types, including maerl (sand containing abundant calcareous algal fragments such as Lithothamium) and debris of gastropods, pelecypods, bryozoans, and other organisms (cf. discussion in Caulet, 1972a, b).

In contrast, some nearshore to neritic regions are veneered by clastic sediment derived primarily from fluvial sources; these include areas in the vicinity of deltas and zones undergoing coastal erosion (Nielsen, 1973). Clastic material is subsequently displaced by longshore transport. Certain areas such

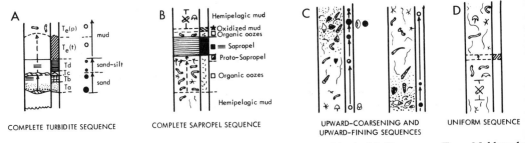

Fig. 4. Graphic representations of major sediment sequences recovered in the Mediterranean. From Maldonado and Stanley (1976a).

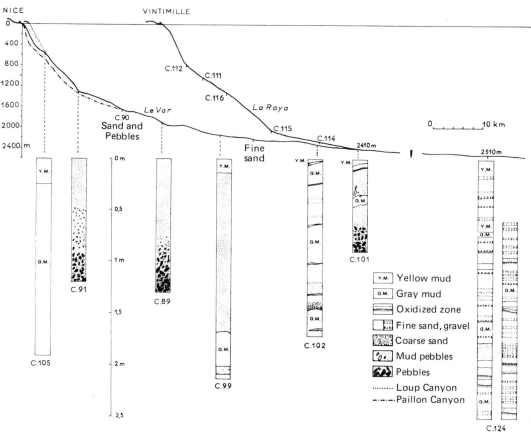

Fig. 5. Logs of cores collected in Côte d'Azur canyons, slope, rise, and Provençal Basin plain showing evidence of downslope movement by turbidite, sand flow, and debris flow processes. From Bourcart (1964).

as the Gulf of Manfredonia are covered by mixes of clastic and biogenic material while others, such as the sector off North Africa, are provided significant amounts of eolian-transported sediments (Fabricius and Schmidt-Thomé, 1972). Nearshore material is sometimes swept into the deeper straits and basins; this is indicated by the occurrence of shallow platform facies in deep-sea piston cores (Kelling and Stanley, 1972a, b; Leclaire, 1972a, b; Rupke and Stanley, 1974; Kelling et al., 1977) and drill cores (Ryan et al., 1973, Chapter 25.2).

Some shallow marine sediment trapped in canyon heads is moved down-slope on relatively steep slopes (3° to 20°; considerably less off deltas) to the basin plains (Fig. 5). Mud (variable mixes of silt and clay), sand, pebbles, and even boulders are transported down canyons and steep escarpments by a plexus of gravitative processes including creep, slump, turbidity currents, and sand and debris flows (these mechanisms are discussed by Middleton and Hampton,

1973). Floods and storms periodically shift material downslope; failure occurs when sediment is transferred directly on metastable material resting on slopes (cf. Ryan *et al.*, 1970; Gennesseaux *et al.*, 1971; Stanley and Unrug, 1972; Got and Stanley, 1974); and in other regions, sediment transport is triggered by frequent and intense earthquakes (Fig. 6). Tar-coated pebbles (Gennesseaux, 1966), man-made objects dumped nearshore, and coastal flora (*Posidonia* grass, etc.) photographed on slopes and basin plains indicate the rapid seaward transfer of material. Deposits tend to be lens-, tongue-, or otherwise irregularly shaped on slopes (pebbly sand in canyons, slump, etc., cf. Bourcart, 1964) and fanlike at the base of some slopes, particularly off deltas (Menard *et al.*, 1965).

The prevailing textural type recovered in piston cores on slopes and in basin plains is mud in which are incorporated minor amounts ($< 2-15\%$) of coarse silt and sand. The coarse silt-and-sand fraction in mud provides a useful indication of sediment mechanism(s) involved. This accessory sand fraction most often consists of planktonic foraminifera or pteropods, or both, while the silt includes wind-blown terrigenous material and coccoliths; the finest fraction of the mud includes clay minerals (Chamley, 1972) and clay-organic

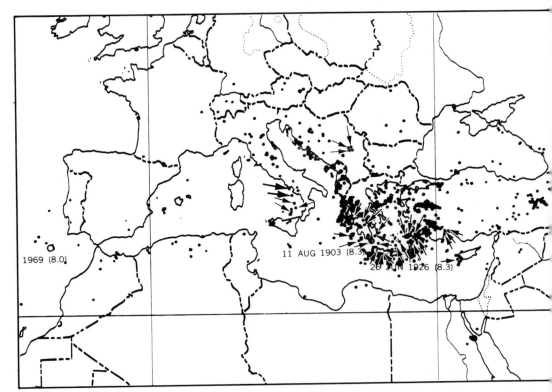

Fig. 6. Seismicity map showing shallow (dots), intermediate (light arrows), and deep (heavier arrows) earthquake epicenters during the period 1963 to 1972. Modified from Tarr (1974).

aggregates (Pierce and Stanley, 1975). Mud-rich sediment of this type is attributed a hemipelagic origin [Fig. 4A, $T_e(p)$]. In other mud types, the accessory sand fraction includes terrigenous or terrigenous/benthic bioclastic mixes; these muds, associated with a distinct suite of sedimentary structures (graded bedding, lamination) revealed by x-radiography, are identified as mud turbidites [Rupke and Stanley (1974); Fig. 4A, $T_e(t)$]. Mud hemipelagites and mud turbidites account for the bulk of the Late Quaternary sequences on slopes and in basins (Fig. 3). Graded sand turbidites also have been recovered in the different basins (Ryan et al., 1965; Horn et al., 1972; Huang and Stanley, 1972). These, and sand flows and slumps, may be responsible for cable breaks (Heezen and Ewing, 1955; Ryan and Heezen, 1965). An average rate of about three fine-grained turbidite flows per 2000 years has been calculated in the more central portions of the Algéro-Balearic Basin (Rupke and Stanley, 1974).

Fine-grained terrigenous, wind-blown, and organic particles are entrained for considerable distances by circulating water masses. Less saline surface water moves differentially above the denser Levantine (or Intermediate) and deep-water masses (Wüst, 1961; Miller 1972a, b); sediment may also be entrained by vertical flow (Stommel, 1972). Suspensates thus can effectively reach all sectors of the Mediterranean (Venkatarathnam and Ryan, 1971; Chamley, 1972; Emelyanov and Shimkus, 1972). Topographically isolated sectors, such as small clefts in the Mediterranean Ridge (Fig. 7) and some of the small basins in the Hellenic Arc, are blanketed by an unconsolidated sediment cover which comprises a larger proportion of hemipelagic mud and organic-rich oozes than is found in more open basins. Suspensate concentrations are generally moderate to quite low except off deltas (the Po and Nile, prior to construction of the Aswan Dam, are good examples) and in straits where erosion of the sea floor and turbulence are significant factors (Kelling and Stanley, 1972a, b; Maldonado and Stanley, 1976a).

The importance of contourites in *Glomar Challenger* drill cores has been suggested by some workers (Nesteroff et al., 1972, 1973). A preliminary survey of suspended sediment in the Algéro-Balearic Basin, using water sampling and optical techniques (Fig. 8), revealed no evidence of a deep nepheloid layer (Pierce and Stanley, 1975). It is quite possible, however, that nepheloid layers and suspensate-rich near bottom water moved by contour-following currents were important in the past when fluvial input was greater and stratification and circulation patterns were different than at present. Identification of contourites in the Mediterranean sequences is not obvious and considerably more detailed study is needed before arriving at firm conclusions about this matter.

Examination of sediment by scanning electron microscope shows that the amount of wind-borne sediment, particularly of silt size, is regionally significant but volumetrically accounts for only a small proportion of the total Pliocene–Quaternary section. Analyses of suspensates and of bottom sediments

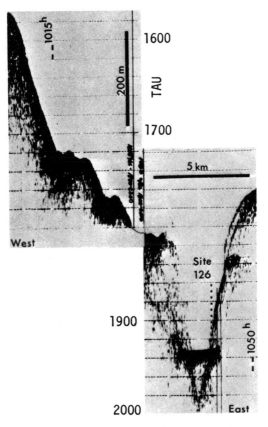

Fig. 7. Fathogram showing small sediment pond in a cleft on the Mediterranean Ridge in the vicinity of *Glomar Challenger* Leg 13, Site 126 (see Fig. 1). From Ryan *et al.* (1973).

reveal eolian material off North African deserts as well as in regions in the northern Mediterranean seasonally influenced by strong winds such as the Bora, Tramontane, and Mistral (Eriksson, 1965; Fierro and Ozer, 1974). Perhaps more obvious are the distinct volcanic ash (tephra) layers recorded in Holocene and Quaternary cores collected in the Tyrrhenian (Ryan *et al.*, 1965), Strait of Sicily (Maldonado and Stanley, 1976a), and throughout much of the eastern basin and Aegean Sea (Ninkovich and Heezen, 1965). Many of these ash layers cover large areas and serve as precise stratigraphic markers (F. W. McCoy, personal communication). Some volcanic-rich sands mixed with biogenic elements are attributed a turbiditic, and not ash flow, origin [cf. examples in the Tyrrhenian Sea and Strait of Sicily examined respectively by Sarnthein and Bartolini (1973) and Maldonado and Stanley (1976a)]. Older (Miocene and possibly Pliocene) tephra and coarser volcanic debris have also been recovered in areas such as the Alboran Sea and Valencia Trough (Ryan *et al.*, 1973).

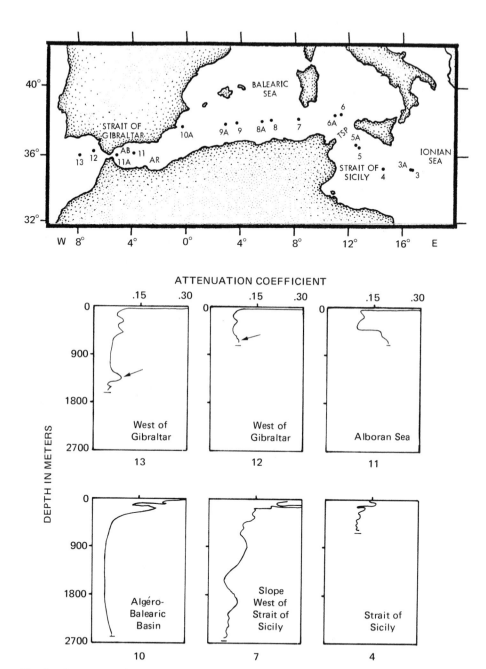

Fig. 8. Light attenuation profiles in the southern Mediterranean. Deflection to the right on graphs indicates increased attenuation (less light transmission due to increased amount of suspended sediment). Note low values in the Algéro-Balearic Basin. Modified from Pierce and Stanley (1975).

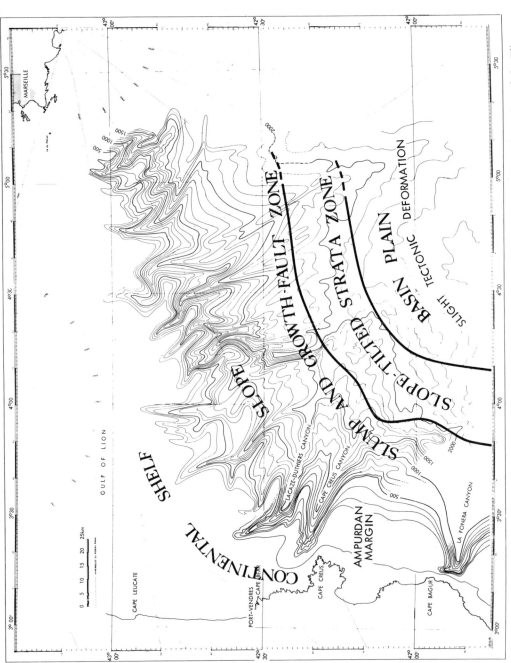

Fig. 9. Chart showing the outer Catalonian Margin major structural zones west of the Rhône Cone; deformation of the Pliocene–Quaternary sequences on the slope and base of slope are due to vertical tectonics and possible salt withdrawal phenomena. From Stanley

The widespread importance of salt tectonics (halokinesis) in the Algéro-Balearic Basin, Provençal–Balearic Basin, and in the Tyrrhenian and Ionian Seas has been recorded (Finetti and Morelli, 1973; see also Table I), but its effect on Pliocene to Recent sedimentation patterns is insufficiently explored. Salt withdrawal along basin margins, as well as faulting (Got, 1973), has resulted in large-scale slumping of sedimentary series on slopes (Fig. 9). Salt domes and related rim-syncline and fault structures, on the other hand, have modified the surface topography of basin plains (Fig. 10). Domes serve as dams (Fig. 10C) which divert or trap dense turbid flows moving from slopes onto basin plains; thus they directly control the aereal distribution of turbidites. Turbid flows spread across a salt-dome-modified surface much as fluvial sediments are diverted across a braided channel system, and as a result, individual turbidites on such a plain cannot be correlated for any considerable distance (Rupke and Stanley, 1974; Stanley et al., 1974a). This is in contrast to well-

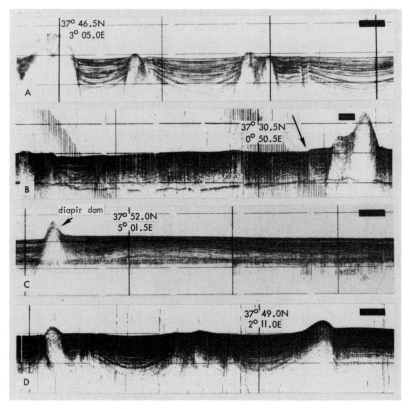

Fig. 10. Examples of salt diapirism and other forms of salt tectonism on 3.5-kHz profiles in the Algéro-Balearic Basin plain. Note how diapir in profile C has acted as a dam behind which sediment is trapped. Horizontal scale (thick bars) is 1.0 km; thin horizontal lines are 20 fm (36.6 m) apart in water. From Stanley et al. (1974a).

stratified units in smaller basins not influenced by salt tectonism or volcanism where it is possible to correlate individual units over a broader region (Ryan *et al.*, 1965, 1970; Bartolini *et al.*, 1972; Huang and Stanley, 1972).

Present depositional patterns are not necessarily representative of those prevalent throughout post-Miocene time. Marked climatic changes in the Pliocene and Quaternary considerably modified sedimentation. The effects of sea level oscillations are particularly well recorded in neritic environments by fining-upward or coarsing-upward cycles (Strait of Sicily region, Fig. 4C; cf. Maldonado and Stanley, 1976*a*); more consistent and widespread changes, however, are recorded by the dark, organic rich sapropel strata (Fig. 4B) found throughout the eastern Mediterranean, including the Adriatic Sea, and pyrite-rich zones in the Strait of Sicily and the western Mediterranean. The youngest sapropel layers are dated between 7500 and 9000 years B.P. (van Straaten, 1972), and these accumulated during the warming trend of the climatic curve in the Early Holocene (Ryan, 1972). The origin of sapropels has been related to stratification of water masses resulting in anaerobic conditions unlike those in effect at present (but perhaps similar to those in the modern Black Sea). Accentuated stratification and anaerobic conditions of the sea floor resulted from

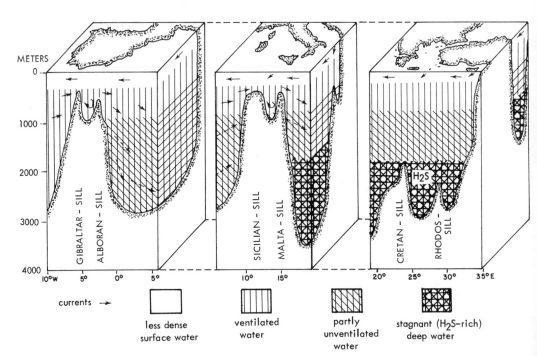

Fig. 11. Schematic showing possible changes of the Early Holocene water mass in the Mediterranean, including stratification and reversal of currents, during the warming phase of the climatic curve. From Stanley *et al.* (1975*b*); topographic base after Wüst (1961).

several factors: (1) possible increased outflow of low-salinity waters from the Black Sea into the eastern Mediterranean coupled with decreased evaporation rates (Olausson, 1961; Ryan, 1972); or (2) surface-water warming (van Straaten, 1972); or (3) excess inflow of fresh water from rivers and melting ice and related short-term current reversals at the Straits of Gibraltar (Mars, 1963; Huang and Stanley, 1972; Nesteroff, 1973c) and Sicily (Stanley et al., 1975b). A schema showing hypothetical stratification and current reversal conditions under which sapropels formed in the eastern and central Mediterranean is depicted in Fig. 11.

The accumulation of sediment in enclosed, silled basins where there is necessarily a close relation between biogenic–terrigenous depositional patterns and climatic and oceanographic factors has in some instances resulted in cyclic sedimentation such as the Late Quaternary Nile Cone deposits (cyclothems) directly related to the Quaternary dynamics (Maldonado and Stanley, 1975, 1976b). Recent deep sea drilling (Glomar Challenger Cruise 42A) also indicates that sapropel layers, related to climatic–oceanographic oscillations in the eastern Mediterranean, were deposited as early as the Pliocene (Hsü et al., 1975).

V. GEOLOGICALLY RECENT MOBILITY OF MARGINS AND BASINS

Depositional patterns reflect a marked tectonic overprint as well as sedimentation mechanisms of the type discussed above. Most geological and geophysical studies of the Mediterranean show that, to varying degrees, this is an "ocean in motion," and displacement of the type affecting the sea floor during the recent geological past appears to continue at present. Seismic subbottom surveys show that most Mediterranean margins are tectonically controlled and that the structural configuration is often complex, i.e., the result of superposition of younger displacement on older trends. A brief review of the manner in which Pliocene and Quaternary sequences have been structurally displaced is considered before progressing with a region-to-region analysis of post-Miocene depositional patterns. It should be recalled that marked vertical exaggeration (times 10 to 30, or more) of the selected CSP illustrated in this chapter distorts the subbottom configuration.

Mauffret et al. (1973) have distinguished two basic types of margins in the western Mediterranean Basin on the basis of seismic lines obtained with deep penetration systems: abrupt and intermediate. The first includes steep (to 20°) continental slopes where the boundary between the margin and basin plain is sharp; subbottom sequences in the basin pinch out within a narrow restricted zone. The intermediate margin is considerably wider and defined by a broader, more gradual transition between the shelf-edge and deep basin; strata are

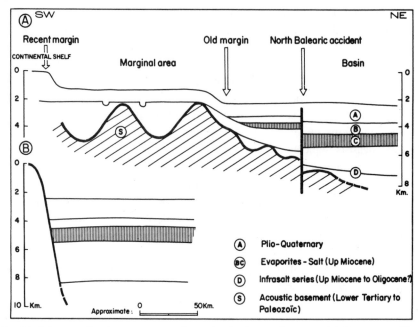

Fig. 12. Diagram showing two types of margins in the northwestern Mediterranean. From Mauffret *et al.* (1973).

offset in a series of steps (Fig. 12A). Further detailing with higher-resolution systems (air-gun and sparker) reveals three types of margins in the western Mediterranean (Stanley *et al.*, 1976): *progressive, intermediate* (or steplike), and *abrupt* (see Fig. 31).

The progressive margin is characterized by a smooth, convex-up sea-floor physiography and lateral seaward continuity of gently dipping Pliocene and Quaternary sediments and underlying M reflectors. This latter unit is slightly offset by faults, which are attenuated in the Pliocene series above it. Growth-faults and slumping have modified the Upper Pliocene and Quaternary sequences. Most of the Catalonian margin off the eastern Pyrenees is of this type (Fig. 9) as is the eastern Betic region north and east of Cape San Antonio (Fig. 13B) and north of the Ibiza margins.

The intermediate type presents a marked, steplike physiography between the outer shelf and base of the slope. The relief of each step or escarpment may be as high as 300–400 m or more; these reflect a series of important vertical offsets of the Pliocene–Quaternary series, M reflectors, and underlying series, all of which are discontinuous between the shelf-edge and the base of the slope (Figs. 13E–F, 14H–I, 20). The downthrown blocks are frequently tilted (Fig. 13E). This structural type, which includes growth-faults, is characteristic of many Balearic Basin margins; examples along the Balearic Platform include sectors southeast of Cape San Antonio, north and northeast of the Mallorca–Menorca Shelf, southeast and south of the Menorca Shelf (including the Balearic Rise), and southwest of the Ibiza–Formentera Shelf. The intermediate margin is probably the most common type of margin encountered in the Mediterranean (Fig. 31).

The third, or abrupt, type is the steepest and presents the greatest relief within a restricted zone; it defines the sharpest and most distinct boundary between the shelf platform sequences and those of the basin plain. The near-horizontal basin plain strata abut sharply against the base of the steep escarpment, and the M reflectors cannot be traced upslope (Fig. 25); the Pliocene–Quaternary sequences are reduced or absent (Figs 14F–G, 19). This margin type defines three prominent slopes in the eastern Betic–Balearic sector: the Menorca, Emile Baudot, and Mazzaron Escarpments. This type, recognized

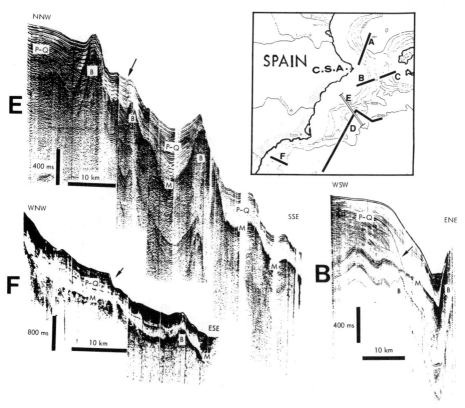

Fig. 13. Air-gun profiles off the eastern Betic chain of Spain, showing examples of progressive (B and northwestern section of E) and intermediate (E, F) margins. P–Q = Pliocene–Quaternary; M = Messinian M reflectors; B = acoustic basement; C.S.A. = Cape San Antonio. Vertical displacement has offset even the uppermost sediment sequences (arrows). Modified from Stanley *et al.* (1976).

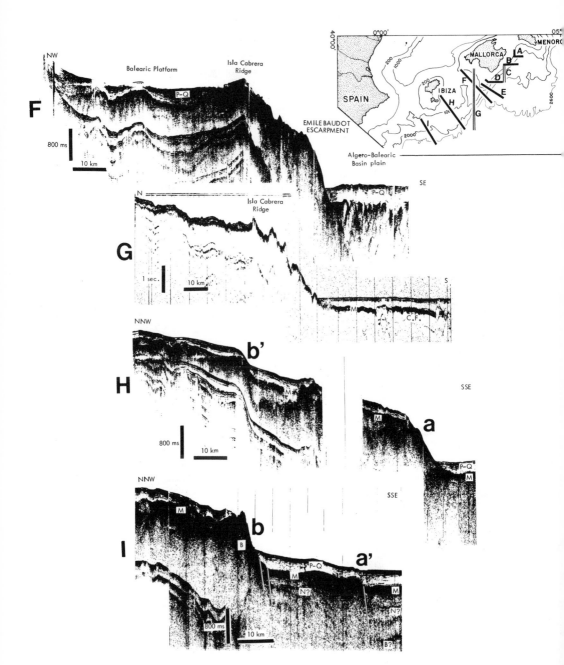

Fig. 14. Seismic profiles along the southern sector of the Balearic Platform, showing examples of abrupt (F, G)- intermediate (H, I)-type margins. Note transition between the two types along the Emile Baudot Escarpment, evidence of post-M (Messinian) displacement. Modified from Stanley et al. (1976).

elsewhere in the Balearic Basin as well as in other sectors of the Mediterranean (cf. chart compiled by Biju-Duval *et al.*, 1974), is associated with steep, narrow margins.

The interrelation of these three margin types is demonstrated in the eastern Betic–Balearic Platform region. A transition from abrupt to intermediate type occurs in the region extending from the south of Mallorca to the south of Formentera (Fig. 14). A lateral transition from progressive to intermediate type is mapped between the Cape San Antonio margin and the Ibiza–Mallorca sectors north of the Balearic Platform. A succession of intermediate to abrupt margins is defined in the regions east and south of Menorca, and extending toward the Cape Palos–Vera margin.

Submarine canyons are present on all three types of margins and appear well developed on the progressive and intermediate types. On abrupt margins, canyons are "perched" and sometimes offset by faulting subsequent to their formation (i.e., the Cabrera Canyon, which has a steep gradient on the Emile Baudot Escarpment). All canyons examined in detail reveal complex tectonic as well as sedimentary origins; they generally follow or are in some way initially controlled by fault trends, and their subsequent development is usually modified to some extent by the Quaternary eustatic events which affected the shallow platforms and influenced sediment transport across the margins toward the basin plain (Got, 1973).

Unconformities and hiatuses prevail in almost all Pliocene and Quaternary sections covering margins and basins, and their importance in terms of Mediterranean evolution have been emphasized (Fabbri and Selli, 1972; Ryan *et al.*, 1970, 1973). The vertical displacement and subsidence of basins related to the three types of margins described above are related to extension, or in some cases, to strike-slip motion. In contrast, there are examples of Pliocene and Quaternary deposits offset by compression. Thrust faults are identified on seismic profiles (Ryan *et al.*, 1973; Finetti and Morelli, 1973), possible allochthonous units (mélanges and/or olistostromes) have been penetrated by deep-sea drilling (*Glomar Challenger* Sites 127–129, Ryan *et al.*, 1973), and small-scale offsets are evident in piston cores (Hieke *et al.*, 1973). Large-scale reverse faulting and thrusts have been recognized offshore in the Hellenic region south and west of the island arc, in sectors of the Adriatic and Tyrrhenian Seas, and in the region south of the Taranto-instep of Italy (Fig. 15). Furthermore, thick accumulations of sediment deformed into large-scale folds as a result of compression are recognized in seismic sections across the Algerian Trench (Auzende and Pautot, 1970; see Fig. 16), and in the region between the islands of Corsica and Elba (Morelli, 1975; see Fig. 17). More detailed examples of concurrent folding and sedimentation in the eastern Mediterranean are illustrated by Ryan *et al.* (1970, their Figs. 17–18) and Got *et al.* (1977).

Post-Miocene to Recent sediment deformation by volcanism and halo-

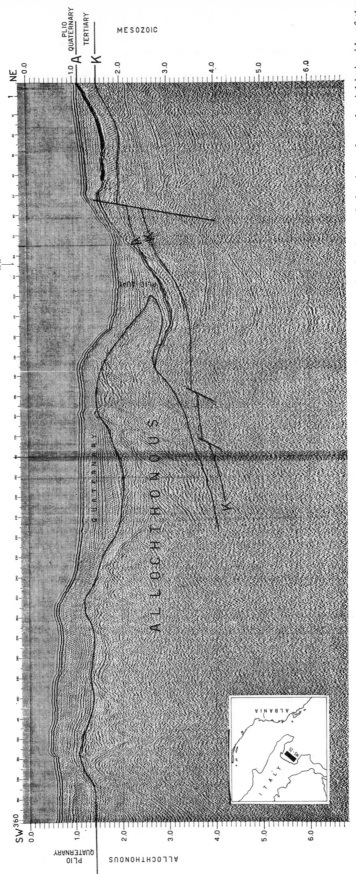

Fig. 15. Seismic reflection transect in the Gulf of Taranto. Note interpreted thrust mass; this type of phenomenon also is observed elsewhere along the Adriatic side of the Apennine system. From Finetti and Morelli (1973).

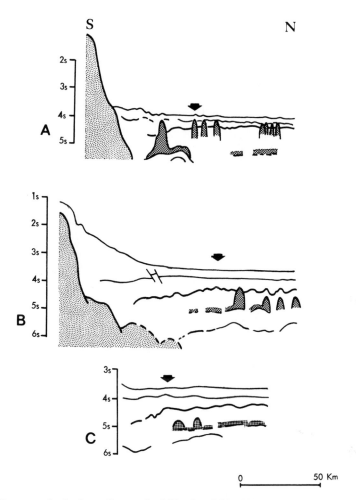

Fig. 16. Interpreted seismic profiles north of Algeria off Alger (A), Bougie (B), and Collo (C). The heavy arrows denote the northern edge of the trenchlike fill. (This feature is also. highlighted by the subbottom reflector above the evaporites shown as a heavy wavy line.) From Auzende *et al.* (1972*b*).

kinesis, also reflecting larger-scale mobility of the sea floor, are amply recorded on seismic records. Salt tectonics in particular has affected the Holocene surficial sediment cover in the western basin (Watson and Johnson, 1968; Hinz, 1972*b*; Stanley *et al.*, 1974*a*; see Profile F, Fig. 14), Tyrrhenian and Ionian Seas (Finetti and Morelli, 1973), and possibly the Nile Cone (Kenyon *et al.*, 1975; Ross *et al.*, 1976).

Neotectonics, which affected recent sedimentation and physiography, are surface indications of large-scale deep-seated events (sea-floor spreading, regional rifting, oceanization, or other events) discussed in other chapters of

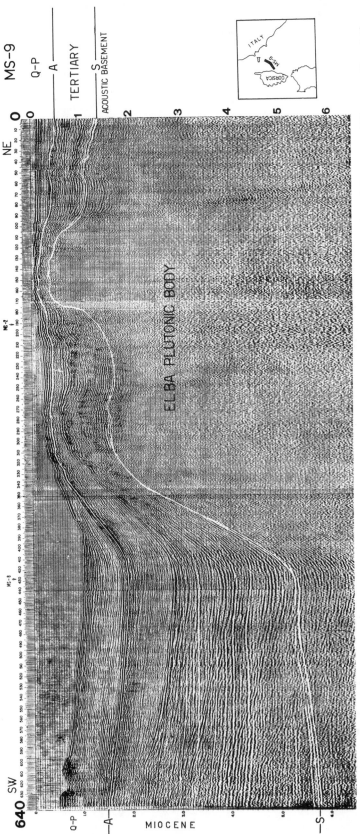

Fig. 17. Seismic reflection transect in the northwestern Tyrrhenian Sea between Corsica and Elba shows about 1 sec of penetration of Quaternary–Pliocene sediment above a thick (at least 8000 m) sediment section of probable Miocene age. From Finetti and Morelli (1973).

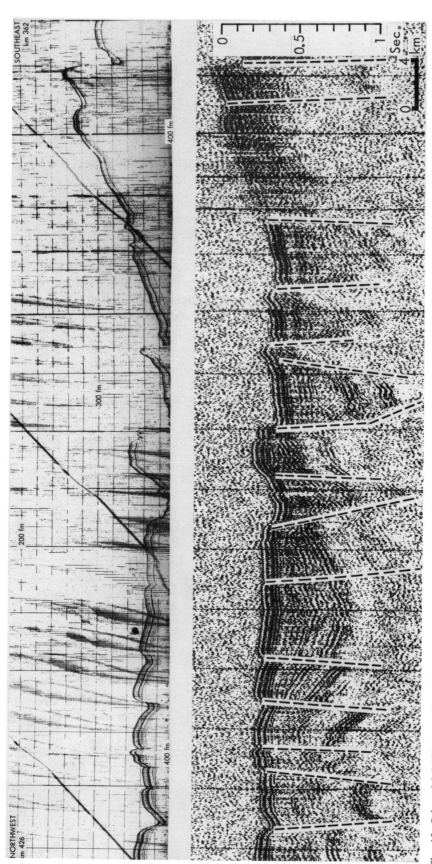

Fig. 18. Selected 3.5-kHz (upper) and 30,000-J (lower) sparker profiles along the same track line in the Strait of Sicily, showing a horst-and-graben network which has offset the uppermost sediment cover. From Maldonado and Stanley (1976a).

this volume. In the following sections, margin configuration and depositional patterns will be considered more specifically in terms of post-Miocene divergent (or extensional), strike-slip, and convergent movements.

VI. DEPOSITIONAL PATTERNS MODIFIED BY DIVERSE NEOTECTONICS

The different styles in which the Pliocene and Quaternary sediment cover has been displaced reflect, in part, the diversity of Mediterranean structural settings. Normal and growth-faults and horst-and-graben systems prevail in divergent and extensional settings. These include, from west to east, the Strait of Gibraltar, much of the Balearic Basin (including the southern Algéro and northern Provençal sectors, Valencia Trough, Ligurian Sea, Balearic Rise, and margins off Catalonia, Corsica, and Sardinia), the Tyrrhenian Sea, Strait of Sicily (Fig. 18), western part of the Ionian Sea (Fig. 19), North Cretan Trough and other Aegean (possibly back-arc) basins (Fig. 20), the large depressions in the northeastern Mediterranean (Rhodes, Antalya, and Cilicia Basins), perhaps small pockets and clefts on the Mediterranean Ridge (Fig. 7), and probably some areas of the Nile Cone. It has also been suggested that recent

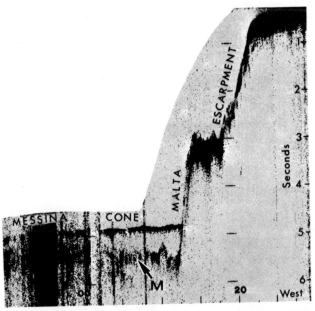

Fig. 19. The Malta Escarpment, an abrupt margin south of Sicily; note broken nature of the M reflectors underlying the Messina Cone. Compare with profiles F and G in Figs. 14 and 25. From Ryan *et al.* (1973).

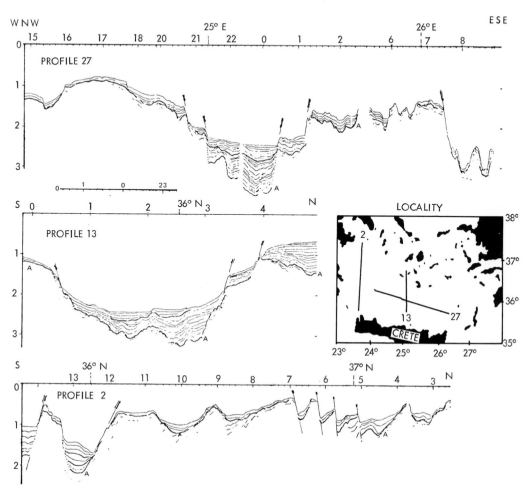

Fig. 20. Line drawings of CSP records in the southern Aegean Sea; profiles 13 and 27 cross the North Cretan Trough. Vertical displacement and marked offset of upper sediment sequences are highlighted. Vertical scale in two-way travel time. Modified from Jongsma *et al.* (1975).

vertical movement recorded in the Nile Cone (Fig. 21) has been activated by solution collapse and other forms of salt tectonics (Kenyon *et al.*, 1975; Ross *et al.*, 1976), as well as by more deep-seated structures (Neev, 1975). The steplike structurally controlled intermediate type margin is prevalent throughout the areas cited above, although excellent examples of both abrupt (Biju-Duval *et al.*, 1974) and progressive margins also are observed. The reader is directed to Table I for a list of published geological, geophysical, and sedimentological studies pertaining to the areas cited above, and to those mentioned in the following sections.

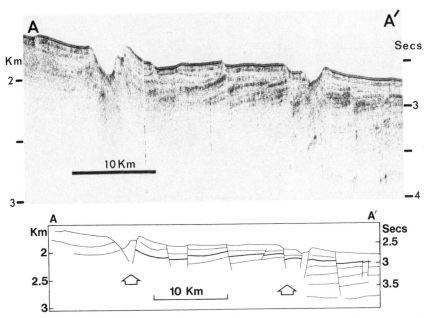

Fig. 21. Air-gun record and interpretation of a Nile Cone section showing horst-and-graben system and normal faults whose throw increases with depth. Note offset of surficial sequences. Deformation has been attributed to diapirism. Modified from Kenyon *et al.* (1975).

Detailed examination of the Catalonian, eastern Betic, and Balearic Platform margins in the Balearic Basin provides evidence of substantial lowering of the Messinian M reflectors and overlying sediment — by perhaps as much as 1500–2000 m — between the upper slope and basin plain (Stanley *et al.*, 1974*a*, 1976). Of particular interest is the Balearic Rise, the large, fan-shaped region at a depth of 1600–2600 m south of the islands of Mallorca and Menorca. This feature is interpreted as a large, detached continental block which formed in the Miocene and subsided significantly in the post-Miocene, but has not yet merged completely with the Balearic Basin plain. Seismic profiles show that the Balearic Rise is completely broken by faults, and its irregular topography reflects a complex network of horsts and grabens (Fig. 22); it is dislocated, perhaps as are other Tertiary and older sections presently buried by the thick Pliocene–Quaternary cover of the Balearic Basin plain (Mulder, 1973; Stanley *et al.*, 1976).

The hypothesis favoring a considerable amount of vertical movement from the Late Miocene to the present, to which this author holds, goes counter to a presently popular concept suggesting that the western Mediterranean Basin was already deep (> 2500 m) in the Late Miocene (Hsü, 1972*b*, Chapter 2, this volume; Cita, 1973; Ryan *et al.*, 1973; Ryan, 1976). Further to the east, Fabbri and Selli (1972) have calculated a mean rate of Tyrrhenian foundering of 1.1 mm/year and suggested that this basin may well be the youngest in the world. While the age and rates of sea-floor wandering remain points of contention, geologists readily agree that the amount of recent vertical movement on the adjacent peri-Mediterranean land has been significant (Fairbridge, 1972; Angelier, 1975; Conchon, 1975; Kidd, 1976). Evidence is mounting for equally significant amounts and rates of vertical displacement in the submarine realm during this same period.

Geological and geophysical studies on land, coupled with those offshore, help pinpoint sectors which have been affected by strike–slip movement. These include, among others, a large part of the Alboran Sea (related to movement of western North Africa relative to the Iberian Peninsula, cf. Olivet *et al.*, 1973*a*, *b*), possibly the eastern Betic–Balearic Platform sector of the Balearic Basin

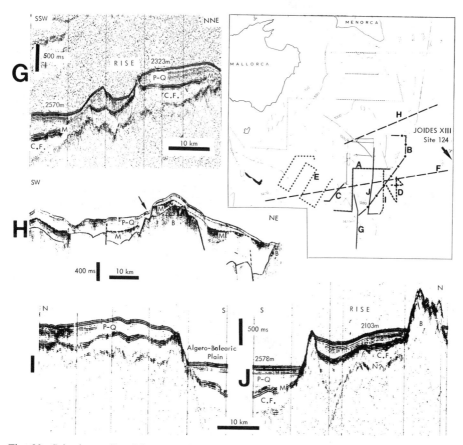

Fig. 22. Seismic profiles (air-gun and sparker) across the Balearic Rise, showing its structurally broken nature and the fault contact with the Balearic Basin plain (profiles G, I, J). Some ridges (B) may be volcanic. Note offset of the Quaternary and surficial sediment series in profile H. From Stanley *et al.* (1976).

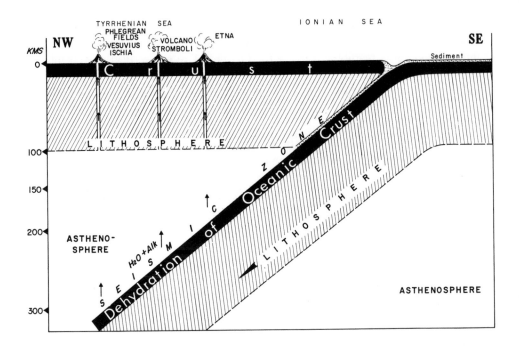

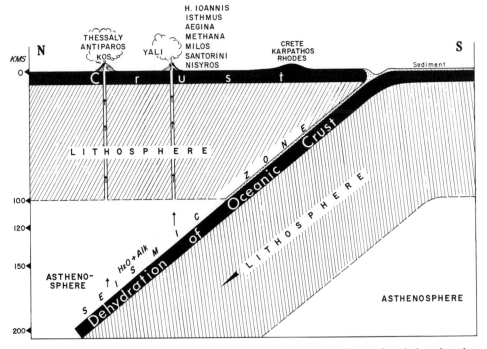

Fig. 23. Underthrusting in the Calabrian (upper figure) and Hellenic (lower figure) Arcs based on seismicity, volcanism, and igneous petrology. From Ninkovich and Hays (1972).

(Auzende *et al.*, 1973*a*, *b*), and the zone between Tunisia, the western Strait of Sicily, and Sardinia (Auzende *et al.*, 1974). Major transcurrent faults also have been delineated in the eastern Leventine Basin–Nile Cone region by Neev *et al.* (1973) and Neev (1975). Margins here, as well as those cited in the previous section, display examples of progressive, intermediate, and locally abrupt margins and provide evidence of important vertical displacement of Pliocene and Quaternary sediment sections.

In recent years, much emphasis has been paid to regions influenced by compression. Of these, the Calabrian Arc, Hellenic Arc, Adriatic Sea–Apulian–eastern Ionian sectors (Belderson *et al.*, 1974*a*), the Algerian Trench, the Corsican–Elba zone, and the long arcuate Mediterranean Ridge have received most attention. The compressive origin of the Mediterranean Ridge, related to underthrusting of the Levantine Basin under the Aegean (Ryan *et al.*, 1970, 1973), has been contested (Sancho *et al.*, 1973); the latter authors provide geophysical evidence favoring intense vertical faulting of mainly Tertiary series by post-Miocene age movements. Seismicity, volcanism, and petrology (Fig. 23) are used to interpret the Calabrian and Hellenic Arcs in light of the convergent plate model (McKenzie, 1970; Rabinowitz and Ryan, 1970; Comninakis and Papazachos, 1972; and Ninkovich and Hays, 1972). However, some geophysicists indicate that the simpler plate boundary–island arc models do not fully account for some observations, and that the Calabrian and Hellenic Arcs, in fact, are distinct from the classic Pacific island arcs (Papazachos, 1974). Although problems of interpretation remain, the evidence of thrustfaults and repetition of section (evidence of compressive motion, cf. Ryan *et al.*, 1973; Finetti and Morelli, 1973; Sorel *et al.*, 1976) as well as of vertical displacement (Wong *et al.*, 1971; Belderson *et al.*, 1972; Angelier, 1975, 1976; Got *et al.*, 1977) are apparent in these seismically active regions.

VII. DEPOSITION IN MOBILE SETTINGS: APPLICATION TO THE GEOLOGICAL RECORD

A. Sediment Dispersal

Sedimentation studies in the various small basins undergoing structural deformation help us unravel Mediterranean evolution and also are valuable for interpreting geological sections in some ancient mobile belts (Menard, 1967; Dickinson, 1974; Dott and Shaver, 1974). The investigations of Mediterranean sediment patterns provide the geologist with some specific examples with which to understand the origin of marine series exposed in the circum-Mediterranean mountain belts and to interpret their paleogeography (cf. Kuenen, 1959*b*; Hersey, 1965*b*; Ryan *et al.*, 1970; Stanley, 1969, 1972; Stanley

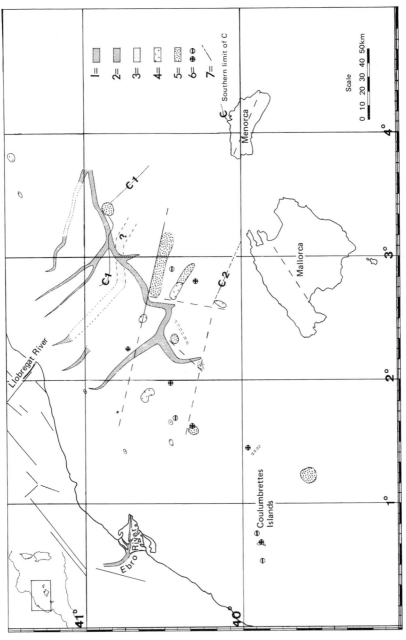

Fig. 24. Structure and subbottom morphology in the Valencia Trough provide evidence of longitudinal plus lateral dispersal: 1 = rivers, 2 = canyons and channels, 3 = buried canyon, 4 = abyssal high, 5 = buried basement high, 6 = magnetic anomalies, 7 = fault. From Mauffret and Sancho (1970).

and Mutti, 1968; Stanley and Unrug, 1972; Stanley *et al.*, 1970, 1972; Hsü, 1972*a*; Hsü and Schlanger, 1971; Hsü and Jenkyns, 1974; and Wezel, 1975).

The transport path followed between source (usually fluvial, but in some cases coastal sections or shallow marine platforms and shelves undergoing erosion) and depositional site is highly variable from region to region. Three distinct types of dispersal patterns are recognized: *longitudinal* (i.e., essentially transport from one end of an elongate depression), *lateral* (primarily entry from one side or margin), and *multiple* (transport from several sides or margins). Examples of these in the western Mediterranean have been schematized by Ryan *et al.* (1970) and Stanley and Unrug (1972, their Fig. 37). In most instances the transport pattern is not strictly of one type, but a composite, usually with one dispersal type predominating. For example, sediment entry into the larger basins (Balearic, Tyrrhenian, and Ionian) is from many sides and thus, in a general sense, these basins fit the multiple source model. However, on closer examination, sediment transport trends are considerably more complex. Sediment entry into the Balearic Basin, the largest, is asymmetrical: the entire northern part of the Basin is dominated by terrigenous sediments primarily from the Rhône, which form the Rhône Cone (Menard *et al.*, 1965) and the thick sediment wedge south of it (Fig. 1); longitudinal dispersal (from the Ebro source) occurs in the wedge-shaped Valencia Trough (Fig. 24) and in the Ligurian Sea; and lateral dispersal prevails on the Algerian margin (filling of the adjacent Algerian Trench, Fig. 16) and the western Corsican–Sardinian margin (Fig. 25).

The Alboran Sea to the west, with its myriad fluvial sources draining the

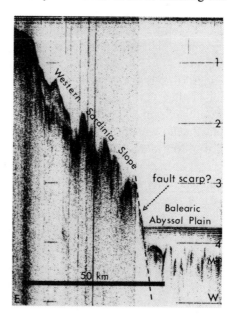

Fig. 25. Seismic profile across the western Sardinian slope–Balearic Basin plain contact. This transect, about 45 km south of *Glomar Challenger* Sites 133 and 134, reveals a possible fault scarp at the base of the slope. Compare with Figs. 14 (profiles F, G) and 19. From Ryan *et al.* (1973).

Betic and Rif chains along its length, would also appear to fit the multiple source model, but detailing of sediment patterns in this oval, mountain-bounded region again reveals a more complex pattern. The numerous rivers and torrents do in fact provide sediments to the coast, but these are then moved on the narrow shelves in a westward direction toward the Strait of Gibraltar and then intercepted by submarine canyons and transported basinward toward the east (Fig. 26; cf. Bartolini et al., 1972; Stanley et al., 1970, 1975a). In contrast, the multiple dispersal pattern in the Tyrrhenian Sea is modified by the structurally and topographically complex margins; ridges have formed structural dams behind which sediments are trapped in peri-Tyrrhenian depressions (Fig. 27). As a result, thick depositional pods are concentrated along the periphery of the Tyrrhenian Sea. Thus, unlike the Balearic Basin where sediment has by-passed the margins, deep sectors of the Tyrrhenian Sea remain only partially buried.

The deposits of the Adriatic Sea record longitudinal dispersal in a much shallower setting. Material is fed largely from the Po at the northwestern end, and additional sediment has been supplied laterally from the Apennine side and, to a lesser extent, from the Adriatic margin (van Straaten, 1965; Pigorini, 1968). CSP records show that structural displacement of the sea floor during deposition has resulted in asymmetry of the sediment wedges: the thickest sections occur near the Italian sector, locally there is a marked reduction of the sediment due to basement highs, and strata dip anomalously northward in the central Adriatic Basin (Hurley, 1971). In contrast, dispersal in the deeper southeastern Adriatic Basin (van Straaten, 1967) is more similar to that of small multi-source basins.

Examples of the three dispersal models are noted in the eastern Mediterranean: lateral, off eastern North Africa where sediments, fed primarily from the Nile River, form the Nile Cone dammed south of the Mediterranean Ridge (this is the largest single sedimentary wedge in the Mediterranean); longitudinal, in the Cilicia Basin between Cyprus and Turkey, derived largely from the Seyhan and other rivers of Turkey to the northeast; and multiple, as illustrated by many smaller basins in the Aegean Sea (Fig. 20) and possibly the Sea of Marmara.

Dispersal in straits (Gibraltar, Sicily, Otranto, etc.) is essentially longitudinal in response to moderate or intense bottom currents in a channelized setting (Fig. 28). Bed load and suspended sediment is swept out of shallow narrows onto slopes and adjacent deep basins. However, some material is trapped in lows within the straits [cf. Strait of Sicily fault-bound troughs (Maldonado and Stanley, 1976a); Strait of Gibraltar small depressions and canyon heads (Kelling and Stanley, 1972a, b)].

It is expected that all of the above examples have counterparts in the ancient rock record. Some comparable dispersal patterns, interpreted from

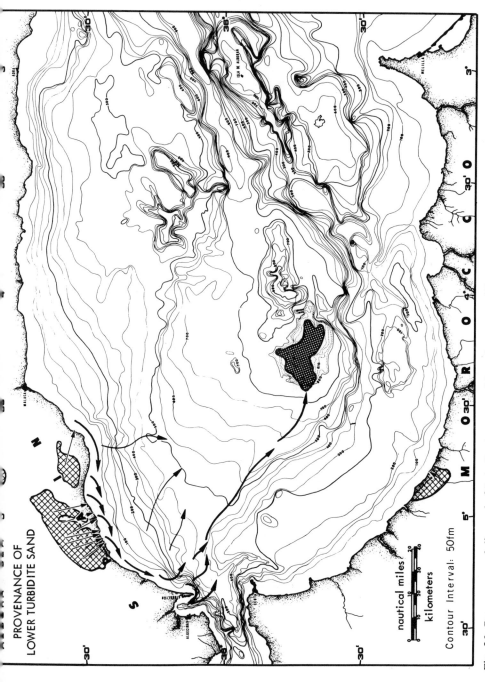

Fig. 26. Provenance and dispersal of Late Quaternary sand into the western Alboran Basin. Note complex pattern: lateral fluvial entry, westward reworking along the narrow Costa del Sol margin, and transfer southeastward via submarine valley to the basin plain. From Stanley *et al.* (1975a).

ancient marine terrigenous formations in geosynclinal settings, have been illustrated (Kuenen, 1959*b*; Dott and Shaver, 1974; and Dickinson, 1974). The rapidly growing interest in ancient basin analysis warrants further elucidation of the complex transport pattern network in the diverse Mediterranean settings.

B. Turbidite Successions and Flysch

Are there analogs of the mud-rich Mediterranean Pliocene and Quaternary facies in the ancient rock record? The thick sequences of alternating turbidite

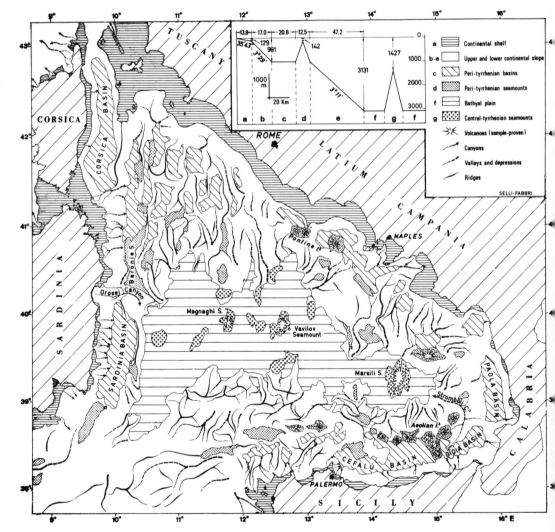

Fig. 27. Tyrrhenian Sea physiographic chart, showing canyons and peri-Tyrrhenian basins along the margins. Fr# Fabbri and Selli (1972).

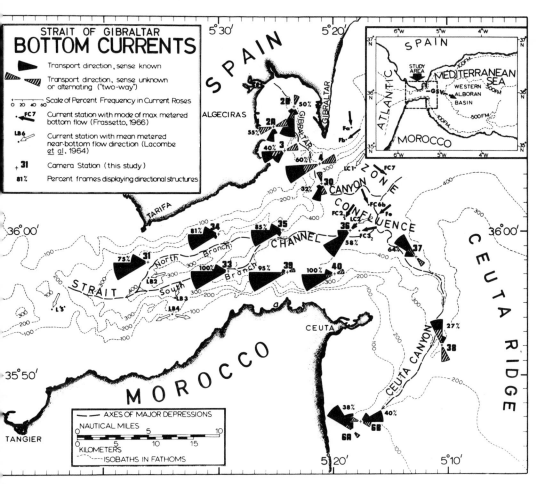

Fig. 28. Longitudinal (westward) transport in the Strait of Gibraltar as determined from orientation of directional features in sea-bed photographs. From Kelling and Stanley (1972a).

and hemipelagic muds interbedded with relatively minor amounts of sand (Fig. 3) are similar to some muddy flysch formations in the circum-Mediterranean mountain chains and mobile belts elsewhere in the world (Hsü, 1972a; Stanley, 1969, 1974c). The rather low proportion of sand turbidites in cores, even those recovered in deltaic cones and off rivers carrying a coarse bed load, is surprising. However, one should recall that a reduction of mud sections to about one-third of their present thickness during consolidation would result in an increased sand–shale ratio; after consolidation, these basin apron and basin plain sequences would resemble some fine-grained units of the type first termed flysch by Studer (1827) in the last century. Possible analogs would include the Ultrahelvetic and other flysch of the Swiss Alps (Hsü, 1970). The more localized tongues of sand and pebbly mud channelized by sand flow and debris flow processes in Mediterranean canyons and valleys, and slumps on the lower slopes and basin aprons also have counterparts in ancient flysch formations (Stanley, 1972, 1974*b*, *c*). The sandier facies within Pliocene and Quaternary sequences, more closely comparable to classic flysch formations, record intense sedimentation associated with phases of structural upheaval, accelerated denudation and/or periods of eustatic low sea level stands.

The Mediterranean Pliocene–Quaternary facies are identified as a type of flysch on the basis of physical characteristics. Although the tectonic and paleogeographic connotations of the term remain controversial, the unconsolidated terrigenous units discussed throughout this chapter are comparable to flysch as descriptively defined (Reading, 1972), i.e., a thick succession of mud, sand, and gravel deposited largely by turbidity currents and mass flow in deep water within a geosynclinal belt. The "geosynclinal" context of the modern Mediterranean has been considered (Stanley, 1969; Ryan *et al.*, 1970). The unconsolidated turbidite–hemipelagite successions now may be evaluated more precisely in terms of a variable geographic–tectonic framework, including divergent, convergent, and strike-slip settings which have undergone variable degrees of deformation since the end of the Miocene.

If flexibly defined, Mediterranean flysch need not be limited to the orogenic–compressive–island arc–cordillera framework of the type originally emphasized by Alpine geologists (cf. Tercier, 1947; Trümpy, 1960; and Hsü, 1970). The largest and thickest deposits — as well as the least deformed and perhaps the most likely to be preserved — are those in extension basins and not in the more stereotyped compressive ones such as the Hellenic, and possibly Calabrian, Arcs. The volume of turbidite successions is much more restricted in the Hellenic Arc setting with its trenches, high seismicity, volcanism, mélanges, olistromes, etc. (Ryan *et al.*, 1973; Stanley, 1974*c*). In some instances, it may be appropriate to interpret the Mediterranean Pliocene–Quaternary sediments as flysch accumulating at plate boundary junctions (cf. Dewey and Bird, 1970; Reading, 1972), but much more sedimentological and geophysical data are needed to confirm such models.

The turbidite-rich sequences in the present Mediterranean occupy diverse geographic and structural settings and a highly variable depth range. While most deposits are found at depths in excess of 1000 m, there are examples of turbidite successions at upper bathyal depths (< 300 m) in the Aegean Sea (Stanley, 1973; Stanley and Perissoratis, 1977). Turbidites also are ponded in the shallower Strait of Sicily fault troughs (Maldonado and Stanley, 1976*a*) and Adriatic Sea basins (van Straaten, 1967), as well as in the deepest (to about 5000 m) Hellenic Arc trenches. It is apparent that the concept of "deep water" now associated with ancient flysch is much too vague, and that other criteria (benthic microfossils, ichnofossils, etc.) are needed to refine paleodepth interpretations.

The Pliocene–Quaternary deposits also record bottom current activity and fine-grained suspended sediment transport in some flysch-type basins. These mechanisms are difficult to evaluate in ancient turbidite successions because of the logistical difficulties of analysing mudstone and shale in the field and laboratory. As a result, much more is known of sand turbidites and sandflow deposits. The "pelagic rain" concept (Aubouin, 1965) which has been used to explain the presence of fine-grained deposits between sand turbidites may be applicable in large ocean basins, but is of questionable use in the smaller, current-influenced basins of the Mediterranean type.

The Mediterranean depositional patterns substantiate indirect evidence provided by ancient basin studies (Stanley, 1969; Stanley and Unrug, 1972) that deltas play a major role in flysch sedimentation. The influence of fluvial input and the close relation between deltas and turbidite–hemipelagite successions is demonstrated by the sediment thickness charts (Fig. 1). Only in the vicinity of large deltas is sediment input a more important factor than tectonic framework in defining depositional trends. The Rhône River debouches in an extensional, the Ebro in an extensional (and/or strike-slip), and the Nile in a compressional setting, but in all cases the margins and adjacent basin floors are masked by a thick, largely terrigenous mantle. The importance of fluvial–deltaic contributions in ancient molasse deposition in orogenic settings also is recognized. A possible example of a modern molasse analog is the thick deposit, derived largely from the Po River, which has filled the shallow northwest part of the Adriatic Sea. As in the case of some ancient molasse, it is possible to trace continental and shallow marine to deeper marine flysch-type (in the southeastern Adriatic basin) deposits. The emplacement of nappes, reduction of the sediment sequences by folds and faults, and other evidence of contemporaneous sedimentation and structural activity are observed in the Adriatic Sea. However, more work in this region, particularly with detailed CSP and drilling, is needed before it can be reliably established that the Adriatic is comparable with some of the classic molasse foredeeps.

C. Depositional Patterns and Paleogeography

Sedimentation studies can be used to resolve larger-scale problems of peri-Mediterranean paleogeography. In several instances, investigations of terrigenous formations on land have shed light on offshore geology. An example is the paleocurrent–petrographic investigation of Paleogene formations in the Maritime Alps and northern Apennines, which has shown that these regions received material from a large area once subaerially exposed in the Ligurian Sea (Kuenen, 1959a; Stanley and Mutti, 1968). This approach, complementing other geologic evidence, provides information on post-Eocene rifting and rotation of Corsica and Sardinia and on the opening of the Ligurian Sea (Fig.

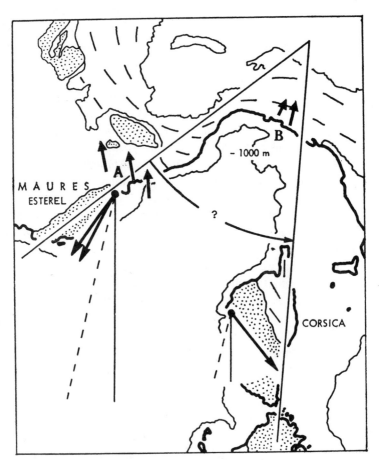

Fig. 29. Map showing opening of Ligurian Sea and rotation of Corsica away from the French Provençal margin, based on geological and geophysical evidence and supplemented by sedimentological studies of Upper Eocene–Oligocene sandstone formations in France and Italy. These units, at A and B, show provenance of material from a source area subaerially exposed in what is now the Ligurian Sea. From Stanley and Mutti (1968).

29). In like manner, it should be possible to apply sedimentation investigations of the offshore Pliocene–Quaternary sequences to problems of post-Miocene Mediterranean paleogeography and evolution.

A case in point is provided by *Glomar Challenger* Deep-Sea Drill Site 130; analysis of its petrology sheds light on the age of formation of the Mediterranean Ridge. Terrigenous layers derived from the Nile and deposited in large part by turbidity currents were recovered in cores (Fig. 1) on the southern part of the ridge. Both fine- and coarse-grained fractions show that until quite recently, this sector occupied a seaward portion of the Nile Cone and that uplift isolated the ridge from Nile clastics approximately one-half million years ago (Ryan *et al.*, 1973).

Other examples of recent geographic changes are revealed by deep-sea drilling coupled with seismic surveys. Sites 127 and 128 in the Hellenic Trench west of Crete show a progressive dipping of the Quaternary strata toward the landward wall of the trench at a rate of about 1°/million years and a stratigraphic inversion of Lower Cretaceous above Pliocene pelagic ooze. The cores penetrated either (1) a tectonic mélange in a zone of underthrusting (possible subduction), or (2) a large olistostrome (Ryan *et al.*, 1973). A rapid rate of Quaternary deformation and uplift also is recorded in the Strabo Trench (Site 129) farther to the east. Furthermore, the majority of drill sites occupied in the eastern, central (Tyrrhenian), and western sectors reveals important discontinuities within the Pliocene series and between the Pliocene and Quaternary units (also apparent on CSP). These hiatuses record changes in sea-floor configuration and sediment regime related to uplift or subsidence, and perhaps also marked a periodic alteration of the circulatory regime, including development of strong thermohaline circulation which may have intensified sea-floor erosion.

The stratigraphic configuration of submarine canyons and their fill has been used to evaluate the Neogene tectonic evolution of some Mediterranean basins. Earlier studies emphasized submarine canyon formation in the Miocene (Bourcart, 1959), and this information has been used as one of the arguments in support of the "deep, dry Mediterranean at the end of the Miocene" hypothesis (Hsü, 1972*b*). While some canyons undoubtedly were cut at that time, detailed seismic surveys also show that the major phase of canyon cutting in some sectors occurred in the Pliocene and again in the Quaternary (Got, 1973); this observation is in support of the "major subsidence in the Pliocene" hypothesis (Biju-Duval *et al.*, 1974; Stanley *et al.*, 1974*b*). Inasmuch as the age of submarine canyon formation is variable along the different margins of the western basin, I feel that too much weight has been given to a Mediterranean-wide canyon cutting phase closely associated with the Messinian as proposed by Ryan *et al.* (1973). Some canyons located on present structural highs appear to have been cut in the Upper Miocene; erosional evidence is best recorded on the continental shelf and upper slope. Pliocene and Quaternary canyon cutting phases are better developed between the outer continental shelf and base of the slope (Stanley *et al.*, 1976). Careful evaluation of the age and style of canyon formation on each margin is needed to more precisely determine the regional changes of sea-floor configuration from the Messinian to the Quaternary and to verify the possibility of accelerated basin foundering in the Pliocene.

In addition to helping resolve paleogeographic problems within the offshore realm and immediately adjacent land, an evaluation of Pliocene–Quaternary sedimentation patterns can be used to interpret (or speculate on) aspects of pre-Messinian peri-Mediterranean geology. One such problem pertains to the Paleogene depositional trends in the Alpine chain. In this respect, it has been shown that the Hellenic Arc, in terms of overall shape and dimensions, is broadly similar to the Alpine belt of western and central Europe (Stanley, 1974a). The comparison can be made by rotating the Hellenic Arc counter-clockwise about 90° and superposing the eastern end of the Arc (the part which extends toward the southeastern Turkey margin) over the French–Italian Maritime Alps. In this position, the Hellenic Arc follows rather closely the arcuate Alpine trend from the Riviera, along the French and Swiss Alps, across to Austria. The reduced width of the Alpine belt arc is a function of the compression which prevailed during the late Mesozoic–Tertiary orogeny.

A comparison of these two arcuate belts (Fig. 30) sheds some light on the origin of the various deep-water clastic formations deposited in the Late Eocene to lowermost Oligocene sea (the Lattorfian Sea of Boussac, 1912) which occupied what is now the western Alps. The different sandstone–shale successions have been mapped as distinct flysch formations—grès d'Annot, grès du Champsaur, grès de Tavayannaz, etc. (Fig. 30, lower left). These turbidite-rich sequences are distributed discontinuously along a belt about 300 km long, which represents only a small part of the Hellenic Arc; this distance is comparable to the one between southern Crete and western Peloponnesus (Fig. 30, lower right). Each of the Paleogene deposits is defined as a separate mappable formation on the basis of lithology and basin analysis, and lithofacies differences reflect distinct geographic and geologic conditions that prevailed along the length of the Lattorfian Sea. It appears that each flysch formation is the result of deposition in a geographically distinct isolated depression within the same sea.

If the same field and lithofacies arguments applicable to the Alpine formations are applied to the Hellenic Arc deposits, it would be possible to designate each of the Pliocene–Quaternary sand–silt–clay sections filling the different Hellenic Arc basins as separate mappable formations. Some basins, like the North Cretan Trough, have received greater volcanic input than others; a comparable analog may be the Late Eocene Champsaur and Tavayannaz Formations, which present volcanic facies in contrast to other formations of equivalent age (such as the Annot Sandstone) that do not. In terms of surface area, some Hellenic basin and apron fills are considerably larger than those of the flysch formations exposed in the western Alps. However, the total surface area of all the Paleogene Alpine flysch exposures represents well under 10% of the external Alpine Arc (Fig. 30) and the distance separating the different flysch formations ranges from about 20 to 50 km; this is remarkably similar to the present Hellenic Basin dispersion pattern (Fig. 30, top).

Why are the various Paleogene turbidite successions separated from each other? Until recently it has been assumed that this discontinuity is the result of removal by erosion and structural displacement during the Alpine surrection.

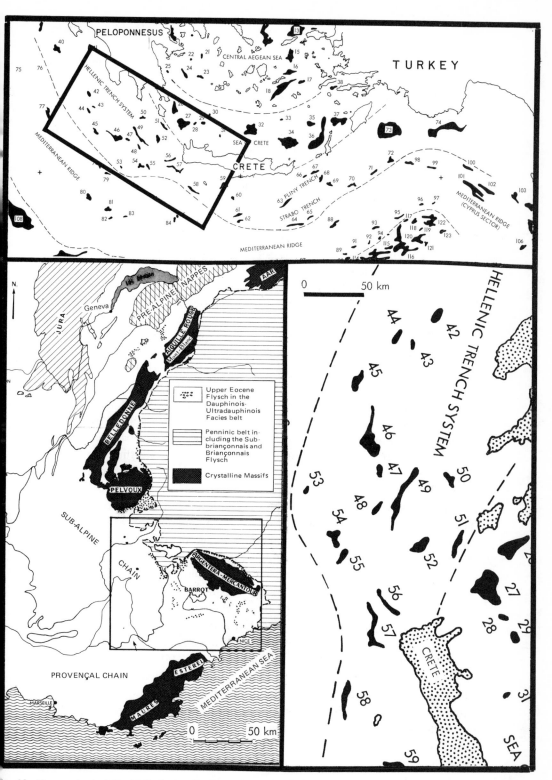

Fig. 30. Upper, chart of the Hellenic Arc (basins and trenches in black); lower left, Upper Eocene–Lower Oligocene flysch formations (dots) in the western Alps; lower right, basin plains in area between Crete and southern Peloponnesus, reproduced at same scale. From Stanley (1974a).

If, however, the Hellenic Arc is an acceptable model, it can be postulated that the geographic discontinuity between the different Paleogene flysch units essentially reflects the original separation of the structural basins. It was shown earlier that the distance between these Alpine formations is roughly comparable to that between modern basin plains in the Hellenic Arc as noted in Fig. 30. Additional light on the matter is provided by seismic profiling which shows the marked regional variability of Pliocene–Quaternary sediment thickness on both sides of the Hellenic island chain (Wong *et al.*, 1971; Jongsma et al., 1975): thicker sections most commonly occur in major depressions and reduced ones on tectonic ridges and topographic highs (Fig. 20). It is likely that a similar pattern prevailed in the Alpine Lattorfian Sea, and that the thick Paleogene turbidite localities define regionally distinct submarine fan and basin plain sequences. It is also probable that the originally thicker basin plain fills would have had a better chance of preservation after surrection and erosion than the thinner, intervening time-equivalent slope and topographic high sequences, which presumably have been removed during and after Alpine formation.

Attributes of the Hellenic Arc also may be applied to the study of ancient deep-water formations in other regions. For instance, many of the well-known flysch formations distributed along elongate mobile belts, such as the Tertiary Macigno and Marnoso Flysch of the Apennines and the Ouachita Flysch units of Arkansas and Oklahoma, show predominant longitudinal dispersal patterns. However, inasmuch as the upper- to mid-slope deposits of ancient turbidite basins are only rarely preserved, the mapped paleocurrent trends identify movement primarily along the major basin axes and thus provide an incomplete record of sediment entry and dispersal. At present, there is insufficient sedimentological data to use readily the Hellenic Arc as a model insofar as dispersal patterns are concerned. Nevertheless, it is noted that most Hellenic basins and trenches are oriented parallel or sub-parallel to the major trend of the arc (Stanley, 1974a). Thus, a running average of transport paths within basin plains conceivably may be oriented along the major regional tectonic trend and, as in some ancient flysch, show a predominant longitudinal dispersal along the arc. The sea-floor topography and preliminary core analysis, however, indicate that sediment in many of these depressions have been introduced laterally from marginal sources by way of canyons and steep scarps (Got *et al.*, 1977). Any deflection of turbid flow trends during downslope transport would probably result from the change in sea-floor orientation between the slope and basin plain axis, i.e., a change in dispersal from one that is normal to one subparallel to the regional tectonic trend of the arc. Bottom currents flowing parallel to the basin axis also may deviate sediment during transport. The probability of a change from transverse to longitudinal flow patterns between slope and plain in each of the individual Hellenic basins is strong. Ongoing research in this area may well shed light on the subject.

It is apparent that many of the basic problems of the type highlighted in this chapter await solution. Detailing of Mediterranean depositional trends in light of sedimentation and structural considerations is certain to provide scientific returns profitable to both land-bound and marine geologists working in this complex region.

VIII. SUMMARY

This synthesis of Pliocene and Quaternary depositional patterns and post-Miocene structural displacement in the Mediterranean highlights the following:

1. The modern Mediterranean is not strictly a relict Tethyan feature. Although influenced to varying degrees by Mesozoic and Paleogene structural trends, the present configuration of margins and basins and their cover of unconsolidated Pliocene and Quaternary sediments clearly record evidence of major tectonic events which have taken place since the end of the Miocene.

2. A new chart of the Pliocene–Quaternary sequences depicts the remarkable regional variation of thickness and major depositional trends. The thickest, most extensive deposits occur near major delta point-sources (Nile, Rhône, Po, Ebro) and in or near zones largely influenced by circulating suspensate-rich water masses (Alboran Sea, straits), and in sectors which have been subject to concurrent sedimentation and structural deformation (Algerian and Corsica–Elba Trenches).

3. The high, overall depositional rates reflect the importance of seasonal fluvial input and suspended sediment transport into relatively small catchment basins. Pliocene and Quaternary rates are roughly comparable to those measured in Late Quaternary sections except in regions, such as the Mediterranean Ridge, which have undergone substantial displacement in the post-Miocene.

4. Mud turbidites and mud hemipelagites, attributed respectively to gravity processes and deposition from circulating water masses, account for the bulk of the Late Quaternary sequences on slopes and in basins. Topographically isolated sectors display enhanced proportions of hemipelagic mud and organic ooze as a result of isolation from turbidite and other mass gravity-influenced flow; all sectors of the Mediterranean have access to fine-grained sediments which are transported for considerable distances.

5. Marked climatic changes in the Pliocene and Quaternary have left their mark on the sediment record. Climatic fluctuations are recorded by cyclic sedimentation patterns and sapropels in the eastern and pyrite-rich zones in the western Mediterranean; some discontinuities may reflect intensified sea-floor erosion at times of thermohaline circulation and current reversal.

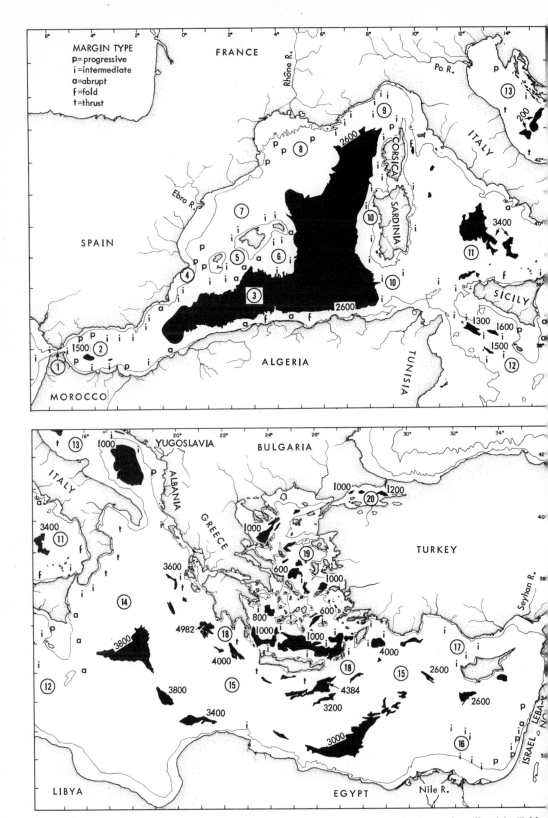

Fig. 31. Chart of Mediterranean, showing geographic areas discussed in text (code numbers listed in Table Margin types: p = progressive; i = intermediate; a = abrupt; f = fold; and t = thrust. Depth of basin pla (in black) in meters.

Levantine Basin, southeastern sector, including Nile Cone and Herodotus Abyssal Plain	16	Emery et al., 1966 Carter et al., 1972 Goedicke, 1972	Oren, 1969 Miller et al., 1970 Morcos, 1972	Shukri, 1950 Einsele, 1967 Wong and Zarudzki, 1969 Venkatarathnam and Ryan, 1971 Emelyanov and Shimkus, 1972 Herman, 1972 Keller and Lambert, 1972 Milliman and Müller, 1973 Nielsen, 1973 Ryan et al., 1973 Maldonado and Stanley, 1975, 1976b Summerhayes and Marks, 1975 Wezel, 1975	131 Ryan et al.
Levantine Basin, northeastern sector	17	Emery et al., 1966 Ryan et al., 1970 Carter et al., 1972	Wüst, 1961 Burman and Oren, 1970 Miller et al., 1970	Emery et al., 1966 Wong and Zarudzki, 1969 Evans, 1971 Venkatarathnam and Ryan, 1971 Emelyanov, 1972 Horn et al., 1972 Ryan, 1972 Stanley, 1973, 1974a, c	375, 376 Hsü et al.
Hellenic (or Aegean or Cretan) Arc, including Ionian Island and southern Aegean sectors	18	Hersey, 1965a Emery et al., 1966 Pareyn, 1968 Maley and Johnson, 1971	Wüst, 1961 Mosetti et al., 1969 Burman and Oren, 1970 Miller et al., 1970 Miller, 1972a, b, 1973	Emery et al., 1966 Pérès, 1968 Wong and Zarudzki, 1969 Pastouret, 1970 Rossignol and Pastouret, 1971 Herman, 1972 Hsü, 1972a Opdyke et al., 1972 Ryan, 1972 Milliman and Müller, 1973 Müller and Staesche, 1973 Ryan et al., 1973 Stanley, 1974a, c Wezel, 1975	127, 128, 12 Ryan et Hsü et al.
Aegean Sea, central and northern sectors	19	Carter et al., 1972 Maley and Johnson, 1971	Wüst, 1961 Bruce and Charnock, 1965 Miller, 1972a, b	U.S. Naval Oceanographic Office, 1965 Wong and Zarudzki, 1969 Frazer et al., 1970 Chamley, 1972 Blanc, 1972 Emelyanov, 1972 Emelyanov and Shimkus, 1972 Mistardis, 1972 Stanley, 1973, 1974a, c Schwartz and Tziavos, 1975 Stanley and Perissoratis, 1977	None
Sea of Marmara	20	Carter et al., 1972 Gunnerson and Ozturgut, 1974	Anderson and Carmack, 1974 Gunnerson and Ozturgut, 1974 Ostlund, 1974 Scholten, 1974	U.S. Naval Oceanographic Office, 1965 Emelyanov, 1972	None

1973	Said, 1962 Soliman and Faris, 1964 Karcz and Kafri, 1973 Neev *et al.*, 1973 Freund *et al.*, 1975 Neev, 1975 Wezel, 1975	Vogt and Higgs, 1969 Papazachos and Comninakis, 1971 Lort, 1973 Ben-Avraham *et al.*, 1976	Ninkovich and Heezen, 1965	Kenyon *et al.*, 1975 Sonnenfeld, 1975 Smith, 1976	Neev *et al.*, 1966 Ryan *et al.*, 1970, 1973 Wong *et al.*, 1971 Biju-Duval *et al.*, 1974 Ginzburg *et al.*, 1975 Kenyon *et al.*, 1975 Morelli, 1975 Ben-Avraham *et al.*, 1976 Ross *et al.*, 1976 Smith, 1976
1975	Aubouin, 1965 Gass and Smewing, 1973 Ozgül and Arpat, 1973 Biju-Duval, 1974 Morelli, 1975 Morgan and Evans, 1975	Vogt and Higgs, 1969 Rabinowitz and Ryan, 1970 Lort, 1973 Lort *et al.*, 1974 Papazachos, 1974	Ninkovich and Heezen, 1965	Lort and Gray, 1974	Hersey, 1965*b* Ryan *et al.*, 1970 Wong *et al.*, 1971 Biiu-Duval *et al.*, 1974 Lort and Gray, 1974
378 , 1973 1975	Aubouin, 1965 Aubouin and Dercourt, 1970 Rabinowitz and Ryan, 1970 Drooger and Meulenkamp, 1973 Biju-Duval, 1974 Biju-Duval *et al.*, 1974 Papazachos, 1974 Angelier, 1975, 1976 Mercier *et al.*, 1976	Drake and Delauze, 1969 Vogt and Higgs, 1969 McKenzie, 1970 Papazachos and Comninakis, 1971 Makris, 1973 Jongsma, 1974	Ninkovich and Heezen, 1965 Marinatos, 1971 Ninkovich and Hays, 1972 Pe and Piper, 1972	Braune and Heimann, 1973	Ryan *et al.*, 1970, 1973 Wong *et al.*, 1971 Finetti and Morelli, 1973 Sancho *et al.*, 1973 Jongsma *et al.*, 1975 Sorel *et al.*, 1976 Got *et al.*, 1977
	Aubouin, 1965 Maley and Johnson, 1971 Makris, 1973 Schuiling, 1973 Biju-Duval *et al.*, 1974 Papazachos, 1974 Smith and Moores, 1974 Keraudren, 1976	Vogt and Higgs, 1969 McKenzie, 1970 Comninakis and Papazachos, 1972 Needham *et al.*, 1973 Jongsma, 1974 Morelli *et al.*, 1975	Ninkovich and Heezen, 1965 Ninkovich and Hays, 1972 Pe and Piper, 1972		Watson and Johnson, 1969
	Maley and Johnson, 1971 Brinkmann, 1974 Papazachos, 1974 Scholten, 1974	Comninakis and Papazachos, 1972 Makris, 1973	Pe and Piper, 1972		

6. The dense network of seismic profiles reveal marked offset of Pliocene and Quaternary sequences along basin margins and in basin plains as a result of geologically recent vertical displacement. A number of margin types are identified, and these are related to extension, strike-slip, and compressive motion, and to salt tectonics (diapirism, solution collapse, etc.).

7. In the western Mediterranean, seismic transects indicate subsidence along margins of the order of the 1500–2000 m between the upper slope and basin plain, suggesting that the basin sea floor occupied bathyal depths (in the range of 1000 m) at the beginning of the Pliocene. Vertical movement is suggested for the origin of the Balearic Rise, interpreted as a large, detached continental block which formed in the Miocene and subsided in the post-Miocene, but has not yet merged completely with the Balearic Basin plain; in contrast, the uplift of the Mediterranean Ridge during this period is demonstrated by deep-sea drilling.

8. In regions not influenced by major deltas, geologically recent structural mobility, and not just sedimentation *per se*, is largely responsible for the marked regional differences in depositional thicknesses, margin configuration, and basin plain depths.

9. The Mediterranean Sea serves as an excellent natural sedimentation laboratory in that it provides a marked diversity of depositional conditions for interpreting the geological record. Its use is demonstrated in the formulation of transport dispersal models, in providing examples of deep-sea deposits analogous to ancient facies such as flysch, and its application to problems of Mediterranean paleogeography and the study of depositional patterns in ancient mobile belts in the circum-Mediterranean region and elsewhere.

10. Many of the major problems pertaining to Mediterranean evolution await solution, and the detailing of depositional patterns is a productive endeavor and one providing a scientific challenge to marine geologists.

ACKNOWLEDGMENTS

Some concepts presented in this chapter are outgrowths of stimulating exchanges with the many colleagues who have participated with me in the Mediterranean Basin (MEDIBA) Project since 1970. Particular appreciation is expressed to Drs. A. Brambati, University of Trieste; H. Got, University of Perpignan; T.-C. Huang, University of Rhode Island; G. Kelling, University of Wales at Swansea; A. Maldonado, University of Barcelona; F. W. McCoy, Lamont–Doherty Geological Observatory; N. A. Rupke, Oxford University; and Y. Weiler, Israel Institute of Petroleum. I also thank Drs. H. Got, University of Perpignan, and J. M. Lort and I. Price, British Petroleum Company, for reviewing the manuscript. Financial support for marine geological and sedimentological studies within the framework of the MEDIBA Project has been provided by the Smithsonian Research Foundation, grant 460132, and the National Geographic Society, grant 156780.

REFERENCES

Akal, T., 1972, The general geophysics and geology of the Strait of Sicily, in: *Oceanography of the Strait of Sicily*, Allan, T. D., Akal, T., and Molcard, R., eds., SACLANT ASW Res. Cent., La Spezia, Conf. Proc. 7, p. 177–192.

Alinat, J., Cousteau, J.-Y., Giermann, G., *et al.*, 1969, Levée de la carte bathymétrique de la mer Ligure, *Bull. Inst. Océanogr. Monaco*, v. 68, p. 1–12.

Alinat, J., Leenhardt, O., and Hinz, K., 1970, Quelques profils en sondage sismique continu en Méditerranée occidentale, *Rev. Inst. Franç. Pétrole*, v. 25, p. 305–326.

Alla, G., 1970, Etude sismique de la plaine abyssale au sud de Toulon, *Rev. Inst. Franç. Pétrole*, v. 25, p. 291–304.

Alla, G., Dessolin, D., Leenhardt, O., and Pierrot, S., 1972, Données du sondage sismique continu concernant la sédimentation Plio-Quaternaire en Méditerranée nord-occidentale, in: *The Mediterranean Sea — A Natural Sedimentation Laboratory*, Stanley, D. J., ed., Stroudsburg: Dowden, Hutchinson, and Ross, p. 471–487.

Alla, G., and Leenhardt, O., 1971, Découverte d'un dôme de sel (Dôme T, Sud Toulon), avec le bathyscaphe, *C. R. Acad. Sci. Paris*, v. 272 (D), p. 1347–1349.

Allain, C., 1964, L'hydrologie et les courants du détroit de Gibraltar pendant l'été de 1959, *Rev. Trav. Inst. Tech. Maritime*, v. 28, p. 3–102.

Allan, T. D., 1966, Underwater photographs in the Strait of Gibraltar, *SACLANT ASW Res. Cent., La Spezia, Tech. Mem.* 116, 19 pp.

Allan, T. D., Akal, T., and Molcard, R., eds., 1972, *Oceanography of the Strait of Sicily*, SACLANT ASW Res. Cent., La Spezia, Conf. Proc. 7, 227 pp.

Allan, T. D., and Morelli, C., 1971, A geophysical study of the Mediterranean, *Bol. Geof. Teor. Appl.*, v. 13, p. 99–142.

Allan, T. D., and Pisani, M., 1970, Gravity, magnetic and depth measurements in the Ligurian Sea, *Rapp. C.I.E.S.M.*, v. 18, p. 907–909.

Alvarez, W., 1972, Rotation of the Corsica–Sardinia microplate, *Nature Phys. Sci.*, v. 235, p. 103–105.

Alvarez, W., and Gohrbandt, K. H. A., 1970, *Geology and History of Sicily*, Tripoli: Petrol. Expl. Soc. Libya.

Alvarez, W., Cocozza, T., and Wezel, F. C., 1974, Fragmentation of the Alpine orogenic belt by microplate dispersal, *Nature*, v. 248, p. 309–314.

Anderson, J. J., and Carmack, E. C., 1974, Observations of chemical and physical fine-structure in a strong pycnocline, Sea of Marmara, *Deep-Sea Res.*, v. 21, p. 877–886.

Andrieux, J., Fontboté, J. M., and Mattauer, M., 1971, Sur un modèle explicatif de l'arc de Gibraltar, *Earth Planet. Sci. Lett.*, v. 12, p. 191–198.

Angelier, J., 1975, Sur les plateformes marines quaternaires et leurs déformations: les rivages méridionaux de la Crète orientale (Grèce), *C. R. Acad. Sci. Paris*, v. 281 (D), p. 1149–1152.

Angelier, J., 1976, Sur la néotectonique de l'arc égéen externe: les failles normales est–ouest et l'extension subméridienne crétoise, *C. R. Acad. Sci. Paris*, v. 282 (D), p. 413–416.

Araña, V., and Vegas, R., 1973, Plate tectonics and volcanism in the Gibraltar Arc, *Tectonophysics*, v. 24, p. 197–212.

Arthaud, F., and Mattauer, M., 1972, Présentation d'une hypothèse sur la genèse de la virgation pyrénéene du Languedoc et sur la structure profonde du Golfe du Lion, *C. R. Acad. Sci. Paris*, v. 274 (D), p. 524–527.

Aubouin, J., 1965, *Geosynclines, Developments in Geotectonics 1*, Amsterdam: Elsevier Publ. Co., 335 pp.

Aubouin, J., 1973, Paléotectonique, tectonique, tarditectonique et néotectonique en Méditerranée moyenne: à la recherche d'un guide pour la comparaison des données de la géophysique et de la géologie, *C. R. Acad. Sci. Paris*, v. 276 (D), p. 457–460.

Aubouin, J., and Dercourt, J., 1970, Sur la géologie de l'Egée: Regard sur le Dodécanèse méridional (Kasos, Karpathos, Rhodes), *Bull. Soc. Géol. France*, v. 12 (7), p. 453–572.

Auffret, G.-A., Pastouret, L., *et al.*, 1974, Influence of the prevailing current regime on sedimentation in the Alboran Sea, *Deep-Sea Res.*, v. 21, p. 839–849.

Auzende, J.-M., 1971, La marge continentale tunisienne: résultats d'une étude par sismique réflexion: sa place dans le cadre tectonique de la Méditerranée occidentale, *Mar. Geophys. Res.*, v. 1, p. 162–177.

Auzende, J.-M., and Olivet, J.-L., 1974, Structure of the western Mediterranean Basin, in: *The Geology of Continental Margins*, Burk, C. A., and Drake, C. L., eds., New York: Springer-Verlag, p. 723–731.

Auzende, J.-M., and Pautot, G., 1970, La marge continentale algérienne et le phénomène de subsidence: exemple du Golfe de Bougie, *C. R. Acad. Sci. Paris*, v. 271 (D), p. 1945–1948.

Auzende, J.-M., Olivet, J.-L., *et al.*, 1972a, La dépression nord-baléare (Espagne), *C. R. Acad. Sci. Paris*, v. 274 (D), p. 2291–2294.

Auzende, J.-M., Olivet, J.-L., and Bonnin, J., 1972b, Une structure compressive au nord de l'Algérie? *Deep-Sea Res.*, v. 19, p. 149–155.

Auzende, J.-M., Olivet, J.-L., and Pautot, G., 1973a, Balearic Islands: southern prolongation, in: *Initial Reports of the Deep Sea Drilling Project*, Ryan, W. B. F., Hsü, K. J., *et al.*, eds., Washington, D. C.: U. S. Govt. Print. Office, v. 13, p. 1441–1447.

Auzende, J.-M., Bonnin, J., and Olivet, J.-L., 1973b, The origin of the western Mediterranean basin, *J. Geol. Soc. London*, v. 129, p. 607–620.

Auzende, J.-M., Olivet, J.-L., and Bonnin, J., 1974, Le détroit sardano-tunisien et la zone de fracture nord-tunisienne, *Tectonophysics*, v. 21, p. 357–374.

Auzende, J.-M., Rehault, J.-P., *et al.*, 1975, Les bassins sédimentaires de la mer d'Alboran, *Bull. Soc. Géol. France*, v. 17, p. 98–107.

Azema, J., 1966, Géologie des confins des provinces d'Alicante et de Murcie (Espagne), *Bull. Soc. Géol. France*, v. 7, p. 80–86.

Bailey, E. B., 1952, Notes on Gibraltar and the northern Rif., *Quat. J. Geol. Soc. London.*, v. 108, p. 157–176.

Barberi, F., Innocenti, F., Ferrara, G., Keller, J., and Villari, L., 1974, Evolution of Eolian arc volcanism (Southern Tyrrhenian Sea), *Earth Planet. Sci. Lett.*, v. 21, p. 269–276.

Bartolini, C., and Gehin, C., 1970, Evidence of sedimentation by gravity assisted bottom currents in the Mediterranean Sea, *Mar. Geol.*, v. 9, p. M1–M5.

Bartolini, C., Gehin, C., and Stanley, D. J., 1972, Morphology and recent sediments of the western Alboran Basin, *Mar. Geol.*, v. 13, p. 159–224.

Belderson, R. H., Kenyon, N. H., and Stride, A. H., 1970, 10-km views of the Mediterranean deep sea floor, *Deep-Sea Res.*, v. 17, p. 267–270.

Belderson, R. H., Kenyon, N. H., and Stride, A. H., 1972, Comparison between narrow-beam and conventional echo-soundings from the Mediterranean Ridge, *Mar. Geol.*, v. 12, p. M11–M15.

Belderson, R. H., Kenyon, N. H., and Stride, A. H., 1974a, Calabrian Ridge, a newly discovered branch of the Mediterranean Ridge, *Nature*, v. 247, p. 453–545.

Belderson, R. H., Kenyon, N. H., and Stride, A. H., 1974b, Features of submarine volcanoes shown on long range sonographs, *J. Geol. Soc.*, v. 130, p. 403–410.

Bellaiche, G., 1968, Reconnaissance du socle sous-marin des Maures et de sa couverture sédimentaire par sismique continue ("air-gun"), *C. R. Acad. Sci. Paris*, v. 266 (D), p. 994–996.

Bellaiche, G., 1970, Géologie sous-marine de la marge continentale au large du massif des Maures (Var, France) et de la plaine abyssale ligure, *Rev. Géogr. Phys. Géol. Dyn.*, v. 12, p. 403–440.

Bellaiche, G., 1972, Prélèvement par 2000 m de profondeur d'un réflecteur acoustique d'âge Pliocène inférieur à faciès peu profond (canyon des Stoechades, Méditerranée nord-occidentale), *C. R. Acad. Sci. Paris*, v. 275 (D), p. 321–324.

Bellaiche, G., and Mauffret, A., 1971, Résultats des dragages réalisés sur un des dômes situés au large de Toulon (Dôme T1 sud-Toulon), *C. R. Soc. Géol. France*, v. 3, p. 187–188.

Bellaiche, G., Gennesseaux, M., Mauffret, A., and Rehault, J. P., 1974, Prélèvements systématiques et caractérisation des réflecteurs acoustiques: nouvelle étape dans la compréhension de la géologie de la Méditerranée occidentale, *Mar. Geol.*, v. 16, p. M47–M56.

Ben-Avraham, Z., Shoham, Y., and Ginzburg, A., 1976, Magnetic anomalies in the eastern Mediterranean and the tectonic setting of the Eratosthenes Seamount, *Geophys. J. R. Astr. Soc.*, v. 45, p. 105–123.

Benson, R., 1972, Ostracodes as indicators of threshold depth in the Mediterranean during the Pliocene, in: *The Mediterranean Sea — A Natural Sedimentation Laboratory*, Stanley, D. J., ed., Stroudsburg: Dowden, Hutchinson, and Ross, p. 63–73.

Benson, R. H., 1973, An ostracodal view of the Messinian salinity crisis, in: *Messinian Events in the Mediterranean*, Drooger, C. W., ed., Amsterdam: North-Holland Publ. Co., p. 235–242.

Benson, R. H., and Ruggieri, G., 1974, The end of the Miocene, a time of crisis in Tethys–Mediterranean history, *Ann. Geol. Surv. Egypt*, v. 4, p. 237–250.

Berthois, L., 1968, Sur la présence d'affleurements calcaires organogènes mio-pliocènes au large de Gibraltar. Ses conséquences tectoniques, *C. R. Acad. Sci. Paris*, v. 267 (D), p. 1186–1189.

Biely, A., Burollet, P. F., and Ladjmi, T., 1972, Etude géodynamique de la Tunisie et des secteurs voisins de la Méditerranée, *23ᵉ Congrès C.I.E.S.M., Athènes, 3–11 Novembre 1972*, 9 pp.

Biju-Duval, B., 1974, Carte géologique et structurale des bassins tertiaires du domaine méditerranéen: commentaires, *Rev. Inst. Franç. Pétrole*, v. 28, p. 607–639.

Biju-Duval, B., Letouzey, J., Montadert, L., *et al.*, 1974, Geology of the Mediterranean Sea basins, in: *The Geology of Continental Margins*, Burk, C. A., and Drake, C. L., eds., New York: Springer-Verlag, p. 695–721.

Biscaye, P. E., Ryan, W. B. F., and Wezel, F. C., 1972, Age and nature of the Pan-Mediterranean subbottom Reflector M, in: *The Mediterranean Sea — A Natural Sedimentation Laboratory*, Stanley, D. J., ed., Stroudsburg: Dowden, Hutchinson, and Ross, p. 83–90.

Bizon, G., Bizon, J.-J., *et al.*, 1973, Présence aux îles Baléares (Méditerranée occidentale) de sédiments "messiniens" déposés dans une mer ouverte, à salinité normale, *C. R. Acad. Sci. Paris*, v. 277 (D), p. 985–988.

Bizon, G., Bizon, J.-J., and Mauffret, A., 1975, Présence de Miocène terminal et de Pliocène inférieur au large de Minorque (Baléares, Espagne), *Rev. Inst. Franç. Pétrole*, v. 30, p. 713–727.

Blanc, J. J., 1954, Erosion et sédimentation littorale actuelle dans le détroit de Messine, *Bull. Inst. Océanogr. Monaco*, v. 51, p. 1–11.

Blanc, J. J., 1958, Sédimentologie sous-marine du détroit Siculo-Tunisien: Campagne du "Calypso," *Bull. Inst. Océanogr. Monaco*, p. 92–126.

Blanc, J. J., 1972, Observations sur la sédimentation bioclastique en quelques points de la marge continentale de la Méditerranée, in: *The Mediterranean Sea — A Natural Sedimentation Laboratory*, Stanley, D. J., ed., Stroudsburg: Dowden, Hutchinson, and Ross, p. 225–240.

Blanc-Vernet, L., Chamley, H., Froget, C., Le Boulicaut, D., Monaco, A., and Robert, C., 1975, Observations sur la sédimentation marine récente dans la région siculo-tunisienne, *Géol. Médit.*, v. 2 (1), p. 31–48.

Boccaletti, M., 1975, Plate tectonics model for the evolution of the western Mediterranean, *Geol. Balcanica*, v. 5, p. 19–28.

Boccaletti, M., and Guazzone, G., 1974, Remnant arcs and marginal basins in the Cainozoic development of the Mediterranean, *Nature*, v. 252, p. 18–21.

Bonnin, J., Olivet, J.-L., and Auzende, J.-M., 1975, Structure en nappe à l'ouest de Gibraltar, *C. R. Acad. Sci. Paris*, v. 280 (D), p. 559–562.

Borsetti, A. M., Colantoni, P., *et al.*, 1974, Preliminary data on the geology of the Balearic Sea, *XXIVᵉ Congrès–Assemblée de Monaco, Comm. Int. Expl. Sci. Médit.*, 3 pp.

Bourcart, J., 1950, La théorie de la flexure continentale, *C. R. XIV Congr. Int. Géogr.*, *Lisbonne*, p. 167–190.

Bourcart, J., 1959, Morphologie du précontinent des Pyrénées à la Sardaigne, in: *La Topographie et la Géologie des Profondeurs Océaniques*, Bourcart, J., ed., Paris: Coll. Int. Cent. Nat. Rech. Sci., Nice–Villefrance, p. 33–52.

Bourcart, J., 1960a, Carte topographique du fond de la Méditerranée occidentale, *Bull. Inst. Océanogr. Monaco*, no. 1163, 20 pp.

Bourcart, J., 1960b, La Méditerranée et la révolution pliocène, in: *Livre à la Mémoire du Professeur P. Fallot, Mém. Soc. Géol. France*, p. 103–118.

Bourcart, J., 1964, Les sables profonds de la Méditerranée occidentale, in: *Turbidites, Developments in Sedimentology*, Bouma, A. H., and Brouwer, A., eds., Amsterdam: Elsevier Publ. Co., v. 3, p. 148–155.

Bourgeois, J., Bourrouilh, R., *et al.*, 1970, Données nouvelles sur la géologie des cordillères bétiques, *Ann. Soc. Nord*, v. 90, p. 347–393.

Bourrouilh, R., 1973, *Stratigraphie, Sédimentologie et Tectonique de l'Ile de Minorque et du Nord-est de Majorque*, Thesis, Univ. Paris, 822 pp.

Bousquet, J. C., and Montenat, C., 1974, Présence de décrochements nord-est–sud-ouest plio-quaternaires dans les cordillères bétiques orientales (Espagne). Extension et signification générale, *C. R. Acad. Sci. Paris*, v. 278 (D), p. 1321–1324.

Boussac, J., 1912, Etudes stratigraphiques sur le Nummulitique alpin, *Mém. Carte Géol. France*, 662 pp.

Brambati, A., 1968, Mixing and settling of fine terrigenous material (16 μm) in the Northern Adriatic Sea between Venice and Trieste, *Studi Trentini Sci. Nat.*, v. 45, p. 103–117.

Brambati, A., and Venzo, G. A., 1967, Recent sedimentation in the northern Adriatic Sea between Venice and Trieste, *Studi Trentini Sci. Nat.*, v. 44, p. 202–274.

Braune, K., and Heimann, K. O., 1973, Miocene evaporites on the Ionian islands (Greece), *Geol. Soc. Greece Bull.*, v. 10, p. 25–30.

Brinkmann, R., 1974, Geologic relations between Black Sea and Anatolia, in: *The Black Sea — Geology, Chemistry and Biology*, Degens, E. T., and Ross, D. A., eds., Amer. Assoc. Petrol. Geol., Mem. 20, p. 63–76.

Brozolo, F. R. di, and Giglia, G., 1973, Further data on the Corsica–Sardinia rotation, *Nature*, v. 241, p. 389–391.

Bruce, J. G., and Charnock, H., 1965, Studies of winter sinking of cold water in the Aegean Sea, *Comm. Int. Expl. Sci. Mer. Médit.*, v. 18, p. 773–778.

Burman, I., and Oren, O. H., 1970, Water outflow close to bottom from the Aegean, *Cah. Océanogr.*, v. 22, p. 775–799.

Burollet, P. F., 1967, General geology of Tunisia, in: *Guidebook to the Geology and History of Tunisia*, Petrol. Expl. Soc. Lybia, 9th Ann. Field Conf., p. 51–58.

Burollet, P. F., and Byramjee, R., 1974, Evolution géodynamique néogène de la Méditerranée occidentale, *C. R. Acad. Sci Paris*, v. 278 (D), p. 1321–1324.

Burollet, P. F., and Dufaure, P., 1972, The Neogene series drilled by the Mistral No. 1 well in the Gulf of Lion, in: *The Mediterranean Sea — A Natural Sedimentation Laboratory*, Stanley, D. J., ed., Stroudsburg: Dowden, Hutchinson, and Ross, p. 91–98.

Caire, A., 1974, Tectonique spirale en Méditerranée centrale, *C. R. Acad. Sci. Paris*, v. 278 (D), p. 3165–3167.

Caputo, M., Pieri, L., and Unguendoli, J., 1970, Geometric investigation of the subsidence in the Po Delta, *Boll. Geof. Teor. Appl.*, v. 13, p. 187–208.

Caputo, M., Panza, G. F., and Postpichl, D., 1972, New evidences about the deep structure of the Lipari arc, *Tectonophysics*, v. 15, p. 219–231.

Carey, S. W., 1958, *A Tectonic Approach to Continental Drift*, Hobart: Univ. Tasmania, 251 pp.

Carter, T. G., Flanagan, J. P., *et al.*, 1972, A new bathymetric chart and physiography of the Mediterranean Sea, in: *The Mediterranean Sea — A Natural Sedimentation Laboratory*, Stanley, D. J., ed., Stroudsburg: Dowden, Hutchinson, and Ross, p. 1–23.

Castillejo, F. F. de, 1975, Variaciones estacionales de temperatura a lo largo del litoral meridional de la Península Ibérica, *Bol. Inst. Esp. Oceanogr.*, v. 187, 53 pp.

Caulet, J.-P., 1972*a*, Les sédiments organogènes du précontinent algérien, *Mém. Mus. Nat. Hist. Nat. Paris*, 289 pp.

Caulet, J.-P., 1972*b*, Recent biogenic calcareous sedimentation on the Algerian Continental Shelf, in: *The Mediterranean Sea — A Natural Sedimentation Laboratory*, Stanley, D. J., ed., Stroudsburg: Dowden, Hutchinson, and Ross, p. 261–277.

Cela, V. S., 1968, Estudio petrológico de las sucesiones volcánicas del sector central de la formación del Cabo de Gata (Almeria), *Est. Geol.*, v. 24, p. 1–38.

Chamley, H., 1972, Sur la sédimentation argileuse profonde en Méditerranée, in: *The Mediterranean Sea—A Natural Sedimentation Laboratory*, Stanley, D. J., ed., Stroudsburg: Dowden, Hutchinson, and Ross. p. 387–399.

Chamley, H., 1976, Sédimentation argileuse en mer Ionienne au Plio-Pléistocène d'après l'étude des forages 125 DSDP, *Bull. Soc. Géol. France*, v. 12, p. 1131–1143.

Charnock, H., Rees, A. I., and Hamilton, N., 1972, Sedimentation in the Tyrrhenian Sea, in: *The Mediterranean Sea — A Natural Sedimentation Laboratory*, Stanley, D. J., ed., Stroudsburg: Dowden, Hutchinson, and Ross, p. 615–629.

Chassefière, B., and Monaco, A., 1973, Relations entre sédimentogenèse, propriétés mécaniques et minéralogie. Application au détroit siculo-tunisien, *C. R. Acad. Sci. Paris*, v. 277 (D), p. 141–144.

Cita, M. B., 1973, Mediterranean evaporites: paleontological arguments for a deep-basin dessication model, in: *Messinian Events in the Mediterranean*, Drooger, C. W., ed., Amsterdam: North-Holland Pub. Co., p. 206–228.

Cocco, E., 1976, The Italian North Ionian coast: tendency towards erosion, *Mar. Geol.*, v. 21, p. M49–M57.

Colantoni, P., and Borsetti, A. M., 1973, Some notes on geology and stratigraphy of the Strait of Sicily, *Bull. Geol. Soc. Greece*, v. 10, p. 31–32.

Colantoni, P., and Zarudzki, E. F. K., 1972, Some principal seafloor features in the Strait of

Sicily, *23ᵉ Congrès C.I.E.S.M., Athènes, 3–11 Novembre 1972*, Comité de géologie et géophysique marines, 3 pp.

Collette, B. J., 1970, La marge continentale au large de l'Esterel (France) et les mouvements verticaux Pliocènes, *Mar. Geophys. Res.*, v. 1, p. 61–84.

Comninakis, P. E., and Papazachos, B. C., 1972, Seismicity of the eastern Mediterranean and some tectonic features of the Mediterranean Ridge, *Bull. Geol. Soc. Amer.*, v. 83, p. 1093–1102.

Conchon, O., 1975, *Les Formations Quaternaires de Type Continental en Corse Orientale*, Thesis, Univ. Paris VI, 514 pp.

Cravatte, J., Dufaure, P., Prim, M., and Rouaix, S., 1974, Les sondages du Golfe du Lion: stratigraphie et sédimentologie, *Compagnie Franç. Pétrole Notes Mém.*, Paris, v. 11, p. 209–274.

Dangeard, L., Rioult, M., Blanc, J.-J., and Blanc-Vernet, L., 1968, Résultats de la plongée en soucoupe no. 421 dans la vallée sous-marine du Planier, au large de Marseille, *Bull. Inst. Océanogr. Monaco*, v. 67, 21 pp.

Dewey, J. F., and Bird, J. M., 1970, Mountain belts and the new global tectonics, *J. Geophys. Res.*, v. 75, p. 2625–2647.

Dewey, J. J., Pitman III, W. C., Ryan, W. B. F., and Bonnin, J., 1973, Plate tectonics and evolution of the Alpine system, *Bull. Geol. Soc. Amer.*, v. 84, p. 3137–3180.

Dickinson, W. R., 1974, Plate tectonics and sedimentation, in: *Tectonics and Sedimentation*, Dickinson, W. R., ed., Soc. Econ. Paleont. Mineral., Sp. Publ. 22, p. 1–27.

Dott, R. H., Jr., and Shaver, R. H., ed., 1974, *Modern and Ancient Geosynclinal Sedimentation*, Soc. Econ. Paleont. Mineral., Sp. Publ. 19, 380 pp.

Drake, C. L., and Delauze, H., 1969, Gravity measurements near Greece from the Bathyscaphe *Archimède*, *Ann. Inst. Océanogr.*, v. 46 (1), p. 71–77.

Drooger, C. W., ed., 1973, *Messinian Events in the Mediterranean*, Amsterdam: North-Holland Publ. Co., 272 pp.

Drooger, C. W., and Meulenkamp, J. E., 1973, Stratigraphic contributions to geodynamics in the Mediterranean area: Crete as a case history, *Geol. Soc. Greece Bull.*, v. 10, p. 193–200.

Duplaix, S., 1972, Les minéraux lourds de sables de plages et de canyons sous-marins de la Méditerranée française, in: *The Mediterranean Sea — A Natural Sedimentation Laboratory*, Stanley, D. J., ed., Stroudsburg: Dowden, Hutchinson, and Ross, p. 293–303.

Egeler, C. G., and Simon, O. J., 1970, Sur la tectonique de la zone bétique (cordillères bétiques, Espagne), *Rev. Géogr. Phys. Géol. Dyn.*, v. 12, p. 173–182.

Einsele, G., 1967, Sedimentary processes and physical properties of cores from the Red Sea, Gulf of Aden and off the Nile Delta, in: *Marine Geotechnique*, Richards, A. F., ed., Urbana: Univ. Illinois Press, p. 154–169.

Emelyanov, E. M., 1972, Principal types of recent bottom sediments in the Mediterranean Sea: their mineralogy and geochemistry, in: *The Mediterranean Sea — A Natural Sedimentation Laboratory*, Stanley, D. J., ed., Stroudsburg: Dowden, Hutchinson, and Ross, p. 355–386.

Emelyanov. E. M., and Shimkus, K. M., 1972, Suspended matter in the Mediterranean Sea, in: *The Mediterranean Sea — A Natural Sedimentation Laboratory*, Stanley, D. J., ed., Stroudsburg: Dowden, Hutchinson, and Ross, p. 417–439.

Emery, K. O., Heezen, B. C., and Allan, T. D., 1966, Bathymetry of the eastern Mediterranean Sea, *Deep-Sea Res.*, v. 13, p. 173–192.

Eriksson, K. G., 1961, Granulométrie des sédiments de l'île d'Alboran, Méditerranée occidentale, *Bull. Geol. Inst. Univ. Uppsala*, v. 15, p. 269–284.

Eriksson, K. G., 1965, The sediment core no. 210 from the western Mediterranean Sea, *Rep. Swed. Deep Sea Exp.*, 1947–1948, v. 8, p. 395–594.

Escowitz, E. C., 1970, Reconnaissance of oceanographic conditions in the Strait of Gibraltar region — May 1970, *Naval Fac. Engin. Comm., Chesapeake Div.*, 15 pp.

Evans, G., 1971, The recent sedimentation of Turkey and the adjacent Mediterranean and Black Seas: a review, in: *Geology and History of Turkey*, Tripoli: Petrol. Expl. Soc. Libya, p. 385–406.

Fabbri, A., and Gallignani, P., 1972, Ricerche geomorfologiche e sedimentologiche nell'Adriatico meridionale, *Giorn. Geol.*, v. 38, p. 453–498.

Fabbri, A., and Selli, R., 1972, The structure and stratigraphy of the Tyrrhenian Sea, in: *The Mediterranean Sea — A Natural Sedimentation Laboratory*, Stanley, D. J., ed., Stroudsburg: Dowden, Hutchinson, and Ross, p. 75–81.

Fabbri, A., Marabini, F., and Rossi, S., 1973, Lineamenti geomorfologici del Monte Palinuro e del Monte delle Baronie, *Giorn. Geol.* v., 39 (1), p. 133–156.

Fabricius, F., and Schmidt-Thomé, P., 1972, Contribution to recent sedimentation on the shelves of the southern Adriatic, Ionian and Syrtis seas, in: *The Mediterranean Sea — A Natural Sedimentation Laboratory*, Stanley, D. J., ed., Stroudsburg: Dowden, Hutchinson, and Ross, p. 333–343.

Fahlquist, D. A., and Hersey, J. B., 1969, Seismic refraction measurements in the western Mediterranean Sea, *Bull. Inst. Océanogr. Monaco*, v. 67, 52 pp.

Fairbridge, R. W., 1972, Quaternary sedimentation in the Mediterranean region controlled by tectonics, paleoclimates and sea level, in: *The Mediterranean Sea — A Natural Sedimentation Laboratory*, Stanley, D. J., ed., Stroudsburg: Dowden, Hutchinson, and Ross, p. 99–113.

Fanucci, F., Fierro, G., Gennesseaux, M., Rehault, J.-P., and Tabbo, S., 1974a, Indagine sismica sulla piattaforma litorale del Savonese (Mar Ligure), *Boll. Soc. Geol. Ital.*, v. 93, p. 421–435.

Fanucci, F., Fierro, G., Rehault, J.-P., and Terranova, R., 1974b, Le plateau continental de la mer Ligure de Portofino à La Spezia: étude structurale et évolution plioquaternaire, *C. R. Acad. Sci. Paris*, v. 279 (D), p. 1151–1154.

Farrand, W. R., 1971, Late Quaternary paleoclimates of the eastern Mediterranean area, in: *The Late Cenozoic Ages*, Turekian, K. K., ed., New Haven: Yale Univ. Press, p. 529–564.

Fernex, F., Jaworsky, G., *et al.*, 1973, Comparaison entre l'évolution géodynamique de la région du n. de la mer d'Alboran et celle des bordures de la mer Ligurienne depuis l'Eocène, *Comm. Int. Expl. Sci. Médit. Monaco, Rapp. Proc.-Verb. Réunions* (Athènes), v. 22, p. 53–54.

Fierro, G., 1970, Minerali pesanti nei sedimenti marini del golfo dell'Asinara e delle bocche di Bonifacio, *Atti Soc. Ital. Sci. Nat. Mus. Civ. St. Nat. Milano*, v. 110, p. 155–194.

Fierro, G., and Ozer, A., 1974, Relations entre les dépôts éoliens quaternaires et les sédiments marins du golfe de l'Asinara et des bouches de Bonifacio (Sardaigne, Italie), *Mem. Inst. Ital. Paleont. Umana*, v. 2, p. 1–9.

Fierro, G., Gennesseaux, M., and Rehault, J.-P., 1973, Caractères structuraux et sédimentaires du plateau continental de Nice à Gênes (Méditerranée nord-occidentale), *Bull. Bur. Rech. Géol. Mineral.*, v. 4, p. 193–208.

Finetti, I., and Morelli, C., 1971, Ricerche sismiche a riflessione nella Laguna e nel Golfo di Venezia, *Boll. Geof. Teor. Appl.*, v. 13 (49), p. 44.

Finetti, I., and Morelli, C., 1973, Geophysical exploration of the Mediterranean Sea, *Boll. Geof. Teor. Appl.*, v. 15, p. 261–341.

Finetti, I., Morelli, C., and Zarudzki, E., 1970, Reflection seismic study of the Tyrrhenian Sea, *Boll. Geof. Teor. Appl.*, v. 13 (48), p. 311–346.

Follador, U., 1967, Il Pliocene ed il Pleistocene dell'Italia Centro-Meridionale, Versante Adriatico, *Boll. Soc. Geol. Ital.*, v. 86, p. 565–584.

Frassetto, R., 1966, Discussion on the distribution of sea floor types at the eastern entrance to the Strait of Gibraltar, as revealed by grabs, *SACLANT ASW Res. Cent., La Spezia, Tech. Mem.* 118, 18 pp.

Frazer, J. Z., Arrhenius, G., Hanor, J. S., and Hawkins, D. L., 1970, *Surface Sediment Distribution — Mediterranean Sea*, La Jolla: Scripps Inst. Oceanography, 9 pp.

Freeman, T., and Simancas, R., 1971, *Guidebook, Tenth International Field Institute, Spain*, Washington, D. C.: Amer. Geol. Inst., 242 pp.

Freund, R., Goldberg, M., Weissbrod, T., Druckman, Y., and Derin, B., 1975, The Triassic–Jurassic structure of Israel and its relation to the origin of the Eastern Mediterranean, *Geol. Surv. Israel Bull.*, v. 65, 26 pp.

Furnstein, J., and Allain, C., 1962, L'hydrologie algérienne en hiver, *Rev. Tr. Inst. Sci. Tech. Pêches Mar. Paris*, v. 26, p. 277–309.

Fuster, J. M., Ibarrola, E., and Martin, J., 1967, Las andesitas piroxénicas de la mesa de Roldán (Almeria, SE de España), *Est. Geol.*, v. 23, p. 1–13.

Gaibar-Puertas, C., 1967, Investigación sistemática de las corrientes océanicas superficiales en el litoral mediterráneo español, *Rev. Cienc. Aplicada*, v. 115, p. 128–147.

Gaibar-Puertas, C., 1969, Estudio geológico de la Isla de Alborán (Almeria). I: Las rocas eruptivas, *Acta Geol. Hispánica*, v. 4, p. 72–80.

Gaibar-Puertas, C., 1973, El campo de pesantez y la estructura geológica del Estrecho de Gibraltar, *Inst. Geogr. Catastral, Madrid*, 85 pp.

Gaibar-Puertas, C., and Ruiz Lopez, J., 1970, Las anomalías geomagnéticas de la Isla del Alborán (Almeria), *Bol. Geol. Mineral.*, v. 91, p. 378–393.

Garala, J., 1929, La geología del Estrecho de Gibraltar, *Bol. Inst. Geol. Mineral.*, v. 51, p. 1–35.

Gass, I. G., and Smewing, J. D., 1973, Intrusion, extrusion and metamorphism at constructive margins: evidence from the Troodos Massif, Cyprus, *Nature*, v. 242, p. 26–29.

Gennesseaux, M., 1966, Prospection photographique des canyons sous-marins du Var et du Paillon (Alpes-Maritimes) au moyen de la Troïka, *Rev. Géogr. Phys. Géol. Dyn.*, v. 8, p. 3–38.

Gennesseaux, M., 1972, La structure occidentale des Bouches de Bonnifacio (Corse), *C. R. Acad. Sci. Paris*, v. 275, p. 2295–2297.

Gennesseaux, M., and Glaçon, G., 1972, Essai de stratigraphie du Pliocène sous-marin en Méditerranée nord-occidentale, *C. R. Acad. Sci. Paris*, v. 275 (D), p. 1863–1866.

Gennesseaux, M., and Thommeret, Y., 1968, Datation par le radiocarbone de quelques sédiments sous-marins de la région niçoise, *Rev. Géogr. Phys. Géol. Dyn.*, v. 10, p. 375–382.

Gennesseaux, M., Guibout, P., and Lacombe, H., 1971, Enrégistrement de courants de turbidité dans la vallée sous-marine du Var (Alpes-Maritimes), *C. R. Acad. Sci. Paris*, v. 273 (D), p. 2456–2459.

Gennesseaux, M., Auzende, J.-M., Olivet, J.-L., and Bayer, R., 1974, Les orientations structurales et magnétiques sous-marines au sud de la Corse et la dérive corso-sarde, *C. R. Acad. Sci. Paris*, v. 278, p. 2003–2006.

Giesel, W., and Seibold, E., 1968, Sedimentechogramme vom ibero-marokkanischen Kontinentalrand, *"Meteor" Forsch.*, v. C, p. 53–75.

Giermann, G., 1961, Erläuterungen zur bathymetrischen Karte der Strasse von Gibraltar, *Bull. Inst. Océanogr. Monaco*, no. 1218, 28 pp.

Giermann, G., 1962, Erläuterungen zur bathymetrischen Karte des Westlichen Mittelmeeres, *Bull. Inst. Océanogr. Monaco*, no. 1254, 24 pp.

Giermann, G., 1969a, Morphologie et tectonique du plateau continental entre le cap Cavallo et Saint-Florent (Corse), *Bull. Inst. Océanogr. Monaco*, no. 1397, 6 pp.

Giermann, G., 1969b, The Eastern Mediterranean Ridge, *Rapp. Comm. Int. Mer. Médit.*, v. 19 (4), p. 605–607.

Giermann, G., Pfannenstiel, M., and Wimmenauer, W., 1968, Relations entre morphologie, tectonique et volcanisme en mer d'Alboran (Méditerranée occidentale), résultats préliminaires de la campagne Jean Charcot (1967), *C. R. Somm. Sci. Soc. Géol. France*, v. 4, p. 116–117.

Ginzburg, A., Cohen, S. S., Hay-Roe, H., and Rosenzweig, A., 1975, Geology of Mediterranean Shelf of Israel, *Bull. Amer. Assoc. Petrol. Geol.*, v. 59 (11), p. 2142–2160.

Giorgetti, F., and Mosetti, F., 1969, General morphology of the Adriatic Sea, *Boll. Geof. Teor. Appl.*, v. 11, p. 49–56.

Glaçon, G. and Rehault, J.-P., 1973, Le Messinien marin de la pente continentale ligure (− 1750 m) (Italie), *C. R. Acad. Sci. Paris*, v. 277 (D), p. 625–628.

Glangeaud, L., 1962, Paléogéographie dynamique de la Méditerranée et de ses bordures. Le role des phases ponto-plio-quaternaires, in: *Océanographie, Géologique et Géophysique de la Méditerranée Occidentale*, Bourcart, J., ed., Coll. Nat. Cent. Nat. Rech. Sci., Villefranche-sur-Mer, p. 125–165.

Glangeaud, L., 1970, Les structures mégamétriques de la Méditerranée. Méditerranée occidentale, *C. R. Acad. Sci. Paris*, v. 270, p. 3184–3289.

Glangeaud, L., 1971a, Evolution géodynamique de la mer d'Alboran et ses bordures. La phase messino-plioquaternaire (résumé), *C. R. Somm. Soc. Géol. France*, Fasc. 8, p. 431–432.

Glangeaud, L., 1971b, Les phases tertiaires de la mer d'Alboran, *C. R. Acad. Sci. Paris*, v. 273 (D), p. 2435–2440.

Glangeaud, L., and Rehault, J.-P., 1968, Evolution ponto-plio-quaternaire du Golfe de Gênes, *C. R. Acad. Sci. Paris*, v. 266 (D), p. 60–63.

Glangeaud, L., Alinat, J., Polvèche, J., Guillaume, A., and Leenhardt, O., 1966, Grandes structures de la mer Ligure, leur évolution et leurs relations avec les chaînes continentales, *Bull. Soc. Géol. France*, v. 8, p. 921–937.

Glangeaud, L., Bobier, C., and Bellaiche, G., 1967, Evolution néotectonique de la mer d'Alboran et ses conséquences paléogéographiques, *C. R. Acad. Sci. Paris*, v. 265 (D), p. 1672–1675.

Glangeaud, L., Bellaiche, G., Gennesseaux, M., and Pautot, G., 1968, Phénomènes pelliculaires et épidermiques du rech Bourcart (golfe du Lion) et de la mer hespérienne, *C. R. Acad. Sci. Paris*, v. 267 (D), p. 1079–1083.

Glangeaud, L., Bobier, C., and Szep, B., 1970, Les structures mégamétriques de la Méditerranée: la mer d'Alboran et l'"arc" de Gibraltar, *C. R. Acad. Sci. Paris*, v. 271 (D), p. 473–478.

Goedicke, T. R., 1972, Submarine canyons on the central continental shelf of Lebanon, in: *The Mediterranean Sea — A Natural Sedimentation Laboratory*, Stanley, D. J., ed., Stroudsburg: Dowden, Hutchinson, and Ross, p. 655–670.

Gostan, J., 1967, Etude du courant géostrophique entre Villefranche-sur-Mer et Calvi, *Cah. Océanogr.*, v. 19, p. 329–345.

Got, H., 1973, *Etude des Corrélations Tectonique-Sédimentation au cours de l'Histoire Quaternaire du Précontinent Pyrénéo-Catalan*, Thesis, Univ. Perpignan, Perpignan, France, 294 pp.

Got, H., and Stanley, D. J., 1974, Sedimentation in two Catalonian canyons, northwestern Mediterranean, *Mar. Geol.*, v. 16, p. M91–M100.

Got, H., Monaco, A., and Leenhardt, O., 1971, Grand traits structuraux et sédimentaires du précontinent pyrénéen au large du Roussillon et de l'Ampurdan, *Rev. Inst. Franç. Pétrole*, v. 26, p. 355–368.

Got, H., Stanley, D. J., and Sorel, D., 1977, Northwestern Hellenic Arc: concurrent sedimentation and deformation in a compressive setting, *Mar. Geol.* v. 24, p. 21–36.

Gunnerson, C. G., and Ozturgut, E., 1974, The Bosporus, in: *The Black Sea — Geology, Chemistry and Biology*, Degens, E. T., and Ross, D. A., eds., Amer. Assoc. Petrol. Geol., Mem. 20, p. 99–114.

Heezen, B. C., and Ewing, M., 1955, Orleansville earthquake and turbidity currents, *Bull. Amer. Assoc. Petrol. Geol.*, v. 39, p. 2505–2514.

Heezen, B. C., and Hollister, C. D., 1971, *The Face of the Deep*, New York: Oxford Univ. Press, 659 pp.

Heezen, B. C., and Johnson, G. L., 1969, Mediterranean undercurrent and microphysiography west of Gibraltar, *Bull. Inst. Océanogr. Monaco*, v. 67, 95 pp.

Heezen, B. C., Gray, C., Segre, A. G., and Zarudzki, E. F. K., 1971, Evidence of foundered continental crust beneath the Central Tyrrhenian Sea, *Nature*, v. 229, p. 327–329.

Herman, Y., 1972, Quaternary eastern Mediterranean sediments: micropaleontology and climatic record, in: *The Mediterranean Sea — A Natural Sedimentation Laboratory*, Stanley, D. J., ed., Stroudsburg: Dowden, Hutchinson, and Ross, p. 129–147.

Hernández Pacheco, F., 1961, Origen y relieve submarino del Estrecho de Gibraltar, *Bol. Inst. Esp. Oceanogr.*, v. 105, 26 pp.

Hersey, J. B., 1965a, Sediment ponding in the deep sea, *Bull. Geol. Soc. Amer.*, v. 75, p. 1251–1260.

Hersey, J. B., 1965b, Sedimentary basins of the Mediterranean Sea, in: *Submarine Geology and Geophysics*, Whittard, W. F., and Bradshaw, R. D., eds., London: Butterworths, v. 17 of the Colston Papers, p. 75–91.

Hertweck, G., 1971, Der Golf von Gaeta (Tyrrhenisches Meer), *Senckenberg. Mar.*, v. 3, p. 247–276.

Hesse, R., and von Rad, U., 1972, Undisturbed large-diameter cores from the Strait of Otranto, in: *The Mediterranean Sea — A Natural Sedimentation Laboratory*, Stanley, D. J., ed., Stroudsburg: Dowden, Hutchinson, and Ross, p. 645–653.

Hieke, W., Sigl, W., and Fabricius, F., 1973, Morphological and structural aspects of the Mediterranean Ridge SW off the Peloponnesus (Ionian Sea), *Bull. Geol. Soc. Greece*, v. 10 (1), p. 109–126.

Hinz, K., 1972a, Results of seismic refraction investigations (Project Anna) in the western Mediterranean Sea, south and north of the island of Mallorca, *Bull. Cent. Rech. Pau-SNPA*, v. 6, p. 405–426.

Hinz, K., 1972b, Zum Diapirismus im westlichen Mittelmeer, *Geol. Jb.*, v. 90, p. 389–396.

Hinz, K., 1974, Results of seismic refraction and seismic reflection measurements in the Ionian Sea, *Geol. Jb.*, v. E-2, p. 33–65.

Hirschleber, H., Rudloff, R., and Snoek, M., 1972, Preliminary results of seismic measurements in the Gulf of Lion, *Bull. Cent. Rech. Pau-SNPA*, v. 6, p. 395–402.

Horn, D. R., Ewing, J. I., and Ewing, M., 1972, Graded-bed sequences emplaced by turbidity currents north of 20°N in the Pacific, Atlantic and Mediterranean, *Sedimentology*, v. 18, p. 247–275.

Hsü, K. J., 1970, The meaning of the word flysch — a short historical search, in: *Flysch Sedimentology in North America*, Lajoie, J., ed., Geol. Assoc. Canada, Sp. Publ. 7.

Hsü, K. J., 1972a, Alpine flysch in a Mediterranean setting, *24th Int. Geol. Congr.*, Sect. 6, p. 67–74.

Hsü, K. J., 1972b, When the Mediterranean dried up, *Sci. Amer.*, v. 227, p. 27–36.

Hsü, K. J., 1973, The dessicated deep-basin model for the Messinian events, in: *Messinian Events in the Mediterranean*, Drooger, C. W., ed., Amsterdam: North-Holland Publ. Co., p. 60–67.

Hsü, K. J., and Jenkyns, H. C., eds., 1974, *Pelagic Sediment: On Land and Under the Sea*, Oxford: Blackwell Scientific Publ., Int. Assoc. Sediment., Sp. Publ. 1, 447 pp.

Hsü, K. J., and Schlanger, S. O., 1971, Ultrahelvetic flysch sedimentation and deformation related to plate tectonics, *Bull. Geol. Soc. Amer.*, v. 82, p. 1207–1218.

Hsü, K. J., Montadert, L., *et al.*, 1975, *Glomar Challenger* returns to the Mediterranean Sea, *Geotimes*, v. 20, p. 16–19.

Hsü, K. J., Montadert, L., *et al.*, *Initial Reports of the Deep Sea Drilling Project*, Washington, D.C.: U.S. Govt. Print. Office, v. 42 (in preparation).

Huang, T.-C., and Stanley, D. J., 1972, Western Alboran Sea: sediment dispersal, ponding and reversal of currents, in: *The Mediterranean Sea — A Natural Sedimentation Laboratory*, Stanley, D. J., ed., Stroudsburg: Dowden, Hutchinson, and Ross, p. 521–559.

Huang, T.-C., Stanley, D. J., and Stuckenrath, R., 1972, Sedimentological evidence for current reversal at the Strait of Gibraltar, *Mar. Tech. Soc. J.*, v. 6, p. 25–33.

Hurley, R. J., 1971, Seismic reflection studies in the Adriatic Sea, *Rev. Géogr. Phys. Géol. Dyn.*, v. 13 (5), p. 429–438.

Jong, K. A. de, and van der Voo, R., 1970, Rotation of Sardinia: palaeomagnetic evidence from Permian rocks, *Nature*, v. 226, p. 933–934.

Jong, K. A. de, Manzoni, M., Stavenga, T., van der Voo, R., van Dijk, F., and Zijderveld, D. A., 1972, Early Miocene age of rotation of Sardinia: paleomagnetic evidence, *23e Congrès–Assemblée, C.I.E.S.M., Athènes, 3–11 Novembre 1972*, 3 pp.

Jongsma, D., 1974, Heat flow in the Aegean Sea, *Geophys. J. R. Astr. Soc.*, v. 37, p. 337–346.

Jongsma, D., Wissman, G., *et al.*, 1975, The southern Aegean Sea: an extensional marginal basin without spreading? *Bundes. Geowissench. Rohstoffe*, 41 pp.

Julivert, M., Fontboté, J. M., *et al.*, 1972, Mapa tectónico de la Península Ibérica y Baleares, *Inst. Geol. Miner. España* (map in one sheet).

Karcz, I., and Kafri, U., 1973, Recent vertical crustal movements between the Dead Sea Rift and the Mediterranean, *Nature*, v. 242, p. 42–44.

Keller, G. H., and Lambert, D. N., 1972, Geotechnical properties of submarine sediments, Mediterranean Sea, in: *The Mediterranean Sea — A Natural Sedimentation Laboratory*, Stanley, D. J., ed., Stroudsburg: Dowden, Hutchinson, and Ross, p. 401–415.

Keller, J., and Leiber, J., 1974, Sedimente, Tephra-Lagen und Basalte der südtyrrhenischen Tiefsee-Ebene im Bereich des Marsili-Seeberges. *"Meteor" Forsch.-Ergebnisse*, v. C (19), p. 62–76.

Keller, J., Ryan, W. B. F., Ninkovich, D., and Altherr, R., 1974, The deep-sea record of Quaternary volcanism in the Mediterranean, *XXIVe Congrès–Assemblée plénière de Monaco, 6–14 Decembre 1974, Comité Géol. Géophys. Mar.*, 4 pp.

Kelling, G., and Stanley, D. J., 1972*a*, Sedimentary evidence of bottom current activity, Strait of Gibraltar region, *Mar. Geol.*, v. 13, p. M51–M60.

Kelling, G., and Stanley, D. J., 1972*b*, Sedimentation in the vicinity of the Strait of Gibraltar, in: *The Mediterranean Sea — A Natural Sedimentation Laboratory*, Stanley, D. J., ed., Stroudsburg: Dowden, Hutchinson, and Ross, p. 489–519.

Kelling, G., Maldonado, A., and Stanley, D. J., 1977, Sedimentation on the Balearic Rise, a foundered block in the western Mediterranean, *Smithson. Contrib. Mar. Sci.* (in press).

Kenyon, N. H., and Belderson, R. H., 1973, Bed forms of the Mediterranean undercurrent observed with side-scan sonar, *Sed. Geol.*, v. 9, p. 77–99.

Kenyon, N. H., Stride, A. H., and Belderson, R. H., 1975, Plan views of active faults and other features on the lower Nile Cone, *Bull. Geol. Soc. Amer.*, v. 86, p. 1733–1739.

Keraudren, B., 1976, Essai de stratigraphie et de paléogéographie du Plio-Pléistocène égéen, *Bull. Soc. Géol. France*, v. 12, p. 1110–1120.

Kermabon, A., Gehin, C., Blavier, P., and Tonarelli, 1969, Acoustic and other physical properties of deep-sea sediments in the Tyrrhenian abyssal plain, *Mar. Geol.*, v. 7, p. 129–145.

Kidd, R. B., 1976, In Sicily: the Messinian stage, *Geotimes*, v. 21, p. 20–22.

Kruit, C., 1955, Sediments of the Rhône delta. Grain size and microfauna, *Nederlandsch. Geol. Mijnb. Genoot. Verh., Geol. Ser.*, v. 15, p. 357–514.

Kuenen, Ph. H., 1959a, L'âge d'un bassin méditerranéen, in: *La Topographie et la Géologie des Profondeurs Océaniques*, Bourcart, J., ed., Cent. Nat. Rech. Sci., Paris, p. 157–163.

Kuenen, Ph. H., 1959b, Transport and sources of marine sediments, *Geol. Mijnb.*, v. 21, p. 191–196.

Lacombe, H., 1971, Le détroit de Gibraltar, océanographie physique, *Notes Mém. Serv. Géol. Maroc*, v. 222, p. 111–146.

Lacombe, H., 1975, Aperçus sur l'apport à l'océanographie physique des recherches récentes en Méditerranée, *Bull. Etude Commun. Médit.*, UNESCO, Monaco, p. 5–25.

Lacombe, H., and Tchernia, P., 1960, Quelques traits généraux de l'hydrologie méditerranéene d'après diverses campagnes hydrologiques récentes en Méditerranée et dans le proche Atlantique, *Cah. Océanogr.*, v. 12, p. 527–547.

Lacombe, H., and Tchernia, P., 1972, Caractères hydrologiques et circulation des eaux en Méditerranée, in: *The Mediterranean Sea — A Natural Sedimentation Laboratory*, Stanley, D. J., ed., Stroudsburg: Dowden, Hutchinson, and Ross, p. 25–36.

Lacombe, H., Tchernia, P., Richez, C., and Gamberoni, L., 1964, Deuxième contribution à l'étude du détroit de Gibraltar, *Cah. Océanogr.*, v. 16, p. 283–314.

Legaaij, R., and Kopstein, F. P. H. W., 1964, Typical features of a fluviomarine offlap sequence, in: *Deltaic and Shallow Marine Deposits*, van Straaten, L. M. J. U., ed., Amsterdam: Elsevier Publ. Co., p. 216–226.

Lambert, R. St. J., and McKerrow, W. S., 1975, Subsidence and deep earthquakes in the Tyrrhenian Sea, *J. Geol.*, v. 83, p. 387–388.

Lanoix, F., 1972, *Etude Hydrologique et Dynamique de la Mer d'Alboran*, Thèse de spécialité, *Univ. Paris* VI, 57 pp.

Leclaire, L., 1968, Contribution à l'étude géomorphologique de la marge continentale algérienne. Note de présentation de 10 cartes topographiques du plateau continental algérien, *Cah. Océanogr.*, v. 20, p. 451–521.

Leclaire, L., 1972a, Aspects of late Quaternary sedimentation on the Algerian precontinent and in the adjacent Algiers–Balearic Basin, in: *The Mediterranean Sea — A Natural Sedimentation Laboratory*, Stanley, D. J., ed., Stroudsburg: Dowden, Hutchinson, and Ross, p. 561–582.

Leclaire, L., 1972b, La sédimentation holocène sur le versant méridional du bassin algéro-baléares, *Mém. Mus. Nat. Hist. Nat. Paris*, v. 24, 391 pp.

Leenhardt, O., 1970, Sondages sismiques continus en Méditerranée occidentale, *Mém. Inst. Océanogr. Monaco*, v. 1, 120 pp.

Leenhardt, O., Pierrot, S., Rebuffatti, A., and Sabatier, R., 1969a, Etude sismique de la zone de Planier (Bouches-du-Rhône), *Rev. Inst. Franç. Pétrole*, v. 24, p. 1261–1287.

Leenhardt, O., Rebuffatti, A., et al., 1969b, Profil sismique dans le bassin Nord-Baléares, *C. R. Somm. Soc. Géol. France*, v. 7, p. 249–251.

Leenhardt, O., Rebuffatti, and Sancho, J., 1970, Carte du Plioquaternaire entre Ibiza et le Cap San Antonio (Méditerranée occidentale), *Rev. Inst. Franç. Pétrole*, v. 25, p. 165–173.

Leonardi, P., Morelli, C., Norinelli, A., and Tribalto, G., 1973, Sintesi geologica e geofisica riguardante l'area Veneziana e zone limitrofe, *Serv. Geol. Italia*, p. 4–17.

Le Pichon, X., Auzende, J. M., et al., 1971a, Deep sea photographs of an active seismic fault zone near Gibraltar straits, *Nature*, v. 230, p. 110–111.

142   Daniel Jean Stanley

Le Pichon, X., Pautot, G., *et al.*, 1971*b*, La Méditerranée occidentale depuis l'Oligocène. Schéma d'évolution, *Earth Planet. Sci. Lett.*, v. 13, p. 145–152.
Le Pichon, X., Pautot, G., and Weill, J. P., 1972, Opening of the Alboran Sea, *Nature Phys. Sci.*, v. 236, p. 83–85.
Le Pichon, X., Hekinian, R., Francheteau, J., and Carré, D., 1975, Submersible study of lower continental slope–abyssal plain contact, *Deep-Sea Res.*, v. 22, p. 667–670.
Letolle, R., and Vergnaud-Grazzini, C., 1973, Essai sur l'évolution générale de la Méditerranée pendant les époques glaciaires, *Coll. Int. C.N.R.S.*, v. 219, p. 231–238.
Loomis, T. P., 1975, Tertiary mantle diapirism, orogeny, and plate tectonics east of the Strait of Gibraltar, *Am. J. Sci.*, v. 275, p. 1–30.
Lort, J. M., 1972, *The Crustal Structure of the Eastern Mediterranean*, Ph. D. Thesis, Univ. Cambridge, Cambridge, 117 pp.
Lort, J. M., 1973, Summary of seismic studies in the eastern Mediterranean, *Bull. Geol. Soc. Greece*, v. 10, p. 99–108.
Lort, J. M., and Gray, F., 1974, Cyprus: seismic studies at sea, *Nature*, v. 248, p. 745–747.
Lort, J. M., Limond, W. Q., and Gray, F., 1974, Preliminary seismic studies in the eastern Mediterranean, *Earth Planet. Sci. Lett.*, v. 21, p. 355–366.
Lucayo, N. C., 1968, Contribución al conocimiento del Mar de Alborán I. Superficie de referencia, *Bol. Inst. Esp. Oceanogr.*, v. 135, 28 pp.
McKenzie, D. P., 1970, Plate tectonics of the Mediterranean region, *Nature*, v. 226, p. 239–243.
Makris, J., 1973, Some geophysical aspects of the evolution of the Hellenides, *Geol. Soc. Greece Bull.*, v. 10, p. 206–213.
Maldonado, A., 1975, Sedimentation, stratigraphy, and development of the Ebro Delta, Spain, in: *Deltas*, Boussard, M. L. S., ed., Houston Geol. Soc., p. 311–338.
Maldonado, A., and Stanley, D. J., 1975, Nile Cone lithofacies and definition of sediment sequences, *9th Int. Sediment. Congr.*, Thème 6, p. 185–191.
Maldonado, A., and Stanley, D. J., 1976*a*, Late Quaternary sedimentation and stratigraphy in the Strait of Sicily, *Smithson. Contrib. Earth Sci.*, v. 16, 73 pp.
Maldonado, A., and Stanley, D. J., 1976*b*, The Nile Cone: submarine fan development by cyclic sedimentation, *Mar. Geol.*, v. 20, p. 27–40.
Maley, T. S., and Johnson, G. L., 1971, Morphology and structure of the Aegean Sea, *Deep-Sea Res.*, v. 18, p. 109–122.
Mars, P., 1963, Les faunes et la stratigraphie du Quaternaire méditerranéen, *Rec. Trav. Station Mar. Endoume, Marseille*, v. 28, p. 61–97.
Marinatos, S., ed., 1971, *Acta of the 1st International Scientific Congress on the Volcano of Thera*, Archaeol. Serv. Greece, Athens, 436 pp.
Mascle, J., 1971, Géologie sous-marine du canyon de Toulon, *Cah. Océanogr.*, v. 23, p. 241–250.
Massuti, M., 1967, Carta de pesca de la región surmediterránea española (desde Estepona a Adra), *Trab. Inst. Esp. Oceanogr.*, v. 33, 24 pp.
Mauffret, A., 1969, Les dômes et les structures "anticlinales" de la Méditerranée occidentale au nord-est des Baléares, *Rev. Inst. Franç. Pétrole*, v. 24, p. 953–960.
Mauffret, A., and Sancho, J., 1970, Etude de la marge continentale au nord de Majorque (Baléares, Espagne), *Rev. Inst. Franç. Pétrole*, v. 25, p. 714–730.
Mauffret, A., Auzende, J., *et al.*, 1972, Le bloc continental baléare (Espagne) — extension et évolution, *Mar. Geol.*, v. 12, p. 289–300.
Mauffret, A., Fail, J. P., *et al.*, 1973, Northwestern Mediterranean sedimentary basin from seismic reflection profile, *Bull. Amer. Assoc. Petrol. Geol.*, v. 57, p. 2245–2262.

Mellis, O., 1954, Volcanic ash-horizons in deep-sea sediments from the eastern Mediterranean, *Deep-Sea Res.*, v. 2, p. 89–92.

Menard, H. W., 1967, Transitional types of crust under small ocean basins, *J. Geophys. Res.* v. 72, p. 3061–3073.

Menard, H. W., Smith, S. M., and Pratt, R. M., 1965, The Rhône deep-sea fan, in: *Submarine Geology and Geophysics*, Whittard W. F., and Bradshaw, R. D., eds., London: Butterworths, v. 17 of the Colston Papers, p. 271–285.

Mercier, J.-L., Carey, E., Philip, H., and Sorel, D., 1976, La néotectonique plio-quaternaire de l'arc égéen externe et de la mer Egée et ses relations avec la séismicité, *Bull. Soc. Géol. France*, v. 13, p. 367–384.

Middleton, G. V., and Hampton, M., 1973, Sediment gravity flows: mechanics of flow and deposition, in: *Turbidites and Deep Water Sedimentation*, Middleton G. V., and Bouma, A. H., eds., Los Angeles: S.E.P.M., Pacific Section, p. 1–38.

Miller, A. R., 1972a, Speculations concerning bottom circulation in the Mediterranean Sea, in: *The Mediterranean Sea — A Natural Sedimentation Laboratory*, Stanley, D. J., ed., Stroudsburg: Dowden, Hutchinson, and Ross, p. 37–42.

Miller, A. R., 1972b, The Levantine Intermediate water mass from Sardinia to Rhodes, in: *Oceanography of the Strait of Sicily*, Allan, T. D., Akal, T., and Molcard, R., eds., SACLANT ASW Res. Cent., La Spezia, Conf. Proc. 7, p. 108–118.

Miller, A. R., 1973, Deep convection in the Aegean Sea, *Coll. Int. C.N.R.S.*, v. 215, p. 1–9.

Miller, A. R., Tchernia, P., and Charnock, C., 1970, *Mediterranean Sea Atlas of Temperature, Salinity, Oxygen, Profiles and Data from Cruises of R.V. Atlantis and R.V. Chain*, Woods Hole Oceanogr. Atlas Ser. 3, 190 pp.

Milliman, J. D., and Müller, J., 1973, Precipitation and lithification of magnesian calcite in the deep-sea sediments of the eastern Mediterranean Sea, *Sedimentology*, v. 20, p. 29–45.

Milliman, J. D., Weiler, Y., and Stanley, D. J., 1972, Morphology and carbonate sedimentation on shallow banks in the Alboran Sea, in: *The Mediterranean Sea — A Natural Sedimentation Laboratory*, Stanley, D. J., ed., Stroudsburg: Dowden, Hutchinson, and Ross, p. 241–259.

Mistardis, G., 1972, Investigations of the geology and mineral wealth of the Aegean Sea area, *24th Int. Geol. Congr.*, Sect. 8, p. 167–181.

Monaco, A. A., 1973, The Roussillon continental margin (Gulf of Lions): Plio-Quaternary paleogeographic interpretation, *Sediment. Geol.* v. 10, p. 261–284.

Montadert, L., Sancho, J., *et al.*, 1970, De l'âge tertiaire de la série salifère responsable des structures diapiriques en Méditerranée occidentale (nord-est des Baléares), *C.R. Acad. Sci. Paris*, v. 271 (D), p. 812–815.

Montenat, C., 1970, Sur l'importance des mouvements orogéniques récents dans le sud-est de l'Espagne (Provinces d'Alicante et de Murcie), *C.R. Acad. Sci. Paris*, v. 270 (D), p. 3194–3197.

Montenat, C., Bizon, G., and Bizon, J.-J., 1975, Remarques sur le Néogène du forage Joides 121 en mer d'Alboran (Méditerranée occidentale), *Bull. Soc. Géol. France*, v. 17, p. 45–51.

Morcos, S. A., 1972, Sources of Mediterranean Intermediate Water in the Levantine Sea, in: *Studies in Physical Oceanography*, Gordon, A. L., ed., New York: Gordon and Breach, v. 2, p. 185–206.

Morelli, C., 1970, Physiography, gravity and magnetism of the Tyrrhenian Sea, *Boll. Geof. Teor. Appl.*, v. 12, p. 276–309.

Morelli, C., 1972, Bathymetry, gravity and magnetism in the Strait of Sicily, in: *Oceanography in the Strait of Sicily*, Allan, T. D., Akal, T., and Molcard, R., eds., SACLANT ASW Res. Cent., La Spezia, Conf. Proc. 7, p. 193–207.

Morelli, C., 1975, Geophysics of the Mediterranean, *Bull. Etude Commun. Médit. Monaco*, Sp. Publ. No. 7, p. 27–111.

Morelli, C., Carrozzo, M. T., Ceccherini, P., Finetti, I., Gantar, C., Pisani, M., and Schmidt di Friedberg, P., 1969, Regional geophysical study of the Adriatic Sea, *Boll. Geof. Teor. Appl.*, v. 10, p. 3–56.

Morelli, C., Pisani, M., and Gantar, C., 1975, Geophysical studies in the Aegean Sea and in the eastern Mediterranean, *Boll. Geof. Teor. Appl.*, v. 18, p. 127–167.

Morgan, P., and Evans, G., 1975, Recent tectonics of the Cilicia Plateau and the northeastern Mediterranean, *EOS*, v. 56 (12), p. 1058.

Mosby, H., 1962, Current measurements, meteorological observations and soundings of the *M/S "Helland-Hansen"* near the Strait of Gibraltar May–June 1961, *Bergen Geophys. Inst.*, v. 4, 29 pp.

Mosetti, F., 1968, Considerazioni preliminari sulla dinamica dell'Adriatico settentrionale, *Arch. Oceanogr. Limnol. Suppl.*, v. 15, p. 237–244.

Mosetti, R., Accerboni, E., and Lavenia, A., 1969, Ricerche oceanografiche nel mare Mediterraneo orientale, *Atti Ist. Veneto Sci., Let. Arti*, v. 127, p. 95–102.

Moskalenko, V. N., 1965, Study of the sedimentary series of the Mediterranean Sea by seismic methods, in: *Basic Features of the Geological Structure of the Hydrologic Regime and Biology of the Mediterranean Sea*, Translation and Interpretation Division, Institute of Modern Languages, p. 60–72.

Mulder, C. J., 1973, Tectonic framework and distribution of Miocene evaporites in the Mediterranean, in: *Messinian Events in the Mediterranean*, Drooger, C. W., ed., Amsterdam: North-Holland Publ. Co., p. 44–59.

Müller, J., and Staesche, W., 1973, Precipitation and diagenesis of carbonates in the Ionian deep-sea, *Bull. Geol. Soc. Greece*, v. 10, p. 145–151.

Needham, H. D., Le Pichon, X., Melguen, M., et al., 1973, North Aegean Sea trough: 1972 *Jean Charcot* cruise, *Geol. Soc. Greece Bull.*, v. 10, p. 152–153.

Neev, D., 1975, Tectonic evolution of the Middle East and the Levantine Basin (easternmost Mediterranean), *Geology*, v. 3, p. 683–686.

Neev, D., Edgerton, H. E., Almagor, G., and Bakler, N., 1966, Preliminary results of some continuous seismic profiles in the Mediterranean shelf of Israel, *Israel J. Earth Sci.* v. 15, p. 170–178.

Neev, D., Almagor, G., Arad, A., Ginzburg, A., and Hall, J. K., 1973, The Geology of the Southeastern Mediterranean Sea, *Israel Geol. Surv. Rep.* MG/73/5, 43 pp.

Nelson, B., 1970, Hydrography, sediment dispersal, and recent historical development of the Po River delta, Italy, in: *Deltaic Sedimentation Modern and Ancient*, Morgan J. P., and Shaver, R. H., eds., Soc. Econ. Paleont. Mineral., Sp. Publ. 15, p. 153–184.

Nesteroff, W. D., 1960, Les sédiments marins entre l'Esterel et l'embouchure du Var, *Rev. Géogr. Phys. Géol. Dyn.*, v. 1, p. 17–18.

Nesteroff, W. D., 1973a, Mineralogy, petrography, distribution and origin of the Messinian Mediterranean evaporites, in: *Initial Reports of the Deep Sea Drilling Project*, Ryan, W. B. F., Hsü, K. J., et al., eds., Washington, D.C.: U. S. Govt. Print. Office, v. 13, p. 673–694.

Nesteroff, W. D., 1973b, Un modèle pour les évaporites messiniennes en Méditerranée: des bassins peu profonds avec dépôt d'évaporites lagunaires, in: *Messinian Events in the Mediterranean*, Drooger, C. W., ed., Amsterdam: North-Holland Publ. Co., p. 68–81.

Nesteroff, W. D., 1973c, Petrography and mineralogy of sapropels, in: *Initial Reports of the Deep Sea Drilling Project*, Ryan, W. B. F., Hsü, K. J., et al., eds., Washington, D.C.: U. S. Govt. Print. Office, v. 13, p. 713–720.

Nesteroff, W. D., Ryan, W. B. F., et al., 1972, Evolution de la sédimentation pendant le

Néogène en Méditerranée d'après les Forages JOIDES–DSDP, in: *The Mediterranean Sea — A Natural Sedimentation Laboratory*, Stanley, D. J., ed., Stroudsburg: Dowden, Hutchinson, and Ross, p. 47–62.

Nesteroff, W. D., Wezel, F. C., and Pautot, G., 1973, Summary of lithostratigraphic findings and problems, in: *Initial Reports of the Deep Sea Drilling Project*, Ryan, W. B. F., Hsü, K. J., *et al.*, eds., Washington, D.C.: U. S. Govt. Print. Office, v. 13, p. 1021–1040.

Nielsen, E., 1973, Coastal erosion in the Nile delta, *Nature and Resources*, v. 9, p. 14–18.

Ninkovich, D., and Hays, J. D., 1972, Mediterranean island arcs and origin of high potash volcanoes, *Earth Planet. Sci. Lett.*, v. 16, p. 331–345.

Ninkovich, D., and Heezen, B. C., 1965, Santorini tepha, in: *Submarine Geology and Geophysics*, Whittard W. F., and Bradshaw, R. D., eds., London: Butterworths, v. 17 of the Colston Papers, p. 413–452.

Norin, E., 1958, The sediments of the Central Tyrrhenian Sea, *Rep. Swed. Deep-Sea Exped.*, 1947–1948, v. 8, 136 pp.

Ogniben, L., Martinis, B., Rossi, P. M., *et al.*, 1973, Structural model of Italy, North and South; lithostratigraphic–tectonic map, bathymetric map, *C.N.R.*, Map.

Olausson, E., 1961, Sediment cores from the Mediterranean Sea and Red Sea, *Rep. Swed. Deep-Sea Exped., 1947–1948*, v. 8, p. 335–391.

Olivet, J.-L., Auzende, J.-M., and Bonnin, J., 1973*a*, Structure et évolution tectonique du bassin d'Alboran, *Bull. Soc. Géol. France*, v. 15, p. 108–112.

Olivet, J.-L., Pautot, G., and Auzende, J.-M., 1973*b*, Alboran Sea, in: *Initial Reports of the Deep Sea Drilling Project*, Ryan, W. B. F., Hsü, K. J., *et al.*, eds., Washington, D.C.: U. S. Govt. Print. Office, v. 13, p. 1417–1430.

Oomkens, O., 1970, Depositional sequences and sand distribution in the post-glacial Rhône delta complex, in: *Deltaic Sedimentation Modern and Ancient*, Morgan J. P., and Shaver, R. H., eds., Soc. Econ. Paleont. Mineral., Sp. Publ. 15, p. 198–212.

Opdyke, N. D., Ninkovich, D., Lowrie, W., and Hays, J. D., 1972, The paleomagnetism of two Aegean deep-sea cores, *Earth Planet. Sci. Lett.*, v. 15, p. 145–159.

Oren, O. H., 1969, Oceanographic and biological influence of the Suez Canal, the Nile and the Aswan Dam on the Levant Basin, in: *Progress in Oceanography*, Sears, M., ed., Oxford: Pergamon Press, v. 5, p. 161–167.

Ostlund, H. G., 1974, Expedition "Odysseus 65": radiocarbon age of Black Sea deep water, in: *The Black Sea — Geology, Chemistry and Biology*, Degens, E. T., and D. A., Ross. eds., Amer. Assoc. Petrol. Geol., Mem. 20, p. 127–132.

Ozgül, N., and Arpat, E., 1973, Structural units of the Taurus orogenic belt and their continuation in neighboring regions, *Geol. Soc. Greece Bull.*, v. 10, p. 156–164.

Papazachos, B. C., 1974, Seismotectonics of the Eastern Mediterranean Area, in: *Engineering Seismology and Earthquake Engineering*, Solnes, J., ed., NATO Advanced Study Institute Series (E), Applied Sciences, No. 3, p. 1–32.

Papazachos, B. C., and Comninakis, P. E., 1971, Geophysical and tectonic features of the Aegean Arc, *J. Geophys. Res.*, v. 76, p. 8517–8533.

Paquet, J., 1969, Etude géologique de la province de Murcie (Espagne), *Mém. Soc. Géol. France*, v. 43, 270 pp.

Pareyn, C., 1968, Observations géologiques et sédimentologiques dans la Fosse Ouest de la mer Ionienne, *Ann. Inst. Océanogr. N. S.*, v. 46, p. 53–69.

Pastouret, L., 1970, Etude sédimentologique et paléoclimatique de carottes prélevées en Méditerranée orientale, *Téthys*, v. 2, p. 227–266.

Pautot, G., 1972, Histoire sédimentaire de la région au large de la Côte d'Azur, in: *The Mediterranean Sea — A Natural Sedimentation Laboratory*, Stanley, D. J., ed., Stroudsburg: Dowden, Hutchinson, and Ross, p. 583–613.

Pe, G. G., and Piper, D. J. W., 1972, Vulcanism at subduction zones: the Aegean area, *Bull. Geol. Soc. Greece*, v. 9, p. 133–144.

Pérès, J. M., 1968, Observations effectuées à bord du bathyscaphe *Archimède* dans la fosse située au S.-W. de l'île de Sapientza (mer Ionienne), *Ann. Inst. Océanogr.*, v. 46, p. 41–46.

Pettersson, H., 1957, The voyage of the ALBATROSS, *Swed. Deep-Sea Exped., 1947–1948*, v. 1, p. 84.

Picard, F., 1971, Etude de la fraction lourde des turbidites dans une carotte profonde de Méditerranée nord-occidentale, *Téthys*, v. 3, p. 949–954.

Pierce, J. W., and Stanley, D. J., 1975, Suspended-sediment concentration and mineralogy in the central and western Mediterranean and mineralogic comparison with bottom sediment, *Mar. Geol.* v., 19, p. M15–M25.

Pierrot, S., 1969, Carte bathymétrique de la mer Ligure (map in 16 sheets, scale 1/50,000), *Musée Océanogr. Monaco.*

Pigorini, B., 1968, Sources and dispersion of recent sediments of the Adriatic Sea, *Mar. Geol.*, v. 6, p. 187–229.

Poizat, C., 1970, Hydrodynamisme et sédimentation dans le Golfe de Gabès (Tunisie), *Téthys*, v. 2, p. 267–296.

Rabinowitz, P. D., and Ryan, W. B. F., 1970, Gravity anomalies and crustal shortening in the eastern Mediterranean, *Tectonophysics*, v. 10, p. 585–608.

Rangheard, Y., 1970, Principales données stratigraphiques et tectoniques des îles d'Ibiza et de Formentera (Baléares); situation paléogéographique et structurale de ces îles dans les Cordillères bétiques, *C.R. Acad. Sci. Paris*, v. 270 (D), p. 1227–1230.

Razavet, C. D., 1956, Contribution à l'étude géologique et sédimentologique du delta du Rhône, *Soc. Géol. France Mém.* 26, 234 pp.

Reading, H. G., 1972, Global tectonics and the genesis of flysch successions, *24th Int. Geol. Congr.*, Sect. 6, p. 59–66.

Rehault, J. P., Olivet, J.-L., and Auzende, J.-M., 1974, Le bassin nord-occidental méditerranéen: structure et évolution, *Bull. Soc. Géol. France*, v. 16, p. 281–294.

Reig, F., 1958, La estructura del Estrecho de Gibraltar y la posibilidad de las obras del cruce del mismo, *Arch. Inst. Est. Africanos, Madrid*, v. 4, 48 pp.

Ritsema, A. R., 1970, Seismo-tectonic implications of a review of European earthquake mechanisms, *Geol. Rundschau*, v. 59, p. 36–57.

Rœver, W. P. de, ed., 1969, Symposium on the problem of oceanization in the western Mediterranean, *Trans. Roy. Geol. Mineral. Soc. Netherlands*, v. 26, 165 pp.

Rosfelder, A., 1955, Carte provisoire au 1/500.000 de la marge continentale algérienne. Note de présentation, *Publ. Serv. Carte Géol. Algérie*, v. 5, p. 57–106.

Ross, D. A., Uchupi, E., and Summerhayes, C., 1976, Recent sedimentation on the Nile Delta and Nile Cone, *Bull. Amer. Assoc. Petrol. Geol.*, v. 60, p. 716.

Rosello Verger, V. M., 1971, Notas sobre la geomorphologia litoral del sur de Valencia (España), *Quatern.*, p. 121–143.

Rossi, S., Mosetti, F., and Cescon, B., 1968, Morfologia e natura del fondo nel golfo di Trieste, *Boll. Soc. Adriat. Sci. Trieste*, v. 56, p. 187–206.

Rossignol, M., and Pastouret, L., 1971, Analyse pollinique de niveaux sapropeliques post-glaciaires dans une carotte en Méditerranée orientale, *Rev. Palaeobotan. Palynol.*, v. 11, p. 227–238.

Rotschy, F., Vergnaud-Grazzini, C., Bellaiche, G., and Chamley, H., 1972, Etude paléoclimatique d'une carotte prélevée sur un dôme de la plaine abyssale ligure ("Structure Alinat"), *Palaeogeogr. Palaeoclimatol. Palaeoecol.*, v. 11, p. 125–145.

Rousset, C., 1975, Etat de la question de la pénéplanation miocène et de ses effets de la phase de compression ponto-pliocène, en Provence occidentale (France), *Géol. Médit.*, v. 2, p. 191–195.

Rupke, N. A., and Stanley, D. J., 1974, Distinctive properties of turbiditic and hemipelagic mud layers in the Algéro-Balearic Basin, western Mediterranean Sea, *Smithson. Contrib. Earth Sci.*, v. 13, 40 pp.

Rupke, N. A., Stanley, D. J., and Stuckenrath, R., 1974, Late Quaternary rates of abyssal mud deposition in the western Mediterranean Sea, *Mar. Geol.*, v. 17, p. M9–M16.

Russell, R. J., 1942, Geomorphology of the Rhône delta, *Assoc. Amer. Geogr. Ann.*, v. 32, 149–254.

Ryan, W. B. F., 1969, *The Floor of the Mediterranean Sea*, Ph.D. Thesis, Columbia Univ., New York, 196 pp.

Ryan, W. B. F., 1972, Stratigraphy of Late Quaternary sediments in the eastern Mediterranean, in: *The Mediterranean Sea — A Natural Sedimentation Laboratory*, Stanley, D. J., ed., Stroudsburg: Dowden, Hutchinson, and Ross, p. 149–169.

Ryan, W. B. F., 1976, Quantitative evaluation of the depth of the Western Mediterranean before, during and after the Late Miocene salinity crisis, *Sedimentology*, v. 23, p. 791–813.

Ryan, W. B. F., and Gustafson, T. B., 1973, Underway geophysical measurements obtained on the *Glomar Challenger* in the eastern North Atlantic and Mediterranean Sea, in: *Initial Reports of the Deep Sea Drilling Project*, Ryan, W. B. F., Hsü, K. J., *et al.*, eds., Washington, D.C.: U. S. Govt. Print. Office, v. 13, p. 517–580.

Ryan, W. B. F., and Heezen, B. C., 1965, Ionian Sea submarine canyons and the 1908 Messina turbidity current, *Bull. Geol. Soc. Amer.*, v. 76, p. 915–932.

Ryan, W. B. F., Workum, F., Jr., and Hersey, J. B., 1965, Sediments of the Tyrrhenian abyssal plain, *Bull. Geol. Soc. Amer.*, v. 76, p. 1261–1282.

Ryan, W. B. F., Stanley, D. J., *et al.*, 1970, The tectonics and geology of the Mediterranean Sea, in: *The Sea*, Maxwell, A. E., ed., New York: Wiley–Interscience, v. 4 (part 2), p. 387–492.

Ryan, W. B. F., Hsü, K. J., *et al.*, eds., 1973, *Initial Reports of the Deep Sea Drilling Project*, Washington, D.C.: U. S. Govt. Print. Office, v. 13, 1447 pp.

Said, R., 1962, *The Geology of Egypt*, Amsterdam: Elsevier Publ. Co., 377 pp.

Sancho, J., Letouzey, J., Biju-Duval, B., Courrier, P., Montadert, L., and Winnock, E., 1973, New data on the structure of the eastern Mediterranean basin from seismic reflection, *Earth Planet. Sci. Lett.*, v. 18, p. 189–204.

Sankey, T., 1973, The formation of deep water in the northwestern Mediterranean, in: *Progress in Oceanography*, Warren, B. A., ed., Oxford: Pergamon Press, v. 6, p. 159–179.

Sarnthein, M., and Bartolini, C., 1973, Grain size studies on turbidite components from Tyrrhenian deep sea cores, *Sedimentology*, v. 20, p. 425–436.

Scholle, P. A., 1970, The Sestri–Voltaggio Line: a transform fault-induced tectonic boundary between the Alps and the Apennines, *Amer. J. Sci.*, v. 269, p. 343–359.

Scholten, R., 1974, Role of the Bosporus in Black Sea chemistry and sedimentation, in: *The Black Sea — Geology, Chemistry and Biology*, Degens E. T., and Ross, D. A., eds., Amer. Assoc. Petrol. Geol., Mem. 20, p. 115–126.

Schreiber, B., Tassi Pelati, L., Mezzadri, M. G., and Motta, G., 1968, Gross Beta radioactivity in sediments of the North Adriatic Sea: a possibility of evaluating the sedimentation rate, *Arch. Oceanogr. Limnol.*, v. 16, p. 45–62.

Schuiling, R. D., 1973, The Cyclads: an early stage of oceanization, *Geol. Soc. Greece Bull.*, v. 10, p. 174–176.

Schwartz, M. L., and Tziavos, C., 1975, Sedimentary provinces of the Saronic Gulf system, *Nature*, v. 257, 573–575.

Selli, R., ed., 1970, Ricerche geologiche preliminari nel Mar Tirreno, *Giorn. Geol.* v., 37, 249 pp.

Selli, R., 1973, An outline of the Italian Messinian, in: *Messinian Events in the Mediterranean*, Drooger, C. W., ed., Amsterdam: North-Holland Publ. Co., p. 150–171.

Selli, R., and Fabbri, A., 1971, Tyrrhenian: a Pliocene deep sea, *Acc. Naz. dei Lincei, Roma Rend. Cl. Sci. Fis. Mat. e Nat.*, v. 8, p. 104–116.

Serra-Raventos, J., and Got, H., 1974, Resultados preliminares de la campana marina realizada en el precontinente catalán entre los cañones de la Fonera y Blanes, *Acta Geol. Hisp.*, v. 9, p. 73–80.

Shukri, N. M., 1950, The mineralogy of some Nile sediments, *Quat. J. Geol. Soc. London*, v. 105, p. 511–534, and v. 106, p. 466–467.

Smith, A. G., and Moores, E. M. III, 1974, Hellenides, in: *Mesozoic–Cenozoic Orogenic Belts*, Spencer, A. M., ed., Edinburgh: p. 159–185.

Smith, S. G., 1976, Diapiric structures in the eastern Mediterranean Herodotus Basin, *Earth Planet. Sci. Lett.* (in press).

Soliman, S. M., and Faris, M. I., 1964, General geologic setting of the Nile Delta province, and its evaluation for petroleum prospecting, *Fourth Arab Petrol. Congr., Beirut*, v. 23, p. 1–11.

Sonnenfeld, P., 1974, The Upper Miocene evaporite basins in the Mediterranean Region — a study in paleo-oceanography, *Geol. Rundschau*, v. 63, p. 1133–1172.

Sonnenfeld, P., 1975, The significance of Upper Miocene (Messinian) evaporites in the Mediterranean Sea, *J. Geol.*, v. 83, p. 287–311.

Sorel, D., Lemeille, F., Limond, J., Nesteroff, W. D., and Sebrier, M., 1976, Mise en évidence de structures compressives sous-marines plio-pléistocènes dans l'arc égéen externe au large de Levkas (Iles Ioniennes, Grèce), *C.R. Acad. Sci. Paris* (in press).

Stanley, D. J., 1969, Turbidites, non-turbidites and outer continental margin paleogeography, in: *The New Concepts of Continental Margin Sedimentation*, Stanley, D. J., ed., Amer. Geol. Inst., Washington, D.C., p. 13.1–13.13.

Stanley, D. J., ed., 1972, *The Mediterranean Sea — A Natural Sedimentation Laboratory*, Stroudsburg: Dowden, Hutchinson, and Ross, 765 pp.

Stanley, D. J., 1973, Basin plains in the eastern Mediterranean: significance in interpreting ancient marine deposits, I. Basin depth and configuration, *Mar. Geol.* v., 15, p. 295–307.

Stanley, D. J., 1974a, Basin plains in the eastern Mediterranean: significance in interpreting ancient marine deposits, II. Basin distribution, *Bull. Cent. Rech. Pau-SNPA*, v. 8, p. 373–388.

Stanley, D. J., 1974b, Dish structures and sand flow in ancient submarine valleys, French Maritime Alps, *Bull. Cent. Rech. Pau-SNPA*, v. 8, p. 351–371.

Stanley, D. J., 1974c, Modern flysch sedimentation in a Mediterranean island arc setting, in: *Modern and Ancient Geosynclinal Sedimentation*, Dott, Jr., R. H., and Shaver, R. H., eds., Soc. Econ. Paleont. Mineral., Sp. Publ. 19, p. 240–259.

Stanley, D. J., and Mutti, E., 1968, Sedimentological evidence for an emerged land mass in the Ligurian Sea during the Paleogene, *Nature*, v. 218, p. 32–36.

Stanley, D. J., and Perissoratis, C., 1977, Sediment entrapment in the Aegean Sea, *Mar. Geol.*, v. 24, p. 97–107.

Stanley, D. J., and Unrug, R., 1972, Submarine channel deposits, fluxoturbidites and other indicators of slope and base-of-slope environments in modern and ancient marine basins, in: *Recognition of Ancient Sedimentary Environments*, Rigby J. K., and Hamblin, W. K., eds., Soc. Econ. Paleont. Mineral., v. 16, p. 287–340.

Stanley, D. J., Gehin, C. E., and Bartolini, C., 1970, Flysch-type sedimentation in the Alboran Sea, western Mediterranean, *Nature*, v. 228, p. 979–983.

Stanley, D. J., Cita, M. B., *et al.*, 1972, Guidelines for future sediment-related research in the Mediterranean Sea, in: *The Mediterranean Sea — A Natural Sedimentation Laboratory*, Stanley, D. J., ed., Stroudsburg: Dowden, Hutchinson, and Ross, p. 723–741.

Stanley, D. J., McCoy, F. W., and Diester-Haass, L., 1974*a*, Balearic abyssal plain: an example of modern basin plain deformation by salt tectonism, *Mar. Geol.*, v. 17, p. 183–200.

Stanley, D. J., Got, H., Leenhardt, O., and Weiler, Y., 1974*b*, Subsidence of the Western Mediterranean Basin in Pliocene–Quaternary time: further evidence, *Geology*, v. 2, p. 345–350.

Stanley, D. J., Kelling, G., Vera, J.-A., and Sheng, H., 1975*a*, Sands in the Alboran Sea: A model of input in a deep marine basin, *Smithson. Contrib. Earth Sci.*, v. 15, 51 pp.

Stanley, D. J., Maldonado, A., and Stuckenrath, R., 1975*b*, Strait of Sicily depositional rates and patterns, and possible reversal of currents in the Late Quaternary, *Palaeogeogr. Palaeoclimatol. Palaeoecol.*, v. 18, p. 279–291.

Stanley, D. J., Got, H., *et al.*, 1976, Catalonian, eastern Betic and Balearic margins: structural types and geologically recent foundering of the western Mediterranean Sea, *Smithson. Contrib. Earth Sci.*, v. 20, p. 1–67.

Stommel, H., 1972, Deep winter-time convection in the western Mediterranean Sea, in: *Studies in Physical Oceanography*, Gordon, A. L., ed., New York: Gordon and Breach, v. 2, p. 185–206.

Storetvedt, K. M., 1973, Genesis of west Mediterranean basins, *Earth Planet. Sci. Lett.*, v. 21, p. 22–28.

Studer, B., 1827, Remarques géognostiques sur quelques parties de la chaîne septentrionale des Alpes, *Ann. Sci. Nat. Paris*, v. 11, p. 1–47.

Summerhayes, C. P., and Marks, N., 1975, Nile Delta: nature, evolution and collapse of continental shelf sediment system, in: *Proceedings, UNESCO Seminar on Nile Delta Sedimentology*, UNESCO Tech. Rept. 2, Alexandria, 29 pp.

Tarr, A. C., 1974, World seismicity map, *U. S. Geol. Surv.* (chart).

Tercier, J., 1947, Le flysch dans la sédimentation alpine, *Eclogae Geol. Helv.*, v. 40, p. 164–198.

Trümpy, R., 1960, Paleotectonic evolution of the Central and Western Alps, *Bull. Geol. Soc. Amer.*, v. 71, p. 843–908.

Udías, A., and Lopez Arroyo, A., 1972, Plate tectonics and the Azores–Gibraltar region, *Nature Phys. Sci.*, v. 237, p. 67–69.

U. S. Naval Oceanographic Office, 1965, *Oceanographic Atlas of the North Atlantic Ocean. Section V, Marine Geology*, U. S. Naval Oceanogr. Office, Publ. 700, 71 pp.

Valette, J. N., 1972, Etude minéralogique et géochimique des sédiments de la mer d'Alboran; résultats préliminaires, *C.R. Acad. Sci. Paris*, v. 275 (D), p. 2289–2290.

van Straaten, L. M. J. U., 1959, Littoral and submarine morphology of the Rhône delta, in: *Proc. 2nd Coastal Geogr. Conf.*, Russell, R. J., ed., Natl. Acad. Sci., Natl. Res. Council, p. 223–264.

van Straaten, L. M. J. U., 1965, Sedimentation in the north-western part of the Adriatic Sea, in: *Submarine Geology and Geophysics*, Whittard, W. F., and Bradshaw, R. D., eds., London: Butterworths, Colston Res. Soc., p. 143–162.

van Straaten, L. M. J. U., 1967, Turbidites, ash layers and shell beds in the bathyal zone of the southeastern Adriatic Sea, *Rev. Géogr. Phys. Géol. Dyn.*, v. 9, p. 219–240.

van Straaten, L. M. J. U., 1972, Holocene stages of oxygen depletion in deep waters of the Adriatic Sea, in: *The Mediterranean Sea — A Natural Sedimentation Laboratory*, Stanley, D. J., ed., Stroudsburg: Dowden, Hutchinson, and Ross, p. 631–643.

Venkatarathnam, K., and Ryan, W. B. F., 1971, Dispersal patterns of clay minerals in the sediments of the eastern Mediterranean, *Mar. Geol.*, v. 11, p. 261–282.

Vergnaud-Grazzini, C., and Bartolini, C., 1970, Evolution paléoclimatique des sédiments würmiens et post-würmiens en mer d'Alboran, *Rev. Géogr. Phys. Géol. Dyn.*, v. 12, p. 325–334.

Villari, L., 1969, On particular ignimbrites of the island of Pantelleria (Channel of Sicily), *Bull. Volcanol.*, v. 33, p. 1–12.

Vogt, P. R., and Higgs, R. H., 1969, An aeromagnetic survey of the eastern Mediterranean Sea and its interpretation, *Earth Planet. Sci. Lett.*, v. 5, p. 439–448.

Vogt, P. R., Higgs, R. H., and Johnson, G. L., 1971, Hypotheses on the origin of the Mediterranean basin: magnetic data, *J. Geophys. Res.*, v. 76, p. 3207–3228.

Watkins, N. D., and Richardson, A., 1970, Rotation of the Iberian peninsula, *Science*, v. 167, p. 209.

Watson, J. A., and Johnson, G. L., 1968, Mediterranean diapiric structures, *Bull. Amer. Assoc. Petrol. Geol.*, v. 52, p. 2247–2249.

Watson, J. A., and Johnson, G. L., 1969, The marine geophysical survey in the Mediterranean, *Int. Hydrogr. Rev. v.*, 46, p. 81–107.

Wezel, F. C., 1970, Numidian flysch: an Oligocene–early Miocene continental rise deposit off the African platform, *Nature*, v. 228, p. 275–276.

Wezel, F. C., 1974, Flysch successions and the tectonic evolution of Sicily during the Oligocene and Early Miocene, in: *Geology of Italy*, Squyres, C., ed., Tripoli: Petrol. Expl. Soc. Libya, p. 1–23.

Wezel, F. C., 1975, Diachronism of depositional and diastrophic events, *Nature*, v. 253, p. 255–257.

Wong, H.-K., and Zarudzki, E. F. K., 1969, Thickness of unconsolidated sediments in the Eastern Mediterranean Sea, *Bull. Geol. Soc. Amer.*, v. 80, p. 2611–2614.

Wong, H.-K., Zarudzki, E. F. K., Phillips, J. D., and Giermann, G. K. F., 1971, Some geophysical profiles in the eastern Mediterranean, *Bull. Geol. Soc. Amer.*, v. 82, p. 91–100.

Wüst, G., 1961, On the vertical circulation of the Mediterranean Sea, *J. Geophys. Res.*, v. 66, p. 3261–3271.

Zarudzki, E. F. K., 1972, The Strait of Sicily — a geophysical study, *Rev. Géogr. Phys. Géol. Dyn.*, v. 14, p. 11–28.

Chapter 4

GEOPHYSICS OF THE MEDITERRANEAN SEA BASINS

Jennifer M. Lort

Deepwater Studies Group
British Petroleum Co., Ltd.
London, England

I. INTRODUCTION

In recent years, the Mediterranean has been the subject of much exploration in the fields of geology and geophysics, both on land and at sea. Reinterpretation and synthesis of these data provide a basis for discussion and define numerous problems still to be resolved. In this chapter the writer presents a summary of the most recent data collected in the Mediterranean basins and attempts a synthesis of the results, referring the reader to other contributions contained in this volume and Volume 4B for detailed regional studies.

The 4000-km-long depression occupied by the Mediterranean Sea is by no means uniform, and for the purposes of this discussion will be subdivided into several regions: the western Mediterranean comprises the Alboran Sea, and the Balearic Sea (Algero-Provençal Basin, South Balearic Basin, and Valencia Basin); the central Mediterranean comprises the Tyrrhenian Sea, and the Adriatic Sea; and the eastern Mediterranean comprises the Ionian Sea, the Aegean and related seas, and the Levantine Sea.

An early "physiographic panorama" was published by Heezen *et al.* (in Ryan *et al.*, 1970). The detailed physiography of these regions is described in an article by Carter *et al.* (1972), which contains references to data sources, and accompanies a new bathymetric chart at the scale 1:2,849,300 compiled at the U.S. Naval Oceanographic Office. A recent international geological map of Europe and the adjacent Mediterranean regions at 1:5,000,000 has been published by UNESCO (1972) and is used together with the tectonic map of Europe and the Mediterranean regions at 1:2,500,000 (1964) in a geological and structural map at 1:2,500,000, which was published by l'Institut Français du Pétrole (IFP), le Centre National pour l'Exploitation des Océans (CNEXO), and l'Institut National d'Astronomie et de Géophysique (INAG), and is described by Biju-Duval (1974); the geological and available marine data are integrated to locate the main Tertiary basins.

Several recent review articles present a synthesis of the marine geological and geophysical data—for example, Ryan *et al.* (1970, 1973), Allan and Morelli (1971), Lort (1971, 1972), Stanley (1972), Biju-Duval *et al.* (1974), and Morelli (1975)—while the Mediterranean region is set in a global tectonics scheme by Dewey *et al.* (1973) after a preliminary geometric fit by Smith (1971). Two Deep-Sea Drilling Project (DSDP) cruises, Legs 13 and 42 in 1970 and 1975, respectively, provide additional lithological and biostratigraphical information from Drill Sites 120–132 and 371–381 (Ryan *et al.*, 1973; Hsü *et al.*, 1975).

The age, structure, and formation of the constituent basins of the Mediterranean Sea are very different. The western, central and some eastern Mediterranean basins, superimposed on or adjacent to the Alpine folded belts, are thought to be Cenozoic in age, while those of the eastern Mediterranean south of the Sicily–Crete–Cyprus Arcs, which are slightly affected by Alpine folding, are probably older, originating in the Mesozoic–Cenozoic (Biju-Duval, 1974; Biju-Duval *et al.*, 1974). Geophysical studies of these basins provide several topics of special interest which are mentioned in the last section of the chapter.

The geophysical data available for the western, central, and eastern Mediterranean are now presented, and are followed by a description of the principal results deduced by various authors.

II. WESTERN MEDITERRANEAN

A. Available Geophysical Data

The western Mediterranean comprises three major basins: the Alboran Basin, the South Balearic Basin, and the Algero-Provençal Basin. Early bathymetric charts of the area include those of Bourcart (1960), Giermann (1961, 1962), Goncharov and Mikhaylov (1964), Mikhaylov (1965), and Allan and Morelli (1971), which are incorporated together with new data from the U.S. Naval Oceanographic Office in the recent map of Carter *et al.* (1972) for the entire Mediterranean Sea. Detailed physiographic charts of the basins are contained in their article. An earlier description of the physiography was given by Ryan *et al.* (1970).

The structural and schematic geological setting of the western and central Mediterranean are shown in Fig. 1, together with the physiographic provinces. Further structural considerations and their implications for the formation of the western basins are contained in Le Pichon *et al.* (1971), Auzende *et al.* (1973*a*), Olivet *et al.* (1973*b*), and Auzende and Olivet (1974).

The first pendulum gravity measurements were made by Dutch, French, and Italian submarines in the years 1930 to 1940, and the results contoured by Coster (1945). The detailed gravity surveys carried out by the Osservatorio Geofisico Sperimentale Trieste (OGST) in the western Mediterranean (Allan and Morelli, 1971) have been included in their comprehensive report of geophysical exploration for the whole Mediterranean Sea (Finetti and Morelli, 1973), which includes charts of bathymetry, gravity (free-air anomalies and Bouguer anomalies), total intensity magnetic field, and results of seismic reflection profiling using 24-channel digital recording and processing (involving CDP gather and stack, deconvolution, TV filtering, and automatic gain control); a wide-angle reflection profile was carried out in the Provençal Basin.

Early magnetic observations were made by the U.S. Naval Oceanographic Office and an anomaly map was produced by Vogt *et al.* (1971); these data were supplemented by the OGST survey, and a total magnetic field intensity map produced (Finetti and Morelli, 1973).

Other magnetic surveys exist in the western Mediterranean, which were flown over a 10-km grid spacing, and total magnetic field and magnetic anomaly maps are presented by Le Mouel and Le Borgne (1970) in southeastern France and the western Mediterranean, Le Borgne *et al.* (1971) in southwestern Europe, including the Provençal Basin, Le Borgne *et al.* (1972) in the southern part of the Algero-Provençal Basin, and Galdeano *et al.* (1974) in the Alboran Basin and western part of the Algero-Provençal Basin.

Paleomagnetic measurements have been described by Nairn and Westphal (1967, 1968), Van der Voo (1969), Zijderveld *et al.* (1970), de Jong *et al.* (1973), and Zijderveld and Van der Voo (1973).

Heat-flow measurements were made by Erikson (1970) and Ryan *et al.* (1970), and compared with those obtained in the east. DSDP Leg 42A reports down-hole temperature measurements (Hsü *et al.*, 1975). Volcanic activity in the western Mediterranean is observed in the Alboran Sea, associated with high magnetic anomalies ($+250\ \gamma$) (Olivet *et al.*, 1973*a*) and close to the Balearic Islands (Pautot *et al.*, 1973). The particular significance of volcanism in the Gibraltar Arc in a plate tectonics scheme is described by Araña and Vegas (1974).

Continuous seismic reflection profiles were first carried out in the Mediterranean by *RV Chain* in 1959 and subsequently in 1961, 1964, and 1966 (Hersey, 1965*a*, *b*; Ryan *et al.*, 1965, 1966; Ryan, 1969). Leenhardt (1969, 1970)

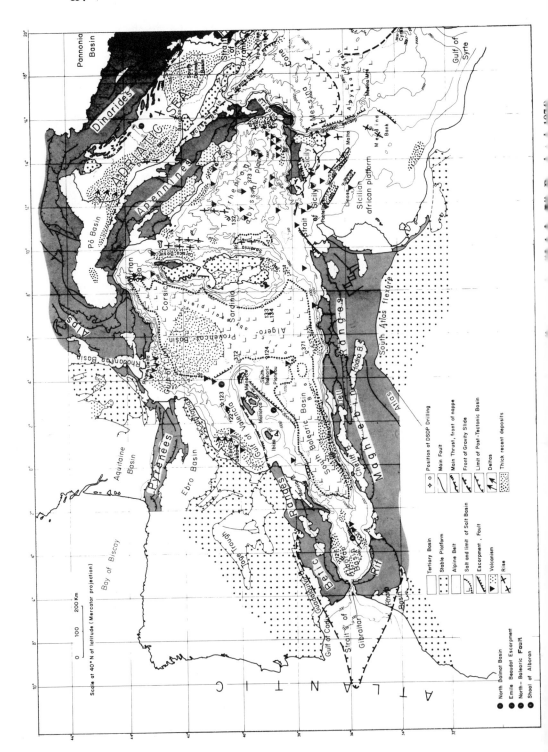

published additional profiles from the Balearic Basin, and Watson and Johnson (1969) describe sparker profiles obtained with cores and echo-sounding profiles during a U.S. Naval Oceanographic Survey which covers the whole of the Mediterranean Sea. Other early profiling was carried out by the Geodynamics Center, Villefranche-sur-mer (Glangeaud *et al.*, 1967), and the Monaco Oceanographic Institute (Alinat *et al.*, 1966). Further high-quality data, often multitrace, has been supplied by IFP (Montadert *et al.*, 1970) and the French oil companies (CFP, Elf-Re, SNPA) (Alla *et al.*, 1972; Delteil *et al.*, 1972), CNEXO (Auzende *et al.*, 1971; Le Pichon *et al.*, 1971; Auzende and Olivet, 1974; Rehault *et al.*, 1974), and the OGST Survey previously mentioned (Finetti and Morelli, 1973). The seismic reflection records have been examined with the additional results of DSDP Leg 13, Sites 121, 122, 123, and 124 by the scientists of CNEXO (Olivet *et al.*, 1973*a*; Pautot *et al.*, 1973; Auzende *et al.*, 1973*b*); those of Leg 42 will appear in the Initial Reports, DSDP, volume 42.

The most extensive early refraction measurements in the western Mediterranean were carried out by Fahlquist and Hersey (1969), and several French scientists have also made crustal studies in the adjacent onland areas (e.g., Muraour *et al.*, 1966; Recq, 1972, 1973*a*, 1974; Bellaiche and Recq, 1973; Bellaiche *et al.*, 1973*a*). Several experiments were conducted during the *Anna* cruise in 1970, by French and German scientists who cooperated to produce a final report by Leenhardt *et al.* (1972) containing separate papers with results of refraction profiles located in the Gulf of Lion (Hirschleber *et al.*), the Balearic Sea (Hinz), and a discussion of the latter results with the DSDP information from the adjacent Site 124 (Hsü).

The use of phase velocities of surface waves generated by earthquakes by Berry and Knopoff (1967) and Payo (1969) provides information on the deep crust and upper mantle structure in the Balearic Sea. Among studies of the adjacent land areas are those of Recq (1973*b*) and Chaudhury *et al.* (1971). Earthquake studies use plots of the locations of earthquake epicenters from ESSA National Earthquake Information Center, Washington, ISS Bulletins, and USCGS listings; the seismicity of the region is discussed by Gutenberg and Richter (1954), Barazangi and Dorman (1969), Karnik (1969), Caputo *et al.* (1970), and Papazachos (1973). Fault plane solutions in the area have been deduced by Banghar and Sykes (1969), Ritsema (1969), and more recently by McKenzie (1972), who delineates the plate boundaries and deduces relative plate motions in the Mediterranean.

B. Principal Results

The western Mediterranean is bounded and intersected by several orogenic belts (Fig. 1). The Alboran Sea contains an eastern and a western depression, divided by the northeast–southwest-trending Alboran Ridge (Carter *et al.*,

1972); Giermann *et al.* (1968) postulate its sedimentary origin and the existence of volcanic piercements. It terminates in the volcanic Alboran Island, the trend corresponding to one of the faults postulated by Bourcart (1960), and to the Alboran marginal fracture of Le Pichon *et al.* (1972).

The Balearic Sea, comprising the South Balearic and Algero-Provençal Basins, has a large abyssal plain, which passes to the northeast into the Ligurian Basin and is flanked to the north and west by the Valencia Basin. The knoll-like topography of the floor of the Balearic plain was interpreted by Hersey (1965*a*) to result from diapirs (Allan and Morelli, 1971; Ryan *et al.*, 1970). The Ebro and Rhône Deltas affect the coastal sediment discharge, broadening the otherwise narrow and steep continental slopes (Menard *et al.*, 1965), which are often dissected, notably in the Ligurian Canyon and off Corsica and Sardinia.

The data collected and interpreted in the years up to 1969 are reviewed by Ryan *et al.* (1970). Some more recent interpretations of data are discussed here and in an article by Morelli (1975). The free-air gravity field in the western Mediterranean shows values in the range -100 to $+100$ mgal, with a value of zero over the deep-water Balearic abyssal plain, and a negative anomaly (-50 mgal) at the foot of the continental slope (Allan and Morelli, 1971) and inside the Gibraltar Arc (Morelli, 1975). The Balearic Islands and volcanic Alboran Islands are associated with low positive free-air anomalies ($+70$ mgal). The Bouguer anomalies, obtained after the application of three-dimensional water terrain correction (Fig. 2) show a positive anomaly ($+200$ mgal) over the deepest part of the Balearic Basin, and a smaller positive anomaly ($+20$ mgal) over the central Alboran Basin, which becomes negative (-50 mgal) inside the Gibraltar Arc and through the Straits, parallel to the trend of the Betic and Rif. Bonini *et al.* (1973) describe the onland data of surrounding areas.

The magnetic field in the Alboran Basin is described by Bayer *et al.* (1973) and by Galdeano *et al.* (1974), several volcanic mounts and banks being correlated with magnetic anomalies. A strong linear magnetic anomaly lies to the north of the Alboran ridge, its source being located below the deep part of the western Basin, not under the ridge itself (Galdeano *et al.*, 1974). In the northern Provençal Basin, Auzende *et al.* (1973*a*) recognize a series of strongly positive anomalies ($+150\,\gamma$) oriented northwest–southeast, which they interpret as transform faults, describing small circles about a pole located at 54°N 24°E about which Corsica and Sardinia have moved relative to Europe during the Oligocene.

This northwest–southeast trend is observed in structural features on the continental margins from the Gulf of Genoa to the region of the Balearic Islands. A correlation between the sea floor topography and magnetic anomalies was suggested by Vogt *et al.* (1971), while Le Borgne *et al.* (1972) and Gonnard *et al.* (1975) further suggest that a magnetic basement and volcanic expression account for many of the anomalies.

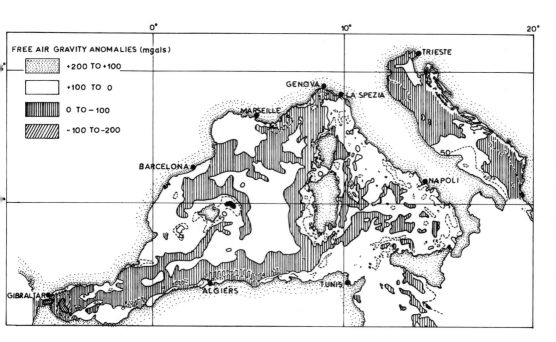

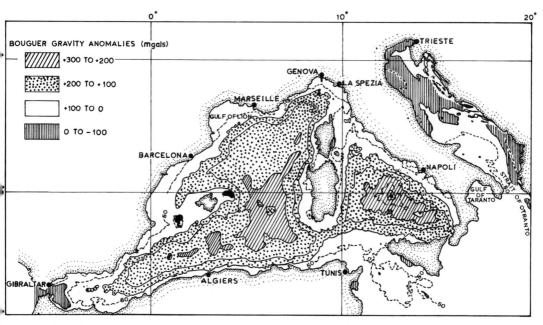

. 2. Free-air and Bouguer gravity anomalies in the western and central Mediterranean Sea from OGST survey
data (after Allan and Morelli, 1971).

In the Valencia Sea, correlations between basement features and strong magnetic anomalies have been made by Vogt *et al.* (1971) and Auzende *et al.* (1972); the Balearic fracture zone, lying to the north of the Balearic Islands and associated with a northwest–southeast magnetic anomaly, is detected by Auzende *et al.* (1973*a*). Volcanoes visible on seismic records have been dated as Upper Miocene in age, and DSDP drilling at Site 123 on the side of a volcano obtained volcanic ash dated as Lower Miocene (21 m.y.) in age.

Heat-flow measurements made by Erikson (1970) reveal a mean heat flow of 2.14 HFU, with maximum values in the deep Balearic Sea, corresponding to the zone of positive Bouguer anomalies; these are well above the global average and are supported by values from Site 372A, DSDP Leg 42A. Values of less than 1.5 HFU were calculated from sites on continental margins. These values support hypotheses which propose that the western Mediterranean is a young oceanic basin.

Numerous reflection profiles carried out in the western Mediterranean define in detail the sedimentary series (Finetti and Morelli, 1972, 1973; Auzende *et al.*, 1973*b*; Mauffret *et al.*, 1973; Rehault *et al.*, 1974; Mulder, 1973; Bellaiche *et al.*, 1973*b*; Biju-Duval *et al.*, 1974). The first studies of seismic reflection data revealed diapiric features on the Balearic abyssal plain (Ryan *et al.*, 1966, 1970; Watson and Johnson, 1969; Leenhardt, 1970; Wong *et al.*, 1970; Allan and Morelli, 1971). Several different ages were attributed to the salt which was assumed to be responsible for the diapirism (Montadert *et al.*, 1970) and DSDP drilling during Legs 13 and 42A confirmed that the evaporite series is of Upper Messinian age (Ryan *et al.*, 1973).

The sedimentary succession present in the western Mediterranean has been interpreted using data from available seismic reflection profiles and from boreholes of the DSDP (Biju-Duval *et al.*, 1974). The main units are as follows: (1) Upper Series, comprising Plio-Quaternary sediments; interval velocities range from 1.7–2.8 km/sec. DSDP boreholes encountered Pliocene muds, with turbidite and contourite components, the series showing halokinetic disturbance, as previously described, on the deep abyssal plain (Fig. 3). (2) Evaporite series, of Upper Miocene age (Messinian), and comprising two series: the upper includes marls alternating with evaporites (anhydrite, gypsum, dolomite, and possibly halite) which has been penetrated by DSDP drilling; the lower is a layer of salt (halite or potash) which is responsible for the halokinetic phenomena observed (Fig. 3). Velocities of these two series are 3.5 km/sec and 4.5 km/sec, respectively, the salt being very variable in thickness (Biju-Duval *et al.*, 1974). (3) Pre-evaporite series, here of Cenozoic age. Miocene sediments (Aquitanian to Tortonian) have been obtained in oil company boreholes in the Gulf of Valencia and Gulf of Lion, and adjacent onshore deposits of the same age are known. A thickness of 4000 m has been calculated for this series by Mauffret *et al.* (1973), with tentative velocities of 4.5–5.5 km/sec (Finetti and Morelli, 1973).

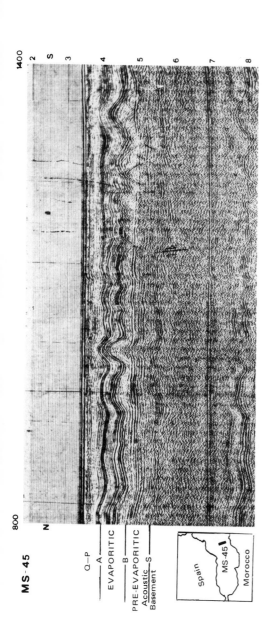

(a)

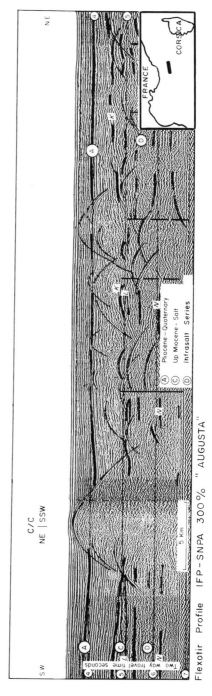

(b)

Fig. 3. (a) Seismic reflection profile in Alboran Sea to show halokinetic features (OGST profile MS-45, after Finetti and Morelli, 1973). (b) Seismic reflection profile in Balearic Sea to show piercement structures in the Algero-Provençal Basin (IFP-SNPA) (after Mauffret *et al.*, 1973).

The extent of the evaporite basin is shown in Fig. 1, practically corresponding to the Balearic abyssal plain. The pre-evaporite deposits pinch out at the basin edges, showing erosion effects which may have been synchronous with the time of formation of the salt on the margins. The overlying evaporites and Pliocene sediments are transgressive on this erosion surface.

The acoustic basement recognized from reflection profiles may be of very different origin—of oceanic type in the abyssal plain, or of volcanic, ancient crystalline, or sedimentary nature at the margins. The degree of tectonic disturbance also varies (Fig. 3).

Ewing and Ewing (1959) in the first seismic refraction experiments in the western Mediterranean, detected the evaporites (4.8 km/sec) below 1.3 km of sediments in the North Algero-Provençal Basin. Fahlquist and Hersey (1969) show that these sediments (1.8–2.9 km/sec) thin from 2 km below the Rhône Cone to 500 m at the basin margins, lying above evaporites (4.1–4.9 km/sec) and a mantle at 11-km depth (8.0 km/sec) in the South Ligurian Sea, which varies southwards to a "soft" mantle of velocity 7.7 km/sec at 12-km depth (Fig. 4). Leenhardt et al. (1972) obtain similar results from the Anna cruise in the Ligurian Sea, their 7.0 km/sec layer occurring at 11-km depth (Fig. 5).

To the south of the Balearic Islands, however, Hinz (in Leenhardt et al., 1972; Hinz, 1973) obtains a mantle velocity of 8.0 km/sec at 12-km depth below an "oceanic type" crustal structure, with sediments and evaporites overlying 5 km of Layer 3 material of velocity 6.0–7.4 km/sec. To the north, a thicker development of material of velocity 4.8–6.0 km/sec and also 7.4–7.8 km/sec increases the crustal thickness to 14 km or more (Fig. 5b); this suggests an "intermediate" type of crust. Wide-angle reflection–refraction profiles were carried out by Finetti and Morelli (1973) in the Algero-Provençal Basin, using digital recording and sonobuoys. They obtain a velocity profile with sediments of velocity 1.7–2.3 km/sec up to 1.2 km below the sea bottom, underlain by evaporites at 4.4 km/sec to 3.7-km depth, pre-evaporites at 4.6–5.5 km/sec to 8.5-km depth, and material at 6.0–7.0 km/sec to a depth of 13.2 km. Malovitskiy et al. (1975) determine a similar sedimentary series above a crystalline basement, which they interpret everywhere as continental platform; they employ vertical displacement of lithospheric blocks to explain the distribution and nature of the crust.

Surface wave studies under the Gulf of Genoa (Recq, 1973b) agree with the results of Fahlquist and Hersey (1969). Berry and Knopoff (1967), Berry et al. (1969), and Payo (1969) all find a low-velocity channel in the upper mantle and an oceanic type crust 11–14 km thick of velocity 7.7 km/sec extending to this channel, which according to Berry and Knopoff is underlain by a mantle of 8.2 km/sec at about 30-km depth. This low-velocity "mantle" at 7.7 km/sec has been discussed by Menard (1967).

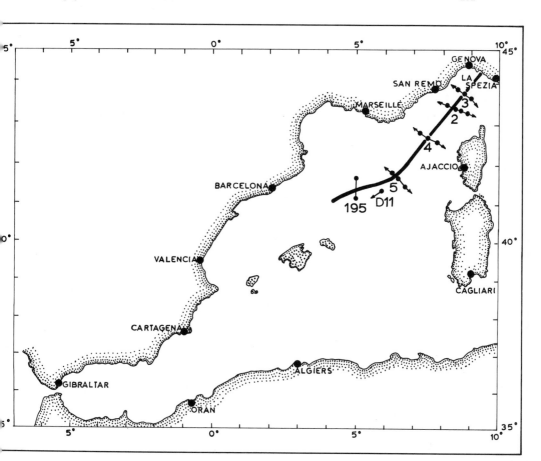

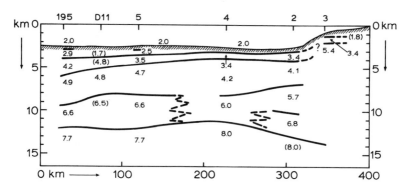

4. Location and interpretation of seismic refraction profiles in the North Balearic Basin (after Fahlquist and Hersey, 1969). D11 is a refraction station of Ewing and Ewing (1959).

The western Mediterranean is the site of shallow-focus earthquakes, this active seismic zone extending from the Mid-Atlantic Ridge close to the Azores through the Straits of Gibraltar across North Africa to Sicily (Gutenberg and Richter, 1954). Figure 6 shows the seismicity of the Mediterranean area. Mc-Kenzie (1972) describes the northward motion of Africa relative to Europe, which is taken up on the compressive plate boundary through North Africa.

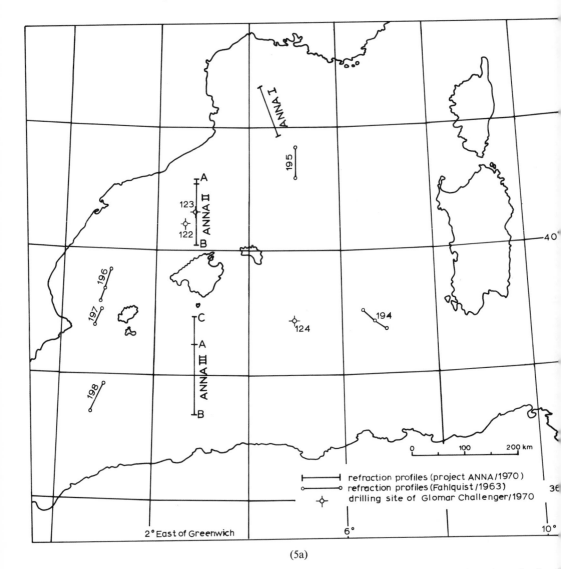

(5a)

Fig. 5. (a) Location of refraction profiles *Anna* 1970 (after Leenhardt *et al.*, 1972). (b) Interpretation of refraction profiles *Anna* 1970. Profile I is after Leenhardt (in Leenhardt *et al.*, 1972); profiles II and III are after Hinz (in Leenhardt *et al.*, 1972).

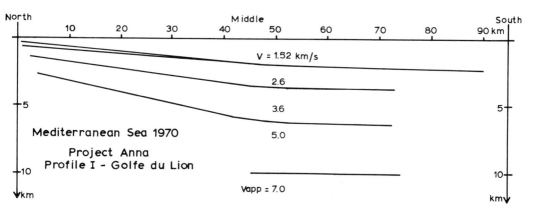

North · Middle · South

V = 1.52 km/s

2.6

3.6

5.0

Mediterranean Sea 1970

Project Anna
Profile I - Golfe du Lion

Vapp = 7.0

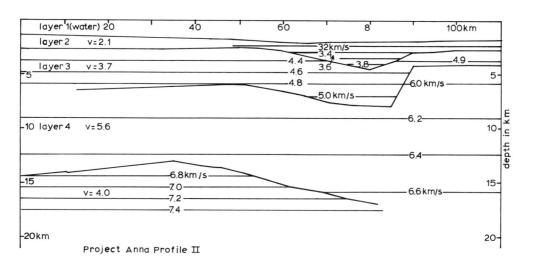

layer 1(water)   20   40   60   8   100 km
layer 2   v = 2.1
layer 3   v = 3.7
32 km/s
3.4
4.4   3.6   3.8   4.9
4.6
4.8
5.0 km/s
6.0 km/s
6.2
layer 4   v = 5.6
6.4
6.8 km/s
7.0
v = 4.0   7.2   6.6 km/s
7.4

depth in km

Project Anna Profile II

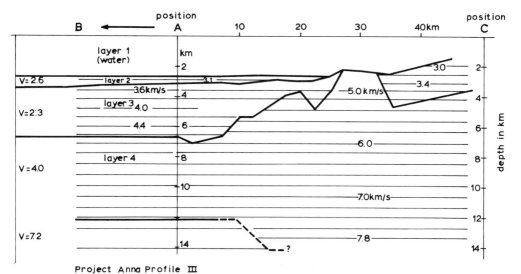

position   position
B ←──────→ A   10   20   30   40 km   C

layer 1
(water)   km
2
V = 2.6   layer 2   3.1   3.0
3.6 km/s   4   3.4
V = 2.3   layer 3 4.0   5.0 km/s
4.4   6
6.0
V = 4.0   layer 4   8
7.0 km/s
10
12
7.8
V = 7.2   14   ?   14

depth in km

Project Anna Profile III

(5b)

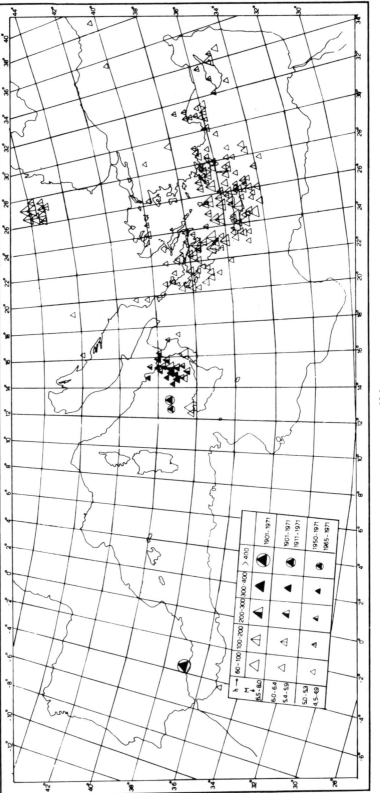

(6a)

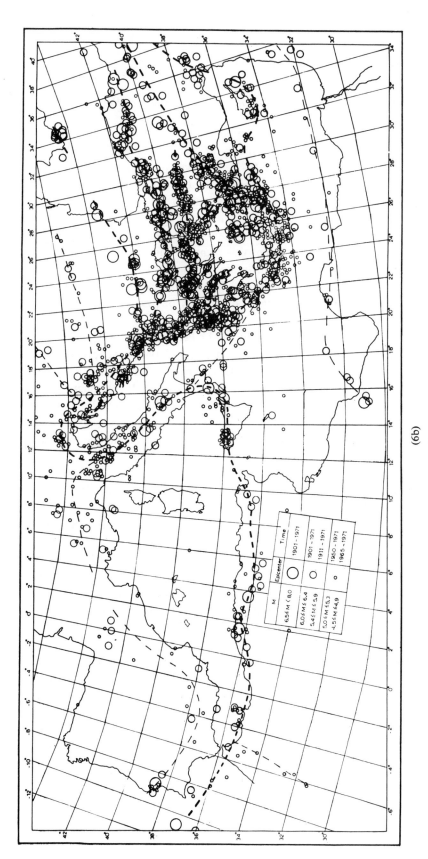

(6b)

Fig. 6. (a) Location of earthquake epicenters of shallow focus ($h = 60$ km) for the period 1901–1971, magnitudes > 4.5; data from Barazangi and Dormann (1969) and Gutenberg and Richter (1954) (after Papazachos, 1973). (b) Location of earthquake epicenters of intermediate to deep focus ($60 < h > 400$ km) (after Papazachos, 1973).

C. Interpretations and Implications

1. *Mechanisms and Models for the Formation of the Western Basins*

Studies of seismicity and earthquake mechanisms provide important information about the present tectonics of the entire Mediterranean area (Ritsema, 1969; McKenzie, 1970, 1972); however, the earlier movements of plates cannot be deduced directly, and the motions have been very variable. The northward movement of Africa relative to Europe is well established and has affected much of the southern part of Europe, producing compression. Africa underwent a counterclockwise rotation between the Early Cretaceous and the Eocene with Iberia and the Alps rotating during the Late Jurassic to Eocene and the Middle Triassic to Middle Eocene, respectively (Van der Voo, 1969; Zijderveld and Van der Voo, 1973), as a consequence of the Upper Jurassic to Lower Cretaceous opening of the Bay of Biscay (Le Pichon *et al.*, 1972).

According to several authors (including Le Pichon *et al.*, 1971; Olivet *et al.*, 1973*b*; Auzende *et al.*, 1973*a*) the age of opening of the Provençal Basin and Algerian Basin is the Middle to Upper Oligocene and Lower to Middle Miocene, respectively. Seismic reflection data (Auzende *et al.*, 1971), magnetic data (Le Borgne *et al.*, 1971; Galdeano *et al.*, 1974), and seismic refraction data (Fahlquist and Hersey, 1969) are consistent with this hypothesis, giving a sedimentary thickness of 5–6 km, which indicates a reasonable sedimentation rate for the period of time involved since the Oligocene.

It is generally proposed that the Algero-Provençal Basin originated by migration of the Corsica–Sardinia block (Argand, 1924; Carey, 1958; Stanley and Mutti, 1968; Ryan *et al.*, 1970; Smith, 1971; Alvarez, 1972).

Auzende *et al.* (1973*a*) propose a double model for the western basins: one of migration of Corsica–Sardinia to form the Algero-Provençal Basin, and that of the dislocation of an "internal zone plate" in the Alboran Basin behind the Betic–Rif Arc, with formation of an inter-arc marginal basin (Le Pichon *et al.*, 1972). The magnetic and structural evidence for this rotation has already been described in Section IIB (Auzende *et al.*, 1973*a*; Bayer *et al.*, 1973) and the palaeomagnetic evidence by Nairn and Westphal (1967, 1968), de Jong *et al.* (1973), Zijderveld *et al.* (1970), and Zijderveld and Van der Voo (1973).

In early fits, Alvarez (1972) rotates Corsica and Sardinia as one unit to lie against Provence and to the north of the Balearic Islands, respectively. He then improved the continental margin fit and developed a differential rotation of the two islands, the relative movement occurring along a boundary through the Straits of Bonifacio (Alvarez, in Ryan *et al.*, 1973, p. 472; Alvarez, 1973, 1974) (Fig. 7). A similar model was achieved by computer fit (Westphal *et al.*, 1973), but Auzende *et al.* (1973*a*) and Bayer *et al.* (1973), while taking

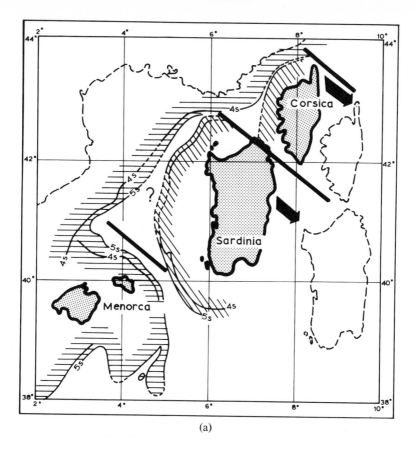

(a)

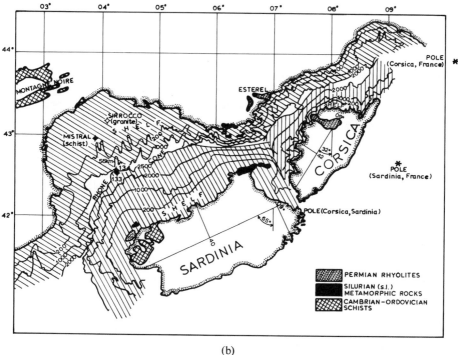

(b)

Fig. 7. Reconstructions of blocks in the western Mediterranean before the formation of the western basins. (a) Thick arrows show the displacement; dashed lines show present position of Europe, Corsica, and Sardinia; shaded areas are continental basement (after Auzende *et al.*, 1973*a*). (b) Angles of rotation are shown for Corsica relative to Europe, and Sardinia relative to Europe, indicating a relative motion between the islands themselves (after Alvarez, 1974).

account of this differential motion, obtain different angles and poles of rotation. Smith (1971) on the other hand, places the islands to the east of the Balearic Islands, off the coast of Spain.

In conclusion, magnetic, structural, and paleomagnetic data confirm counterclockwise rotation of several blocks in the western Mediterranean Sea relative to stable Europe, and the subsequent development of the present constituent basins.

2. Origin and Evolution of the Basins

Several different origins have been considered for the basins of the western Mediterranean, and early ideas embraced all hypotheses: (a) the basin is a relic of the ancient Tethyan seaway, in which sediments of the Alpine mountain belts were deposited; (b) crustal opening occurred during the Tertiary, so that the present crustal structure is young oceanic, similar to that of the Mid-Atlantic Ridge; (c) an intermediate type of crust exists similar to that of inter-arc basins; (d) the basin consists of a foundered continental block, possibly of collapsed Alpine structures covered by recent sediments and affected by oceanization processes; (e) the crust is a mixture of continental and oceanic blocks, similar to the rifted continent of the Red Sea–Afar area.

A description of these hypotheses can be found in de Roever (1969), Van Bemmelen (1969), Ritsema (1969), Ryan et al. (1970), Le Pichon et al. (1971), Auzende et al. (1973a), Olivet et al. (1973b), and Morelli (1975).

The acquisition of recent data has greatly modified thoughts on the problem. Gravity, refraction, and surface wave studies indicate the existence of crust of an "oceanic" type in the center of the deep Balearic Sea, but the magnetics are generally flat with few lineations (Auzende et al., 1973a; Bayer et al., 1973) (Section IIB).

The Valencia Basin, north of the Balearic Islands, has an "intermediate" type of thinned continental crust attributed to the downward movement of the oceanic lithosphere below the continental Spanish block during Cretaceous to Palaeogene times, and associated with volcanic activity and Betic–Balearic mountain building processes (Hinz, 1973).

The direction of the magnetic anomalies seems to differ between the Alboran Basin and the Algero-Provençal Basin. The trend is north-northwest–south-southeast (Bayer et al., 1973; Galdeano et al., 1974) in the Alboran Basin, and northwest–southeast in the Provençal Basin (Auzende et al., 1973a) and in the southern basin to the south of Corsica (Gennesseaux et al., 1974). Bayer et al. (1973) favor an interpretation of linear anomalies of the Vine and Matthews type in the south Balearic Basin, with displacement by fracture zones. However, Gonnard et al. (1975) correlated the observed anomalies with volcanic events; the volcanism occurred after the opening of the basin, which suggests that the crust here is of "inter-arc" rather than "oceanic" type.

Recent geological and structural constraints in the area are discussed by Auzende and Olivet (1974) and Biju-Duval et al. (1974), the post-Messinian events being well defined from seismic profiles and DSDP results. Sedimentary features include those of halokinetic activity in the Plio-Quaternary, often with considerable faulting; subsidence occurred along the margins of the basin whose formation dates back to Upper Miocene, when salt was deposited; below the salt, the sediments fill the deep basin, but correlation is difficult between the center and the margins as tectonic activity has affected the series. Further discussion is contained in Section V.

To summarize, the western Mediterranean basins owe their formation to an Oligo-Miocene phase of distension, accompanied by counterclockwise rotation of landmasses when oceanic crust was created in the deep Algero-Provençal Basin; in the adjacent Valencia and South Balearic Seas, the crust is probably of thinned continental origin, corresponding to formation in a marginal basin.

III. CENTRAL MEDITERRANEAN

A. Available Geophysical Data

The central Mediterranean is divided, for this discussion, into the areas of the Tyrrhenian Sea and the Adriatic Sea. Early bathymetric charts by Segré (1958), Pfannenstiel (1960), Mikhaylov (1965), and Morelli et al. (1969) for the Adriatic Sea, and Morelli (1970) for the Tyrrhenian Sea, were incorporated in recent publications by Carter et al. (1972) and Finetti and Morelli (1972, 1973). Results of narrow-beam side-scan sonor exploration are published by Belderson et al. (1974b).

The structural setting of the central Mediterranean is shown in Fig. 1. Maps available for the area are published by the C.M.R. Italieni, at 1:1,000,000, including bathymetry, structure, Bouguer gravity, total magnetic field, and seismicity. The geology and structure of surrounding land areas is well studied; authors include Castany (1956), Burollet (1967), Martin (1967), Wezel (1967, 1970), Alvarez and Gohrbandt (1970), Aubouin et al. (1970), Auzende (1971), Wezel and Ryan (1971), Auzende et al. (1972), Barberi et al. (1973), Scandone et al. (1974), Bishop (1975), and Chabrier and Mascle (1975).

References to early gravity and magnetic observations have been given in Section IIA. OGST survey results are published by Morelli et al. (1969), Morelli (1970), Finetti et al. (1970), and Finetti and Morelli (1972, 1973). Many articles have appeared on the relationship between topography, chiefly seamounts and volcanic islands, and the magnetic anomalies in the Tyrrhenian Sea (Hersey, 1965a; Ryan et al., 1965; Finetti et al., 1970; Heezen et al., 1971; Selli and Fabbri, 1971; Allan and Morelli, 1971; Barberi *et al.*, 1973; Belderson *et al.*, 1974*b*; Gonnard *et al.*, 1975).

Paleomagnetic data are published by Zijderveld and de Jong (1969), Zijderveld *et al.* (1971), Soffel (1972), de Jong *et al.* (1973), Channell and Tarling (1975), Klootwijk and Van den Berg (1975), and Lowrie and Alvarez (1975).

Heat-flow values in the Tyrrhenian and Adriatic Seas are recorded by Erikson (1970) and DSDP Leg 42A (1975), and studies of volcanic rocks have been made by Gasparini and Adams (1969), Locardi (1968), and Maccarrone (1970). Information about marine sediments is given by Ryan *et al.* (1965), Kermaboni *et al.* (1969), and from coring by DSDP Leg 13 at Site 132 and Leg 42A at Site 373.

Seismic reflection profiles have been described by Ludwig *et al.* (1965) and Ryan *et al.* (1965) for the Tyrrhenian Sea, and more recently by Finetti *et al.* (1970), Finetti and Morelli (1972, 1973), and by Morelli *et al.* (1969) in the Adriatic Sea. Refraction techniques were used by Ewing and Ewing (1959), Moskalenko (1967), Fahlquist and Hersey (1969) and Colombi *et al.* (1973).

Phase-velocity studies already described for the western Mediterranean in Section IIA cover the Tyrrhenian Sea. The seismicity of the Calabrian Arc is described by Peterschmitt (1956), Ritsema (1969), Caputo *et al.* (1970), Bousquet (1972), and Papazachos (1973), and the delineation of plates is described by McKenzie (1970, 1972); paleomagnetic and volcanological evidence from Sicily in support of the location of the plate boundary is presented by Barberi *et al.* (1974). The seismicity of Italy is described by Giorgetti and Iaccarino (1971), and the distribution of earthquakes is shown in Fig. 6.

B. Principal Results

The results will be discussed separately for the Tyrrhenian Sea, Straits of Sicily, and the Adriatic Sea.

1. *Tyrrhenian Sea*

The Tyrrhenian Sea occupies a triangular-shaped depression between the ancient Hercynian schists of Sardinia, and the Alpine belt comprising the Apennine, Calabrian, and Sicilian chains which form the Calabrian Arc (Fig. 1). The deep abyssal plain at 3000–3500-m depth is enclosed by elongate ridges or mountains parallel to the present coastlines. Peripheral basins exist on many of the margins, e.g., the Calabrian Basin (Finetti and Morelli, 1972, 1973; Morelli, 1975). The ridges form effective sediment barriers and determine the extent of the continental slope and shelf.

The abyssal plain has a very different character from that of the western basin, being studded with seamounts which mainly trend north–south, some of them lying at extremely shallow depths (Morelli, 1970). Some of these have given samples of Paleozoic and Triassic age (Heezen *et al.*, 1971; Selli and Fabbri, 1971). The topographic relief is closely related to the observed free-air gravity anomalies in the Tyrrhenian Sea (Allan and Morelli, 1971) (Fig. 8). A weak positive free-air anomaly over most of the basin indicates that the area is out of isostatic equilibrium. The Bouguer anomalies reach 200 mgal and more in areas of deeper water (Fig. 2), indicating the existence of high-density material at shallow depths. Neither the western Sardinia to Elba ridge, nor the Italian mainland have prominent associated gravity anomalies.

The magnetic field reflects the location of the volcanic arc, and basaltic activity seen in the seamounts of the abyssal plain, the Straits of Sicily, and in eastern Sardinia is related to tension. Potassic trachytes are found along the Tyrrhenian coast of Italy, associated with continental collapse structures. Volcanic activity took place either subsequent to the collapse of the basin after the Pliocene, according to Selli and Fabbri (1971) and Morelli (1970), or before the Middle Miocene, and thus before the salinity crisis (Ryan *et al.*, 1973). Gasparini and Adams (1969) propose that volcanism has migrated southeastward along a zone of activity behind the Calabrian Arc. Basalts have been dredged from some seamounts (Maccarrone, 1970) and also phyllites, schists, and marbles of Paleozoic and Triassic age (Heezen *et al.*, 1971).

Seismic reflection profiles in the abyssal plain reveal thin (200 m), unconsolidated sediments (Ryan *et al.*, 1965, 1970) above a series including the Messinian evaporites, with only restricted development of a salt layer, and thus limited halokinetic effects (Fig. 9). A basin with thick salt development exists in the southwest (Finetti and Morelli, 1972), which is closed by the Straits of Sardinia–Tunisia. Detailed interpretation of records with respect to the geological setting are contained in Finetti and Morelli (1972, 1973) and Biju-Duval *et al.* (1974). DSDP Leg 13 at Site 132 drilled into Plio-Quaternary sediments before reaching salt at a depth of 223 m (Ryan *et al.*, 1973). DSDP Leg 42A at Site 373 drilled through a Plio-Quaternary sequence of marls, ashes, and volcanic sediments on the side of a seamount, and into basalt breccias with a limestone matrix above basalt flows (Hsü *et al.*, 1975). Biju-Duval *et al.* (1974) summarize the geology and tectonic events; their significance and effects on the evolution of the basin will be discussed in Section IIIC.

Early refraction measurements by Ewing and Ewing (1959) detected evaporites (4.9–5.7 km/sec) beneath a thin sediment cover in the abyssal plain, and Moskalenko (1967) observed evaporites (4.0 km/sec) 0.4–1.2 km thick above 1.8–2.4 km of material of velocity 5.0 km/sec. Fahlquist and Hersey (1969) found a layer of velocity 6.9–7.5 km/sec at a depth of 6 km, while Finetti and Morelli (1973) interpreted wide-angle seismic reflection profiles to show a thin layer of 7.0 km/sec material above an 8.2 km/sec mantle at 12-km depth. Thus, the crust in the Tyrrhenian Sea is relatively thin; the structure may be

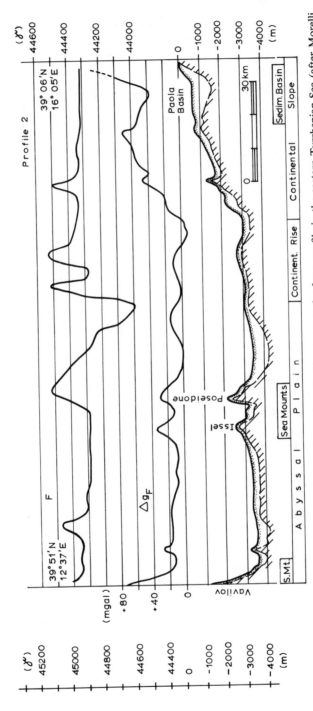

Fig. 8. Free-air gravity anomalies, magnetic intensity, and sea-bottom topography for a profile in the eastern Tyrrhenian Sea (after Morelli, 1970).

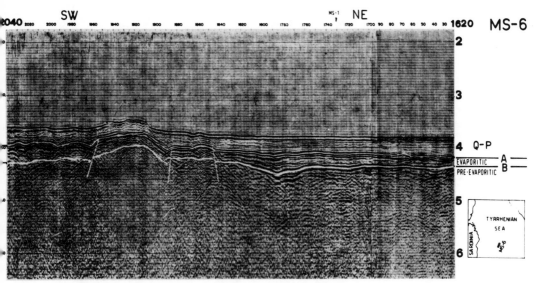

Fig. 9. Seismic reflection profile in abyssal plain, Tyrrhenian Sea (OGST profile MS-6, after Finetti and Morelli, 1973).

compared with that of the Mid-Atlantic Ridge (Finetti and Morelli, 1973; Morelli, 1975). Explosion seismology indicates that between the shelf edge and Gulf of Salerno the crustal thickness is about 20 km (Colombi et al., 1973), increasing to 25 km near the coast. There is some evidence that the Tyrrhenian has been thrust over the crust of the Apennines, as the velocity interpretation can be used in a crustal shortening model, and southwest–northeast movements are seen in geological structures on land.

The Calabrian Arc is associated with the only deep earthquakes found in the Mediterranean Sea (300–700-km epicenter depth) (Fig. 6). The earthquakes are known to define a Benioff zone, a conical slab, which dips from the Ionian Sea under Calabria to the northwest at an angle of 60° (Peterschmitt, 1956; Caputo et al., 1970; Ritsema, 1969; Bousquet, 1972). Papazachos (1973) draws isodepths in the Calabrian Arc to demonstrate the structure, and plots the focal depths against distance from the arc (Fig. 10a, b). He notes that discontinuities in the dipping seismic zone occur, and attributes this to variations in the physical conditions. Ritsema (1969) presents an alternative interpretation, that instead of the Ionian Sea being thrust below the Calabrian Arc, the arc overrides the Ionian Sea in an almost east–west direction. The existence of a volcanic arc is indicative of a back-arc basin environment. The seismicity is used to define the boundary between the European and African plates (McKenzie, 1972) which curves northwards into the Italian peninsula and around the Adriatic (Fig. 11).

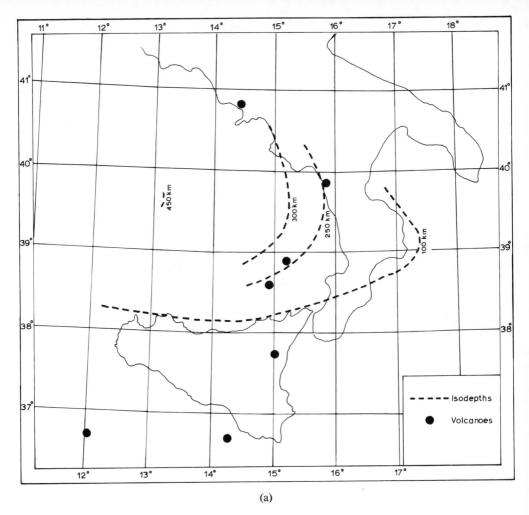

(a)

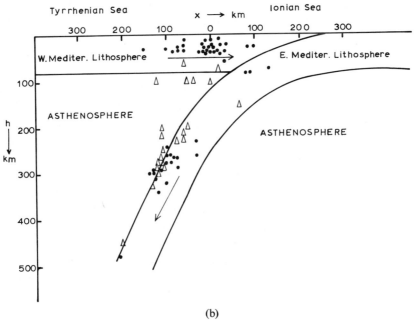

(b)

Fig. 10. (a) Features of the Calabrian Arc (after Papazachos, 1973). Proposed isodepths (lines of equal focal depth) in the Calabrian Arc and Tyrrhenian Sea. (b) Plot of focal depth of epicenters *vs.* distance from Calabrian Arc; triangles represent intermediate or deep-focus earthquakes for the period 1901–1964; dots represent earthquakes occurring 1964–1971, including data from Caputo *et al.* (1970).

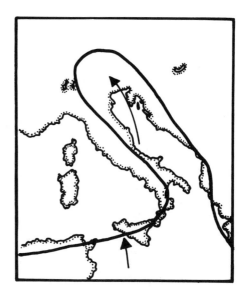

Fig. 11. Plate boundaries in the central Mediterranean; definition of Apulian plate is uncertain; arrow denotes movement relative to Europe (after McKenzie 1972).

2. *Straits of Sicily and Sardinia*

The straits between Sicily, Sardinia, and the African coast divide the western Mediterranean from the eastern Mediterranean and two areas of different morphological and geophysical characteristics. The passage is described in detail by Finetti and Morelli (1973) and Morelli (1975). Bathymetrically, the straits form an epicontinental bridge between Sicily and Africa, bounded to the east by a north–south fault system, the "Malta escarpment," which has foundered in the center to form three main grabens, also dissected by a northwest–southeast fault system; communication with the eastern and western Mediterranean basins is by shallow sills, lying at depths of 500 and 400 m, respectively (Carter *et al.*, 1972).

Free-air anomalies are positive, and the magnetic field is generally flat, corresponding to thick sediments (Finetti and Morelli, 1972) (Fig. 2). Magnetic anomalies are associated with volcanic instrusions along faults in a north–south direction, and northwest–southeast aligned volcanic islands and submerged volcanoes. Burollet (1967) describes the structural elements of the region, which are shown in Fig. 12.

Seismic reflection profiles indicate that the Palaeozoic basement is continuous between Sardinia and Africa (Auzende *et al.*, 1971) and that a thick Mesozoic to Tertiary platform exists, with a basal layer dated as the top of the Triassic, undisturbed by tectonics, except for Pliocene foundering (Morelli, 1975); Plio-Quaternary sediments are extremely thin, due to emergence of the feature (Fig. 13). Explosion seismology indicates a 20–21-km-thick crust, thickening to 35 km under Sicily, the Moho separating layer of velocity 6.3 km/sec and 8.0 km/sec.

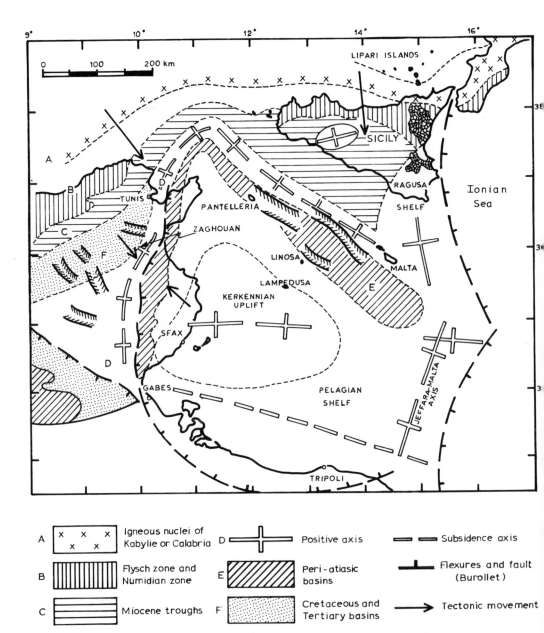

Fig. 12. Tectonic sketch of the Sicily Channel (after Burollet, 1967).

Fig. 13. Seismic reflection profile in the Straits of Sicily (OGST profile MS-34, after Finetti and Morelli, 1973).

3. *Adriatic Sea*

The Adriatic Sea lies between Italy and Greece, a relatively shallow basin in the north, 200 m, deepening to 1200 m in the south; the northern and central basins have the characteristics of a continental shelf area continuing from the Po Plain (Van Straaten, 1965), while the southeast is a small "basin plain" (Emery *et al.*, 1966). The basin is separated from the Ionian Sea by a submarine sill, lying at approximately 700-m depth. A detailed morphological description is given by Giorgetti and Mosetti (1969), and interpretation of geophysical data by Morelli *et al.* (1969).

The free-air gravity indicates that the northern and central basins are approximately in isostatic equilibrium, with a negative anomaly of −100 mgal over the extension of the Po Basin in the northwest, while in the south, anomalies of −50 to +50 mgal are observed sloping from the Albanian coast into the basin. A Bouguer anomaly of +50 mgal lies over the basin. The magnetic field is negative over the length of the Adriatic Sea, reaching a minimum of −160 γ in the south.

Finetti and Morelli (1973) and Morelli (1975) describe reflection profiles carried out in the southern basin, which reveal a thick sequence (2500 m) of Mesozoic and Tertiary sediments, with thickening in the Plio-Quaternary towards the east and pinching out to the west (Biju-Duval *et al.*, 1974) (Fig. 14). The front of the Dinarides folded zone can be interpreted as reverse faults, indicating compression (Aubouin *et al.*, 1970; Mercier *et al.*, 1972). In the north a sedimentary basin contains thick Pliocene and Quaternary sediments; these are at their maximum thickness in the Po Basin (7000 m) and along the Italian coast where subsidence has occurred. A folded Mesozoic sequence is known below the Cenozoic, which flattens out toward the east, where evaporites are found (Biju-Duval *et al.*, 1974). Miocene evaporites are included in gravity nappes offshore of the Apennines. Refraction profiles are described by Dragasevic (1969) to determine the crustal thickness under Yugoslavia.

Seismological data show that a belt of earthquakes exists around the Adriatic, but these are not clearly continuous across the south (Lort, 1972; McKenzie, 1972), so that the present plate margins are not altogether clear (Fig. 11). The Adriatic is surrounded by thrust belts and tectonically emplaced ophiolite bodies, dated as Jurassic and Cretaceous, being remobilized in the Cenozoic (Abbate *et al.*, 1970; Aubouin *et al.*, 1970).

C. Interpretations and Implications

1. *Nature of the Crust*

Hypotheses for the origin of the Tyrrhenian Sea included that of an older extensive ocean basin, possibly Permian (Glangeaud, 1962) or Jurassic (Scan-

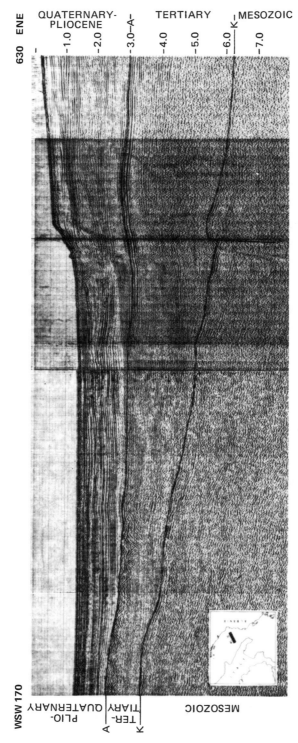

Fig. 14. Seismic reflection profile in the southern Adriatic Sea (OGST profile MS-30, after Finetti and Morelli, 1973).

done, 1975) in age. Other hypotheses include Tertiary foundering of a continental craton (Heezen *et al.*, 1971; Selli and Fabbri, 1971) and formation in a marginal back-arc basin by lithospheric subduction at a plate margin located in the Ionian Sea (Barberi *et al.*, 1973). Several hypotheses and their tectonic implications are discussed in detail by Morelli (1970). In fact, all indications suggest an insular arc structure; high heat flow, gravity and magnetic anomalies, the thin crust, and "soft" mantle structure point toward a marginal sea-type of basin.

The Adriatic Sea has rather the characteristics of a basin on a continental shelf, at least in the north, while in the south, the crust is probably continental.

2. *Subduction Zones*

Crustal subduction is still taking place below the Tyrrhenian Basin, as demonstrated by the intermediate and deep-focus earthquakes and the internal volcanic arc (Caputo *et al.*, 1970; Bousquet, 1972) (Fig. 10). According to Barberi *et al.* (1973) the contact between the African and European plates can be located in the Ionian Sea by a belt of low-gravity anomalies existing to the southeast of Calabria, which is displaced north of Mount Etna on Sicily by a transcurrent fault between Sicily and Calabria. Scandone *et al.* (1974) describe the geological and tectonic evolution of the Apulian–Sahara continental margins and evidence for the past existence of subduction zones.

The ophiolites located around the Adriatic may represent the relics of an ancient oceanic domain, although not the Tethys (Smith, 1973). Subduction or downthrusting has probably occurred along the Apennines and the Dinarides. The axis of the Apennines corresponds to a gravity maximum, and a maximum allochthonous thickness, which may be indicative of an ancient subduction zone (Finetti and Morelli, 1972). The lack of seismicity across the southern Adriatic also precludes the delineation of an Apulian plate, although McKenzie (1972) suggests that a "tongue" became detached from the African plate and moved northwards with the general movement of Africa relative to Europe, following the rotation of Italy when the Tyrrhenian Sea opened up (Fig. 11).

IV. EASTERN MEDITERRANEAN

A. Available Geophysical Data

The eastern Mediterranean may be considered as three distinct units for the purpose of discussion: the Ionian Sea, the Aegean Sea, and the Levantine Sea. (Fig. 15).

Early bathymetric charts include those of the Russians (Goncharov and Mikhaylov, 1964; Mikhaylov, 1965). Emery *et al.* (1966) describe soundings reported in ancient literature and use data collected by the Saclant ASW

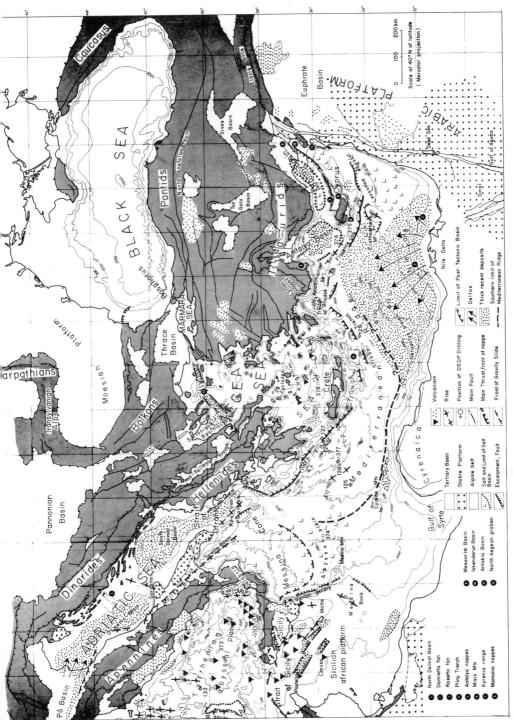

Fig. 15. Structural framework and physiography of the eastern Mediterranean Sea (modified after Biju-Duval *et al.*, 1974).

Research Center, La Spezia, Italy, in 1964 and by the Lamont Geological Observatory and Woods Hole Oceanographic Institution, together with U.S. Naval Survey data, to draw a 500-m contour map and delineate the physiographic provinces. The physiography was also described by Giermann (1969), Ryan (1969), Ryan et al., (1970) and Wong et al. (1971); Watson and Johnson (1969) present the results of a U.S. Naval Oceanographic Survey in the form of a physiographic map. Ryan et al. (1970) describe a physiographic chart compiled by Heezen et al. (in Ryan et al., 1970). Recent, more detailed bathymetric charts have been drawn by Allan and Morelli (1971) for the western Levantine Sea, by Finetti and Morelli (1972) for the Ionian Sea, by Maley and Johnson (1971) for the Aegean Sea, by Ross et al. (1974) for the Black Sea, and by Wright (in Morelli et al., 1975) for the eastern Levantine Sea. The map by Carter et al. (1972) is the most recent version, and the detailed physiography of the region is described in their article (Fig. 16). The results of profiles using a narrow-beam side-scan sonar are published by Belderson et al. (1970, 1972, 1974a) and Kenyon et al. (1975).

The first gravity data obtained were submarine measurements (Cassinis, 1941; Cooper et al., 1952). Other early gravity data were interpreted by Mace (1939), Harrison (1955), Gass and Masson-Smith (1963), Gass (1968), Fleischer (1964), and more recently by Rabinowitz and Ryan (1970), Woodside and Bowin (1970), Allan and Morelli (1971), Finetti and Morelli (1973), and Woodside (1975).

Maps of total magnetic field intensity have been produced by Allan and Morelli (1971), by Finetti and Morelli (1973) from measurements at sea, and by Vogt and Higgs (1969) from aeromagnetic data. A magnetic anomaly map for the eastern Levantine Basin was prepared by Matthews (1974), and an aeromagnetic survey conducted by Huntings Surveys Ltd. over Cyprus is described by Gass (1968) and Vine et al. (1973).

Other discussions of magnetic profiles are contained in Watson and Johnson (1969), Vogt and Higgs (1969), and Wong et al. (1971). Ben-Avraham et al. (1976) described the magnetic field of the eastern Mediterranean, with special reference to the tectonic setting of the Eratosthenes Seamount.

The volcanicity of the region is chiefly located in the Aegean Basin (Nicholls, 1971; Ninkovitch and Hays, 1972; Boccaletti et al., 1974; Vilminot and Robert, 1974). Heat-flow measurements were first made by Erikson (1970) and Ryan, and are described in Ryan et al. (1970). Jongsma (1975) describes several recent results from the Aegean Sea. DSDP Leg 42A (Hsü et al., 1975) reports down-hole temperature measurements.

Early seismic reflection profiling was carried out by Chain cruises in 1959, 1961, 1964, and 1966 (Woods Hole Oceanographic Institution), and interpreted by Hersey (1965a), Ryan et al. (1965, 1966, 1970), Ryan and Heezen (1965), Ryan (1969), and Wong et al. (1971). These were interpreted for determinations

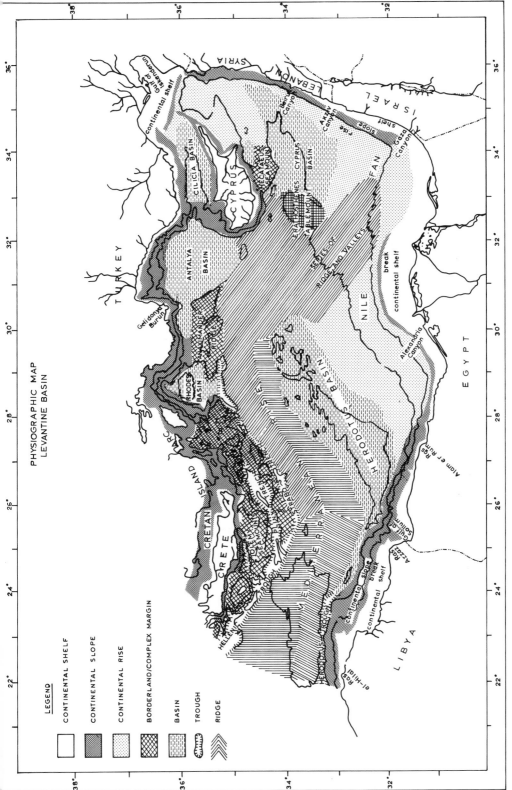

Fig. 16. Physiographic map of the Levantine Sea (after Carter *et al.*, 1972).

of sediment thicknesses by Wong and Zarudski (1969) and Zarudski *et al.* (1969); other determinations have been made by Ludwig *et al.* (1965); Moskalenko (1966); Yelnikov (1966), and more recently by Malovitskiy *et al.* (1975) as a result of the Russian survey in the eastern Mediterranean. Watson and Johnson (1969) describe the results of the U.S. Naval Oceanographic Survey, which used a sparker source. Other seismic reflection profiles were obtained by DSDP Leg 13 (Ryan *et al.*, 1973) and DSDP Leg 42A (Hsü *et al.*, 1975) in the Mediterranean. The Israel Geological Survey covered the eastern Levantine Sea close to the coast (Neev *et al.*, 1973). French oil companies have published data from individual profiles in the Ionian Sea (Sancho *et al.*, 1972), as have the Germans (Hinz, 1974), and this area is covered by an OGST grid survey (Finetti and Morelli, 1972, 1973). Woodside (1975) describes the results of seismic profiling in the Levantine Basin, and Jongsma (1975) interprets profiles in the Aegean Sea, both cruises being organized by the Department of Geodesy and Geophysics, University of Cambridge.

Seismic refraction and point reflection measurements have also been made in the eastern Mediterranean (Gaskell and Swallow, 1953; Gaskell *et al.*, 1958; Ewing and Ewing, 1959; Ludwig *et al.*, 1965; Moskalenko 1967; Moskalenko and Yelnikov, 1966; Yelnikov, 1966; Neprochnov, 1968; Weigel, 1974; Wright, personal communication) during experiments carried out by the Lamont Geological Observatory, the Russians, a cooperative Greek–German team, and by the Department of Geodesy and Geophysics, University of Cambridge, together with l'Institut Français du Pétrole. In addition, surface wave dispersion studies yield deep crustal–lower mantle results (Payo, 1967; Papazachos and Delibasis, 1969). Seismicity studies and fault plane solutions allow the plate boundaries to be delineated and the relative motions to be determined (Constantinescu *et al.*, 1966; Arieh, 1967; Papazachos and Delabasis, 1969; Caputo *et al.*, 1970; Scheidegger, 1970; Papazachos and Comninakis, 1971; Comninakis and Papazachos, 1972; McKenzie, 1970, 1972; Papazachos, 1973).

B. Principal Results

The structural setting of the eastern Mediterranean is shown in Fig. 15. The Ionian Sea is bounded by the Calabrian Arc system of Sicily and southern Italy, the African continent to the south, and the complex Hellenides of Greece to the east. Detailed physiographic charts by Carter *et al.* (1972) delimit the Messina Cone, located outside the Calabrian Arc, the deep Sicilia Basin with abyssal plain, bounded to the east by the Mediterranean Rise and beyond by the Hellenic Trench system, to the south by the continental rise of the Gulf of Syrte (Surt), and to the west by the marginal complex of Sicily, which is dissected by numerous troughs (Ryan and Heezen, 1965).

The Mediterranean Rise, an arcuate structure which extends eastward through the Levantine Basin (Fig. 16) is composed of many small ridges and troughs, but it is not a continuous ridge as previously described (Mikhaylov, 1965; Ryan *et al.*, 1970); it terminates in the west at the Messina Rise and in the east almost reaches the Florence Rise, west of Cyprus. The strongly deformed nature of the rise has been described by Hersey (1965*a*), Giermann (1966, 1969); Emery *et al.* (1966); Watson and Johnson (1969); Allan and Morelli (1971); Ryan *et al.* (1970); Wong *et al.* (1971); Belderson *et al.* (1972); and Sigl *et al.* (1973) who also correlate the different modes of deformation with the physiographic provinces present.

The free-air gravity anomalies through the eastern Mediterranean show

To the north lies the series of deeps commonly known as the Hellenic Trench. Minor basins within the Levantine Sea include the Herodotus Basin, lying on the southern flank of the rise, and the basins associated with Cyprus, lying externally and internally to the arc, which are the Antalya and Cilicia Basins, respectively. The Nile Cone is an important delta feature extending over the Egyptian shelf (Kenyon *et al.*, 1975; Maldonado and Stanley, 1975). Detailed physiography of this complex Levantine Sea is shown in Fig. 16.

The Aegean Basin, lying interior to the Cretan (Hellenic) Arc, comprises several islands and small deeps, most of the associated shelves being extremely narrow and steep (Maley and Johnson, 1971). The Anatolian Trough in the north marks the extension of the North Anatolian Fault (Vogt and Higgs, 1969; Mikhaylov, 1965; Jongsma, 1975).

The free-air gravity anomalies through the eastern Mediterranean show a greater variation than in the west. In the Ionian Sea the range is from $+50$ to -50 mgal, being close to zero, a steep gradient corresponding to the edge of the Malta Escarpment; the values become negative over the Sicilia Basin. Over the Hellenic Trench system, a broad belt of discontinuous high negative anomalies (up to -240 mgal) extends from the west coast of Greece to the Antalya Basin, the most negative values occurring in the north. To the south the negative anomalies also follow the arcuate Mediterranean Rise, but become relatively positive over the Herodotus Basin, east of $27°$, and on the southern flanks of the rise. Both Cyprus and the Aegean Sea are associated with significantly positive anomalies, greater than $+100$ mgal (Makris, 1973). Gass and Masson-Smith (1963) and Gass (1968) describe anomalies of $+280$ mgal over Cyprus, associated with the Troodos Massif, which require high-density material to lie at a shallow depth (Vine *et al.*, 1973).

Bouguer anomalies in the eastern Mediterranean are predominantly positive, but the values are lower than those observed in the west, except in the Ionian Sea, where values of $+300$ mgal are associated with the deep basin (Finetti and Morelli, 1973). A narrow zone of negative Bouguer anomalies extends from the southern Peloponnese to a plateau area, lying to the south of Crete, which terminates at $33°30$ N, $24°30'$E. To the east of this point the Bouguer anomalies become positive and the trend of the rise changes to east-northeast. An arcuate low anomaly transverses Crete and meets another low, which passes around the south of Cyprus. The continuation of gravity anomalies on land is described by Jongsma (1975) for the Aegean Sea and by Woodside (1975) for the eastern Levantine Sea. A positive anomaly of $+100$ mgal is seen over most of the Nile Cone, and an anomaly of $+120$ mgal over the Anaximander Mountains.

Interpretations of crustal structure, based on available gravity data, are given by Rabinowitz and Ryan (1970), Woodside and Bowin (1970), and Woodside (1975), who state that the eastern Mediterranean is out of isostatic equilibrium. The Moho appears to lie at great depth (20–34 km) under the Mediterranean Rise, and the sedimentary column is extremely thick. Rabinowitz and Ryan (1970) compared the rise to an arc-associated sedimentary ridge, formed by a crustal shortening process, while Woodside and Bowin (1970) attributed its origin to an accumulation of sediment.

The magnetic field intensity in the eastern Mediterranean is extremely undisturbed (Vogt and Higgs, 1969; Allan and Morelli, 1971; Finetti and Morelli, 1973; Matthews, 1974; Malovitskiy et al., 1974, 1975). The only exceptions are in the central and northern Aegean Sea, associated with volcanic activity and possible rifting (Vogt and Higgs, 1969; Allan and Morelli, 1971) over Cyprus (Wong et al., 1971) and onland Turkey and Syria, associated with ophiolite bodies; the Eratosthenes Seamount, over which there is a distinctive positive anomaly of $+250\,\gamma$ above a regional anomaly of $-100\,\gamma$ to $-150\,\gamma$ (Ben-Avraham et al., 1976), and in the Ionian Sea over the Malta Escarpment and deep basin, where the anomalies reach $+500\,\gamma$. Volcanic activity near Malta and in the Ionian Sea can be correlated with magnetic anomalies.

Palaeomagnetic results from Cyprus made by Vine and Moores (1969) show that a counterclockwise rotation has occurred through 90°, and they suggest that the oceanic floor found here was formed during a quiet period during the Lower or Middle Cretaceous, as there is no evidence of magnetic stripes over the island (Vine et al., 1973).

Heat-flow measurements in the eastern Mediterranean indicate consistently low values (0.74 ± 0.30 HFU) (Erikson, 1970). DSDP Leg 42A measurements gave values of less than one half the global average in the Ionian Sea (Site 374), which suggests that the region is undergoing low-angle regional underthrusting, thus depressing the isotherms and causing a reduction in surface heat flow. At Site 376 on the Florence Rise, the value is extremely low (0.12 HFU), indicating active tectonics over the area. The Aegean Sea is characterized by higher than average heat flow, typical of values in marginal seas behind active island arcs (Hsü et al., 1975). Jongsma (1975) observes values in the range 1.24 to 2.73 HFU to support this hypothesis.

Reflection profiles are described by Hersey (1965a, b), Zarudski et al. (1969), Ryan et al. (1970), Finetti and Morelli (1972, 1973), Lort (1972), Biju-

Duval *et al.* (1974), Hinz (1974), Jongsma (1975), Morelli (1975), and Woodside (1975). Profiles show that the Ionian Sea is dominated by the Messina Cone, now thought to comprise a slipped mass on the outer margin of the Calabrian Arc, recognized from the absence of coherent reflections and the presence of diffractions (Sancho *et al.*, 1973; Biju-Duval *et al.*, 1974; Letouzey *et al.*, 1974) (Fig. 17). A tectonically disturbed zone lies behind the cone; compressional effects are seen throughout the Ionian Sea (Morelli, 1975). The Messina Cone comprises a Mesozoic to Cenozoic sequence of sediments (Finetti and Morelli, 1972, 1973), including Messinian evaporites and salt of variable thickness. In the western Ionian Sea and the Gulf of Syrte area, a thick sequence of sediments with a thin evaporite development is recognized above a Mesozoic (Upper Cretaceous) sequence superimposed on a continental crystalline basement visible only in the Gulf of Syrte (Finetti and Morelli, 1973; Morelli, 1975).

Over the Mediterranean Rise in both the Ionian and Levantine Basins there is great loss of penetration with characteristic diffractions caused by the irregular topography of the surface and of the Plio-Miocene boundary (Closs and Hinz, 1974; Ryan *et al.*, 1970; Lort, 1972; Sigl *et al.*, 1973; Finetti and Morelli, 1973; Hinz, 1974; Morelli, 1975). In some places, in the eastern Ionian Sea, seismic reflection profiles have penetrated the rise, and evidence of fragmentary deep reflectors is seen (Sancho *et al.*, 1973; Morelli, 1975). The origin of the rise is discussed in Section IIIC.

A thick development of sediments, including evaporites, is seen in the Hellenic Trench and the Herodotus Basin. The evaporites may be faulted out against the flanks of the rise, or may continue over it as a thin, highly diffracting cover (Fig. 18). Penetrative diapiric features occur in the North Cyprus Basins and to the southwest and south of Cyprus. DSDP has sampled various stages in the Miocene on the Mediterranean Rise from Messinian to Langhian (Leg 13, Sites 125, 126, 129), and the pre-evaporitic sequence has only been sampled in Site 375 (Leg 42) on the Florence Rise. Finetti and Morelli (1973) recognize a deeper reflector and interpret the sequence between Egypt and the Mediterranean Rise as Tertiary and probably Mesozioc in age, thus indicating a far greater age for the eastern basins than for the west. Thick sediments on the Nile Cone date from at least the Eocene, and DSDP Leg 31 (Site 131) reached Pleistocene clays; the sediments pinch out against the faulted Eratosthenes Seamount. Recent faulting along the Israeli coast is described by Neev *et al.* (1973). Internal basins in the Aegean Sea contain Middle Eocene to Oligocene sediments, which were subsequently tectonically affected during the Miocene (Maley and Johnson, 1971; Jongsma, 1975). The basins are older in the north, but differ according to the tectonic activity.

Until 1971, seismic refraction exploration in the eastern Mediterranean was extremely limited. Gaskell and Swallow (1953) and Gaskell *et al.*, (1958) describe results from three deep-water stations in the Levantine and Ionian

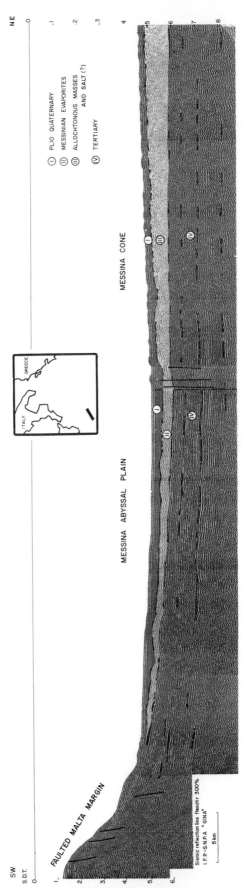

Fig. 17. Seismic reflection profile in the Ionian Sea, across the Messina Cone (after Biju-Duval *et al.*, 1974).

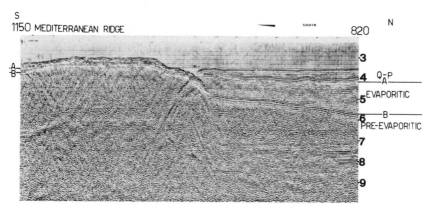

Fig. 18. Seismic reflection profile south of Crete (OGST profile MS-51, after Finetti and Morelli, 1973).

Basins, where they identify evaporites of velocity 4.3–4.7 km/sec below 0.4 km of sediments. Ewing and Ewing (1959) confirm these results, using data from stations in the Ionian Sea. During 1969 to 1971, the *Meteor* investigated the Ionian Sea, using reflection and refraction profiling techniques (Hinz, 1974; Weigel, 1974). Hinz reports that in the west, close to the Sicily–Malta Platform, 0.5 km of sediments overlie material of velocity 6.1 km/sec, but that this layer lies at 10-km depth in the deep basin, indicating that the Ionian Sea has subsided. The evaporites (velocity 4.0–4.5 km/sec) were detected on the Messina abyssal plain, underlain by low velocity material (2.2 km/sec) and with the Moho (8 km/sec mantle velocity) lying at 19-km depth. The sedimentary material is 4–5 km thick over the Mediterranean Rise, while the crystalline basement of 6 km/sec lies at 6–10-km depth. Weigel (1974) discusses a profile from the Malta Shelf into the Ionian Trench, where he correlates his results with those of Hinz (1974) (Fig. 19). He suggests that a process of oceanization of continental crust is occurring because of a deep crystalline layer (6.1 km/sec) and a relatively shallow Moho (18–20 km) in the central Ionian Sea. The Moho descends to 46 km below the Peloponnese.

Makris (1975) combines the seismic results of *Meteor* with gravity and seismic data from the Hellenides (Makris, 1973) in a profile from the Aegean Sea to the Ionian Sea, but his results only indicate a "transitional" type of crust rather than oceanic or continental. A contour map of the Moho discontinuity (Makris, 1973) from seismic and gravity data in the Ionian Sea, Peloponnese, and Aegean Sea indicates that strong compressional forces are responsible for the crustal thickening of the Hellenides (to 46 km below the Peloponnese) and the doming of the Aegean Sea area. The Moho lies at about 30-km depth over the central Aegean and to the south of the Hellenic Arc, but rises attenuated to 22 km over the Ionian Sea.

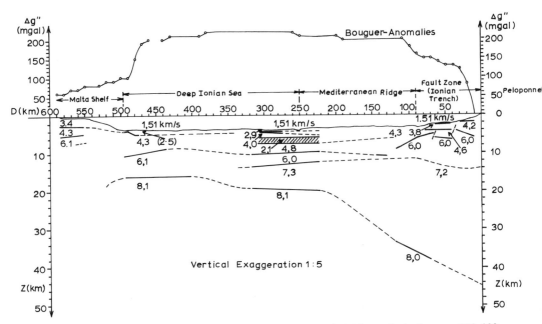

Fig. 19. Crustal section under the Ionian Sea (after Weigel, 1974); results in the range 230–300 km are after Hinz (1973).

In the Levantine Sea, the Russians identify a 7.0 km/sec layer which they attribute to the crystalline African Platform (Moskalenko, 1966). Malovitskiy *et al.* (1975) support this result and state that a crustal thickness of 20–25 km at the margins diminishes to 10–15 km in the central sea. Four seismic refraction profiles in the Levantine Sea are described by Lort *et al.* (1974); the location of the profiles and chief results are summarized in Fig. 20 and Table I. The Moho was located only on an unreversed line at the foot of the Nile Cone, and further results by D. Wright (personal communication) indicate that it may lie at a shallower depth of 20 km. He interprets wide-angle seismic profiles obtained by the Department of Geodesy and Geophysics, University of Cambridge, and I.F.P. The evaporites (velocities 3.5–4.1 km/sec) are found everywhere at approximately the same depth (1.5–2.0 km below the sea bottom). Material of velocity ranges 4.5–5.8 km/sec and 5.4 km/sec underlies the evaporites, the total producing a thick crustal column, which indicates that the crust is not necessarily oceanic. Figure 20 also shows results of a wide-angle seismic profile by Finetti and Morelli (1973) located to the north of the Nile Cone, indicating that the Moho lies at about 25-km depth.

Several seismic refraction profiles in and around Cyprus have been described by Matthews *et al.* (1971); Lort (1972); Lort and Matthews (1972); Khan *et al.* (1972), and laboratory measurements made of the rock velocities (Poster, 1973) in order to compare the velocities obtained in the rocks exposed

TABLE I

Structure and Location of Seismic Profiles[a]

Line	R_1	R_6	R_6R	R_4	R_4A	R_7
Lat. and long. (1)	33°50'N	33°54'N	34°03'N	32°42'N	33°33'N	32°54'N
	22°30'E	29°00'E	30°27'E	27°44'E	28°42'E	32°00'E
Lat. and long. (2)	33°35'N	34°03'N	33°54'N	33°33'N	32°34'N	32°59'N
	24°00'E	30°27'E	29°00'E	28°40'E	28°09'E	33°12'E
v (velocity, km/sec)	1.5	1.5	1.5	1.5	1.5	1.5
d (layer thickness, km)	1.4	3.0	3.1	3.1	2.9	1.6
v (unconsol.)	1.8, 2.3	2.8	2.0, 3.1	2.6	2.1, 2.3, 2.6	2.0
d (sediments)	0.3, 7.5	1.3	0.5, 1.0	2.7	0.3, 0.6, 0.9	0.5
v (3.5–4.1)	3.5	3.7	3.6	3.8	3.8	4.1
d	2.0	11.3	9.5	0.3	1.0	5.9
v (4.5–5.8)	5.8	4.8	5.5	4.5, 5.4	4.5, 5.4	5.1
d				9.0, 1.9	4.0, 0.9	7.5
v (6.4–6.5)				6.5	6.4	6.4
d						12.9
v (mantle)						8.4
d						

[a] Position (1): buoy position; position (2): last shot. Layer thickness (d) is calculated under buoys for reversed lines. R_4 and R_7 are reversed profiles; R_1, R_4A, R_6, and R_6R are unreversed.

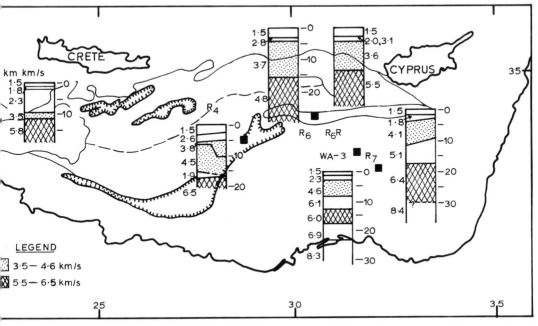

20. Seismic refraction results in the eastern Mediterranean (after Lort, 1972); additional data at WA-3 after Morelli (1975). Figures adjacent to the columns are velocities in km/sec and the vertical scale shows depth in km.

in Cyprus with those obtained in typical oceanic areas. The results were consistent with the hypothesis that Cyprus is an upthrust slice of oceanic crust, though due to the shattered nature of the surface rocks, surface measurements obtained rather lower velocities than anticipated (Lort and Matthews, 1972). The rocks exposed on Cyprus may also be traced into Morphou and Famagusta Bays (Gaskell et al., 1958; Lort, 1972).

The deep structure of the eastern basins is described from surface wave studies by Papazachos (1969) and Payo (1969), who suggest that the crust is 20–25 km thick, with lower crustal velocities than in the west, and a low-velocity channel in the mantle at 100–150-km depth.

The eastern Mediterranean is more seismically active than the west (Gutenberg and Richter, 1954; Ergin et al., 1967; Karnik, 1969; Papazachos and Comninakis, 1971; Papazachos, 1973), the Aegean Sea being totally delineated by a zone of shallow-focus earthquakes which surrounds the sea, the northern boundary connecting with the highly active Anatolian fault system (Galanopoulos, 1969; McKenzie, 1972). The deep bathymetric trench noted previously in association with a magnetic anomaly in the north Aegean Sea lies to the north of the major seismicity (Vogt and Higgs, 1969). Scattered seismic activity indicates that this rift does not mark the site of major movement at present and according to Papazachos (1973), the lithosphere may be discontinuous here. Fault plane solutions show variable mechanisms, due to the complex tectonic activity of the region. A feature of interest is the episodic nature of the earthquakes (McKenzie, 1972; Papazachos, 1973).

The Hellenic Arc is well defined eastwards from 39°N in Greece, across Crete, Karpathos, and Rhodes into Turkey, the intense seismic activity occurring over the Hellenic Trench. Depth distribution of earthquake foci indicates that this boundary is a seismic plane which dips to the northeast (Nicholls, 1971). Earthquakes within the Aegean do not indicate the existence of a well-defined Benioff zone similar to that of the Calabrian Arc, although there is an internal Tertiary–Quaternary volcanic arc (Nicholls, 1971; Papazachos and Comninakis, 1971; Vilminot and Robert, 1974) below which the slab lies at about 150-km depth. At the western end, there is no evidence for the continuation of the seismic zone across the Adriatic Sea (McKenzie, 1972); it probably passes north through Yugoslavia and Turkey. A diffuse zone of earthquakes marks the outer Cyprus Arc (Arieh, 1967; McKenzie, 1972).

McKenzie (1972) uses the seismicity to delimit a complex of small plates in the eastern Mediterranean (Fig. 21a) and shows how the motion between Africa and Europe is taken up by these plates. He attributes most of the seismicity of the region to the relatively fast-moving Aegean plate. The underthrusting of the African plate below the European plate is now confirmed (Caputo et al., 1970; McKenzie, 1972), although the complicated tectonics of the region is reflected by the close proximity of zones of thrusting and zones of extension

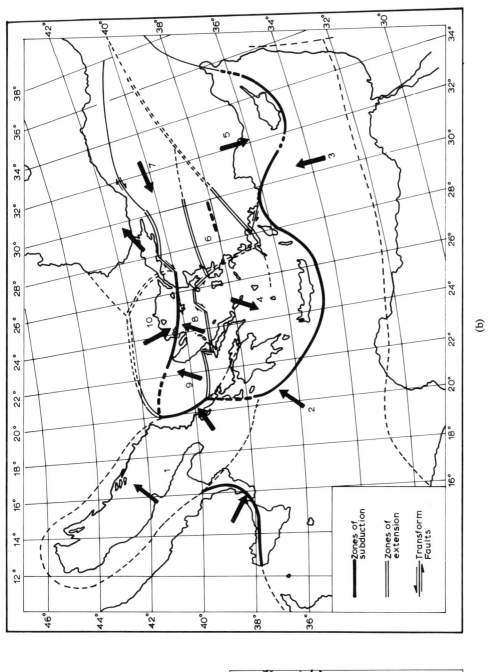

(b)

(a)

Fig. 21. (a) Sketch of plate boundaries and motion in the eastern Mediterranean (after McKenzie, 1972). (b) Tectonic features of lithospheric blocks in the eastern Mediterranean (after Papazachos, 1974).

(Papazachos, 1974), which are also observed by the author in seismic reflection profiles across the Hellenic Arc (OGST survey, Finetti and Morelli, 1973; University of Paris, *Medor* 1975 survey, to be published). Ritsema (1969) concludes that plate motion is also accompanied by deep-rooted mantle flow, and interprets a dextral motion on faults in the eastern Hellenic Arc, and sinistral motion in the west, which he explains by compressive forces acting east–west along the arc. Ozelçi (1971) discusses the motions along the Anatolian fault in relation to plate tectonics of the region. A further interpretation of the "seismotectonics" of the eastern Mediterranean is given by Papazachos (1974) in a synthesis article, and he presents a configuration of lithospheric aseismic "blocks" as shown in Fig. 21b. However, these models do not account for all the earthquake phenomena observed, indicating that the region is extremely complex.

C. Interpretations and Implications

1. *Eastern Mediterranean*

The present tectonics of the eastern Mediterranean are largely dominated by the relative northward movement of Africa with respect to Europe, and the consequent subduction of the African plate at the Hellenic Trench (McKenzie, 1972). The Aegean plate moves rapidly southeastward and the Anatolian plate southwestward relative to Africa (Fig. 21a); the stress field is complex, both compressional and tensional stresses are recognized in close proximity in the area, and in considering the evolution of the basins, several associated topographic features are of special interest.

a. *Mediterranean Rise.* The Mediterranean Rise was originally interpreted as a crustal thickening in the sediments due to piling up in front of the subduction zone at the Hellenic Arc (Woodside and Bowin, 1970) or to crustal shortening (Rabinowitz and Ryan, 1970). Recent ideas propose that the Mediterranean Rise comprises several allochthonous masses which have been emplaced in the Tertiary as olistostromes or flysch nappes; Mulder (1973) favors a Late Cenozoic displacement, while Biju-Duval *et al.* (1974) suggest that the Upper Miocene salt has acted as a surface of decollement. Finetti and Morelli (1973), on the other hand, suggest that the rise comprises Mesozoic to Tertiary material, originating from Alpine foundering.

It is difficult to determine the nature of the rise from geophysical evidence, as in reflection profiles the penetration beneath the rise is generally very poor, due to an extremely dissected surface topography and a strongly diffracting basal Pliocene horizon (Giermann, 1969; Hieke, 1972; Weigel, 1974; Closs and Hinz, 1974). There is some evidence of discontinuous contorted reflectors in the western Levantine Basin on the southern flank of the rise. The rise was originally described as a linear, arcuate feature, with distinctive "cobblestone" topography (Hersey, 1965a; Giermann, 1969; Ryan et al., 1970), following the trend of the Hellenic Arc from the Messina Cone through the eastern Mediterranean to Cyprus. However, it is not continuous, being composed of a series of depressions and elevations of varying trend, nor is it well defined bathymetrically. Toward the east it narrows considerably; the deformation and thicker sedimentary series due to contributions from the Nile Cone mask the true extent. The author suggests that it may disappear at about 26°E, far to the west of the Florence Rise and Cyprus Arc features. The evaporite series is seen over the rise (OGST and Cambridge *Shackleton* 1972 profiles) although it is not everywhere continuous; a thicker development exists to the north and south. This suggests that the rise existed before the evaporite deposition commenced. Several collapse structures are visible in the rise, which are due to salt tectonics and dissolution effects. DSDP has not succeeded in solving the problem of the nature of the rise, though Site 125 indicates the presence of evaporites and Hsü et al. (1973) formulated a dissolution model to explain the "cobblestone" topography. It is possible that brittle fracturing of carbonates on the flanks of the domed rise, followed by dissolution, produced this distinctive texture (Al-Chalabi, personal communication).

The gravity field over the far eastern Mediterranean, including the rise, shows negative free-air anomalies and generally positive Bouguer anomalies, which indicate a thicker sedimentary section rather than a basement high under the rise. Only in the Ionian Sea is the associated Bouguer anomaly strongly positive. The feature has no associated magnetic anomaly, no associated earthquake activity, and furthermore, the heat flow is low; this evidence conclusively refutes the early hypothesis that the "Rise" was tectonically similar to a mid-ocean ridge. Refraction measurements (Lort, 1972) on the Mediterranean Rise (Fig. 20 and Table I) indicate that thick sediments exist (9–11 km), a 5.0–5.8 km/sec layer occurring below the evaporite series. There is no evidence of the occurrence of high velocity layers at shallow depths, which would indicate a basement ridge. This evidence leads Finetti and Morelli (1973) and Morelli (1975) to suggest that at least in the Ionian Sea, a complete Cenozoic and Mesozoic section exists under the rise; it is a foundered continental crust. The contorted sequence is possibly metamorphosed, so that the layering is often not visible on reflection records (Al-Chalabi, personal communication).

On the other hand, Mulder (1973) and Biju-Duval et al. (1974) compare the position of the Mediterranean Rise, on a frontal arc, with recognizable slipped and allochthonous masses in other areas of the Mediterranean; the Messinian Cone in front of the Calabrian Arc (Letouzey et al., 1974), the Antalya and Lycian nappes in Turkey (Brunn et al., 1971), the nappes in front of the Gibraltar Arc (Bonnin et al., 1975; Lajat et al., 1975), and the nappes present on Cyprus (Lapierre, 1972; Biju-Duval et al., 1974). It should be mentioned that a hypothesis involving allochthony of the whole Mediterranean Rise necessitates the transport of a great volume of material over long distances.

On the evidence of existing data, these hypotheses cannot be proved or disproved. It is probable that several mechanisms are involved in the formation of the Mediterranean Rise, which although named as a single feature has diverse characteristics, as previously described.

b. *Eratosthenes Seamount.* The Eratosthenes Seamount, situated to the south of Cyprus, and delineated by fault scarps, is described in early bathy-

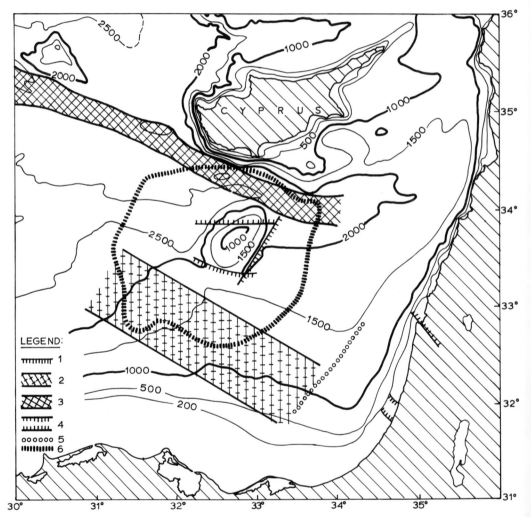

Fig. 22. Tectonic setting of the Eratosthenes Seamount (after Ben-Avraham *et al.*, 1976): (1) normal faults, (2) fault belts, (3) Giermann's fault (1969), (4) graben, (5) axis of magnetic high, (6) boundary of magnetic anomaly (50 γ contour).

metric studies (Emery *et al.*, 1966) of the eastern Mediterranean. Its present tectonic setting is considered in detail by Woodside (1975) and Ben-Avraham *et al.* (1976) (Fig. 22). It is described as a "single massive peak," but the author observes from seismic profiles (OGST; University of Cambridge, *Shackleton*; University of Paris, *Medor* 1975) a feature which is faulted and complex. Ben-Avraham *et al.* (1976) notes that small channels or valleys surround the mount, and that the sides are a series of step faults from a relatively flat crest; it is traversed by a left-lateral strike-slip fault and has normal faulting on the north side. Active tectonics have caused the vertical displacement of blocks relative to one another, and affected the adjacent Herodotus Basin.

The associated magnetic anomalies cannot be explained by the shape of the seamount alone, but require a larger source. Ben-Avraham *et al.* (1976) interpret the mount to be normally magnetized, in agreement with results from the pillow lavas in Cyprus (Vine *et al.*, 1973). Most of the seamount itself is composed of non-magnetic material, probably sediments, and the actual magnetic structure occupies a large part of the Levantine Basin (Ben-Avraham *et al.*, 1976; Malovitskiy *et al.*, 1975). There is no directly associated Bouguer anomaly, but indications are that the crust is slightly thinner under the Eratosthenes Seamount than in the surrounding area, where the Bouguer Anomaly is generally positive. They also compare the Eratosthenes Seamount with the Troodos Massif and deduce, from respective angles of rotation of 90° (Vine *et al.*, 1973) and 22° (Ben-Avraham *et al.*, 1976), that the blocks have moved independently, the shear zone between the two features being the fault originally observed by Giermann (1966).

It is apparent that important recent faulting has affected the seamount and this may result from an earlier rotation of crustal blocks, producing magnetic anomalies, and possibly explaining the low slip rates for the crust observed by North (1974) in this area. The present mode of deformation is a northwest–southeast compression, associated with the north-northeast movement of the Sinai plate (Ben-Avraham *et al.*, 1976; Neev *et al.*, 1973).

c. *Hellenic Trough*. This feature is generally considered to represent a presently active subduction zone, where the African plate moves below the Aegean plate. It comprises a series of depressions, the deepest of which are the Strabo and Pliny Trenches lying to the southeast of Crete. Some of the holes contain up to 1500 m of sediments; others are without sediment infill. There is no equivalent trough in front of the Cyprus Arc, indicating that the conditions differ here. The Florence Rise marks the western boundary of the arc, recognized from bathymetry and seismic reflection profiles, where the sediments of the Antalya Basin, including a thick evaporite series which can be traced from land (Brunn *et al.*, 1971; Magné and Poisson, 1974), wedge out against the rise. It is thought to represent the front of southward overthrust structures known from Cyprus, the presence of ophiolites possibly being associated with the observed magnetic anomalies, and corresponding in the west to the fault recognized by Giermann (1969).

2. *Cyprus*

As previously discussed, the position of the Cyprus Arc can be deduced, from the geological and geophysical data, to stretch from the Anaximander Mountains to the Levantine coast, a diffuse zone of seismicity occurring along its length (Lort, 1972; McKenzie, 1972). It is composed of nappes that are recognized on land at either end of the arc in Oman and Turkey (Biju-Duval et al., 1974). Drilling at Site 375 (DSDP Leg 42A) on the Florence Rise penetrated Messinian evaporites, bottoming in marls similar to the Packhna Formation on Cyprus.

The island of Cyprus itself has a special role in the history of the eastern Mediterranean, and the presence of the allochthonous Mammonia thrust sheets is now established (Lapierre, 1972). Biju-Duval et al. (1974) discuss a possible allochthonous origin of the Troodos Massif. This Cretaceous oceanic crust (Gass and Masson-Smith, 1963; Gass and Smewing, 1973) is interpreted in relation either to mid-ocean ridge activity (Moores and Vine, 1971; Pearce and Cann, 1973; Vine et al., 1973) or to an insular arc position (Miyashiro, 1973), although formation in a small Cretaceous ocean near the African plate margin (Robertson and Hudson, 1974) is now favored rather than a Tethyan origin.

The Mammonia complex was thrust on to Troodos Ocean floor in Maestrichtian time as the result of a collision, but there is some controversy over the direction of thrusting which was described as from the north by Lapierre (1972) and Biju-Duval et al. (1974); the situation is probably more complex, that is the Mammonia nappes were thrust southward over the Troodos Ocean floor as a result of a collision with a Triassic to Jurassic continental margin, represented by the Mammonia sediments (Robertson and Hudson, 1974). Acid volcanism preceded the emplacement, suggesting subduction and embryonic island arc development. This development was probably linked to plate motions and the relative motion between Africa and Europe; Turkey is known to have rotated counterclockwise, as have Cyprus, Israel, and the Lebanon. According to Van der Voo (1968, 1969) and Zijderveld and Van der Voo (1973), the landmasses lying to the south of the Alpine belt moved and rotated counterclockwise together with the African plate in mid-Cretaceous to Eocene time. However, Van der Voo and French (1974) state that Cretaceous and Tertiary data on Turkish rocks indicate that Turkey was located in a position halfway between Africa and Europe in the Tethys.

Biju-Duval et al. (1974) infer that to the north of Cyprus lies a Late Cretaceous subduction zone, which became passive, and that an autochthonous substratum is developed to the north, an extension of the Late Cretaceous–Early Tertiary African plate.

3. Nature and Age of Formation of the Crust

The division by McKenzie (1972) of the eastern Mediterranean into a series of small, rapidly moving plates has already been discussed (Fig. 21a); however, this concept does not allow deductions to be made about the early history of the area. The northern boundary of the African plate is defined seismically by a system of active arcs: the Calabrian Arc, the Hellenic Arc, and the Cyprus Arc, which enclose to the north marginal basins of a different nature from the Ionian and Levantine Basins. The latter are generally considered to be the oldest existing crust in the Mediterranean Sea, being Mesozoic to Cenozoic in age (Biju-Duval et al., 1974), and to lie exterior to the Alpine folded belt. The Aegean and north Cyprus Basins, which are described as inter-arc or back-arc type basins (Biju-Duval et al., 1974) on the basis of their situation and magnetic and heat-flow data, are younger Cenozoic basins, similar to those in the western Mediterranean.

Ophiolite bodies are very abundant around the eastern Mediterranean, and their tectonic setting in the Alpine Range dates from the Jurassic and Cretaceous: the age attributed to the Troodos Massif in Cyprus is Late Cretaceous, as discussed in the previous Section IVC, while Triassic ophiolites exist in Antalya, southern Turkey (Marcoux, 1970; Brinkmann, 1972), and Syria. They are interpreted as remnants of ocean crust and used by Smith (1971, 1973) as evidence to support a geometric fit of the coastlines of the Mediterranean and to indicate the age of the sea floor as Triassic to Late Cretaceous (Freund et al., 1975).

Geological evidence from the Negev indicates that the eastern Mediterranean Basin was initiated in Triassic time by splitting of the edge of the African continent along the continental margin. The Tethyan continental margin was pushed to the north by this expanding basin (Freund et al., 1975), and the Tethyan ocean crust consumed in a subduction zone which lies to the north of the present Mediterranean Sea (Scandone, 1975) during a compressive phase in the Cretaceous, which is seen as Upper Jurassic to Quaternary compression throughout the Alps. It is possible that the eastern Mediterranean Basin developed between the continent and an orogenic belt as a marginal sea, similar to the western Pacific marginal basins, Sea of Japan (Karig, 1971). Freund et al. (1975) identify a Mesozoic Benioff zone from the high potash igneous rocks of Israel and the Lebanon, which now (after rotation of the landmasses) dips towards the south.

The geophysical data indicate that there is no "normal" oceanic crust in the eastern Mediterranean. Crustal thicknesses vary between 16–19 km in the

Ionian Sea, where a thick sedimentary series of Mesozoic to Tertiary age is recognized, and where in the Gulf of Syrte the basement is clearly continental, to over 20 km in the Levantine Sea, on the basis of gravity and a single seismic refraction profile (Lort, 1972). Refraction results show the widespread occurrence of a "granitic" layer in the velocity range 5.5–6.5 km/sec, and intermediate depths to the Moho, but more data are necessary if this problem is to be solved. In the Levantine Basin a continental margin type of crust is thought to exist, the original ocean having already been consumed.

V. DISCUSSION

The Mediterranean Sea presents diverse characteristics, the chief controls being the location of the individual constituent basins with respect to the Alpine folded belts. As previously described, the basins superimposed on or adjacent to the belt are thought to be Cenozoic in age (western Mediterranean Basin, Tyrrhenian Basin, Aegean Basin, North Cyprus Basins) and they were possibly formed in an island arc situation. The older basins, of Mesozoic to Cenozoic age, have themselves been affected by Alpine folding (Ionian and Levantine Basins, Adriatic Sea).

Features of interest are large offshore nappes, dating from the Upper Cretaceous, which are detected by seismic reflection profiles (Cyprus Arc, Messina Cone, West Gibraltar Arc). These are all situated on active continental margins which surround the entire Mediterranean Sea, the notable exception being the stable margin of Israel.

Evidence of recent tectonic activity is seen throughout the Mediterranean, indicated by the present seismicity, volcanicity, and compressional features, which are recognized from deep seismic studies and from seismic reflection surveys. McKenzie (1972) delineates the plate boundaries in the Mediterranean, showing that in the east between the African and European plates there are sub-plates which are undergoing rapid motion in directions oblique to the north–south approach of the two major plates, with crustal consumption occurring at the Hellenic Arc. In the west the movement is taken up by a lateral translation along the continuation of the Azores–Gibraltar fracture boundary. Palaeomagnetic results indicate that several of the constituent landmasses, particularly in the western Mediterranean, have rotated counterclockwise (Zijderveld and Van der Voo, 1973).

The presence of tectonically emplaced Triassic, Jurassic, and Cretaceous ophiolites through the entire Mediterranean from Oman to Gibraltar indicate the former existence of a Tethyan Ocean, which was thought to lie north of the present eastern Mediterranean Basins, and to have been largely consumed (Smith, 1973). Reconstructions from present plate motions, together with the contribution of geological evidence from land, enable only the recent history of the basins to be deduced. The past history is often masked by recent geological events, which give the different basins a certain uniformity in geological and geophysical character, even though their age and genesis are extremely different.

The crustal structure of the constituent basins of the Mediterranean Sea has already been discussed in Sections II, III, and IVC. In the western Mediterranean the deep Algero-Provençal Basin is oceanic in nature, and the South Balearic and Valencia Basins are of inter-arc basin "intermediate" type with a "soft" mantle. The Tyrrhenian Basin has the thinnest crust in the whole Mediterranean, which is attributed to formation in a marginal sea. In the eastern Mediterranean Sea, the Aegean and North Cyprus Basins have an inter-arc or back-arc "intermediate" crustal structure, while the Levantine Basin has a thick crust, possibly of continental margin development; the Adriatic Sea is continental in nature. The Ionian Basin has an "intermediate" type of crust, thinner than that in the Levantine Sea to the east.

A single common feature through all the basins of the Mediterranean Sea is the widespread occurrence of evaporites in the Upper Miocene. The evaporites are described by Finetti et al. (1970), Finetti and Morelli (1973), Mulder (1973), Biju-Duval et al. (1974), and Sonnenfeld (1975), who also plot their distribution. Salt is not everywhere present, but generally coincides with the areas of deep water abyssal plain, such as the Balearic Basin; halokinetic features are seen particularly in the western Mediterranean Basin, and the Antalya Basin and the Herodotus Basin in the east.

Several hypotheses have been suggested for the formation of the evaporite basins, the depth of deposition being a problem: a deep-water origin (Schmalz, 1969); shallow-water deposition (Ogniben, 1957; Nesteroff, 1973), or deep sea precipitation (Hsü, 1972; Hsü et al., 1973; Ryan, 1973; Ryan et al., 1973). A comprehensive collection of papers on the Messinian events in the Mediterranean is edited by Drooger (1973) and the conflicting ideas described and discussed; Sonnenfeld (1975) also reviews the various evaporite models. It is likely that there existed several basins of deposition which communicated with sources of salt-water supply: the Atlantic through straits similar to the present Straits of Gibraltar, the Red Sea in the east, the Tethys through the Aegean Sea, and so on. Deposition was controlled by topography, which related to structures, and subsequent down-faulting and subsidence probably occurred during the Pliocene and Quaternary. The evolution of pre-evaporite sedimentation is problematic, as the correlation between onshore information and marine seismic reflection profiles is difficult. A contrast between east and west is seen as the sediments in the west are post-tectonic in relation to the surrounding folded belts, while in the east the sediments may include older formations.

It seems likely that evaporite deposition in the Mediterranean differed in its evolution in the west and the east, the alimentation, degree of subsidence, and evaporite facies development varying with the region. The genetic model seems to be that of shallow-water deposition at the margins of shallow to moderately deep basins, there being no catastrophic dessication cycles, but replenishment from a number of different sources and a simultaneous subsidence; inadequate water exchange controls the deposition of the evaporite series (Sonnenfeld, 1975).

In order to establish the evolution of the complex Mediterranean region, it is evident that there still remains much work to be done, in particular, the correlation of additional offshore data with that of geological data on land.

REFERENCES

Abbate, E., Bortoletti, V., Passerini, P., Sagri, M., and Sestini, G., 1970, Development of the northern Apennines Geosyncline, *Sediment. Geol.*, v. 4, p. 3–4.

Alinat, J., Giermann, G., and Leenhardt, O., 1966, Reconnaissance sismique des accidents de terrain en Mer Ligure, *Compt. Rend.*, v. 262 (B), no. 19, p. 1311–1314.

Alla, G., Byramjee, R., and Didier, J., 1972, Structure géologique de la marge continentale du Golfe du Lion, *23rd Congr. CIESM Athens*, Abstract.

Allan, T. D., and Morelli, C., 1971, A Geophysical Study of the Mediterranean Sea, *Boll. Geofis. Teor. Appl.*, v. 13 (50), p. 99–142.

Alvarez, W., 1972, Rotation of the Corsica–Sardinia Microplate, *Nature*, v. 235, p. 103–105.

Alvarez, W., 1973, The application of plate tectonics to the Mediterranean Region, in: *Implications of Continental Drift to the Earth Sciences*, Tarling, D. H., and Runcorn, S. K., eds., London and New York: Academic Press, v. 2.

Alvarez, W., 1974, Sardinia and Corsica, one microplate or two? *Rend. Sem. Fac. Sci., Univ. Cagliari*, p. 1–4.

Alvarez, W., and Gohrbandt, K. H. A., 1970, *Geology and History of Sicily*, Tripoli: Petrol. Expl. Soc. Libya.

Araña, V., and Vegas, R., 1974, Plate tectonics and volcanism in the Gibraltar arc, *Tectonophysics*, v. 24 (3), p. 197–212.

Argand, E., 1924, La tectonique de l'Asie, *C. R. XIIIe Congrès Géologique Internationale*, 1922, p. 171–372.

Arieh, E. J., 1967, Seismicity of Israel and adjacent areas, *Bull. Geol. Surv. Israel.*, v. 43, p. 1–14.

Aubouin, J., Blanchet R., Cadet, J. P., Celet, P., Charuet, J., Chorowitz, J., Cousin, M., and Rampnoux, J. P., 1970, Essai sur la géologie des Dinarides, *Bull. Soc. Géol. France*, Sér. 7, v. 12, p. 1060–1095.

Auzende, J.M., 1971, La marge continentale tunisienne, resultats d'une étude par sismique réflexion, sa place dans le cadre tectonique de la Mediterranée occidentale, *Mar. Geophys. Res.*, v. 1, p. 162–177.

Auzende, J. M., and Olivet, J. L., 1974, Structure of the western Mediterranean Basin in: *The Geology of the Continental Margins*, Burk, C. A., and Drake, C. L., eds., New York: Springer-Verlag, p. 723–731.

Auzende, J. M., Bonnin, J., Olivet, J. L., Pautot, G., and Mauffret, A., 1971, Upper Miocene salt layer in the western Mediterranean Basin, *Nature Phys. Sci.*, v. 230, p. 82–84.

Auzende, J. M., Olivet, J. L., and Bonnin, J., 1972, Une structure compressive au nord de l'Algérie, *Deep Sea Res.*, v. 19, p. 149–155.

Auzende, J. M., Bonnin, J., and Olivet, J. L., 1973a, The origin of the western Mediterranean Basin, *J. Geol. Soc. London*, v. 129, p. 607–620.

Auzende, J. M., Olivet, J. L., and Pautot, G., 1973b, Structural framework of selected regions of the western Mediterranean: Balear Islands, southern prolongation, in: *Initial Reports of Deep Sea Drilling Project*, v. 13, p. 1441–1447.

Baird, D. W., 1970, Review of Mediterranean Alpine tectonics, in: *Geology and History of Turkey*, Campbell, A. G., ed., Petrol. Expl. Soc. Libya, 13th Ann. Field Conf., p. 139–158.

Banghar, A. R., and Sykes, L. R., 1969, Focal mechanisms of earthquakes in the Indian Ocean and adjacent regions, *J. Geophys. Res.*, v. 74, p. 632–649.

Barazangi, M., and Dorman, J., 1969, World seismicity maps compiled from ESSA, crust and geodetic survey, epicenter data 1961–67, *Bull. Seismol. Soc. Amer.* v. 59, p. 369–380.

Barberi, F., Gasparini, P., Innocenti, F., and Villari, L., 1973, Volcanism of the southern Tyrrhenian Sea and its geodynamic implications, *J. Geophys. Res.*, v. 78 (23), p. 5221–5232.

Barberi, F., Civetta, L., Gasparini, P., Innocenti, P., Scandone, R., and Villari, L., 1974, Evolution of a section of the Africa–Europe plate boundary: palaeomagnetic and volcanological evidence from Sicily, *Earth Planet. Sci. Lett.*, v. 22, p. 123–132.

Bayer, R., Le Mouel, J. L., and Le Pichon, X., 1973, Magnetic anomaly pattern in the western Mediterranean, *Earth Planet. Sci. Lett.*, v. 19, p. 168–176.

Belderson, R. H., Kenyon, N. H., and Stride, A. H., 1970, 10 km wide views of Mediterranean deep sea floor, *Deep Sea Res.*, v. 17, p. 267–270.

Belderson, R. H., Kenyon, N. H., and Stride, A. H., 1972, Comparison between narrow-beam and conventional echo-soundings from the Mediterranean Ridge, *Mar. Geol.*, v. 12, p. 11–15.

Belderson, R. H., Kenyon, N. H., and Stride, A. H., 1974a, Calabrian Ridge—a newly associated branch of the Mediterranean Ridge, *Nature*, v. 247, no. 5441, p. 453–454.

Belderson, R. H., Kenyon, N. H., and Stride, A. H., 1974b, Features of submarine volcanoes shown on long-range sonographs, *J. Geol. Soc. London*, v. 130, p. 403–410.

Bellaiche, G., and Recq, M., 1973, Offshore seismological experiments south of Provence, in relation to Glomar Challenger deep sea drillings (JOIDES DSDP, Leg 13), *Mar. Geol.*, v. 15, p. M49–52.

Bellaiche, G., Recq, M., and Rehault, J.-P., 1973a, Nouvelles données sur la structure du haut fond du Méjean obtenues par la "sismique refraction," *C. R. Acad. Sci.*, v. 276 (D), p. 1529–1532.

Bellaiche, G., Gennesseaux, M., Mauffret, A., and Rehault, J.-P., 1973b, Prélèvements systématiques et caractérisation des réflecteurs acoustiques, nouvelle étape dans la compréhension de la géologie de la méditerranée occidentale, *Inst. Franç. Pétrole, Div. Geol.*, Internal Report.

Ben-Avraham, Z., Shoham, Y., and Ginzburg, A., 1976, Magnetic anomalies in the eastern Mediterranean and the tectonic setting of the Eratosthenes Seamount, *J. R. Astr. Soc.*, v. 45, p. 105–123.

Berry, M. J., and Knopoff, L., 1967, Structure of the Upper Mantle under the Western Mediterranean Basin, *J. Geophys. Res.*, v. 72, p. 3613–3626.

Berry, M. J., Knopoff, L., and Mueller, St., 1969, The low velocity channel of the upper mantle under the Mediterranean Basin, *Rapp. Comm. Int. Mer Médit.*, v. 19 (4), p. 669–670.

Biju-Duval, B., 1974, Carte géologique et structurale des bassins tertiares du domaine Méditerranéen, En vente (160 FF), *Rev. Inst. Franç. Pétrole*, v. 29, p. 607–639.

Biju-Duval, B., Letouzey, J., Montadert, L., Courrier, P., Mugniot, J. F., and Sancho, J., 1974, Geology of the Mediterranean Sea Basins, in: *The Geology of Continental Margins*, Burk, C. A., and Drake, C. L., eds., New York: Springer-Verlag, p. 695–723.

Bishop, W. F., 1975, Geology of Tunisia and adjacent parts of Algeria, Libya, *Bull. Amer. Assoc. Petrol. Geol.*, v. 59(3), p. 413-450.

Boccaletti, M., Manetti, P., and Peccarillo, A., 1974, The Balkanids as an instance of back-arc thrust belt; possible relation with the Hellenids, *Bull. Geol. Soc. Amer.*, v. 85, p. 1079–1084.

Boer, J., de, 1965, Palaeomagnetic indications of megatectonic movements in the Tethys, *J. Geophys. Res.*, v. 70, p. 931–944.

Bonini, W. E., Loomis, T. P., and Robertson, J. D., 1973, Gravity anomalies, ultramafic intrusions and the tectonics of the region around the strait of Gibraltar, *J. Geophys. Res.* v. 78, p. 1372–1382.

Bonnin, J., Olivet, J. L., and Auzende, J. M., 1975, Structure en nappe à l'ouest de Gibraltar, *C. R. Acad. Sci. Paris*, v. 280, D(5), p. 559–562.

Bourcart, J., 1960, Carte topographique du fond de la Mediterranée occidentale, *Bull. Inst. Océanogr. Monaco*, no. 1163.

Bousquet, J. C., 1972, *La Tectonique Récente de l'Apennin Calabro-Lunanien dans son Cadre Géologique et Géophysique*, Unpublished thesis, Univ. Montpellier, 172 p.

Brinkmann, R., 1972, Mesozoic troughs and crustal structure in Arabia, *Bull. Geol. Soc. Amer.*, v. 83, p. 819–826.

Brunn, J. H., Dumont, J. F., Ch. de Graciansky, P., Gutnik, M., Juteau, Th., Marcoux, J., Monod, O., and Poisson, A., 1971, Outline of the geology of the Western Taurids, in: *Geology and History of Turkey*, Campbell, A. G., ed., Petrol. Expl. Soc. Libya, 13th Ann. Field Conf., p. 225.

Burollet, P. F., 1967, General geology of Tunisia and tertiary geology of Tunisia, in: *Guidebook to the Geology and History of Tunisia*, Amsterdam: Breumelhof, p. 51–58, 215–225.

Caputo, M., Panza, G. F., and Postpischl, D., 1970, Deep structure of the Mediterranean Basin, *J. Geophys. Res.*, v. 75 (26), p. 4919–4923.

Carey, S. W., 1958, A tectonic approach to Continental drift, in: *Continental Drift, a Symposium*, Hobart: Univ. Tasmania, p. 177–355.

Carter, G. T., Flanagon, J. P., Jones, R. C., Marchant, F. L., Murchison, R. R., Rebman, J. H., Sylvester, J. C., and Whitney, J. C., 1972, A new bathymetric chart and physiography of the Mediterranean Sea, in: *The Mediterranean Sea, a Natural Sedimentation Laboratory*, Stanley, D. J., ed., Stroudsburg: Dowden, Hutchinson and Ross.

Cassinis, G., 1941, La crociera gravimetrica del R. Sommergibile "Des Geneys" anno 1935, *R. Accad. Ital., Red. Fis.*, v. 12 (7), p. 11.

Castany, G., 1956, Essai de synthèse géologique du territoire Tunisie–Sicile, *Ann. Min. Geol.*, v. 16.

Chabrier, G., and Mascle, J., 1975, Comparaison des évolutions géologiques de la Provence et de la Sardaigne, *Rev. Géogr. Phys. Géol. Dyn.* (2), v. 17 (2), p. 121–136.

Channell, J. E. T., and Tarling, D. H., 1975, Palaeomagnetism and the rotation of Italy, *Earth Planet. Sci. Lett.*, v. 25, p. 177–188.

Chaudhury, M., Giese, P., and Visininni, G. de, 1971, Crustal structures of the Alps; some general features from explosion seismology, *Boll. Geofis. Teor. Appl.*, v. 13, p. 211–240.

Closs, H., and Hinz, K., 1974, Seismische und bathymetrische Ergebnisse von Mediterranean Rucken und Hellenischen Graben, *Aeg. Symposium*, Hannover, 1974.

Colombi, B., Giese, P., Luongo, G., Morelli, C., Riuscetti, M., Scarascia, S., Schutte, K. G., Strowald, J., and Visintini, G., 1973, Preliminary report on the seismic refraction profile Gargano–Salerno–Palermo–Pantelleria, *Boll. Geofis. Teor. Appl.*, v. 15, p. 225–254.

Comninakis, P. E., and Papazachos, B. C., 1972, Seismicity of the eastern Mediterranean and some tectonic features of the Mediterranean Ridge, *Bull. Geol. Soc. Amer.*, v. 83, p. 1093–1102.

Constaninescu, L., Ruprechtova, L., and Enescu, D., 1966, Mediterranean earthquake mechanisms and their seismotectonic implications, *Geophys. J. R. Astr. Soc.*, v. 10, p. 347–368.

Cooper, R. I. B., Harrison, J. C., and Willmore, P. L., 1952, Gravity measurements in the eastern Mediterranean, *Phil. Trans. Soc. London*, v. 244, p. 533–559.

Coster, H. P., 1945, The gravity field of the western and central Mediterranean, Wolters, J. B., ed., Groningen-Batavia.

Delteil, J. R., Durand, J., Semichon, P., Montadert, L., Fondeur, C., and Mauffret, A., 1972, Structure géologique de la marge continentale catalone, *23 Congr. CIESM, Athens*, Abstract.

Dewey, J. F., Pitman, W. C., Ryan, W. B. F., and Bonnin, J., 1973, Plate tectonics and the evolution of the Alpine system, *Bull. Geol. Soc. Amer.*, v. 84, p. 3137–3180.

Dragasevic, T., 1969, Investigation of the structural characteristics of the Mohorovicic discontinuity in the area of Yugoslavia, *Boll. Geofis. Teor. Appl.*, v. XI (41–42), p. 57–69.

Drooger, C. W., ed., 1973, *Messinian Events in the Mediterranean*, Amsterdam: North-Holland Publ. Co.

Emery, K. O., Heezen, B. C., and Allan, T. D., 1966, Bathymetry of the eastern Mediterranean Sea, *Deep Sea Res.*, v. 13, p. 173–174.

Ergin, K., Guclu, U., and Uz, Z., 1967, A catalogue of earthquakes for Turkey and surrounding areas, *Publ. Tech. Univ. Istanbul*, v. 24, no. 1.

Erikson, A. J., 1970, *The Measurement and Interpretation of Heat Flow in the Mediterranean and Black Sea*, Unpublished thesis, Massachussetts Institute of Technology.

Ewing, J., and Ewing, M., 1959, Seismic refraction measurements in the Atlantic Ocean basins, in the Mediterranean Sea, on the Mid-Atlantic Ridge and in the Norwegian Sea, *Bull. Geol. Soc. Amer.*, v. 70, p. 291–305.

Fahlquist, D. A., and Hersey, J. B., 1969, Seismic refraction measurements in the western Mediterranean Sea, *Bull. Inst. Océanogr. Monaco*, v. 67, no. 1386, 52 p.

Finetti, I., and Morelli, C., 1972, Wide scale digital seismic exploration of the Mediterranean Sea., *Boll. Geofis. Teor. Appl.*, v. 14 (56), p. 291–342.

Finetti, I., and Morelli, C., 1973, Geophysical exploration of the Mediterranean Sea, *Boll. Geofis. Teor. Appl.*, v. 15 (60), p. 263–340.

Finetti, I., Morelli, C., and Zarudski, E., 1970, Reflection seismic study of the Tyrrhenian Sea, *Boll. Geofis. Teor. Appl.*, v. 12, p. 311–346.

Fleischer, U., 1964, Schwerestörungen im östlichen Mittelmeer nach Messungen mit einem Askania-Secgravimeter, *Son. Deutschen Hyd. Zeitschr.*, v. 17 (14), p. 153–164.

Freund, R., Goldberg, M., Weissbrod, T., Druckmann, Y., and Derin, B., 1975, The Triassic–Jurassic structure of Israel and its relation to the origin of the eastern Mediterranean, *Bull. Geol. Surv. Israel*, v. 65, p. 1–26.

Galanopoulos, A. G., 1969, The seismotectonic regime in Greece, *Phys. Solid Earth*, v. 7, p. 455–460.

Galdeano, A., Courtillot, V., Le Borgne, E., Le Mouel, J. L., and Rossignol, J.-C., 1974, Aeromagnetic survey of the Mediterranean: description and tectonic implications, *Earth Planet. Sci. Lett.*, v. 23 (3), p. 323–336.

Gaskell, J. F., and Swallow, J. C., 1953, Seismic refraction experiments in the Mediterranean and Indian Ocean, *Nature*, v. 172, p. 535.

Gaskell, T. F., Hill, M. N., and Swallow, J. C., 1958, Seismic measurements made by *H.M.S. Challenger* in the Atlantic, Pacific and Indian Oceans and in the Mediterranean Sea 1950–1953, *Phil. Trans. Roy. Soc. London*, v. A.251, p. 23–83.

Gasparini, P., and Adams, J. A. S., 1969, K–Ar dating of Italian Plio-Pleistocene volcanic rocks, *Earth Planet. Sci. Lett.*, v. 6, p. 225–230.

Gass, I. G., 1968, Is the Troodos Massif, Cyprus, a fragment of Mesozoic ocean floor? *Nature*, v. 220, p. 39–42.

Gass, I. G., and Masson-Smith, D., 1963, The geology and gravity anomalies of the Troodos Massif, Cyprus, *Phil. Trans. Roy. Soc. London*, v. A.255, p. 417–467.

Gass, I. G., and Smewing, J. D., 1973, Intrusion, extrusion and metamorphism at constructive margins: evidence from the Troodos Massif, Cyprus, *Nature*, v. 242, p. 26–29.

Gennesseaux, M., Auzende, J. M., Olivet, J. L., and Bayer, R., 1974, Les orientations structurales et magnétiques sous-marines au Sud de la Corse et la dérive corse–sarde, *C. R. Acad. Sci. Sér. D*, v. 278 (16), p. 2003–2006.

Giermann, G., 1961, Erläuterungen zur bathymetrischen Karte der Strasse von Gibraltar, *Bull. Inst. Océanogr. Monaco*, no. 1218.

Giermann, G., 1962, Erläuterungen zur bathymetrischen Karte des westlichen Mittelmeers, *Bull. Inst. Océanogr. Monaco*, no. 1254.

Giermann, G., 1966, Gedanken zur Ostmediterranean Schwelle, *Bull. Inst. Océanogr. Monaco*, v. 66, no. 1362.

Giermann, G., 1969, The Eastern Mediterranean Ridge, *Rapp. Comm. Int. Mer Médit.*, v. 19, no. 4, p. 605–607.

Giermann, G., Pfannenstiel, M., and Wimmenauer, W., 1968, Relations entre morphologie, tectonique et volcanisme en mer d'Alboran (Méditerranée occidentale). Résultats préliminaires de la campagne *Jean Charcot* (1967), *C. R. Somm. Séances Soc. Géol. France*, v. 4, p. 116–117.

Giorgetti, F., and Iaccarino, E., 1971, Seismicity of the Italian region, *Boll. Geofis. Teor. Appl.*, v. 13 (50), p. 143–154.

Giorgetti, F., and Mosetti, F., 1969, General morphology of the Adriatic Sea., *Boll. Geofis. Teor. Appl.*, v. 11, p. 49–56.

Glangeaud, L., 1962, Palaeogéographie de la Méditerranée et de ses bordures. Le rôle des phases ponto-plio-quaternaires, *Océanogr. Géol. Géogr. Médit., Cent. Nat. Rech. Sci.*, p. 125.

Glangeaud, L., Bobier, C., and Bellaiche, G., 1967, Evolution néotectonique de la mer d'Alboran et ses conséquences paléogéographiques, *C. R. Acad. Sci. Paris*, v. 265, p. 1672–1675.

Goncharov, V. P., and Mikhaylov, O. V., 1964, New data concerning the bottom relief of the Mediterranean, *Deep Sea Res.*, v. 11, p. 625.

Gonnard, R., Letouzey, J., Biju-Duval, B., and Montadert, L., 1975, Apports de la sismique réflexion aux problèmes du volcanisme en Méditerranée et à l'interprétation des données magnétiques, *Sciences de la Terre, Montpellier*, Abstract.

Gutenberg, B., and Richter, C. E., 1954, *Seismicity of the Earth and Associated Phenomena*, Princeton, N. J.: Princeton Univ. Press, 2nd ed.

Harrison, J. C., 1955, An interpretation of gravity anomalies in the Eastern Mediterranean, *Phil. Trans. Roy. Soc. London*, v. A.248, p. 283–325.

Heezen, B. C., Gray, C., Segré, A. G., and Zarudski, E. F. K., 1971, Evidence of a foundered continental crust beneath the Central Tyrrhenian Sea, *Nature*, v. 229, p. 327–329.

Hersey, J. B., 1965a, Sedimentary basins of the Mediterranean Sea, in: *Submarine Geology and Geophysics*, Whittard, W. F., and Bradshaw, R., eds., London: Butterworths, Proc. 17th Symp. Colston Res. Soc. 1965, p. 75–91.

Hersey, J. B., 1965b, Sediment ponding in the deep sea, *Bull. Geol. Soc. Amer.*, v. 76, p. 1251–1260.

Hieke, W., 1972, Erste Ergebnisse von stratigraphischen, sedimentologischen und morphologishtektonischen Untersuchungen auf dem Mediterranean Rucken (Ionischen Meer), *Zeitschr. Deutsch. Geol. Ges.*, v. 123, p. 567–570.

Hinz, K., 1970, A low velocity layer in the upper crust of the Ionian Sea, *Rapp. Comm. Int. Mer Médit.*, v. 21 (11), p. 865.

Hinz, K., 1973, Crustal structure of the Balearic Sea, *Tectonophysics*, v. 20, p. 295–302.

Hinz, K., 1974, Results of seismic refraction and seismic reflection measurements in the Ionian Sea, *Geol. J.*, v. 2, p. 35–65.

Hsü, K. J., 1972, Origin of saline geants: a critical review after the discovery of the Mediterranean evaporite, *Earth Sci. Rev.*, v. 8, p. 371–396.

Hsü, K. J., 1973, The desiccated deep-basin model for the Messinian events, in: *Messinian Events in the Mediterranean*, Drooger, C. W., ed., Amsterdam: North-Holland Publ. Co., p. 60-67.

Hsü, K. J., Cita, M. B., and Ryan, W. B. F., 1973, The origin of the Mediterranean evaporites, in: *Initial Reports of Deep Sea Drilling Project*, v. 13 (II), p. 1203–1232.

Hsü, K. J., Montadert, L., Garrison, R. E., Fabricius, F. H., Bernouilli, D., Melieres, F., Kidd, R. B., Muller, C., Cita, M., Bizon, G., Wright, R. and Erikson, A., 1975, Summary of Deep Sea Drilling Project Leg 42A, Scripps Inst. Oceanogr., April 14, 1975.

Jong, K. A., de, Manzoni, M., Stauenga, T., Van Der Voo, R., Van Dijk, F., and Zijderveld, J. D. A., 1973, Rotation of Sardinia: palæomagnetic evidence for rotation during the early Miocene, *Nature*, v. 243, p. 281–283.

Jongsma, D., 1975, *A Marine Geophysical Study of the Hellenic Arc*, Unpublished Ph.D. thesis, Univ. of Cambridge.

Karig, D. E., 1971, Origin and development of marginal basins in the Western Pacific, *J. Geophys. Res.*, v. 76, no. 11, p. 2542–2561.

Karnik, V., 1969, *Seismicity of the European Area, Part 1*, Dordrecht, Netherlands: D. Reidel Publ.

Kenyon, N. H., Stride, A. H., and Belderson, R. H., 1975, Plan views of active faults and other features on the Lower Nile Cone, *Bull. Geol. Soc. Amer.*, v. 86, p. 1733–1739.

Kermaboni, A., Gehin, C., Blavier, P., and Tonarelli, B., 1969, Acoustic and other physical properties of deep-sea sediments in the Tyrrhenian abyssal plain, *Marine Geol.*, v. 7, p. 129–145.

Khan, M. A., Summers, C., Bamford, S. A. D., Chroston, P. N., Poster, C. K., and Vine, F. J., 1972, A reversed seismic refraction line on the Troodos Massif, Cyprus, *Nature Phys. Sci.*, v. 238, no. 87, p. 134–136.

Klootwijk, C. T., and Van den Berg, J., 1975, The rotation of Italy: Preliminary palaeomagnetic data from the Umbrian sequence, Northern Apennines, *Earth Planet. Sci. Lett.*, v. 25, p. 263–273.

Lajat, D., Gonnard, R., and Letouzey, J., 1975, Prolongation dans l'Atlantique de la partie extrême de l'arc béticorifain, *Bull. Soc. Géol. France*, v. 7, no. 17, p. 481–485.

Lapierre, H., 1972, *Les Formations Sédimentaires et Éruptives des Nappes de Mamonia et leurs Relations avec le Massif de Troodos* (Chypre), Doctorat d'état, Univ. Nancy, 420 p.

Le Borgne, E., Le Mouel, J. L., and Le Pichon, X., 1971, Aeromagnetic survey of south-western Europe, *Earth Planet. Sci. Lett.*, v. 12, p. 287–299.

Le Borgne, E., Bayer, R., and Le Mouel, J. L., 1972, La cartographie magnétique de la partie sud du bassin algéro-provençal, *Compt. Rend.*, v. 274, p. 1291–1294.

Leenhardt, O., 1969, Analysis of continuous sounding profiles, *Int. Hydrogr. Rev.*, v. 46, p. 51–80.

Leenhardt, O., 1970, Sondages sismiques continus en Méditerranée occidentale—enrégistrement, analyse, interprétation, *Mém. Inst. Océanogr. Monaco*, No. 1.

Leenhardt, O., Gobert, B., Hinz, K., Hirn, A., Hirschleber, H., Hsü, H. J., Rebuffatti, A., Rudant, J.-P., Rudloff, R., Ryan, W. B. F., Snoek, M., and Steinmetz, L., 1972, Results of the *Anna* cruise. Three north–south profiles through the western Mediterranean Sea, *Bull. Cent. Rech. Pau, SNPA*, v. 6 (2), p. 365–452.

Le Mouel, J., and Le Borgne, E., 1970, Les anomalies magnétiques du Sud-Est de la France et de la Méditerranée Occidentale, *C. R. Acad. Sci. Paris, Sér. D*, v. 271, p. 1348–1350.

Le Pichon, X., Pautot, G., Auzende, J. M., and Olivet, J., 1971, La Méditerranée occidentale depuis l'oligocène, Schèma d'évolution, *Earth Planet. Sci. Lett.*, v. 13, p. 145.

Le Pichon, X., Pautot, G., and Weill, J. P., 1972, Opening of the Alboran Sea, *Nature Phys. Sci.*, v. 236, p. 83–85.

Letouzey, J., Biju-Duval., B., Courrier, P., and Montadert, L., 1974, Nappes de glissement actuelles au front de l'arc calabrais en mer ionienne (d'après la sismique réflection), *2nd Reunion Sciences de la Terre* (*Nancy*), Abstract.

Locardi, E., 1968, Uranium and thorium in the volcanic processes, *Bull. Volcanol.*, v. 31, p. 235–260.

Lort, J. M., 1971, The tectonics of the Eastern Mediterranean: A geophysical review, *Rev. Geophys. Space Phys.*, v. 9 (2), p. 189–216.

Lort, J. M., 1972, *The Crustal Structure of the Eastern Mediterranean*, Unpublished thesis, Univ. Cambridge, 117 p.

Lort, J. M., and Matthews, D. H., 1972, Seismic velocities measured in rocks of the Troodos Igneous complex, *Geophys. J. R. Astr. Soc.*, v. 27 (4), p. 383.

Lort, J. M., Limond, W. Q., and Gray F., 1974, Preliminary seismic studies in the eastern Mediterranean, *Earth Planet. Sci. Lett.*, v. 21, p. 355–366.

Lowrie, W., and Alvarez, W., 1975, Palaeomagnetic evidence for rotation of the Italian peninsula, *J. Geophys. Res.*, v. 80 (11), p. 1579–1592.

Ludwig, W. J., Gunturi, B., and Ewing, M., 1965, Sub bottom reflection measurements in the Tyrrhenian and Ionian Seas, *J. Geophys. Res.*, v. 70 (18), p. 4719–4723.

McKenzie, D. P., 1970, Plate tectonics of the Mediterranean region, *Nature*, v. 226, p. 239–243.

McKenzie, D. P., 1972, Plate tectonics of the Mediterranean region, *Geophys. J. R. Astr. Soc.*, v. 30, p. 109–181.

Maccarrone, E., 1970, Petrography and chemistry of the submarine lava from Seamount 4, south Tyrrhenian Sea, *Boll. Soc. Geol. Ital.*, v. 89, p. 159–180.

Mace, C., 1939, Gravity measurements on Cyprus, *Roy. Astr. Soc., Geophys. Suppl.*, v. 4 (7), p. 473.

Magné, J., and Poisson, A., 1974, Présence de niveaux oligocènes dans les formations sommitales du Massif de Bey Daglary près de Korkuteli et de Bucak (autochtone du Taurus, Lycien, Turquie), *Compt. Rend.*, v. 278, p. 205–208.

Makris, J., 1973, Some geophysical aspects of the evolution of the Hellenides, *Bull. Geol. Soc. Greece*, v. 10, p. 206–213.

Makris, J., 1975, Crustal structure of the Aegean Sea and the Hellenides obtained from geophysical surveys, *J. Geophys.*, v. 41, p. 441–443.

Maldonado, A., and Stanley, D. J., 1975, Nile Cone lithofacies and definition of sediment sequences, *IX Int. Congr. Sed.*, Nice, 1975.

Maley, T. S., and Johnson, G. L., 1971, Morphology and structure of the Aegean Sea, *Deep Sea Res.*, v. 18, p. 109–122.

Malovitskiy, J. P., Emelyanov, E. M., Kazakov, O. V., Moskalenko, V. N., Osipov, G. V., Shimkus, K. M., and Chumakov, I. S., 1975, Geological structure of the Mediterranean Sea floor (based on geological–geophysical data), *Mar. Geol.*, v. 18, p. 231–261.

Marcoux, J., 1970, Age Carnien de termes effusifs du cortège ophiolitique des nappes d'Antalya (Taurus lycien oriental, Turquie), *C. R. Acad. Sci. Paris*, v. 271, p. 285–287.

Martin, L., 1967, *Geology and History of Libya*, Petrol. Expl. Soc. volume, Amsterdam: Breumelhof N. V.

Matthews, D. H., 1974, Bathymetry and magnetic anomalies of the Eastern Mediterranean, *CIESM Congr., December, 1974, Monaco*, Abstract.

Matthews, D. H., Lort, J. M., Vertue, T., Poster, C. K., and Gass, I. G., 1971, Seismic velocities at the Cyprus autocrop, *Nature Phys. Sci.*, v. 231, no. 26, p. 200–201.

Mauffret, A., Fail, J. P., Montadert, L., Sancho, J., and Winnock, E., 1973, Northwestern Mediterranean sedimentary basin from a seismic reflection profile, *Bull. Amer. Assoc. Petrol. Geol.*, v. 57, p. 2245–2262.

Menard, H. W., 1967, Transitional types of crust under small ocean basins, *J. Geophys. Res.*, v. 72, p. 2061–3073.

Menard, H. W., Smith, S. M., and Pratt, R. M., 1965, The Rhone deep-sea fan, in: *Submarine Geology and Geophysics*, Whittard, W. F., and Bradshaw, R., eds., London: Butterworths, Proc. 17th Symposium Colstan Res. Soc., p. 271–285.

Mercier, J., Bousquet, B., Delibasis, N., Drakapoulos, I., Keraudren, B., Lemeille, F., and Sorel, D., 1972, Déformations en compression dans le Quaternaire des rivages ioniens (Céphalonie, Grèce) données néotectoniques et sismiques, *Compt. Rend.*, v. 275D, p. 2307–2310.

Mikhaylov, O. V., 1965, The relief of the Mediterranean sea-bottom, in: *Basic Features of the Geological Structures of the Hydrological Regime and Biology of the Mediterranean*, Fomin, L. M., ed., Moscow: Nauka, 224 p.

Miyashiro, A., 1973, The Troodos ophiolite complex was probably formed in an island arc, *Earth Planet. Sci. Lett.*, v. 19, p. 218–224.

Montadert, L., Sancho, J., Fail, J., Debyser, J., and Winnock, E., 1970, De l'âge tertiaire de la série salifère responsable des structures diapiriques en Mediterranée Occidentale (Nord-Est des Baléares), *C. R. Acad. Sci. Paris*, v. 271, p. 812–815.

Moores, E. M., and Vine, F. J., 1971, The Troodos Massif, Cyprus, and other ophiolites as oceanic crust: evaluation and implications, *Phil. Trans. Roy. Soc. London*, v. A.268, p. 443–466.

Morelli, C., 1970, Physiography, gravity and magnetism of the Tyrrhenian Sea, *Boll. Geofis. Teor. Appl.*, v. 12 (48), p. 274–308.

Morelli, C., 1975, Geophysics of the Mediterranean, *Bull. CIESM*, v. 7, p. 27–111.

Morelli, C., Carozzo, M. T., Ceccherini, P., Finetti, I., Gantar, C., Pisani, M., and Schmidt Di Friedberg, P., 1969, Regional geophysical study of the Adriatic Sea, *Boll. Geofis. Teor. Appl.*, v. 11 (41-42), p. 8–56.

Morelli, C., Pisani, M., and Gantar, C., 1975, Geophysical studies in the Aegean Sea and in the eastern Mediterranean, *Boll. Geofis. Teor. Appl.*, v. 66.

Moskalenko, V. N., 1966, New data on the structure of sedimentary strata and basement in the Levant Sea, *Oceanology*, v. 6, p. 828–836.

Moskalenko, V. N., 1967, Study of the sedimentary series of the Mediterranean Sea by Seismic methods, *Proc. Acad. Sci. USSR Oc. Comm.* (Transl. 1969).

Moskalenko, V. N., and Yelnikov, I. N., 1966, Seismic data about the possible continuation of the African platform in the Crete–African region of the Mediterranean Sea, *Int. Oceanogr. Congr.*, v. 2, p. 259–260.

Mulder, C. J., 1973, Tectonic framework and distribution of Miocene evaporites in the Mediterranean, in: *Messinian Events in the Mediterranean*, Drooger, C. W., ed., Amsterdam: North-Holland Publ. Co., p. 44–59.

Muraour, P., Marchand, J. P., Ducrot, J., and Ceccaldi, X., 1966, Remarques sur la structure profonde du précontinent de la région de Calui (Corse) à la suite d'une étude de sismique réfraction, *Compt. Rend.*, v. 262, p. 17–19.

Nairn, A. E. M., and Westphal, M., 1967, A second virtual pole from Corsica, the Ota Gabbrodiorite, *Palaeogeogr. Palaeoclimat. Palaeoecol.*, v. 3, p. 277–286.

Nairn, A. E. M., and Westphal, M., 1968, Possible implications of the palaeomagnetic study of Late Palaeozoic igneous rocks of north-western Corsica, *Palaeogeogr. Palaeoclimat. Palaeoecol.*, v. 5, p. 179–204.

Neev, D., Almagor, G., Arad, A., Ginzburg, A., and Hall, J. K., 1973, The geology of the southeastern Mediterranean Sea, Israel Geol. Survey, Jerusalem, Rep. MG/73/5, 43 p.

Neprochnov, Y. P., 1968, Structure of the earth's crust of epicontinental seas: Caspian, Black and Mediterranean, *Can. J. Earth Sci.*, v. 5, p. 1037–1043.

Nesteroff, W. D., 1973, Un modèle pour les évaporites messiniennes en Méditerranée: des bassins peu profonds avec dépot d'évaporites lagunaires, in: *Messinian Events in the Mediterranean*, Drooger, C. W., ed., Amsterdam: North-Holland Publ. Co.

Nicholls, I. A., 1971, Santorini Volcano, Greece—Tectonic and petrochemical relationships with volcanoes of the Aegean region, *Tectonophysics*, v. 11, p. 377–385.

Ninkovitch, D., and Hays, J. D., 1972, Mediterranean island arcs and origin of high potash volcanoes, *Earth Planet. Sci. Lett.*, v. 16, p. 331–345.

North, R. G., 1974, Seismic slip rates in the Mediterranean and Middle East, *Nature*, v. 252, no. 5484, p. 560–563.

Ogniben, L., 1957, Petrographia della serie solsifera-siciliana e considerazioni geoteche relative, *Mem. Descrit. Carta. Geol. Ital.*, v. 33, 275 p.

Olivet, J. L., Pautot, G., and Auzende, J. M., 1973a, Structural framework of selected regions of the western Mediterranean: Alboran Sea, in: *Initial Reports of Deep Sea Drilling Project*, v. 13, p. 1417–1430.

Olivet, J. L., Auzende, J. M., and Bonnin, J., 1973b, Structure et évolution tectonique du bassin d'Alboran, *Bull. Soc. Géol. France*, Sér. 7, v. 15, p. 77–90.

Ozelçi, H. F., 1971, Gravity anomalies of the eastern Mediterranean, *Bull. Min. Res. Expl. Inst. Turkey*, v. 80, p. 54–92.

Papazachos, B. C., 1969, Phase velocities of Rayleigh waves in south-eastern Europe and the eastern Mediterranean Sea, *Pure Appl. Geophys.*, v. 75 (4), p. 47–55.

Papazachos, B. C., 1973, Distribution of seismic foci in the Mediterranean and surrounding area and its tectonic implication, *Geophys. J. R. Astr. Soc.*, v. 33, p. 421–430.

Papazachos, B. C., 1974, Seismotectonics of the eastern Mediterranean area, Eng. Seism. & Earthquake Eng., *NATO Adv. Studies Inst. Ser.*, Ser. E, v. 3, p. 1–32.

Papazachos, B. C., and Comninakis, P. E., 1971, Geophysical and tectonic features of the Aegean arc, *J. Geophys. Res.*, v. 76, p. 8517–8533.

Papazachos, B. C., and Delibasis, N. D., 1969, Tectonic stress field and seismic faulting in the area of Greece, *Tectonophysics*, v. 7 (3), p. 231–255.

Pautot, G., Auzende, J. M., Olivet, J. L., and Mauffret, A., 1973, Structural framework of selected regions of the western Mediterranean: Valencia Basin, in: *Initial Reports of Deep Sea Drilling Project*, v. 13, p. 1430–1441.

Payo, G., 1967, Crustal structure of the Mediterranean Sea by surface waves, Part I: Group velocity, *Bull. Seism. Soc. Amer.*, v. 57 (2), p. 151–172.

Payo, G., 1969, Crustal structure of the Mediterranean Sea by surface waves, Part II: Phase velocity and travel times, *Bull. Seism. Soc. Amer.*, v. 59 (1), p. 23–42.

Pearce, J. A., and Cann, J. R., 1973, Tectonic setting of basic volcanic rocks determined using trace element analyses, *Earth Planet. Sci. Lett.*, v. 19, p. 290–300.

Peterschmitt, E., 1956, Quelques donnés nouvelles sur les séismes profond de la mer Tyrrhénienne, *Ann. Geofis. Rome*, v. 9, p. 305–334.

Pfannenstiel, M., 1960, Erläuterungen zu den bathymetrischen Karten des östlichen Mittelmeeres, *Bull. Inst. Océanogr. Monaco*, no. 1192, 60 p.

Poster, C. K., 1973, Ultrasonic velocities in rocks from the Troodos Massif, Cyprus, *Nature Phys. Sci.*, v. 243, p. 2.

Rabinowitz, P. D., and Ryan, W. B. F., 1970, Gravity anomalies and crustal shortening in the eastern Mediterranean, *Tectonophysics*, v. 10, p. 585–608.

Recq, M., 1972, Profils de réfraction en Ligurie, *Pure Appl. Geophys.*, v. 101, p. 155–161.

Recq, M., 1973*a*, Contribution à l'études de la structure profonde de la crôute terrèstre dans la région de Nice, *Boll. Geofis. Teor. Appl.*, v. 15 (58), p. 161–180.

Recq, M., 1973*b*, The P_N velocity under the Gulf of Genoa, *Earth Planet. Sci. Lett.*, v. 20, p. 447–450.

Recq, M., 1974, Contribution à l'étude de l'évolution des marges continentales du Golf de Gènes, *Tectonophysics*, v. 22, p. 363–375.

Rehault, J. P., Olivet, J. L., and Auzende, J. M., 1974, Le bassin nord occidentale Méditerranéen: structure et évolution, *Bull. Géol. Soc. France*, (3), v. 16, p. 281–294.

Ritsema, A. R., 1969, Seismic data of the west Mediterranean and the problem of oceanization, in: *Symposium on the Problems of Oceanisation in the Western Mediterranean, Verhand. Kon. Ned. Geol. Mij. Gen.*, v. 26, p. 105–120.

Robertson, A. H. F., and Hudson, J. D., 1974, Pelagic sediments in the Cretaceous and Tertiary history of the Troodos Massif, Cyprus, in: Hsü, K. J., and Jenkyns, eds., Spec. Publ. Int. Assoc. Sedimentol., v. 1, p. 403–436.

Roever, W. P., de, 1969, Genesis of the western Mediterranean Sea, in: *Symposium on the Problem of Oceanisation in the Western Mediterranean, Verhand. Kon. Ned. Geol. Mij. Gen.*, v. 26, p. 9.

Ross, D. A., Uchupi, E., Prada, K. E., and MacIlvane, J. C., 1974, Bathymetry and microtopography of Black Sea, in: *The Black Sea—Geology, Chemistry and Biology*, Mem. Amer. Assoc. Petrol. Geol. 20, p. 1–11.

Ryan, W. B. F., 1969, *The Stratigraphy of the Eastern Mediterranean, Parts I and II*, Ph.D. thesis, Columbia Univ., New York.

Ryan, W. B. F., 1973, Geodynamic implications of the Messinian crisis of salinity, in: *Messinian Events in the Mediterranean*, Drooger, C. W., ed., Amsterdam: North-Holland Publ. Co., 272 p.

Ryan, W. B. F., and Heezen, B. C., 1965, Ionian Sea submarine canyons and the Messina Turbidity current, *Bull. Geol. Soc. Amer.*, v. 76, p. 915–932.

Ryan, W. B. F., Workum, F., and Hersey, J., 1965, Sediments on the Tyrrhenian Abyssal Plains, *Bull. Geol. Soc. Amer.*, v. 76, p. 1261–1282.

Ryan, W. B. F., Ewing, M., and Ewing, T. I., 1966, Diapirism in the sedimentary basins of the Mediterranean Sea, *Trans. Amer. Geophys. Union*, v. 47, p. 120.

Ryan, W. B. F., Stanley, D. J., Hersey, J. B., Fahlquist, D. A., and Allan, T. D., 1970, The tectonics and geology of the Mediterranean Sea, in: *The Sea*, Maxwell, A. E., ed., New York: Wiley–Interscience, v. IV, Pt. 2.

Ryan, W. B. F., Hsü, K. J., Nesteroff, W., Pautot, G., Wezel, F. C., Lort, J. M., Cita, M. B., Maync, W., Stradner, H., and Dumitrica, P., 1973, *Initial Reports of Deep Sea Drilling Project*, v. 13, Parts I and II, 1447 p.

Sancho, J., Letouzey, J., Biju-Duval, B., Courrier, I. P., Montadert, L., and Winnock, E., 1973, New data on the structure of the eastern Mediterranean Basin from seismic reflection data, *Earth Planet. Sci. Lett.*, v. 18, p. 189–204.

Scandone, P., 1975, The seaways and the Jurassic Tethys Ocean in the central Mediterranean area, *Nature*, v. 256, no. 5513, p. 117–119.

Scandone, P., Giunta, G., and Liguori, V., 1974, The connection between the Apulia and Sahara continental margins in the southern Apennines and in Sicily, *Cong. Ass. CIESM, Monaco*, December 1974, Abstract.

Scheidegger, A. E., 1970, The tectonic stress and tectonic motion direction in Europe, western Asia as calculated from earthquake fault plane solutions, in: *Theoretical Geomorphology*, 2nd ed., Scheidegger, A. E., ed., New York: Springer-Verlag.

Schmalz, R. F., 1969, Deep water evaporite deposition: a genetic model, *Bull. Amer. Assoc. Petrol. Geol.*, v. 53 (4), p. 798–822.

Segré, A. G., 1958, La morfologia del mare Tirreno secondo i più recenti studi, *Riv. Geogr. Ital.*, v. 65, p. 137–143.

Selli, R., and Fabbri, A., 1971, Tyrrhenian: a Pliocene deep sea, *Atti. Accad. Naz. Lincei*, Sér. 8, v. 50, p. 579–592.

Sigl, W., Hinz, K., and Garde, S., 1973, "Hummocky and rolling landscape" in the Ionian Sea—A contribution to the cobblestone problem, *Meteor-Forschungs. Ergebnisse*, Reihe C, No. 14, p. 51–54.

Smith, A. G., 1971, Alpine deformation and the oceanic areas of the Tethys, Mediterranean and Atlantic, *Bull. Geol. Soc. Amer.*, v. 82, p. 2039–2070.

Smith, A. G., 1973, The so-called Tethyan ophiolites, in: *Implications of Continental Drift to the Earth Sciences*, Tarling, D. H., and Runcorn, S. K., eds., London and New York: Academic Press, v. 2, p. 777–986.

Soffel, H., 1972, Anticlockwise rotation of Italy between Eocene and Miocene, palaeomagnetic evidence, *Earth Planet. Sci. Lett.*, v. 17, p. 207–210.

Sonnenfield, P., 1975, The significance of Upper Miocene (Messinian) evaporites in the Mediterranean Sea, *J. Geol.*, v. 83, p. 287–311.

Stanley, D. J., ed., 1972, *The Mediterranean Sea: A Natural Sedimentation Laboratory*, Stroudsburg: Dowden, Hutchinson and Ross, 765 p.

Stanley, D. J., and Mutti, E., 1968, Sedimentological evidence for an emerged land mass in the Ligurian Sea during the Palaeogene, *Nature*, v. 218, p. 32–36.

UNESCO, 1972, International geological map of Europe and adjacent Mediterranean region 1:5,000,000.

Van Bemmelen, R. W., 1969, Origin of the Western Mediterranean Sea, in: *Symposium on the Problem of Oceanisation in the Western Mediterranean*, *Verhand. Kon. Ned. Geol. Mij. Gen.*, v. 26, p. 13–52.

Van der Voo, R., 1968, Palaeomagnetic and the Alpine tectonics of Eurasia, part IV, Jurassic, Cretaceous and Eocene pole positions from N. E. Turkey, *Tectonophysics*, v. 6, p. 251–269.

Van der Voo, R., 1969, Palaeomagnetic evidence for the rotation of the Iberian Peninsula, *Tectonophysics*, 7 (1), p. 5–56.

Van der Voo, R., and French, R. B., 1974, Apparent polar wandering for Atlantic bordering continents: Late Carboniferous to Eocene, *Earth Sci. Rev.*, v. 10, p. 99–120.

Van der Voo, R., and Zijderveld, J. D. A., 1969, Palaeomagnetism in the Western Mediterranean area, *Verhand Kon. Ned. Geol. Mij. Gen.*, v. 26, p. 121–137.

Van Straaten, L. M. J. U., 1965, Sedimentation in the northwestern part of the Adriatic Sea, in: *Submarine Geology and Geophysics*, Whittard, W. F., and Bradshaw, R., eds., London: Butterworths, p. 143–162.

Vilminot, J.-C., and Robert, U., 1974, A propos des relations entre le volcanisme et la tectonique en Mer Eegée, *C. R. Acad. Sci. Paris*, v. 278, p. 2099–2102.

Vine, F. J., and Moores, E. M., 1969, Palaeomagnetic results for the Troodos Igneous Massif, Cyprus, *Trans. Amer. Geophys. Union*, v. 50, p. 131.

Vine, F. J., Poster, C. K., and Gass, I. G., 1973, Aeromagnetic survey of the Troodos Igneous Massif, Cyprus, *Nature Phys. Sci.* v. 244, p. 34–38.

Vogt, P. R., and Higgs, R. H., 1969, An Aeromagnetic survey of the Eastern Mediterranean Sea and its interpretations, *Earth Planet. Sci. Lett.*, v. 5, p. 439–448.

Vogt, P. R., Higgs, R. H., and Johnson, G. L., 1971, Hypothesis on the origin of the Mediterranean Basin: magnetic data, *J. Geophys. Res.*, v. 76 (14), p. 3207–3228.

Watson, J. A., and Johnson, G. L., 1969, The Marine geophysical survey in the Mediterranean, *Int. Hydrogr. Rev.*, v. 46, p. 81–107.

Weigel, W., 1974, Crustal structure under the Ionian Sea, *J. Geophys.*, v. 40, p. 137–140.

Westphal, J. D. A., Bardon, C., Bossert, A., and Hamzeh, R., 1973, A computer fit of Corsica and Sardinia against southern France, *Earth Planet. Sci. Lett.*, v. 18, p. 137–140.

Wezel, F. C., 1967, Sedimentological characteristics of some Italian turbidites, *Estr. Geol. Romana*, v. 6, p. 396–403.

Wezel, F. C., 1970, Numidian Flysch: an Oligocene–early Miocene continental rise deposit off the African platform, *Nature*, v. 228, p. 275–276.

Wezel, F. C., and Ryan, W. B., 1971, Flysch, margini continentali e zolle litosferiche, *Boll. Soc. Geol. Ital.*, v. 90, p. 249–268.

Wong, H. K., and Zarudski, E. F. K., 1969, Thickness of unconsolidated sediments in the eastern Mediterranean Sea., *Bull. Geol. Soc. Amer.*, v. 80, p. 2611–2614.

Wong, H. K., Zarudski, E. F. K., Knott, S. T., and Hays, E. E., 1970, Newly discovered group of diapiric structures in western Mediterranean, *Bull. Amer. Assoc. Petrol. Geol.*, v. 54 (11), p. 2200–2204.

Wong, H. K., Zarudzki, E. F. K., Phillips, J. D., and Giermann, G. K. F., 1971, Some Geophysical Profiles in the eastern Mediterranean, *Bull. Geol. Soc. Amer.*, v. 82 (1), p. 91–100.

Woodside, J., 1975, *The Evolution of the Eastern Mediterranean*, Unpublished Ph.D. thesis, Univ. Cambridge.

Woodside, J., and Bowin, C., 1970, Gravity anomalies and inferred crustal structure in the Eastern Mediterranean Sea, *Bull. Geol. Soc. Amer.*, v. 81, p. 1107–1122.

Wright, D. W., 1975, *Variable Angle Seismic Experiments in the Eastern Mediterranean and Northeastern Atlantic*, Unpublished Ph.D. thesis, Univ. Cambridge.

Yelnikov, I. N., 1966, Seismic reflection surveys in the Mediterranean between Crete and Africa, *Akad. Nauk SSSR, Inst. Okaenol.* (Transl.), v. 5.

Yemel'Yanov, Ye M., Mikhaylov, O. V., Shimkus, K. M., and Moskalenko, V. N., 1968, Some special features of the geomorphological structure and tectonic development of the Mediterranean Sea, *Acad. Sci. USSR, Oc. Comm.* (Transl.), p. 34).

Zarudski, E. F. K., Kin Wong, and Phillips, J. D., 1969, Structure of the eastern Mediterranean, *Trans. Amer. Geophys. Union.*, v. 50, p. 208.

Zijderveld, J. D. A., and Jong, K. A., de, 1969, Palaeomagnetism of some late Palaozoic and Triassic rocks from the eastern Lombardic Alps, Italy, *Geol. Mijb.*, v. 48 (6), p. 559–564.

Zijderveld, J. D. A., and Van der Voo, R., 1973, Palaeomagnetism in the Mediterranean area, in: *Implications of Continental Drift to the Earth Sciences*, Tarling, D. H., and Runcorn, S. K., eds., London and New York: Academic Press, v. 1, p. 133–161.

Zijderveld, J. D. A., Jong, K. A., de, and Van Der Voo, R., 1970, Rotation of Sardinia: Palaeomagnetic evidence from Permian rocks, *Nature*, v. 226, p. 933–934.

Zijderveld, J. D. A., Hazen, G. J. A., Nardin, M., and Van der Voo, R., 1971, Shear in the Tethys and the Permian Palaeomagnetism in the southern Alps, including new results, *Tectonophysics*, v. 10 (5/6), p. 639.

Chapter 5A

THE DINARIC AND AEGEAN ARCS:
THE GEOLOGY OF THE ADRIATIC

Paul Celet

Département de Géologie Dynamique
U. E. R. des Sciences de la Terre
Université de Lille
Villeneuve d'Ascq, France

I. INTRODUCTION

The special character of the Adriatic Sea is conferred by its girdle of Alpine mountains. The axes of two chains mark its edges, the Apennines separating it from the western Mediterranean, and the Dinaric Alps separating it from the Pannonian Basin and Eastern Alps. Closed to the northwest by the arc of the southern Alps and closely linked with this structure, the Adriatic communicates with the Ionian Sea through the narrow Straits of Otranto (Fig. 1). The eastern Mediterranean, to which the Adriatic appears attached like a long appendage, opens southeastward to two abyssal plains, the Ionian Basin off Calabria and Greece and the Levantine Basin between Egypt and Turkey.

The Adriatic Sea which extends for about 800 km between latitudes 40–45°N is relatively narrow, its width lying between 120 and 200 km. Morphologically, the floor of the Adriatic is asymmetrical. Longitudinally, there are two hydrographic domains separated by the Gargano–Dubrovnik Ridge. The northernmost of these is shallow, its depth seldom exceeding a few hundred meters, and is floored by southeasterly dipping alluvial deposits, a continuation of the Po plain sediments, which can be traced along the Italian coast as far as Gargano. Two Dalmatian provinces, according to Giogetti *et al.* (in Morelli *et al.*, 1969), can be recognized along the Yugoslavian littoral between Istria and Dubrovnik, and two Adriatic provinces, a northern and a central, on either side of a transverse line from San Benedetto to Zadar. The deep sub-Adriatic depression opening into the Ionian Sea forms the southern domain. The Adriatic Sea and the peri-Adriatic region form an integral part of the Alpine system; inserted between the Apennines and Dinarides their elongation is controlled by the major structural trend of these chains.

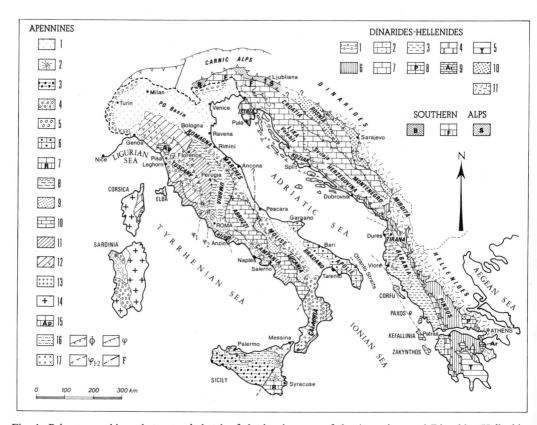

Fig. 1. Paleogeographic and structural sketch of the isopic zones of the Apennines and Dinarides–Hellenides. *Apennines*: 1. Plio-Quaternary basins (Po, Adriatic, and Bradano Basins), 2. recent volcanism, 3. Mio-Pliocene and Quaternary basins of Sicily, 4. Tuscan Neogene intramontane basin, 5. peri-Adriatic Paleogene depressions (Venetian foredeep, Italo-Dinaride foredeep, Polan slope), 6. Apulia and Gargano platform, 7. Ragusa platform (Sicily), 8. Marche–Umbria zone, 9. Lucanian zone, 10. Abruzzi–Campania limestone platform, 11. Tuscan domain, 12. Ligurian domain, 13. Calabrian basement, 14. Corso-Sardinian basement, 15. Apuane window, 16. Sicilian flysch zone, 17. Peloritan crystalline and limestone chains. *Dinarides–Hellenides*: 1. Dalmatian–Albanian foredeep, 2. pre-Apulian zone, 3. Ionian zone, 4. Istro-Dalmatian platform and Gavrovo Ridge, 5. Tripolitza zone, 6. Budva–Cukali zone and Pindus Trough, 7. High Karst platform, 8. Parnassos zone, 9. Argolid Series, 10. Bosnian and Beotian zone, 11. internal Dinarides and Hellenides. *Southern Alps*: B. Belluno Trough, F. Frioulane Ridge, S. Slovene Trough (Aubouin, 1960, 1963a,b; Cousin, 1970). *Tectonic lines*: Φ. Overthrust front, φ. major thrust, φ_1. Falterona–Trasimeno thrust front, φ_2. Tuscan nappe thrust front, F. principal faults.

The birth of the Adriatic is closely related to the history of the Alpine geosyncline and of the Tethys, the ocean which originally separated the European and African plates. This history had its origin at the beginning of the Mesozoic, a time when the Adriatic region may still have been linked to the African continent, forming an embryonic marginal Mesogean Sea (Argyriadis, 1975). The closure of the Tethys brought about by the approach of the continents concomitant with the opening of the Atlantic resulted in compressional and tensional phases after the end of the Jurassic. This radically modified the paleogeography of the Alpine geosyncline (Dewey *et al.*, 1973; Smith 1971).

The Adriatic region is bounded by an ophiolitic belt characteristic of the internal thrust zones of the Apennines and Dinarides. The ophiolites, tectonically emplaced during different orogenic phases of the Jurassic and Cretaceous, were remobilized during Cenozoic tangential movements. If ophiolites are regarded as of oceanic origin, then the Adriatic, which may be regarded as a Mesozoic continental plate or sub-plate, was surrounded by the Tethys. Seismological data indicate the existence of a circum-Adriatic seismic belt (McKenzie, 1972) outlining the present sub-plate, the area of which is probably much less than that of the old continental Adriatic.

In this article the structural and paleogeographic development of the Adriatic will be detailed for the Alpine geosynclinal cycle, that is during the Mesozoic–Tertiary. The Paleozoic cycle and orogenies, although their existence is suggested by deep geophysical exploration, are not considered here.

The configuration of the present Adriatic with respect to the distribution of the isopic zones all around its margins makes possible a limited extrapolation of the geological character of the sea floor. The sea floor, better known now than only a few years ago, continues to yield its secrets to the expanding program of drilling off both the Italian (Vercellino, 1970) and Yugoslav coasts (Simpozij Zadar, 1971). The Dinaric trend parallels the Dalmatian coast, forming a single 600-km zone; the Italian coastline, in contrast, cuts obliquely across several Apennine zones, ranging from the external miogeosynclinal zone to the Apulian foreland.

The first part of the following geological review will be concerned with the survey of the Dinaric and Apennine margins and their extension beneath the Adriatic. The second part will be concerned with the post-geosynclinal development of the basin and the geodynamic character of the Adriatic.

II. STRUCTURAL AND PALEOGEOGRAPHIC EVOLUTION OF THE TETHYSIAN ADRIATIC

Within the Italo-Dinaric area (Aubouin, 1960) at the beginning of the Mesozoic, a vast shallow sea occupied a region bounded by troughs to the northeast and southwest (Bernouilli, 1971). It was an area of dolomitic limestone and evaporite deposition bordered by two oceanic areas of pelagic sedimentation which later became the sites of ophiolite emplacement (internal Serbian and Ligurian zones). The Dinaric carbonate platform underwent a complex evolution during the 200 m.y. from the Triassic to the Miocene (Jelaska, 1973), in the course of which the platforms and basins of an external geosyncline with continuous or nearly continuous sedimentation can be distinguished from the internal ridges and troughs affected by precursory (Jurassic–Cretaceous) tectonism.

The facies distribution has led the majority of authors to consider the Mesozoic–Cenozoic history of the peri-Adriatic region in terms of a number of isopic zones, each characterized by sedimentary features reflecting depositional and environmental conditions. The paleogeography established at the beginning of the Mesozoic shows the development of elongate zones on both sides of the Adriatic. The present trends, NW–SE Dinaric trend in the Balkans and NNW–SSE Apennine trend in peninsular Italy, are not parallel as a result of subsequent deformation. These external zones can be divided from east to west as follows (Fig. 2): in the Dinarides—High Karst zone, Budva zone, Dalmatian zone, Ionian zone; in the Apennines—Apulian zone, Umbria zone, Abruzzi–Campania zone.

All of these zones more or less directly affect the Adriatic. They either form part of the margin or are prolonged under the sea where the nature of the paleogeographic extensions and their structural character are hypothetical. They will be described in their present order of outcrop, symmetrical about an axis which may lie in the Apulian zone in the Adriatic (Fig. 3) itself or in the Molise Trough (Aubouin, 1963a).

A. The Dinaric Geosynclinal Cycle

In the Yugoslavian Dinarides northwest of a line from Pec–Scutari, in Montenegro to Istria, two carbonate regions, the High Karst and the Dalmatian zone, are developed and overlap. They are separated along the littoral southeast of Dubrovnik by a narrow pelagic zone, the Budva zone.

1. *The High Karst Platform* (Fig. 2A)

Well developed in Montenegro, this zone, defined by Kosmatt (1924), has a wide extent in Herzegovina and extends into Croatia (Lika) and Slovenia towards Postojna, where it probably passes into the southern Calcareous Alps (Frioul) (Aubouin, 1963a).*

---

* To the southeast across the Scutari ridge, limestone zones with a similar paleogeographic significance and presumably homologous are known as the Albanian Alps and the Parnassian zone (in Greece).

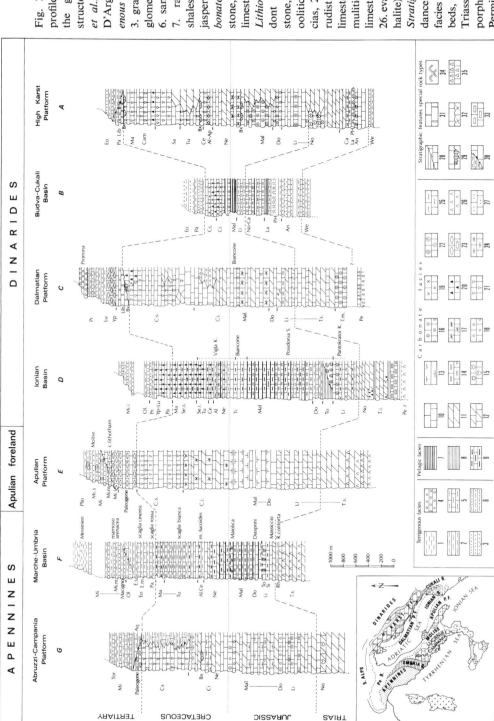

Fig. 2. Simplified stratigraphic profiles and correlations during the geosynclinal phase (reconstructed mostly after Aubouin et al., 1970; Sestini, 1970; and D'Argenio et al., 1971). *Terrigenous facies*: 1. shale, 2. marl, 3. graywacke sandstones, 4. conglomerate, 5. marly limestone, 6. sandy pelites. *Pelagic facies*: 7. radiolarites, 8. radiolarite shales, 9. limestones with chert, jasper, and radiolarites. *Carbonate facies*: 10. compact limestone, 11. dolomite, 12. dolomitic limestone, 13. reef limestone, 14. *Lithiotis* limestone, 15. megalodont limestone, 16. algal limestone, 17. coral limestone, 18. oolitic limestone, 19. microbreccias, 20. bioclastic limestone, 21. rudist limestone, 22. ammonitic limestone, 23. alveoline and nummulitic limestone, 24. orbitoline limestone, 25. nodular limestone, 26. evaporites (anhydrite, gypsum, halite), 27. sandy limestone. *Stratigraphic symbols*: 28. discordance, 29. bauxite, 30. lateral facies passage, 31. Liburnian beds, 32. gap, concordant, 33. Triassic bituminous limestone, 34. porphyry and porphyritic tuff, 35. Permian fusulinid breccia.

Stratigraphy. In the southern part of the High Karst zone detrital red sandstones of Lower Triassic age are found, a lithology also known in the Werfenian of more internal zones. These are followed by limestones or dolomites capped by a nodular Anisian limestone with ammonites (Han–Bulog

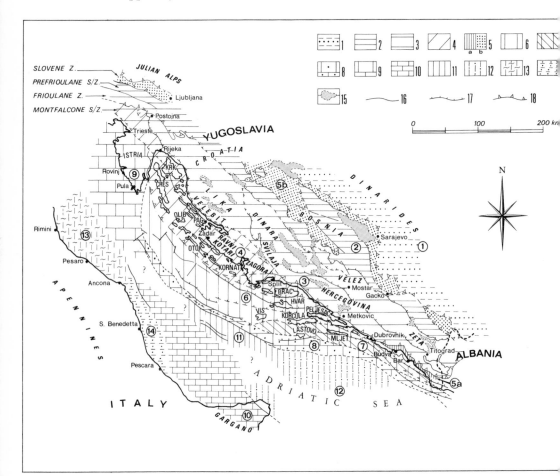

Fig. 3. Principal structural trends in the central Adriatic and the lateral Dinaric units (after Miljuš, 1972, and Turk, 19
1. Internal Dinarides, 2. internal margin of the High Karst (= pre-Karst subzone), 3. High Karst units, 4. units of
external Karst subzone (Chorowicz, 1975) = sub-Dalmatian subzone (Blanchet, 1974) = Ravni–Kotari Syncli
rium (Miljuš, 1972), 5. Budva units (5a) and Bosnian (5b) = Gacko Synclinorium (Miljuš, 1972), 6. northern Dalma
units (Dalmatian zone), 7. southern Dalmatian units (Dalmatian zone), 8. Adriatic Ridge (Turk, 1971; Miljuš, 19
[The existence of such a limestone ridge (mid-Adriatic) crossing the axis of the Adriatic Sea is hypothetical. Accordin
Miljuš, it is a particular structure separating the north Dalmatian depression (6) from the southern Adriatic Basin (12). T
associates it with Istria and the Gargano Ridge and thus to the Mesozoic platform of Apulia. It may also be envisio
as prolonged toward the Dalmatian zone and the most distant zones of Kruja (Albania) and Gavrovo (Greece).], 9. Ist
platform (Dalmatian zone), 10. Apulian platform, 11. central Adriatic depression, ?continuation of the Ionian z
12. Ionian units and the southern Adriatic Tertiary basin, 13. Recent Po depression, 14. Pliocene Marche depress
15. continental subsiding basins. Tectonic lines: 16. normal fault, 17. reversed and thrust faults, 18. overthrust nap

facies), overlain by volcanic beds* associated with radiolarites ("Porphyrit–Hornstein") and fine banded jaspery limestones of Ladinian age.

Carbonate sedimentation on the platform began in Upper Triassic times and continued until the Upper Cretaceous. It involves a Norian and infra-Cretaceous dolomitic and breccia facies which extends laterally towards the margin of the Bosnian Trough in a transitional paleogeographic domain (the pre-Karst subzone of Blanchet *et al.*, 1970). The main Mesozoic sequence consists of Upper Jurassic coralline reef limestones and Upper Cretaceous rudist reefs, with back-reef and peri-reef facies and oolitic limestones during the Dogger and bioclastic, stromatolitic, and algal limestones during the Lias, Malm, and Cretaceous. The continuity of this sequence is broken by some relatively short lacunes, often indicated by bauxite horizons. These occur particularly during the Kimmeridgian–Portlandian, the Cenomanian–Turonian, and at the end of the Cretaceous.

The thickness of Mesozoic carbonates may exceed 2600 m in the central part of the platform in Montenegro. Along the external margin of the platform a Liburnian (Lower–Middle Eocene) laguno-lacustrine transitional facies (Bignot, 1972) is found. These are overlain by Paleocene limestones. In Zagora, Ravni–Kotari, the northern Dalamatian Islands, and Cicaria, the neritic Triassic–Jurassic sequence contains numerous breccias and a volcanic horizon (analogous with the Triassic flysch of Budva). In addition, during the Upper Cretaceous, pelagic influences are felt. The Paleogene transgression began in the Lower to Middle Eocene. These features indicate the existence of the external margin of the High Karst platform (Chorowicz, 1975).

Generally in the High Karst zone the deposition of flysch began in the Paleocene and Lower Eocene, but may be earlier (Maestrichtian–Paleocene) in the pre-Karst, although reduced in thickness in comparison with the surrounding zones. The flysch is discordantly overlain by the Upper Priabonian–Oligocene Promina Molasse.

Paleogeography. During the greater part of the Mesozoic the High Karst zone was a shoal or platform region. The submarine morphology of the platform was doubtless far from uniform. In fact, detailed lithological studies indicate facies changes which show that the platform was subject to synsedimentary deformation related to elevation or depression along a system of faults which cut the platform into a series of horsts and grabens. These fractures, by providing relief, introduce variations into the deposits which even though of shallow-water type, sometimes at sea level (bauxite formation), were formed in a slowly subsiding region. The absence of a marked angular discordance in the thick carbonate sequence is a characteristic feature.

---

* This mixed Ladinian volcanic episode with rhyolites and andesites is indicative of extensive subsidence of the platforms (Bébien *et al.*, 1977).

Tectonics. The allochthonous nature of the High Karst unit can be demonstrated clearly in the coastal region of Montenegro. Here, southwest of the Lovčen Mountains in the imposing cliffs dominating "Kotor's moth," it abuts rocks of the Budva zone. The magnitude of the thrust can be seen for tens of kilometers along the Scutari (Shkoder)–Pec line up to the Albanian border. The Karst nappe is thus widely developed in the southern Dinarides. Its northwestern prolongation raises some problems, particularly because of its overlap on to the Dalmatian zone and the disapperance of the Budva zone (tectonic or paleogeographic?) (Aubouin *et al.*, 1970). The tectonic front of the High Karst, which can be followed along the Dalmatian coastal region from Dubrovnik, Makarska, and Omis to near Split, is continued by faults which bend strongly northward toward Svilaja and Dinara. Beyond this limit typical Karst reappears in Slovenia north of Istria. In southern Croatia the Velebit, which extends to the coast and which shows some Karst affinities, is itself allochthonous. The overlap of the Karst on to the Dalmatian zone, according to Miljuš (1972, Figs. 21, 23), may extend as far as the Krk, Rab, and Pag Islands between Rijeka and Zadar (see Fig. 3), but its prolongation southeast of a line from Zadar to Split beyond the Ravni–Kotari and Zagora Mountains is uncertain (external Karst of Chorowicz, 1975).

The Knin thrust near Svilaja, which does not necessarily represent the maximum extension of the High Karst nappe, underlines the magnitude of the displacement. This abrupt virgation of the major structural trend above Mosor may be linked to the existence of an old structural fault at the latitude of Zrmanja (Chorowicz, 1970). It cannot be excluded, however, that this tectonic overlap, which seems to spare the most external zone, may not be a sure indication of major cover and general allochthony of the Dinaric zones, signs of which appear more and more clearly in the Albanides and the Hellenides.

Whatever may be the explanation, one part of the Dalmatian littoral at the latitude of Velebit belongs to the High Karst zone, which thus advances to the sea and forms part of the peri-Adriatic domain.

2. *The Budva–Cukali Trough* (Fig. 2B)

In Yugoslavia this only crops out along the coast, from south of Lake Scutari (Shkoder) to west of Kotor, near the Albanian border. It passes out to sea between Bar and Budva. It crosses Albania, where it is called the Krasta–Cukali zone, and as the Pindus zone it is well developed in Greece. It is a paleogeographic unit of major importance with a succession comprising three distinct sedimentary units: (1) A sequence of Werfenian arenaceous pelites and reddened sandstones passing up into an Anisian alternation of flyschoid sandstones and shales capped by tuffs and porphyries (Bébien *et al.*, 1977). The volcanic rocks are associated with jasper and radiolarities ("Porphyrit–Hornstein").

(2) A series beginning with platy-bedded Ladinian cherty limestones, Upper Triassic to Jurassic *Halobia* limestone and flaggy jasper, Upper Jurassic to Lower Cretaceous radiolarites and siliceous pelites, and terminating with brecciated, banded limestones with flint. The last limestones contain debris of rudists and a microfauna (*Globotruncana*) of Upper Cretaceous age. (3) A flysch sequence made up of a rhythmic alternation of sandstones, graywackes, and shales. Deposition began in the Maestrichtian and continued into the Lower Eocene.

The Mesozoic sequence of Budva is thus characterized by a thick detrital Lower and Middle Triassic sequence, relatively thin Upper Triassic and Jurassic pelagic beds suggesting deposition in a deep trough, and an early flysch sequence. No indication of a break in sedimentation has been found.

Structurally the Budva zone is thrust over the neritic Dalmatian zone. At the contact the series is broken up into thrust sheets with shearing at the base, the units at outcrop giving the appearance of a thick thrust slice bordering the sea between Ulcinj and Bar and from Budva to Kotor as far south as Orjen where the Budva units front the Upper Eocene Dalmatian flysch (Cadet, 1970).

Southeast of Dubrovnik the Budva zone is tectonically cut out, but its continuation to the northwest below the High Karst zone may be envisaged. Effectively, the Budva thrust extends along the Dalmatian littoral by Dubrovnik and Makarska to end a little north of Split where it continues as the Koziak front (Fig. 3). The work of Chorowicz (1975) seems to indicate that the thrusting marks the junction of this zone whose paleogeographic termination is to be sought further to the northwest, in particular in eastern Istria west of Rijeka at the site of the Ucka and Cicaria faults. This argument finds stratigraphic support; the transitional subzone of the external Karst (cf. above) marked by pelagic influences in the Upper Cretaceous on one side, and the Dalmatian subzone with an equally deep-water Cretaceous on the other.

Nevertheless, if the Budva zone does continue much to the northwest its importance appears to diminish for Ypresian–Lutetian flysch appears at the same time along the borders of the Karst and Dalmatian zones of Istria (Bignot, 1972). The significance of the Budva zone thus diminishes in time and space towards the southern Alps, the two external carbonate platforms joining to form a single unit, the Croatian unit of Aubouin (1963*a*).

3. *The Dalmatian Platform* (Fig. 2C)

In the preceding section, the difficulty of distinguishing the Dalmatian zone from the High Karst zone in northern Dalmatia has been illustrated. As a result of an inflection in the structural trend at the latitude of Split, the axial trends in southern Dalmatia are oriented east–west (at the latitude of Hvar, Brač, Korcula, Lastovo, Mljet, up to the Pelješac Peninsula). It is this typical Dalmatian domain whose edge follows the coast between Makarska, Dubrovnik, and Ulcinj. It reappears in southern Istria south of the Cicaria–Ucka thrust. Only rarely in this neritic zone with its cover of rigid limestone thrown into large amplitude folds and largely under water, are the deeper parts exposed. The deep exploration well, Rovinj 1, on the west coast of Istria passed through more than 700 m of Permian breccias with fusulinids and other Permian micro-fossils, as well as Middle Triassic algal and foraminiferal limestones (Koch-ansky-Devidé, 1964). Carbonate and dolomitic facies, also developed within the Upper Triassic and the Jurassic (*Orbitopsella*, *Lithiotis*, and *Paleodasycladus* limestones) including the Upper Jurassic (*Clypeina* and *Cladocoropsis* lime-stones), are found in Pula 1 well in extreme southwestern Istria and in Lastovo Island, where Dogger dolomites and sub-reefal Malm limestones crop out.

There are important facies variations in the Cretaceous with the appearance of several hundred meters of anhydrite containing rare intercalations of Aptian reef limestones in the deep Ulcinj wells, while in contrast on the Island of Mljet and on the Peljesac Peninsula, there are outcrops of Malm *Clypeina* limestones, Lower Cretaceous dolomites, and Upper Cretaceous limestones with *Globotruncana*, *Pithonella*, and rudists.

In short, the Mesozoic sequence of the Dalmatian zone is indicative of a shallow carbonate platform subject to phases of emersion in the Upper Cre-taceous, which is represented by lagunal facies. Liburnian marly limestones with carbonaceous horizons occasionally capped by bauxite witness a break of sedimentation in the Upper Senonian which in some places extends from the Cenomanian–Turonian to the base of the Lutetian (central and southern Istria). The Eocene is represented by alveolinid bioherms and nummulite horizons of Ypresian and Lower Lutetian age.

A regressive episode marked by the formation of bauxite brought the period of carbonate sedimentation to an end. Flysch with a basal conglomerate invaded the platform at the end of the Lutetian and continued up to the Pri-abonian (and perhaps until Oligocene), a time of tectonic activity as is shown by the thrusting along its internal margin. The thickness of the Dalmatian Mesozoic sequence is considerable. In western Istria the Triassic exceeds 2000 m, the Jurassic more than 1300 m (Rovinj 1 well), and the Cretaceous at least 1500 m. The total thickness of sediments, almost exclusively carbonates, thus approaches 5000 m. This compares with the Mesozoic limestone and dolomite sequence of the High Karst, which may reach and even exceed 6000 m. The lithological character suggests a subsiding Mesozoic platform, probably with irregular relief as in the High Karst, and marked by minor ridges and depres-sions in the center and faults along the margins. This is the Adriatic carbonate platform of Radoičić (D'Argenio *et al.*, 1971).

Margins of the Dalmatian Platform. The existence of an important thrust on the southern flank of Koziak northwest of Split has already been noted.

North of this thrust outcrops of the external Karst (Zagora) are found over-thrust by internal Karst units (Svilaja). To the south, under thrusts of Karst, are a series of tight folds cut by reverse faults (Trogir). In this structural assemblage brecciated Cretaceous limestones extend as high as Maestrichtian, in contrast to the external Karst where Campanian–Maestrichtian is not represented. A Paleogene transgression began in the Middle Lutetian (Liburnian and Nummulitic limestones) with the development of thick flysch of Priabonian age conformably overlain by the Upper Priabonian Promina Molasse. Additionally, pelagic horizons are more abundant in the Cretaceous, which thus differs from the sub-reef facies of the High Karst and the distinctly neritic facies of the Dalmatian Islands (Vis, Hvar).

Chorowicz (1975) interpreted this littoral assemblage, extending from Split to eastern and southern Istria, as a transitional zone having a sub-Dalmatian location between the Budva and Dalmatian zones (the Split subzone). The Dalmatian zone (*sensu stricto*) in this interpretation corresponds to western Istria, the southern islands (Vis, Hvar, Korčula, Mjlet, etc.) and the Peljsač Peninsula, with the Split subzone to a certain extent acting as the internal border of the former. The external border, according to Radoičić (D'Argenio *et al.*, 1971), may be found in the Islands of Lastovo and Susac where above the peri-reefal calcareous and dolomitic horizons of ?Dogger–Malm are reefs of latest Jurassic–infra-Cretaceous age, which contain intercalated sediments with a pelagic fauna.

Tectonics. The structural deformation of the Dalmatian zone is well known only in its southern internal part, where imbricated anticlinal thrust sheets, oriented essentially east–west oblique to the Dinaric trend, are thrust south over one another. These structures replace one another with roots formed by faults or flexures. The tectonism of the Upper Eocene was repeated during the Plio-Quaternary by a network of faults. West of southern Istria, the Dalmatian zone has the form of a broad Cretaceous dome plunging to the east, the Jurassic core of which is exposed along the axis between Rovinj and Poreç. A boxlike style in which anticlines have broad flat crests, and synclines or monoclinal flexures are sharp and locally sheared by vertical faults is characteristic. It seems likely that the greater part of the Dalmatian central platform, now under the Adriatic and connecting western Istria to the southern Dalmatian Islands, has the same structural style.

4. *The Ionian Basin* (Fig. 2D)

This basin emerges in western Albania north of the Otranto Straits between the Adriatic littoral and the Greek border and continues to the southeast bordering the Ionian Sea, from which it derives its name. Toward the northwest, the Ionian Basin disappears below the Adriatic and is not found in Yugoslavia. According to Pieri (1966), eastern Gargano in Italy forms its external flank. There are strong similarities between the Ionian facies of Greece (Epirus) and southern Albania. They belong to the same isopic zone which will be studied in the latter country, for it is there that it disappears under the Adriatic.

Stratigraphy. Upper Triassic (Norian)–Lower Triassic dolomites, with intercalated lenticular bituminous horizons in the lower part, crop out in the center of the Albanian Ionian zone (Mali Gjere Mountains). In Greece, Norian dolomites rest upon black limestones with *Cardita gumbeli* assigned to the Carnian (I.F.P.–I.G.R.S., 1966). It is difficult to assess the nature of the pre-Carnian rocks. On occasion, around tectonic contacts, anhydrite–gypsum–halite appears associated with breccias and cavernous dolomite. Wells at Akarnania and Corfu show that these beds underlie the Triassic–Liassic limestones. They may be Triassic or Permian or could even represent the Hercynian basement and so be unrelated to the Alpine cover (Dalipi *et al.*, 1971). There is a gradual transition from the Triassic to the Liassic, massive dolomites giving way to diagenetic dolomites intercalated with organo-detrital limestones enclosing cherts and dated as Lower–Middle Lias.

From Toarcian times onward the Albanian Ionian zone consists of three axially aligned isopic zones, a central subzone of limestones, and *Posidonia* shales bordered laterally by two red nodular argillaceous limestones (ammonitico rosso). The Dogger is characterized by marly filamentous limestones (with *Posidonia*) alternating with bands of chert and occasional oolitic limestone horizons. This is succeeded by pelagic cherty limestones with a characteristic microfauna. The lower part, Neocomian–Albian, contains marly horizons, while in the upper part, Senonian, there are inclusions of breccia and biodetrital beds (Dalipi *et al.*, 1971).

Calcareous–siliceous sedimentation continued into the Paleocene through the Ypresian–Lutetian and into the Priabonian, with micritic and microbrecciated organic limestones containing a rich foraminiferal assemblage. Detrital sedimentation began with the deposition of Priabonian marls followed by Upper Eocene–Oligocene conglomeratic and sandy argillaceous flysch. This thick sequence (in excess of 2000 m) becomes progressively more shaly with coarse horizons and turbidites. It continues into the Burdigalian and Lower Helvetian with the deposition of marls and clays (Gjata *et al.*, 1971).

The molasse transgression of the Upper Helvetian marks the end of tectogenesis in the Ionian zone. The Upper Miocene is characterized by Tortonian sandstones and conglomerates and Messinian gypsum. Marine sedimentation was continuous through much of Tertiary time, at least in those external synclinal parts of the Ionian zone which form the peri-Adriatic depression, the submerged margin of the present Adriatic Basin.

Paleogeographic Evolution. An important paleogeographic feature is the large subsiding evaporite basin developed during the Triassic–Lower Liassic. More open marine facies developed along the margins of the basin in the internal and external subzones.

A phase of folding and uplift in the Upper Lias led to the uplift along paleofaults of the principal anticlines of which the median horst (central sub-zone) is the most important, bounded by depressions which themselves contain areas of uplift. Unconformities marked by hard grounds formed at this time. This basin far from a continent received pelagic sedimentation consisting essentially of marly limestones with some chert up to the Lower Senonian, indicating the gradual deepening of a starved basin bounded by two platforms, one internal (Kruja ridge = the prolongation of the Istrian, Dalmatian, and Gavrovo zones), the other external (Sazan ridge-margin of the Apulian zone). The influence of these two highs on the intervening Ionian Trough is clearly indicated by the calcareous flank breccias intercalated in the sequences in the marginal subzones.

The Ionian Basin became shallower in the Upper Senonian–Priabonian, but the Liassic ridges persisted though of less importance, the deposition of microbreccias became very important while the amount of silica present diminished. The change of sedimentation which began in the Upper Eocene was linked to the uplift of the Krasta–Cukali chain (Budva zone = Pindus zone) which provided a source of detritus. Several thousand meters of flysch accumulated in the internal Ionian syncline bordering the Kruja Ridge. This trough, which subsided during the Oligocene, contrasts with the external margin of the basin where the flysch is thin and marly–calcareous in character, passing westwards into the condensed neritic limestones of the Sazan (Apulian) zone. At the beginning of the Aquitanian, the Liassic folds and the internal zone were uplifted; however, the flysch sequence of the central and external margins of the basin continued unchanged.

The major tectonic activity dates from the Helvetian. The Ionian Basin was compressed and deformed against the Apulian platform. The anticlinal units compressed by a westerly stress were thrust over the eastern synclinal flanks. During these movements the evaporite beds flowed, producing typical diapiric intrusion and injection. The compressive movements continued until the Quaternary, and thrusting over Neogene and even Pliocene still occurs.

The regression at the end of the Miocene (gypsum) was followed by a Pliocene transgression seen in estuaries along the western coastal margin. In its turn the Ionian zone emerged, and at that time the borders of the continent acquired a configuration approaching that of the present day. In short, the Adriatic–Ionian zone was a remarkably constant domain during the geosynclinal cycle, behaving as an active depression of the marginal sea between the stable Dalmatian Ridge to the east and the Apulian Ridge to the west.

B. The Apennine Geosynclinal Cycle

The external isopic zones of the Apennines trending north–south cut the Italian Peninsula obliquely. As a result of tectonic movements to which they have been subjected, their present distribution is a poor reflection of the geosynclinal paleogeography. The Apulian domain, which retains a stable position, is covered in the west by the Marche–Umbria, Abruzzi–Campania, and Lucania zones. They will be discussed in that order. A fundamental contrast exists between the flysch and terrigenous sediments of the northern Apennines and the calcareous chains of the southern Apennines; the latter are reminiscent of the external Dinarides.

1. *The Apulian Platform* (Fig. 2E)

The Apulian platform in southern Italy consists of the region between Monte Gargano and Brindisi and the Gulf of Taranto. This calcareous massif of Pouilles, Murges, and Monte Gargano dominates the broad, post-geosynclinal, Plio-Quaternary Bradanic depression and the late geosynclinal Molise Molasse basin to the southwest.

The eastern border of the Apulian foreland is found in western Albania, where it is exposed in Sazan Island and the Karaburun Mountains off the Otranto Straits. It is also found in the Ionian Islands of Paxos, Kephalline, and Zakynthos (the pre-Apulian zone of Aubouin). Only the margins of the Apulian foreland are well known for the central part lies under the Adriatic and Ionian Seas off the Greek Islands. One part found in Italy forms the Apulian or Apulo-Garganian zone.

The Apulian foreland was a stable platform which probably received more than 6000 m of reef deposits between the Triassic and Lower Miocene. The stratigraphic sequence begins with dolomites, alternating with evaporitics, which are penetrated by the Foresta Umbra well at Gargano.* Calcareous sedimentation continued into the Liassic with dolomite episodes and then with reefs which appear to take on a major role from the Dogger to the Upper Cretaceous (coral limestones from Malm to Lower Cretaceous; rudist limestones Turonian–Santonian), though dolomites also occur during the Lower Cretaceous. North of Gargano the Middle Cretaceous is absent and a bauxite horizon marks this gap. This horizon does not appear further south. Along the eastern margin of Gargano the Upper Cretaceous reef limestones pass laterally into micritic chalky, cherty limestones of Ionian facies.

The Paleogene is represented by biomicrites, biosparites, and organic

---

* Upper Triassic deposits of marl and evaporites with basic igneous rocks whose structural position is uncertain have been reported north of Gargano (Punta della Pietre Nere, D'Argenio, 1971).

limestones with an abundant planktonic foraminiferal assemblage of Eocene and Oligocene age. The period began with regressive organo-detrital limestones which continue to the Lower Miocene.

A late tectonic sequence followed, with an interval of algal limestone formation in the Helvetian and thick terrigenous deposits in the Upper Miocene and Pliocene in the Molise Basin. The basin is regarded by some as an independent paleogeographic unit evolved during the geosynclinal cycle (Pieri, 1966; D'Argenio et al., 1971) while others attribute it to a molasse trough lying in an immense platform uniting Apulia and the Abruzzi–Campania (Aubouin, 1965).

The Apulian foreland was thus a stable platform of which the eastern part crops out extensively in southern Italy. It is found on the Albanian coast (Sazan zone) (Papa, 1970) where the succession, beginning with the Cretaceous and extending to the Lower Miocene (Aquitanian), has a character comparable with that just described.

The Mesozoic Apulian carbonates show a major development of biostromes with reef limestones common and some dolomitic intervals. The platform may thus be considered a permanent ridge forming the common foreland of the Dinarides and Apennines during the Alpine geosynclinal period. From a structural point of view the ridge was stable although subject to faulting at least in the central part, but without apparently being affected by recent tectogenic phases. This was due to its distal position with respect to the orogenic axes and to the fact that successive waves of tectonic activity reached the area late and were low in intensity.

In eastern Albania the Sazan zone (Papa, 1970), with more than 2000 m of Cretaceous–Oligocene ridge carbonates, is regarded as the prolongation of the Apulian zone. As on Monte Gargano, the eastern flank of the great Karaburun anticlinal fold, the backbone of the region, thrusts over the external synclinal unit of the Ionian zone. This pattern argues in favor of the passage of the centripetal axis of symmetry of the Italo-Dinaric system to the middle of the Adriatic, and, at least in the southern part, in the Ionian zone (Pieri, 1966).

2. *The Marche–Umbria Basin* (Fig. 2F)

An oblique isopic zone crosses the Italian Peninsula from Venetia to the Tyrrhenian Sea south of Rome. It probably forms much of the Adriatic shore at Marche and Romagna. Outcrops occur near Ancona, but elsewhere this zone is masked by post-geosynclinal sediments of the Padan Basin and late geosynclinal deposits of the northeastern margin of the Apennines, a prolongation of the Italo-Dinaric foredeep (Aubouin, 1972). Inland its limit generally follows the course of the Tiber. This border is tectonic (Falterona–Trasimeno line) and is due to the eastward overthrust of the Tuscan nappe, which itself supports the more internal eugeosynclinal units of the Ligurian zone. In the central

Apennines, the Marche–Umbria zone is in contact with and overthrusts the
Abruzzi zone along a line from Ancona to Anzio. The sediments of the zone
are miogeosynclinal and consist of several lithological sequences following the
Hercynian orogenic cycle and the Permian volcanic episode.

The Detrital Subbasement. During the Triassic an arenaceous continental
and lagunal series (Verrucano *s.s.**) developed. Remains of these sediments
exist near Perugia in the Umbria sequence.

The Carbonate Beds. Following the deposition of detrital beds reaching up
to the Carnian, there began a period of widespread carbonate deposition in the
Upper Triassic (Alpine facies). It is well known in the northern, central, and
southern Apennines and in the southern Alps and Dinarides. In this shallow
basin, which on occasion tended to be lagunal, thick Norian anhydrites were
deposited with bands of dolomite, and dolomitic limestone with cargneules
(calcare cavernoso). This formation, which passes laterally into the Hauptdo-
lomit in the northern Apennines, appears to continue in the southern Apennines
where its presence has been recorded in boreholes between the Abruzzi and
Gargano Peninsula (Martinis and Pieri, 1963); but its further continuation
into southern Italy remains hypothetical. Its thickness is variable, ranging from
3 m to more than 1000 m in the Ladarello well.

At the end of Triassic time the marine platform was the site of shallow
water sedimentation, alternating between low energy and high energy, of
turbulent and interdital facies (stromatolites). Micrite (intra, pel, bio, and
oomicrites) and oosparite, sometimes argillaceous and fetid, or marls with
dolomitic horizons and cavernous breccias were also formed. These beds contain a Rhaetic fauna of ammonites, brachiopods, anthozoans, and echinoderms.

In the Lower Liassic neritic sedimentation was established with massive
limestones (calcare massiccio), with dolomitic and epigenetic dolomitic horizons. The limestones, which are micrites, intramicrites, sparites, intrasparites,
and biosparites contain dasyclads and solenopores. In the Umbrian sequence
these beds may begin in the Rhaetic and continue locally to the Middle Lias
(Lower Pliensbachian ammonitic limestones of Monte Cucco and Perugia).
The thickness of the limestone varies from a few to several hundred meters. It
covers formations of differing age. After the end of the Lower Lias the platform
was subject to important deformation. In structural depressions well-bedded
limestones (micrites and calcarenites) with chert nodules, siliceous lenses, or

---

* In the older and broader sense of this term, the Verrucano consisted of three members
(Abbate and Sagri, 1970): (1) clastic Carboniferous sediments, often metamorphosed,
strongly folded during the Hercynian orogeny; (2) clastic, often fossiliferous beds, arkosic
and volcanic rocks of Upper Carboniferous–Permian age (and perhaps even Lower
Triassic); (3) coarse sandstones and conglomerates (Verrucano *s.s.*) Ladinian–Carnian
in age discordant upon or separated by a break in sedimentation from the underlying beds.

thin slabs (Pietra corniolo = Calcare selafero) formed, while on the ridges swept by currents, and toward the top of the series in general, a reddish nodular "ammonitico rosso" limestone facies developed. The last contains an abundant Toarcian–Aalenian fauna and ranges from 200–300 m thick (Bortolotti et al., 1970).

The Pelagic Siliceous Facies. Following the Upper Lias the sea deepened and limestones, marly limestones with chert nodules, and *Posidonia* marls (Dogger) and *Aptychus* marls (calcari Diasprini–Scisti and Aptici) (Malm) were deposited. This pelagic sequence, varying in thickness from 50 to 100 m, is occasionally marked by a reduced thickness or breaks in the succession. In the latter case nodular limestones with Oxfordian–Tithonian ammonite fragments (Grigio ammonitico) replace jaspery limestones and rest discordantly on the massive limestones of the Lower Lias, providing evidence of vertical fault movements during the Middle Lias in the Marche–Umbria zone. The pelagic environment persisted into the Upper Jurassic as is suggested by the presence of banded micritic limestones with chert nodules and siliceous lenses intercalated with bands of dolomite and dolomitic limestone. This formation, known as the Maiolica (Calcare rupestre) in Umbria, where it is of the order of 300 m thick, contains a rich fauna of Tithonian and Neocomian tintinnids. Reworking, solution, and breaks in sedimentation are found with occasional breccias, conglomerates, and stratigraphic discontinuities on paraconformity in the series.

Terrigenous Sedimentation. During the Albian–Cenomanian, deep, open-water marine, marly deposits accumulated over the whole isopic zone. This stage is marked by an alternation of marly limestone, fissile bituminous shales, and pelites containing abundant fucoids and a characteristic microfauna (fucoid marls "Scaglia facies"). Well-bedded limestones, marls and shales or pelites, and interbedded bituminous horizons occur in the Upper Cretaceous. They are from 200 to 400 m thick and are well developed in the subsurface from the Venetian Alps north of Padane to south of Latium. This formation contains a rich foraminiferal fauna, representing 4 biozones characteristic of the Cenomanian to Middle Eocene. Two lithofacies, one white (Scaglia bianca) and the other red (Scaglia rossa) are found.

The open sea deepened and, during the Upper Eocene and Oligocene, received an increasing amount of terrestrial argillaceous debris. Marls and marly limestones of the "Scaglia cinerea" type predominated, their thickness increasing from Umbria towards Marche. They are dated by an abundant microfauna. In the Marche–Umbria zone marly sedimentation terminated in marls and marly limestones. Sequences of nodular chert bands and sandy volcanic tuffs are interbedded in the western, more calcareous part of the zone (Biscario) where the beds are Aquitanian–Langhian in age. In the east (Marche) the marly sequence reaches to the Helvetian and is overlain by Tortonian–Messinian marls (Schlier) directly below molasse. In short, the Scaglia facies reach higher stratigraphic levels in the east, from Oligocene in Umbria to Miocene in Marche.

The Period of Flysch Accumulation. In the Marche–Umbria zone there was a rapid incursion of detrital material into a large subsiding external trough. This is now exposed in Romagna and the Marche and in the Umbrian Apennines. The flysch sequence is a marly–sandy rhythmic alternation of calcareous, sandy turbidites, marly beds with planktonic foraminifera, siltstones, conglomeratic lenses with granite and schist pebbles, and sandy calcarenites. There is evidence of intensive reworking, indicating an unstable sea floor and resedimentation.

The Marche–Umbria Flysch consists of two distinct stratigraphic units, the Middle Miocene (Langhian–Tortonian) "Marnoso–Arenacea" and the Upper Miocene (?Upper Tortonian–Messinian) "Piceno" Flysch. The total thickness may exceed 5000 m. An eastward migration of facies can be observed, the basal detrital beds becoming progressively younger to the east. In addition, the "Marnoso–Arenacea" passes laterally northeastward into a marly and marly limestone Scaglia facies (Biscario–Schlier) in the Romagna and Umbrian foothills. Only in southwestern Umbria and in the Abruzzi Apennines can the "Piceno" Flysch be found resting on the "Marnoso–Arenacea," here on a horizon of pteropod marls. The interesting feature is that in its upper part Messinian gypsiferous sandstone and beds of gypsum are found.

Comment. It is worth noting that the Tuscany series lie to the west of the Marche–Umbria series and pass laterally into them with comparable facies: cavernous Triassic limestones, massive Lower Jurassic limestones and dolomites, a condensed Upper Lias sequence with "ammonitico rosso," cherty and radiolarian limestones of the Malm–Neocomian, and varicolored shales, marls, and calcarenites (Scisti Policromi) of Upper Cretaceous to Oligocene age (= Scaglia of Umbria). The sandy–marly flysch, the Macigno, 2000–3000 m thick, which rests on these beds dates from the Upper Oligocene to Lower Miocene. Thus, the onset of the detrital facies is still earlier in Tuscany, a zone of pelagic miogeosynclinal character.

Stages in Deformation. The Marche–Umbria, as the Tuscan zone, was a subsiding region during the Triassic and Jurassic. Locally the subsidence was interrupted by periods of uplift or variation in the subsidence rate, or simple pauses in subsidence. Differential movements, which began at the end of the Liassic in the Marche–Umbria zone, are marked by lacunas, "ammonitico rosso" facies, etc. Vertical movements developed during the Cretaceous and Lower Eocene with olistostromes and paraconformities.

As a result of Tertiary tectogenesis, the Tuscan zone was thrust over Umbrian units (Hsü, 1967). At the beginning of the Miocene (Aquitanian) the most external of the Ligurian eugeosynclinal units were thrust onto the Tuscan Flysch and accompanied by a shearing of the Tuscan zone during the Lower–Middle Miocene and the decollement of the Mesozoic. The thrust of the Tuscan nappe and flysch over Marche–Umbria must have taken place at the end of the Miocene. The advance of the nappes must have occurred throughout the Miocene and no doubt fed material into the more external miogeosynclinal troughs (olistolites, olistostromes, gravity slides).

At a later date (Messinian–Pliocene) the Marche–Umbria zone itself thrust over the Abruzzi–Latium sequence to the east. Thus the Marche–Umbria Basin is characterized by thin-skin tectonics of the Jura type during a phase of geosynclinal activity which does not exclude fracture. Its relation with Tuscan and Ligurian units is complex (Fig. 6B) as is shown by allochthonous eugeosynclinal units buried in the chaotic San Marino complex and the Canetolo Eocene olistostromes intercalated in the Umbrian Oligo-Miocene flysch.

Lucanian Series. To the southeast in the southern Apennines, the Lucanian Series of Triassic–Cretaceous pelagic marly, calcaro-siliceous beds in front of the thrust limestones of Abruzzi–Latium–Campania represents a basin margin facies (D'Argenio *et al.*, 1971). The Lucanian Trough poses a paleogeographic problem (Pieri, 1966). On the one hand, the persistence of the transitional facies along the external margin of the Apennine Abruzzi zone suggests that the Lucanian sedimentary trough continued to Marche, and in consequence the Marche–Umbria zone connected with that of Lucania. On the other hand, the geological history of the two troughs is quite different. The Lucanian Trough differentiated earlier (Upper Triassic) than the Marche Basin (Lower Liassic), and the appearance of basic eruptive rocks in the Galestrino Flysch would suggest that the Lucanian Trough was eugeosynclinal whereas the Marche–Umbria zone was, as already demonstrated, clearly miogeosynclinal. In the absence of direct proof of one or the other hypothesis, the author has adopted the paleogeographic reconstruction (Fig. 4b) showing continuity of the two troughs (after Pieri, 1966). This reconstruction requires that most of the carbonate zones of the southern Apennines, particularly the internal parts, be allochthonous (Grandjacquet, 1963).

A detailed examination of these structures goes beyond the limits of a review concerned primarily with the Adriatic; however, the magnitude of thrusting is shown by recent deep drilling in southern Italy and it is clear that the limestone zones override east of Grand Sasso and border the coastal post-geosynclinal series between Lanciano, Chieti, and San Benedetto. Only the Abruzzi–Campania sequence part of the neritic Mesozoic of the central–

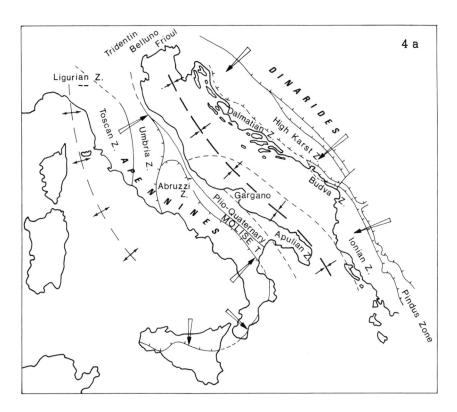

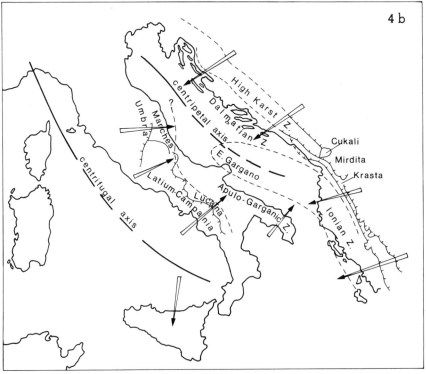

southern Apennines, which forms the Latium–Campania zone of Pieri (1966), will be considered here.

3. *The Abruzzi–Campania Platform* (Fig. 2G)

Carbonate sedimentation predominates in this sequence which is well developed in the central Apennines (Catenacci *et al.*, 1964). The characteristic sequence of the platform is well developed in the Ocre Mountains and Monte Maggiore, while the marginal facies appears along the external borders north of Monte San Franco, Gran Sasso, and Acquasante as well as to the east in Matesi (AGIP Frosolone wells), Mt. Meta, and Maiella.

During the Upper Triassic and Lower Liassic the marine platform deposits commonly found in the neritic peri-Adriatic zones consist of stromatolitic dolomites and dolomitic limestones. At Monte Maggiore, in the center of the platform, carbonate sedimentation continued into the Upper Jurassic and Lower Cretaceous with Liassic *Lithiotis* limestone, Dogger oolitic reef and dolomitic limestone, and *Cladocoropsis*, *Clypeina*, and algal limestone during the Malm–Tithonian (Pescatore and Vallario, 1964). After Neocomian dolomites, reef limestones developed. The sequence is broken by bauxites and calcareous breccias, marking a stratigraphic break from the Albian to the lower Turonian. Reef conditions persisted into the Upper Cretaceous with Turonian–Senonian rudist limestones, which may themselves contain a brief break and rest on a bauxitic horizon. The Cretaceous sequence in the Abruzzi marks an end to the period of subsidence and shows a tendency to emergence (D'Argenio, 1964). This tendency continued into the Tertiary, for in a typical sequence in the central and southern Apennines, the Paleogene is absent or rare. In the Abruzzi,

---

Fig. 4. Paleogeographic and tectonic models attempting to interpret the Apennine–Dinaric system. (a) Model proposed by Aubouin (1963*b*, 1965, 1972). The structure comprises: Apulian foreland (Apulo-Garganic, Abruzzi, and pre-Apulian zones); external miogeosynclinal zones (Ionian and Marche–Umbrian Troughs, Croatian and Tuscan Ridges); internal eugeosynclinal zones (Serbian and Ligurian Ophiolitic Troughs). The Apulian foreland ends north of the Abruzzi under the Adriatic. Note on the loop of the isopic zones of the southern Alps: If the Frioul Ridge is the homologue of the High Karst, the Belluno Trough must occupy a more external position to be considered as a continuation of the Marche–Umbria zone. In this case the Belluno Trough must terminate paleogeographically between the Tridentine and Frioulane Ridges, the two ridges being thus equivalents. (b) Model proposed by Pieri (1966). The Apulo-Garganic zone forms the common foreland and should be correlated with the Dalmation zone across the Adriatic. The Ionian zone terminates paleogeographically off the Gargano Ridge. The paleogeographic location of the Lucanian zone is uncertain, or else it occupies an intraplatform position equivalent to that of the Budva and merges with the Marche–Umbria zone, or alternatively it is in an "ultra" position as the Latium–Campania zone. The centripetal axis of symmetry follows the median line of the present Adriatic and passes from there to the Ionian zone of Albania.

Fig. 5. Distribution of the Adriatic Tertiary basins (after Morelli
et al., 1969; Turk, 1971; and Aubouin, 1965). *Miocene basins*:
A. northern Dalmatian Basin, B. Basin of southern Dalmatia and
Albania, C. Molise Basin. *Pliocene and Quaternary basins*: 1.
Venetian Basin (Pliocene), 2. Po Basin (Plio-Quaternary), 3.
Marche Basin (Pliocene), 4. Central Adriatic Basin (Plio-Quater-
nary), 5. Bradanic Basin (Pliocene), 6. Adriatic–Ionian Basin
(Pliocene). *Mesozoic platforms*: A–P. Apulia, I–P. Istria, R–G.
transverse ridge of Gargano and mid-Adriatic.

transgressive Miocene marls and bands of organic limestone rest directly on
the Upper Cretaceous. East of Maiella, the Casoli well penetrated transgressive
Miocene limestones resting directly on neritic Lower Cretaceous rocks (Pieri,
1966).

Significant changes are found in the Mesozoic rocks along the margins of
this zone. In the Abruzzi's marly limestones and calcareous organo-detrital
marls, the "Scaglia cinerea" invade the Upper Cretaceous succession and
continue into the Paleogene. Along the eastern and northern external front of
the southern calcareous Apennines, transitional sequences mark a passage
from a neritic to a pelagic environment lying to the east and northeast and
particularly marked in the vicinity of Marsica where pelagic facies occur within
the Mesozoic section during the Dogger–Malm and Neocomian (Colacicchi
and Praturlon, 1965). In the section east of Matese cherty limestones, marls,
and carbonate flank breccias also mark the transition to a deep-water facies
lying to the east (the Molise–Sannitica depression of Manfredini, 1964).

Deep drilling along the eastern margin of the centro-southern calcareous
Apennines has demonstrated important easterly displacements of the tectonic

across the Adriatic and adjacent regions. The profiles are based on data from Aubouin *et al.* (1970), Blanchet (1974), Morelli *et al.* (1969), Miljuš (1972), Sestini (1970), and Vercellino (1970). The tectonic style is marked by tangential faults on both sides of the Adriatic. Recent subsidence and marginal deformation are responsible for the huge Plio-Quaternary basin of which the Adriatic is the present representative. It should be noted that late stresses along the eastern margin of the Apennines has resulted in mass slides of pre-Miocene (Ol/M) olistostromes, followed by post-Miocene (Ol/P) olistostromes as a result of Pliocene movements, resedimented during the Plio-Quaternary cycle. Symbols: Pz. undifferentiated Upper Paleozoic, T. Triassic, J. Jurassic, C. Cretaceous, E-O. Eocene–Pliocene, Pg. Paleogene, P. Pliocene, Q. Quaternary, P-Q. Plio-Quaternary, Δ. deep wells. Geographic distribution of the principal phases of deformation (inset): 1. Upper Cretaceous, 2. Eocene, 3. Oligocene, 4. Miocene.

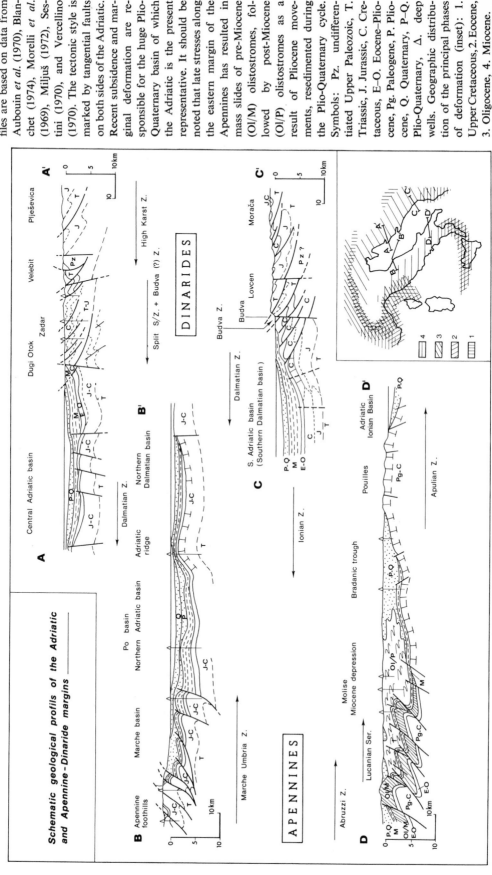

Schematic geological profils of the Adriatic and Apennine-Dinaride margins

units of the Abruzzi–Campania zone over terrigenous Miocene beds (Pieri, 1966). The magnitude of the displacement is still not well defined. In some cases they amount to several kilometers of eastward thrusting of simple scales. The general style, however, does not exclude a greater allochthony of the whole and a major displacement of the calcareous Apennines which could occupy an "ultra" position and comprise a huge nappe whose origin was internal to the Marche–Umbria. Whatever may be the case, there exists an important decollement at the base of the Mesozoic carbonate sequence in the Abruzzi–Campania Series. The sequence is that of a neritic platform whose external margin passes into a basin, the significance of which is still not clearly defined.

C. Apennine–Dinaride Correlation and Relation to the Southern Alps

The isopic zones which surround the Adriatic and whose characteristics have just been described, may be separated into platforms and basins (ridges and troughs) by differences in depth of sedimentation, nature and origin of the sedimentary material, and subsidence. They alternate following a geometric pattern characteristic of a geosynclinal system (Aubouin, 1965).

1. *Relationship between the Apennine–Dinaride Isopic Zones*

The same Mesozoic facies distribution, which permits the recognition of different zones, also reveals certain similarities and analogues on both sides of the Adriatic toward which the external zones of the Apennine–Dinaric system converge. The internal zones lie to the northeast in the Dinarides and to the southwest in the Apennines (Fig. 1). The polarity of the geosyncline shows itself by a migration in time and space of the orogeny and of flysch deposition from the internal toward the external zones, as well as by a centripetal symmetry of the tectonic vergence directed toward the Adriatic, which is apparent on the Yugoslav–Albanian coast and along the coast of peninsular Italy. This raises the question of homology between the principal isopic zones of the Apennines with their symmetrical equivalents in the Dinarides.

In a first model Aubouin (1963a, b) clearly established comparisons between the Apennines and Dinarides. Aubouin's arguments concerning the early stages of orogeny and the internal flysch zone where ophiolites are found will not be discussed here. In the external zone he distinguished the following: in the Apennines—a miogeanticlinal ridge (Tuscan zone), a miogeosynclinal trough (Marche–Umbria zone), a common foreland (Apulian ridge comprising the Abruzzi and Gargano–Murge zone); in the Dinarides—a miogeanticlinal ridge (Croatian zone = High Karst + Dalmatian zone, the Budva zone being considered as absent paleogeographically), a miogeosynclinal trough (Ionian zone).

In this succinct analysis there is demonstrable parallelism between the domains on each side of the Apulian foreland, and Aubouin established the homologous nature of the Ionian and Umbrian geosynclinal zones. There are in fact many striking facies similarities (see Fig. 2D, F), especially in the Jurassic successions, although the one in Umbria is thinner. There are several differences in detail; thus the Upper Cretaceous and Eocene are more calcareous in the Ionian zone* and more marly in the Umbrian zone, a difference attributed by Aubouin to a greater development of terrigenous facies in the Ligurian eugeosyncline (Cretaceous flysch) than in the Dinaric homologue.[†]

Without going into detail, it appears that the characteristics of the Apennines and Dinarides are the same not only in paleogeographic and orogenic history, but also in the migration of the zone of flysch deposition and magmatic evolution. Despite the peculiarities of the internal zones and tectonic styles, the two systems are similar and symmetrical with respect to the Apulian zone. The axis of symmetry formerly may have lain along the median line of the Molise, then along the border of the Pouilles littoral and off Gargano (Aubouin, 1972). Pieri (1966) adopted the principal features of the concept of an Italo-Dinaric unit, but introduced modifications to take into account geological results from petroleum exploratory wells sunk in the central–southern Apennines (Fig. 4b).

In the revised interpretation, the neritic Latium–Campania zone (Abruzzi–Campania zone) is separated from the Garganic (Apulian) Ridge; the Lucania zone between these two ridges may then be considered either as a trough in continuity with the Marche–Umbria miogeosyncline or as independent of it and of eugeosynclinal type. In the latter view the Lucania and Latium–Campania zones may then represent the symmetrical equivalents of the Budva and High Karst zones (Farinacci and Radoicic, 1964). As Pieri (1966) emphasized, these reconstructions are highly problematic and the continuity of the Marche–Umbria trough with that of Lucania remains conjectural, nor is there any argument for the eventual reunion of the Dalmatian and Apulo-Garganian zones under the Adriatic.

The existence of an Upper Jurassic to Lower Eocene sequence of siliceous and cherty limestones of Ionian type east of Gargano, which forms the western margin of the Ionian zone, requires emphasis. In addition, the sense of tectonic movement and the vergence towards the Adriatic seems clearly to indicate that the axis of symmetry of the Apennino-Dinaride system lies under the Adriatic, and passes under the Otranto Straits to the Albanian littoral and perhaps beyond to Karaburun (cf. Fig. 4b).

---

* Foraminiferal calcarenites are intercalated in the succession in both areas.

[†] This requires modification following the recognition of extensive thick Eo-Cretaceous flysch in the median domain in Yugoslavia (Bosnian zone of Blanchet et al., 1969).

2. Extension of the Apennine–Dinaric Zones under the Adriatic and Their Relation to the Southern Alps

The continuation of some of the isopic zones of the peri-Adriatic geosynclinal system is still uncertain. The pattern is clear along the northeastern shore of the Adriatic. The Dalmatian zone, which runs along the Yugoslav coast for most of its length, is unquestionably developed offshore as far as Istria. The role of the stable foreland during the geosynclinal cycle played by the external Apennine–Dinaric zones (Dalmatian, Ionian, Apulian) leads to the suggestion that these zones remained autochthonous and that their extension under the central and southern Adriatic scarcely changed during late and post-geosynclinal times. If this were the case, the external margin of the High Karst (sub-Dalmatian subzone of Blanchet, Ravni–Kotari synclinorium of Miljuš, 1972) spreads out in the coastal region and in the Dalmatian Islands between Rijeka and Split (Fig. 3). It is continued to the southeast by the Dalmatian and Budva zones sandwiched between these two platforms. Further toward the Otranto Straits and the Albanian littoral, the axis of the Ionian zone continues out to sea. Because of its structural orientation it crosses the Adriatic along a slightly oblique course to emerge in eastern Gargano. Its continuation beyond that point, according to Aubouin (1960, 1963a, b) may be the Umbrian Trough as the continuation of the Ionian, the two troughs uniting to form a unique basement of the Italo-Dinaric foredeep of Marche and Romagna. This basement then links with the Belluno zone beyond the lower plain of the Po.

Alternatively Pieri (1966) envisages the possible periclinal closure of the Ionian zone north of Gargano under the central Adriatic. The Apulo-Garganic foreland on the Italian side borders the Murge littoral and extends below the Ionian Sea to include the Islands of Paxos, Cephalonia, and Zante, but its continuation below the sea north of Gargano raises the same problems as the preceding interpretation. In the first hypothesis the Apulian Ridge ends as a paleogeographic feature to the north, joining the centro-southern calcareous Apennines. In the second view it would reach Istria via the northern Adriatic to coalesce with the Dalmatian zone. Richter (1974) recently proposed an alternative interpretation, with facies and structural correlations between Monte Gargano, the southern Dalmatian islands, and the Budva zone, which he suggested grouping as a southern Dalmatian Arc across the Adriatic. There are no submarine data to permit a choice or provide alternatives. In fact, the Adriatic coastal section of Italy between Gargano and Venice is occupied by thick (5000–6000 m) Plio-Quaternary argillaceous–sandy beds which lie upon the external geosynclinal zones of Italy–Yugoslavia. These beds represent the sub-Adriatic continuation of the Po Basin. While there is nothing to indicate the extension of the Lucania zone under these recent deposits, everything points to the continuation of the Upper Cretaceous–Eocene (Scaglia) marls and calcarenites (microbreccias) of the Umbria zone. It is known that the marls (shale) and marly limestone facies rise progressively in the sequence toward the northeast, that is towards the foothills of Romagna and Marche in the Cretaceous and Middle and Upper Miocene. They also extend north in the subsurface of the Padan Basin (Ferrara Hills) and towards the Venetian pre-Alps (Bortolotti *et al.*, 1970). This vertical and lateral migration of the marly facies is accompanied by progressively younger ages of the flysch turbidites towards the external zones.

In short, if the molasse and Plio-Quaternary deposits are excluded, the base of the northern Adriatic probably comprises two parts, one at the latitude of Ancona belonging to the Dalmatian carbonate platform (Istrian) and the other at very great depths at the latitude of Venice and under the Padan Basin, forming the continuation of the Umbrian Basin.

The junction of the Apennines–southern Alps–Dinarides in northern Italy and the Italo-Yugoslav region raises the problem of their relation to the southern Alps, both from a paleogeographic and tectonic standpoint. A link exists between the southern Alps and the Dinarides at the latitude of the Julian Alps. Even though correlations are difficult to establish because of the compression of the Dinaric Arc as a result of the southerly displacement of the southern Alpine nappes, some comparisons may be made.

It is uncertain whether the Montefalcone subzone (Frioulane zone) belongs with the Dalmatian zone (Cousin, 1970), for although characterized by a Mesozoic neritic or reef-ridge sequence together with Liburnian beds and bauxite, these develop later and cover a more restricted interval (Senonian–Thanetian), and in addition flysch appears earlier (end of the Lower Eocene). It is thus more likely that the Montefalcone subzone should be assigned to the external border of the High Karst to which it shows the greatest affinity.

Aubouin *et al.* (1965) and Cousin (1970) defined a Frioulane Ridge and a Frioulane zone which is characterized by Jurassic and Cretaceous platform carbonates and an Upper Paleocene flysch with basal red marly (Scaglia) beds. This sequence has all the characteristics of beds in the High Karst. There is, in addition, an internal flank (pre-Frioulane subzone) of pre-Karst type (intercalations of breccia in a pelago-terrigenous Mesozoic sequence, paraconformities, and breaks in sedimentation). It is overthrust by beds which in some respects recall the Bosnian zone in exhibiting a pelagic Jurassic substrate (cherty, finely banded limestones and radiolarites) overlain by arenaceous horizons and tintinnid limestones, followed by Upper Cretaceous flyschoid beds, a characteristic of the Slovene zone. Without entering into further detailed comparisons it is clear that the paleogeographic and structural disposition of the Julian Alps and of western Slovenia has much in common with the Dinarides *sensu stricto*. If there was no continuity between the different homologous zones, at least they present a striking resemblance.

Comparisons with the Apennines are much more tentative, and in effect, the relationships between the Slovenian isopic zones and zones with Apennine affinity are complex (Fig. 4). Not only are the latter (Belluno, Tridentine, and Lombardy zones of Aubouin *et al.*, 1965) not directly paralleled by Slovene (Julian) zones, but the latter are totally isolated from the northern and southern Apennines by the Po Basin. Nevertheless, recent research has shown that the Frioulane zone closes to the north (Aubouin *et al.*, 1965) and passes progressively westward into the Belluno Trough which developed during the Malm–Cretaceous and thus possesses a pelagic facies (siliceous limestones and radiolarites). These facies are overlain by a terrigenous flysch deposited during the Eocene and Oligocene. A parallel has been drawn between the Belluno sequence and that of the Marche–Umbria zone (Aubouin, 1963*a*). The two zones present some similarities in the Upper Jurassic (Diasprini facies), but the development of the basal Mesozoic is different (i.e., their Triassic–Liassic sequences) and also at the top where there is a calcareous-radiolarite sequence in the Belluno zone and a Scaglia facies in the Umbrian zone (cf. Fig. 2F).

In the case of the Tridentine zone, or rather the internal margin of the zone, this may correspond, according to Aubouin, to the Tuscan zone with the proviso that the nature, thickness, and perhaps even the age of the flysch may differ appreciably.

In summary, the distribution of the isopic zones on both sides of the Adriatic present a symmetry and structural polarity which gives an individuality to the Apennine–Dinaric system and demonstrates a characteristic tectonic style within the Alpine system in general. A common foreland must be sought under the Adriatic, and should be made up of the most external parts of the miogeosyncline and characterized by youthful flysch deposits, late folding, and the absence of major tangential movement.

This stable foreland, solidly anchored to its basement, is presumably represented to a large extent by the Apulian Ridge which crops out in Pouilles and western Gargano. Its extension to the west is masked by the Apennine nappes and its extension under the central Adriatic is still more conjectural. If it terminates there tectonically and paleogeographically, then the floor of the central Adriatic must essentially be Ionian and linked to the Marche–Umbria Trough (Aubouin). If on the other hand the Apulo-Garganic Ridge links with the Dalmatian zone as far as Istria, then the foreland loop must occur further north under the Adriatic off Venice (the Dalmatian zone would then represent in some measure the internal margin of the Apulian zone). In such a case, the Ionian Trough would end north of Gargano (Pieri) and play the secondary role of an intraplatform basin.

It is clear that there is considerable uncertainty concerning the deep geology of the Adriatic Sea. It goes without saying that the closure of the Apennino-Dinaric isopic zones, following the arcs which embrace the marine region, reflect an original paleogeographic disposition accentuated by Alpine tectonic deformation. In fact, many authors have insisted upon the importance of slide movements and translation parallel to the major Dinaric structures leading to differential displacement of the major blocks of the Apennine–Dinaric system, the eastern blocks advancing the greatest distance to the north (Aubouin, 1963b). These movements could have produced a concentric piling up of units, one over the other, and the disappearance of the most external zones towards the southern Alps. They must also be considered in the more general context of the tectogenic evolution of the Mediterranean because of the importance of the consecutive phases of Miocene compression in the opening of the Ligurian Sea and the rotation of the Corso-Sardinian block (Grandjacquet et al., 1972).

D. The Paleogene Peri-Adriatic Depressions

These depressions are elongated along the general axis of the geosynclinal structures and filled at a later date by fine to coarse detrital molasse or flysch. Sedimentologically these rhythmic arenaceous deposits differ little from geosynclinal flysch, and in consequence a distinction betwen the two is often difficult and controversial. Be that as it may, the most obvious characteristic is the accumulation of thick terrigenous beds deposited in subsiding basins on top of earlier formed structures. In consequence, the late geosynclinal sediments which began to be deposited following the earlier deformation of the internal zone are less intensely tectonized by the late Alpine phases than the basement upon which they rest. They are generally discordant on earlier beds.

Such late tectonic basins are known in the Dinarides, Hellenides, and "Apennides." Some are superposed upon internal zones (Liguro-Piedmont backdeep, Pannonian Basin), others upon external zones (Italo-Dinaric molasse foredeep of Aubouin, 1963b). The latter form a fundamental element in the Adriatic area from a Tertiary paleogeographic standpoint. The late geosynclinal Italo-Dinaric basin is elongated along a NW–SE axis and extends from the border of the southern Alps to the Gulf of Tarento (Fig. 1). Within the basin several regions may be recognized.

1. The Venetian Foredeep

This region contains sandy Miocene beds which in the east of the trough rest discordantly on the folded geosynclinal series, while on the western side the beds appear concordant. This asymmetry results from post-Oligocene (and pre-Burdigalian) folding which affected the Frioulane region and its Slovenian margin (Aubouin, 1963a).

2. *Padan Talus Slope*

Along the northern Apennines and the length of the Po plain (Bologna, Modena, and Parma) as well as in the Romagna foothills, Cenozoic clastic terrigenous deposits rest discordantly on the underlying geosynclinal beds. This applies to the sandy turbidites and molasse Ranzano–Bismantova sequence of Upper Eocene to Middle Miocene age deposited along the northern margin of the Apennines.* The Padan late geosynclinal beds were fed by detrital material from the adjacent tectonized regions and from the front of the flysch nappes and ophiolites as they slid over the Apennines.

On the Adriatic side (Fig. 4) near San Marino (in the Marrechia Valley) the Tuscan and Ligurian allochthons underlie biostromal calcareous breccias and calcarenites of Langhian–Helvetian age from which they are separated by a coarse conglomerate. Locally, the top of the sequence is formed of sandstones and glauconitic marls or Tortonian (Montebello) marls. All these sediments have been involved in Late Mio-Pliocene folding. According to Sestini (1970), important horizontal displacements toward the east and northeast may have moved these late geosynclinal sediments even while they were forming. They enclose olistolites, some of which contain xenoliths (clasts) with a Cretaceous–Eocene microfauna. Along the external Apennine margin from Pavia to San Marino, slices of chaotic sandy–calcareous Paleogene material are intercalated (tectonically or sedimentary?) in a Miocene molasse complex.

3. *Italo-Dinaric Foredeep*

South of Romagna, the Italo-Dinaric foredeep continues along the margin of the Marche Province and cuts obliquely across the Gargano Peninsula from Molise and Lucania to Bradano, thus separating the Apennine margin from Pouilles and Gargano. In this Apennine external trough thick conglomerates, sandy and marly molasse, and flysch accumulated. These beds are of varying aspect and age so that diverging interpretations have been proposed (but not considered here). The Molise Series, several thousand meters of Upper Miocene–Pliocene terrigenous sediments, includes older (Upper Cretaceous–Eocene–Oligocene) reworked sediments. The Molise Trough was superposed on the internal basement margin of the Apulian platform. According to Pieri (1966), this superposition was contemporaneous with the uplift of the Apennines which reactivated sliding of the still plastic Miocene beds which themselves represent the reworking of sediments of differing age. The Pliocene of Molise–Lucania is equally involved in Late Pliocene tangential movements (see below).

---

* Alpine as well as other sources of the detrital quartzose material may be envisaged, especially to the west, before the rotation of the Corso-Sardinian massif (Nairn and Westphal, 1968) and the opening of the Ligurian Sea (Grandjacquet *et al.*, 1972).

In short the molasse-flysch trough developed over the external Apennine zones, the Apulian in the south, the Marche–Umbria in the north, as a common, actively subsiding foredeep, the subsidence of which began earlier in the north-west and continued until the end of the Miocene. Although there is no surface continuity of outcrop of the late geosynclinal deposits along the Po Basin margin from Romagna to Emilia, there does not seem to have been a clear limit between the Apennino-Dinaric foredeep and the Liguro-Piedmont back-deep in which equally thick Oligocene and Miocene sandy–marly beds and conglomerates were also accumulating. The distinction is thus relative, a function of their position with respect to the internal and external zones of the geosyncline.*

4. *The Dalmatian–Albanian Intratrough*

This trough corresponds to two Miocene depressions adjacent to the Yugoslav littoral, a northern, Dalmatian Basin from Cres to Split and a south-ern, Albanian–Dalmatian Basin from western Albania to the mouth of the Neretva (Fig. 6) (see Turk, 1971). In Albania the intratrough became distinguish-able following the orogenesis of the Ionian zone. Detrital marine sediments, conglomeratic, arenaceous, and argillaceous, were deposited unconformably upon the folded internal margin of the Ionian zone, but are in continuity in the central part (Aquitanian molasse, Burdigalian clays and marls of the lower Helvetian). To the east, upper Helvetian beds are transgressive and discordant, molassic in character, and reach up to the Tortonian. The Albanian intratrough area is characterized by gentle folds paralleling earlier fold axes (Papa, 1970).

III. RECENT EVOLUTION OF THE ADRIATIC–MEDITERRANEAN DOMAIN

This interval includes the final stage of the Alpine cycle, that is, the post-geosynclinal stage as defined by Aubouin (1963*b*, 1965). It reflects a period of tectonic relaxation in the different isopic geosynclinal zones.

The clastic sedimentation characteristic of this post-orogenic epoch began at the end of the Miocene (Messinian) and continued to the Pleistocene. The

---

* The same is true for the distinction of the different stages in the development of the geo-syncline (*s.s.*), particularly in the northern Apennines. The migration of orogenesis means that the same lithological sequence, i.e., the Messinian gypsum (Gessoso solfifera) may belong to three different stages, miogeosynclinal in the Romagna–Marche piedmont, late geosynclinal in the Modena–Bologna Apennines, and post-geosynclinal in southern Tuscany. Different sequences of the same age may characterize successive stages (Ranzano–Bismanova–Scaglia cinerea–Marnoso–Arenacea, end of Eocene to beginning Miocene, late geosynclinal and geosynclinal, respectively).

post-geosynclinal paleogeography, independent of the earlier paleogeography, is reflected in Neogene basins whose age, facies, and deformation vary according to their position with respect to present relief. Two basic types may be distinguished.

A. Neogene Intramontane Basins

These basins result from tensional stresses. In general, they date from the end of the Miocene (Pontian deformation) and the end of the Pliocene (Villafranchian movements). They may be superposed over the paleogeographic zones of the geosynclinal phase and over tectonic units, thus demonstrating their total independence of the isopic zones and structures.

In this category may be grouped the peri-Adriatic intramontane basins developed over the miogeosynclinal domain. For morphological reasons linked to lithology (carbonate basement) and fracture tectonics, these basins are particularly developed in Yugoslavia, notably in the external Dinarides of the littoral where they form the poljes of the Yugoslav karst. The faulted Livno Basin in western Bosnia, contains more than 2000 m of terrestrial lacustrine sediments. Here, resting discordantly upon Mesozoic and Paleogene limestones, are conglomerates, marls, sands, and limestone with beds of Lower to Upper Miocene coal upon which again rests a discordant sequence of detritals with lignites of Pontian or Mio-Pliocene age. There are many other Mio-Pliocene lacustrine basins spread throughout the Dalmatian region (Fig. 3) and these are related to late geosynclinal tectonics with which are associated a network of normal–subvertical faults with throws of up to several thousand meters. The same neotectonic activity also controlled volcanicity, seen in the more internal basins, and is presumably responsible for the present seismic activity.

The major Neogene internal basins, whether superposed over late geosynclinal troughs or not, have the same tensional origin. This is true not only for the Pannonian Basin within the Balkans, but also of the Tuscan Basin along the Tyrrhenian coast where Neogene beds are transgressive and discordant. The first is a vast interior sea (Paratethys) which was separated from the Tethys in Middle Miocene time and evolved into a lagoon and finally a lake during the Upper Miocene and Pliocene. It received detrital sediments (sandstones, marls, and conglomerates) and is characterized by a specialized lacustrine fauna (Levantine Paludinas, during the Ponto-Pliocene). Dacitic and andesitic tuffs provide evidence of volcanic activity at this time. The second was a subsiding basin in which, during the Messinian, evaporitic and lacustrine conditions reigned. A marine incursion occurred during the Pliocene with a general regression at the end of that interval. Vertical movements (horst–graben structures) predominated during the Plio-Quaternary, affecting the whole region.

Thus, the subsiding Neogene basins are characterized by active detrital sedimentation linked to the erosion of the marginal relief during a period of active tectonism, the Neogene beds being transgressive and discordant upon the older beds. All these movements may be attributed to an unstable crust and strongly influenced the conditions of deposition, and were presumably responsible for the rhyolitic and trachy-basaltic effusive and pyroclastic activity which prevailed in some parts of the Italian Peninsula.

B. Peri-Adriatic Mio-Pliocene and Quaternary Basins

These basins are filled with argillaceous–arenaceous sediments which extend under the Po plain and border the Italian Adriatic coast as far as Bradano. There is also a continuation of the Pliocene basin of the Albanian littoral off the Otranto Straits. This distribution (Fig. 5) shows a Dinaric trend over which the Adriatic Basin is superposed. Several of the basins occupy an important place in the structure of the Adriatic.

1. *The Po Basin and Northern Venetia*

More than 7000 m of Pliocene and Quaternary sediments have accumulated in the Po Basin. This foredeep follows the valley of the Po and in Emilia is the northern Apennine margin and the coastal region of Romagna. It continues southeastward into Marche (Fig. 6, Profiles 2 and 3). Mio-Pliocene beds are poorly represented along the northern edge of the Po Basin while there is a thickening of the Quaternary beds in Venetia, the result of increasing subsidence in this direction. Along the Padan margin, outcrops are small and isolated and only in Romagna–Marche do we find more typical sequences with thicknesses reaching 1000 m.

The Messinian succession rests upon Tortonian marls and consists, from bottom to top, of bituminous marls or sandy marls (Gessoso solfifera) followed by diatomaceous beds overlain in turn by siliceous sulphur-rich beds in which gypsum and dolomite are intercalated. This marine ensemble contains beds of tripoli,* a facies well known in the Mediterranean region. The rapid facies changes underline the tectonic instability at this time. Resting upon these beds are clays with lenticular conglomerate horizons and bands of lagunal limestone.

Deposition of the sandy argillaceous and silty Pliocene beds of the Po Valley began with a marine transgression which ended in the Middle Pliocene, a time in which tectonic movements produced a regression marked by local discordances, by folds and by faults inclined in opposite directions on the two sides of the Po axis.

The top of the Middle Pliocene and the Upper Pliocene are argillaceous and on occasion may reach a thickness of 1000 m. The basal marine clastic

---

* Tripoli, a local term given to diatomites.

(Calabrian) beds, as well as the fluviatile–lacustrine and continental Pleistocene and Holocene (Van Straaten, 1970) are unfolded and variable in thickness, ranging up to a 2000 m. It is worth noting that the Pliocene and Lower Pleistocene beds have been affected by important vertical movements in Middle and Upper Pleistocene time, which uplifted the Apennines as much as 1000 m above sea level, rejuvenating the drainage pattern.

The tectono-sedimentary evolution of the post-geosynclinal cycle along the external margin of the northern Apennines is complex. The Messinian and later formations result from the erosion of eu- and miogeosynclinal beds tectonized by earlier orogenic phases, and are themselves affected by deformation at the end of the Tertiary as a consequence of the general eastward migration of the orogenic wave. The Messinian Sea was thus pushed further and further to the east and northeast as folding reached these regions (Upper Miocene–Lower Pliocene in Emilia, Lower–Middle Pliocene in Marche). A phase of extension which followed the orogenic wave resulted in subsidence and magmatic activity in the internal basins. Very active subsidence characterized the external domain, converting it to a foredeep into which allochthonous masses continued to slide or be horizontally displaced, thus continuing the movements begun in the Miocene.

What was the origin of all the detrital and terrigenous material piled into these troughs and post-geosynclinal foredeeps? In the case of the northern Apennines, there is little doubt that it must be sought in the more internal parts of the chain. In Emilia and Romagna, the Ligurian and Tuscan nappes constitute the prime source of argillaceous–arenaceous sediments as these allochthonous rocks were transported earlier into the area. In Marche, the Umbrian autochthon (Marnoso–Arenacea sandstones, Tortonian marls) probably provided part of the clastic material, with a contribution from the crystalline massifs of the Alpine foreland north of the Adriatic.

2. *Bradanic Trough*

The Plio-Quaternary series of Marche is continued to the southeast along the Adriatic coast of the Abruzzi to Gargano. It then forms a narrow band about 50 km wide between the molasse basin and Miocene flysch of Molise and the carbonate "table" of Pouilles and Murge, and extending to the Gulf of Tarento, following the course of the Bradano River (whence the name). The basement of this asymmetric tectonic trough rises to the north in a series of step faults while the southern margin is abrupt, formed by a series of subvertical faults with throws reaching 2000 m (Ghezzi and Marchetti, 1964). It is bordered to the southeast by a ridge (Dorsale de Rotondella) which separates it from the Miocene Molise Basin. This faulted structure of horsts and grabens is imposed upon a Mesozoic (? Cretaceous) carbonate basement, a westward extension of the Apulian foreland. It is clearly defined by strong negative gravity anomalies over the basins.

The marine marly–argillaceous succession contains coarse, sandy inter-calations at the base of the Pliocene. The sea was progressively pushed to the northeast in a restricted but deep basin. The Lower and Middle Pliocene phase of Alpine orogenic activity uplifted the southwest flank of the basin, triggering gravity slides of chaotic masses of flysch into the center of the basin (olisto-stromes) while at the same time more rapid subsidence of the trough resulted in new marine incursions. In short, the Plio-Quaternary sedimentation in the Bradanic Basin was profoundly influenced by its tectonic history which resulted in discordances, lacunas, and paraunconformities in a dominantly argillo-arenaceous sequence, which in the deeper parts of the basin is in excess of 3000 m thick. Lavas were erupted along the major fault along the western basin margin (Vulture).

3. *The Albanian Peri-Adriatic Basin*

This molasse basin extends over the Albanian littoral covering part of the Ionian zone, particularly in the low hills between Durres, Kavaja, Fier, and Vlore. It continues under the Adriatic and its external margin rests on the Dalmatian–Albanian intrabasin. The Messinian strata which crop out along the margin (Kavaja) are characterized by evaporites (clay and gypsum) which mark the "salinity crisis" between Miocene and Pliocene. Sometimes the Messinian gypsiferous clays are complete and progressively overlain by Plio-cene, on other occasions Pliocene rests directly upon lower members of the series (Pashko, 1973). The Pliocene marine beds are generally transgressive, essentially argillaceous, and usually have a characteristic planktonic microfauna. The upper part, which is conglomeratic, passes up to Pleistocene clastic beds.

C. Plio-Quaternary of the Adriatic Area

In its present form the Adriatic Sea represents the end of the evolution of a Tertiary basin which is becoming better known as the result of geophysical research (Morelli *et al.*, 1969) and offshore drilling (to 6000 m). This submarine exploration has been particularly intense in the last few years in the northern, Italian zone of the northern Adriatic between Venice and Gargano (Vercellino, 1970).

The northern Adriatic Basin (Fig. 6B), the submarine continuation of the Po Basin, is filled with more than 7000 m of sandy and argillaceous Pliocene beds. The filling of the basin occurred as the result of terrestrial material pro-grading along the Apennine margin. The northern Adriatic Basin is asym-metrical. The basement rises and joins the Istrian platform to form a ridge in the central Adriatic off Ancona. The ridge serves to separate the basin from the Miocene basin of northern Dalmatia (Fig. 6B). The western margin of the basin is involved in Apennine overthrusts which affect even the youngest beds. In contrast, the Dalmatian coast is effected only by subsidence and rejuvenation of relief (Fig. 6A). The basin deepens southwards and between Ancona and Pescara the Pliocene thickens to more than 5000 m. It is closed abruptly by the Gargano Ridge which crosses the Adriatic, dividing it into two basins.

The southern Adriatic Basin (Fig. 6C) is much the deeper of the two. To the west it joins with the Apulian platform (Fig. 6D); however, to the east it is limited by reverse faults in front of late Dinaric structures which continued to be active during the Miocene and even into the Pliocene in the external Ionian units (Fig. 6D). In the axial region of the basin where it communicates with the Ionian Sea through the Otranto Straits, considerable thicknesses of Plio-Quaternary sediments (more than 2500 m) have been detected. The sequence thins to the west and shows progradational features near the coast.

In summary, a new paleogeography developed during the Neogene in the Adriatic regions. The Miocene epicontinental seas, little by little, were rolled back toward the common, Apulian foreland of the Dinarides and Apennines. The deposits reflect the continuing tectonic instability of the time and the rapid changes in environment. Neotectonic activity was important (Aubouin, 1970) and corresponds in general to extensional readjustments, although as has been seen, late compressive phases persist to the present day as witnessed by the numerous earthquakes along the fractures and by major displacements along the margins of the basins.

It is interesting to note that in the northern Adriatic Basin the stable foreland lies along the Yugoslav coast, while in the southern Adriatic it is on the Italian shore. This inversion occurs across the Gargano–Split (Velebit) axis, already shown to correspond to a zone of positive gravity anomalies as well as an important offset in the isopic zones (Maksimovic and Sikosek, 1971). This could be explained by a right-lateral shear leading to differential displacement and translation along an arc which follows structures known in the Apennines (Anzio–Ancona line in particular). This situation recalls such major transverse virgations as that of Sarajevo or the Scutari–Pec transverse fault, already interpreted as an ancient transform fault (Aubouin et al., 1970; Aubouin and Dercourt, 1975).

IV. MAJOR STAGES IN THE GEOLOGICAL EVOLUTION OF THE ADRIATIC REGION

In the following section the major steps in the history of the Dinaric and Apennine chains will be retraced in the more general framework, taking into account the evolution of the sedimentary regions on a Tethysian scale.

A. Geotectonic Development

Without entering into details of the numerous recent syntheses of which that of Dewey *et al.* (1973) is the latest, it is apparent that the Alpine edifice of the Mediterranean was formed from a complex of continental fragments and oceanic basement compressed between the major cratonic blocks of Africa and Europe. Within this grand scheme the external Mesozoic domains of the Apennines and Adriatic, and the present peri-Adriatic zone, were part of an Apulian block which at the end of the Paleozoic formed part of the African craton (Argyriadis, 1975). In the semi-enclosed marginal sea which occupied this place during the Permo-Triassic, evaporites as well as clastic and terrigenous deposits were laid down. This was the initial stage.

The separation of the Apulian block and the opening of a Mesogean paleo-Ocean in the early Mesozoic marked the beginning of an important phase in the evolution of the Tethys, the stages of which have been sketched by Dercourt (1970) and defined by Smith (1971), who integrated them into a general scheme linked with the evolution of the Atlantic. The phases of this evolution have been described in detail by Dewey *et al.* (1973). Extension and spreading, resulting in separation of the Apulian block from the African and Rhodopian cratons, occurred during the Triassic and Lower Jurassic (rifting of plate margins). From the Toarcian to the Kimmeridgian (-180 to -140 m.y.) there was rapid accretion of oceanic crust and active subsidence of the continental basement of the Apulian block. The platform, more extensive than at present, had already a complex paleogeography with shoals and basins, some of the latter with a troughlike character (Budva–Cukali), while the whole may be compared, within limits, to the carbonate platform off the coast of Florida and the Bahamas (D'Argenio *et al.*, 1971; Bernouilli, 1972).

The onset of subduction and formation of active trenches along the European cratonic margin following the displacement of the Apulian block at the end of the Jurassic led to shearing, the formation of tectonic thrust sheets, and the emplacement of ophiolites.* Whatever may be the case, a new paleogeography developed in the internal zones following the compressive stress and tangential tectogenesis of the Jurassic–Cretaceous boundary interval. The closure of the Tethys followed in step with the opening of the Atlantic during the Cretaceous and Paleogene and the orogenic crises of the Tertiary. Oceanic material was piled upon the Apulian pre-continent; the magnitude of the cover being unknown, there is nothing against the view that the oceanic zone

---

* In the Dinarides the ophiolites would be the product of that part of the Tethys lying in an internal location about the Vardar zone (Rampnoux, 1973; Blanchet, 1974) or a still more internal location. It may also be supposed that certain platforms with a continental basement (Golja, for example) represent small continents separated by a marginal ocean from the main ocean (Dimitrijevic and Dimitrijevic, 1973).

was completely eliminated on both sides of the Apulian foreland and that the external zones continue great distances under the internal Apennine and Dinaric nappes.

Under these conditions, it appears that the peri-Adriatic region was attached to the Euro-African domain at the end of the Paleozoic and then became decoupled from it to form an independent block (Blanchet, 1974). During the Mesozoic it behaved as a vast submerged platform (Apulian foreland) along the margins of which, bordering the paleo-Tethys Ocean, a number of paleogeographic zones were distinguishable either as platforms or depressions. Among them are: the Apennine platforms and basins (Abruzzi–Latium–Campania and Marche–Umbria–Lucania); the Dinaric platforms and basins (High Karst, Dalmatian, Budva–Ionian). These external Apennine and Dinaric zones during the Jurassic and Cretaceous belonged to a vast geosynclinal domain covered under the term "peri-Adriatic carbonate platforms" (D'Argenio *et al.*, 1971) where each platform perhaps represents a distinct paleogeographic unit surrounded by oceanic basins (intraoceanic carbonate platforms), or else intraplatform basins?

B. Evolution of the Carbonate Platforms and Peri-Adriatic Basins

The stratigraphic reconstructions (Fig. 2) of both the Italian and Yugoslav coasts (D'Argenio *et al.*, 1971) clearly illustrate the homology in facies, age, and distribution of the peri-Adriatic carbonate platforms, suggesting a common origin during the Middle to Upper Triassic. At that time, the broad platforms were flanked by basins receiving shallow-water deposits (dolomites, evaporites).

1. The Platforms

During the Jurassic and Cretaceous, there are striking similarities between the platform reef sequences and the transitional facies at the basin margins (peri-reefal sediments, oolites, flank breccias, etc.) on both sides of the Adriatic. The synchronism and remarkable correspondence can be seen in the lithofacies succession as well as in the principal stratigraphic members—the evaporites and stromatolitic limestones of the Triassic–Lower Liassic, and the bauxitic horizons of the Lower–Middle Cretaceous, for example. Many of the biozones have identical paleontological character, both in the macrofauna (megalodont, *Lithiotis*, *Ellipsactinia*, rudist biostromes, etc.) and in the microfauna (*Orbitopsella*, *Clypeina*, *Cuneolina*, and *Globotruncana* limestones, etc.).

Nonetheless, some differences in the reef regime due to synsedimentary tectonic control of relief can be detected. Some stages of evolution may appear at different times (Jurassic bauxites in the Dinaric domain, Cretaceous evaporites of the Adriatic platform).

The major phases in the paleogeographic evolution of the platforms and adjacent peri-Adriatic basins can be summarized in four principal stages, three of which according to D'Argenio *et al.* (1971) correspond to minor transgressive–regressive cycles. These stages are:

(1) The intial end of the Triassic–infra-Liassic stage, the period during which the Adriatic carbonate platform and its two basins (Budva–Cukali and Lucania) were established. Deposition of evaporites continued over the Apulian Ridge, while on either side a littoral facies alternated with the carbonate facies of the back-reef sublittoral.

(2) The Jurassic cycle was characterized by a shortening of the Triassic platforms and a deepening of the basins. In many regions the ridge with its reefs showed a tendency to emerge at the end of the Malm.

(3) The Upper Malm–Lower Cretaceous cycle began with an expansion of the platforms and reefs, accompanied by a migration of the marginal facies toward the basins. During the Albian–Lower Cenomanian, the reef platform became emergent at several points. This was the time of greatest emergence during the Mesozoic when many of the bauxite deposits formed.

(4) The Upper Cretaceous–Paleogene cycle began with a transgression of the rudist sea over the continental beds. There was a progressive reduction in the size of the platforms and basins with the formation of some new intra-platform depressions (Jelaska, 1973). This breakup of the peri-Adriatic platforms, accompanied by some emergence during the Maestrichtian–Paleocene, was the precursor of a new paleogeography which became established during the Tertiary.

It has been shown that during the Tertiary terrigenous sediments, little by little, invaded the platform region. In the writer's opinion, this had several causes of which the two most important were tectonic and geodynamic. Tectonically, a significant shortening became apparent during the Paleocene, resulting from the progressive orogenesis of the internal zones. This created a new paleo-environment and the erosion of the newly formed relief provided a considerable mass of detrital sediment which gradually invaded the platforms. Geodynamic activity led to a subsidence of the crust along the Apulian foreland margins, which, however, remained stable until the Middle Miocene (see below).

There remains however a fairly distinct delay in the timing of events on the two sides of the Adriatic. The principal phases in the Apennine orogeny and in the appearance of flysch are more recent in the external zones (Lower–Middle Miocene) than the corresponding events in the Dinarides, where the invasion of terrigenous sediments dates from the Middle–Upper Eocene.

2. *The Basins*

The evolution of the peri-Adriatic sedimentary basins poses still more complex problems because their relatively thin, plastic material was often ejected onto the adjacent platforms. Their presence is sometimes revealed only by the existence of marginal basin facies with calcarenites, turbidites, calcirudites, and other bioclastic beds in which metasomatic cherts may be formed. The most difficult question concerns the nature of the substrate. In some cases, where the movement has been small (Marche–Umbria, Ionian Basin), it is probably continental or at least it belonged to the continental margin of the Apulian craton. In other cases, the existence either of oceanic crust (formed by spreading?) or an intermediate, intra-arc or oceanized? continental crust may be imagined. Such would be the case for the Budva–Cukali Basin, an appendage of the Pindus Trough, which could be a marginal oceanic basin between the Apulian shelf and island arc (of High Karst type) or further from the continent (Golijian and others), but whatever the interpretation, distinct from the main Tethys.

The significance of the Lucanian Trough or Basin remains uncertain in view of its present tectonic situation and the major nappe movements of the southern Apennines. The Lucanian zone could well be of a more internal origin than the calcareous zones surrounding it.

In summary the Adriatic and the bounding areas to the northeast and southwest consist of:

(1) The foreland of the Apulian plate, a zone of true continental crust, which expands to the southeast.

(2) The Dinaric margin with its platforms and basins, isolated from the paleo-Tethys by a system of arcs and troughs.

(3) The Apennine margin, also with its own paleogeography, separated from the ocean by ridges and platforms, a large part of which is now below the Tyrrhenian Sea.

C. Geodynamics of the Adriatic Area

To the writer's knowledge no bore has penetrated the crust either under the Adriatic or in the peripheral continental area, despite their depth (Foresta Umbra well, Gargano, 5912 m; Zadar well, 4435 m). Geophysical data, however, permit the construction of crustal profiles (Morelli *et al.*, 1969; Roksandic, 1966). Seismic refraction profiles, for example, indicate a sediment thickness of 10 to 12 km, with local variations to 8 km in the central Adriatic.

1. *Crustal Structure*

A preliminary and still incomplete sketch based upon gravimetric and aeromagnetic data and the distribution of seismicity as developed by Morelli shows: (1) The Mohorovičíc Discontinuity up to 50 km below the Apennines and more than 50 km below the external Dinarides, but rising slightly under the Adriatic. This crustal thickening is in strong contrast with the upwarping of the mantle under the Tyrrhenian Sea, particularly off Sicily. According to Morelli, the tangential movements of the Apennines and Dinarides could be explained by the combined effects of gravity due to the uplift of the Moho, and compression followed by relative displacement. (2) The distribution of gravity anomalies is highly significant. Strong negative isostatic anomalies, perhaps the sign of an active orogenic front, are observed in the Romagna Apennine foreland and along the eastern margin of the southern Dalmatian Basin. In this context, an asymmetry between the two sides of the Adriatic can be noted. The Dalmatian–Istrian littoral is a stable, positive area while the southern, Albanian–Yugoslav margin is unstable. In contrast, on the Italian side it is the northern Adriatic shore off the Po and Adriatic Trough which is unstable with strong negative anomalies, while the Apulian littoral is a stable zone. The region north of the Split–Gargano Ridge is distinctly more active (greater seismic energy release) than the region to the south. The areas of strong negative anomalies are marked by rapid* subsidence and a deeply buried rigid limestone subbasement. Thus to the northwest (extension of the Po Valley) and southeast (western Albania) of the Adriatic, two slopes link the subsiding troughs to the carbonate platforms of Istria on the one hand and Apulia on the other. Each of the two recent depressions just described represents a true basin at the continental shelf margin from a geological standpoint. The thick, plastic evaporites which underlie the rigid carbonate carapace act as a lubricant on both sides of the subsiding Adriatic Sea.

In conclusion, the present Adriatic is a stable region against which two unstable areas, the Apennines and Dinarides, have reacted as a result of convergent compressional forces.

2. *Structural Genesis*

The geophysical data indicate a marked crustal thickening in the peri-Adriatic region, resulting from compression and folding during the emplacement of the Cenozoic nappes. The rise of the Moho can then be regarded as the result of mechanical extension, linked to foundering and neotectonic dislocations, in other words, following post-orogenic isostatic readjustment. The evolution of the Adriatic crust can thus be summarized in retrospect. The present tensional phase was preceded by a major period of Cenozoic compression (closure, see above) which saw the creation of large crustal intumescences at the margins of the Apulian platform (dip of the Moho to 50 km or more). At this stage, intracrustal decollement may have lead to the formation of a

---

* In the Po delta, the rate of subsidence since the Pleistocene is of the order of 2 mm/yr.

basement (Paleozoic) thrust sheet along the margins of the internal Dinaric zone.*

In considering the paleogeographic evolution during the Mesozoic, of the region studied, even after allowing for unfolding of a region reduced to about half its width by Alpine tectonics, it is still necessary to assume crustal stretching and a markedly thinner crust under the peri-Adriatic external zones which had the character of subsiding marginal seas bounding the continental Apulian block (Adriatic plate) on both sides of the Tethys.

V. CONCLUSIONS

Geological knowledge of the Mediterranean in general, and the Adriatic in particular, has made great strides in the last few years, the result of continuing field work, increased offshore and onshore drilling, and geophysical studies which provide a clearer idea of deep structure. The preceding paleogeographic review recounts the basic stages in the evolution of the geosyncline in the Adriatic region, an evolution, which from a quasi-uniform Triassic stage became progressively more complex during the course of the Alpine cycle. This cycle began after an episode of Permo-Triassic evaporite formation and ended with the Messinian salts (now under the Adriatic), marking the beginning of a new cycle of which the present Mediterranean is an evolutionary stage.

Within this general framework, isopic zones, platforms and basins developed along the margins of the stable foreland. Here carbonate (or siliceous) sedimentation which predominated during the Mesozoic gave way, little by little, during the Cenozoic to terrigenous sediments.

Questions still unresolved concern the significance of zones of pelagic sedimentation, e.g., the Budva–Cukali Trough, Lucanian Series, and their structural evolution. It is important also to better understand the mechanism and environment of salt deposition (deep marine or not, lagunal?), for present sedimentation models are often incomplete or do not always clearly reflect the paleogenetic conditions.

The history and distribution of the peri-Adriatic isopic zones with respect to the Apulian foreland suggests that they belong together, justifying to a certain extent the idea of an Apulian plate or sub-plate. The paleogeographic evolution of the region was controlled by the kinematics of the two major cratons, European and African, which, during their approach, led to the complex translational motions of the Apulian microcontinent. Clearly this interpretation remains hypothetical as does the nature of the crust below deep basins where present geophysical data cannot indicate past significance.

---

* Regions of high heat flow under which, according to Scheffer (1963), the crystalline basement became mobile and could deform readily (Sana, Praca, Bijelo-Polje, etc.).

The rate of deformation, generally slow in the peri-Adriatic external zones, also reflects the particular characteristics of the region. The fundamental structural feature is the convergence on both sides of the Adriatic of tectonic stress and its attenuation towards the southern part of the Apulian axis.

Many problems remain, in particular the continuation of the isopic zones under the Adriatic and their interrelations there (does the axis of tectonic symmetry coincide with the axis of symmetry of the isopic zones? In other words does it pass into the Adriatic–Ionian Basin or the Apulian? Or are the paleogeographic and structural frameworks independent?). The questions of eventual links between the Apennines and the southern Alps and Dinarides under the Po plain, of the tectonic relationship of the Italo-Yugoslav ensemble with the northern Alpine Arc also remain. Finally there is the question of the significance of the major Dinaric and Apennine transverse structures.

Despite the complexity of the problems which remain to be solved and the major advances already made, the Adriatic remains on exciting region for research, rich in hope and promise.

ACKNOWLEDGMENTS

This review has benefited from Alpine research, in particular in the Dinarides, of my colleagues at the Laboratoire Associé no. 215, C.N.R.S. Heartfelt thanks are accorded to all those geologists, French and Italian, who have shared the secrets of their field areas and with whom I have had many profitable discussions. My thanks are also due to Dr. A. E. M. Nairn for his translation of this article.

REFERENCES

Abbate, E., and Sagri, M., 1970, Development of the Northern Apennines geosyncline. The eugeosynclinal sequences, *Sed. Geol.*, v. 4, p. 251–340.

Argyriadis, I., 1975, Mésogée permienne, chaîne hercynienne et cassure téthysienne, *Bull. Soc. Géol. France*, v. 17, p. 56–57.

Aubouin, J., 1960, Essai sur l'ensemble italo-Dinarique et ses rapports avec l'arc alpin, *Bull. Soc. Géol. France*, v. 2, p. 487–526.

Aubouin, J., 1963*a*, Essai sur la paléogéographie post-triassique et l'évolution secondaire et tertiaire du versant sud des Alpes orientales (Alpes méridionales; Lombardie et Vénétie, Italie; Slovénie occidentale, Yougoslavie), *Bull. Soc. Géol. France*, v. 5, p. 730–766.

Aubouin, J., 1963*b*, Esquisse paléogéographique et structurale des chaînes alpines de la Méditerranée moyenne, *Geol. Rundsch.*, v. 53, p. 480–534.

Aubouin, J., 1965, *Géosynclines*, Amsterdam: Elsevier Publ. Co., 355 p.

Aubouin, J., 1970, Réflexion sur la tectonique de faille plio-quaternaire, *Geol. Rundsch.*, v. 60, p. 833–848.

Aubouin, J., 1972, Chaînes liminaires (Andines) et chaînes géosynclinales (Alpines), *24th Int. Geol. Congr.*, Sect. 3, p. 438–461.

Aubouin, J., and Dercourt, J., 1975, Les transversales dinariques dérivent-elles de paléofailles transformantes? *C. R. Acad. Sci. Paris, Sér. D*, p. 347–350.

Aubouin, J., Bosellini, A., and Cousin, M., 1965, Sur la paléogéographie de la Vénétie au Jurassique, *Mem. Geopal. Univ. Ferrara*, v. 1, fasc. II, p. 147–158.

Aubouin, J., Blanchet, R., Cadet, J. P., Celet, P., Charvet, J., Chorowicz, J., Cousin, M., and Rampnoux, J. P., 1970, Essai sur la géologie des Dinarides, *Bull. Soc. Géol. France*, v. 12, p. 1060–1095.

Bébien, J., Blanchet, R., Cadet, J. P., Charvet, J., Chorowicz, J., Lapierre, H., and Rampnoux, J. P., 1977, Le volcanisme triassique des Dinarides. Sa place dans l'évolution géotectonique mésogéenne, *Tectonophysics* (in press).

Bernouilli, D., 1971, Redeposited Pelagic sediments in the Jurassic of the central Mediterranean area, *Ann. Inst. Geol. Publ. Hungarica*, v. 54 (2), p. 71–90.

Bernouilli, D., 1972, North atlantic and mediterranean mesozoic facies: a comparison, *Initial Reports of Deep Sea Drilling Project*, v. 11, p. 801–871.

Bignot, G., 1972, *Recherches Stratigraphiques sur les Calcaires du Crétacé Supérieur et de l'Eocène d'Istrie et des Régions Voisines. Essai de Révision du Liburnien*, Thesis, Paris, 353 p.

Biju-Duval, B., Letouzey, J., Montadert, L., Courrier, P., Magniot, J. F., Sancho, J., 1974, Geology of the Mediterranean sea basins, in: *The Geology of the Continental Margins*, Burke, C. A., and Drake, C. L., eds., New York: Springer-Verlag, p. 695–721.

Blanchet, R., 1974, De l'Adriatique au Bassin Pannonique. Essai d'un modèle de chaîne alpine, *Mém. Soc. Géol. France*, v. 53, no. 12, p. 1–172.

Blanchet, R., Cadet, J. P., Charvet, J., and Rampnoux, J. P., 1969, Sur l'existence d'un important domaine de flysch tithonique–crétacé inférieur en Yougoslavie: l'unité du flysch bosniaque, *Bull. Soc. Géol. France*, v. 11 (7), p. 871–880.

Blanchet, R., Cadet, J. P., and Charvet, J., 1970, Sur l'existence d'une unité intermédiaire entre la zone du Haut-Karst et l'unité du flysch bosniaque en Yougoslavie: la sous-zone prékarstique, *Bull. Soc. Géol. France*, v. 12 (7), p. 227–236.

Boccaletti, M., Elter, P., and Guazzone, G., 1971, Plate tectonic model for the development of the western Alps and Northern Apennines, *Nature*, v. 234, p. 108–111.

Bortolotti, V., Passerini, P., Sagri, M., and Sestini, G., 1970, Development of the Northern Apennines geosyncline—The miogeosynclinal sequences, *Sed. Geol.*, v. 4, p. 341–444.

Boskov-Stajner, Z., 1971, Prilog stratigrafiji jadranskog oboda, *Nafta*, Zagreb, v. 22, p. 270–274 (English summary).

Cadet, J. P., 1970, Esquisse géologique de la Bosnie–Herzégovine méridionale et du Monténégro occidental, *Bull. Soc. Géol. France*, v. 12 (7), p. 973–985.

Caire, A., 1970, Tectonique de la Méditerranée centrale, *Ann. Soc. Géol. Nord*, Lille, v. 90, p. 307–346.

Catenacci, E., De Castro, P., and Sgrosso, I., 1964, Complessi-guida del Mesozoico calcareo-dolomitico nella zona orientale del Massiccio del Matese, *Mem. Soc. Geol. Ital.*, Bologna, v. 4, p. 837–856.

Charvet, J., 1970, Aperçu géologique des Dinarides aux environs du méridien de Sarajevo, *Bull. Soc. Géol. France*, v. 12, p. 986–1002.

Chorowicz, J., 1970, La transversale de Zrmanja aux confins de la Croatie et de la Bosnie–Herzégovine, *Bull. Soc. Géol. France*, v. 12, p. 1028–1033.

Chorowicz, J., 1975, Le devenir de la zone de Budva vers le Nord-Ouest de la Yougoslavie, *Bull. Soc. Géol. France*, v. 17, p. 699–706.

Colacicchi, R., and Praturlon, A., 1965, Stratigraphical and paleogeographical investigations on the Mesozoic shelf edge facies in eastern Marsica (Central Apennines, Italy), *Geol. Romana*, v. 4, p. 80–118.

Cousin, M., 1970, Esquisse géologique des confins italo-yougoslaves leur place dans les Dinarides et les Alpes méridionales, *Bull. Soc. Géol. France*, v. 12, p. 1034–1047.

Dalipi, H., Kondo, A., Pejo, I., Ikonomi, J., and Mecaj, B., 1971, Stratigraphy of the Deposits of the Mesozoic in Southern and Western Albania (outer Albanids), *Nafta*, Zagreb, v. 4–5 (XXII), p. 227–253.

D'Argenio, B., 1964, Una transgressione del Cretacico superiore nell'Appennino Campano, *Mem. Soc. Geol. Ital.*, Bologna, v. 4, p. 881–933.

D'Argenio, B., Radoicic, R., and Sgrosso, I., 1971, A paleogeographic section through the italo-dinaric external zones during Jurassic and Cretaceous times, *Nafta*, Zagreb, v. 22, p. 195–207.

Dercourt, J., 1970, L'expansion océanique actuelle et fossile: ses implications géotectoniques, *Bull. Soc. Géol. France*, v. 12, p. 261–317.

Dewey, J. F., and Bird, F. M., 1970, Mountain belts and the new global tectonics, *J. Geophys. Res.*, v. 75, p. 2625–2647.

Dewey, J. F., Pitman, W. C., Ryan, W. B. F., and Bonnin, J., 1973, Plate tectonics and the evolution of the Alpine system, *Bull. Geol. Soc. Amer.*, v. 84, p. 3137–3180.

Dimitrijevic, M. D., and Dimitrijevic, M. N., 1973, Olistostrome melange in the Yougoslavian Dinarides and late Mesozoic plate tectonics, *J. Geol.* v. 81, p. 328–340.

Farinacci, A., and Radoicic, R., 1964, Correlazione fra serie guiresi e cretacee dell'Appennino centrale e delle Dinaridi esterne, *La Ricerca Scientifica*, Roma, Rendiconti A, v. 7, p. 269–300.

Ghezzi, G., and Marchetti, M. P., 1964, Contributo alla conoscenza stratigrafica e sedimentaria del Terziario superiore della Calabria e Basilicata, *Mem. Soc. Geol. Ital.*, v. 4, fasc. II, p. 1155–1174.

Gjata, T., Skela, V., Ylly, J., and Kici, V., 1971, Stratigraphy of Paleogenic Deposits in Western and Southwestern Albania (Outer Albanids), *Nafta*, Zagreb, v. 22, p. 208–217.

Glangeaud, L., 1966, Les méthodes de la géodynamique et leurs applications aux structures de la Méditerranée occidentale, *Rev. Géogr. Phys. Géol. Dyn.*, Paris, v. 10, p. 83–135.

Grandjacquet, C., 1963, Schéma structural de l'Apennin Campano-lucanien (Italie), *Rev. Géogr. Phys. Géol. Dyn.*, Paris, v. 5, p. 185–202.

Grandjacquet, C., Haccard, D., and Lorenz, C., 1972, Essai de tableau synthétique des principaux évènements affectant les domaines alpin et apennin à partir du Trias, *C. R. Somm. S.G.F.*, p. 158–163.

Hsü, J. K., 1967, Origin of large overturned slabs of Apennines, Italy, *Bull. Amer. Assoc. Petrol. Geol.*, v. 51, p. 65–72.

I.G.R.S.–I.F.P. (Institut de Géologie et Recherches du Sous-Sol, Athènes, et Institut Français du Pétrole), 1966, *Etude Géologique de l'Epire (Grèce Nord-occidentale)*, Paris: Editions Technip, 306 p.

Jelaska, V., 1973, Paleogeographic and oil-geologic considerations of the West part of the Dinaride carbonate shelf, *Geol. Vjesnik*, Zagreb, v. 25, p. 57–64.

Kochansky-Devidé, V., 1964, Die Mikrofossilien des Yugoslawischen Perm, *Paleont. Z.*, Stuttgart, v. 38, p. 180–188.

Kondo, A., 1970, De la présence de structures ioniennes à partir du Jurassique inférieur et de leur développement sous-marin, *Bull. U.S.H.T.*, Tirana, no. 4.

Kosmatt, F., 1924, Geologie des zentralen Balkan Halbinsel, in: *Die Kriegsschauplatze*, v. 12, Berlin: Borntraëger Ed.

Laubscher, H. P., 1970, Das Alpen-Dinariden-Problem und die Palinopastik der südlichen Tethys, *Geol. Rundsch.*, v. 60, p. 813–833.

McKenzie, D. P., 1970, Plate tectonics of the Mediterranean region, *Nature*, v. 226, p. 239–243.

McKenzie, D. P., 1972, Active tectonics of the Mediterranean region, *Geophys. J. R. Astr. Soc.*, v. 30, p. 109–185.

Maksimovic, B., and Sikosek, B., 1971, Significance of theoretic conceptions of geotectonic structure of the Adriatic sea belt and of its regional oil exploration pattern, *Nafta*, Zagreb, v. 4–5 (XXII), p. 343–346.

Manfredini, M., 1964, Osservazioni geologiche sul bordo interno della depressione Molisano–Sannitica (Italia meridionale), *Mem. Soc. Geol. Ital.*, Bologna, v. 4, p. 959–999.

Martinis, B., and Pieri, M., 1963, Alcune notizie sulle formazioni evaporitiche del Triassico superiore nell'Italia centrale e meridionale, *Mem. Soc. Geol. Ital.*, v. 4, 30 p.

Miljuš, P., 1972, Geological–tectonic structure and evolution of outer Dinarides and Adriatic area, *Ann. Geol. Pén. Balkanique*, v. 37, p. 19–77 (in Serbo-Croatian, English summary).

Morelli, C., Carrozzo, M. T., Ceccherini, P., Finetti, I., Gantar, C., Pisani, M., and Schmidt di Frieberg, P., 1969, Regional geophysical study of the Adriatic sea (with an appendix by Giogetti, G., and Mosetti, F.: General morphology of the Adriatic sea), *Boll. Geofis. Teor. Appl.*, v. 11, p. 3–56.

Nairn, A. E. M., and Westphal, M., 1968, Possible implications of the paleomagnetic study of the late Paleozoic igneous rocks of northwestern Corsica, *Palaeogeogr. Palaeoclimat. Palaeoecol.*, v. 5, p. 179–204.

Papa, A., 1970, Conceptions nouvelles sur la structure des Albanides (présentation de la carte tectonique de l'Albanie au 500,000e), *Bull. Soc. Géol. France*, v. 12, p. 1096–1109.

Pashko, P., 1973, Le Messinien dans la zone ionienne, *Perm. Studimesh*, Tirana, v. 3, p. 51–69 (French summary).

Pescatore, T., and Vallario, A., 1964, La serie Mesozoica nel gruppo del Monte Maggiore (Caserta), *Mem. Soc. Geol. Ital.*, Bologna, v. 4, p. 699–709.

Pieri, M., 1966, Tentativo di ricostruzione paleogeografico-strutturale dell'Italia centro-meridionale. *Geol. Romana*, v. 5, p. 407–424.

Polsak, A., 1965, Géologie de l'Istrie méridionale spécialement par rapport à la biostratigraphie des couches crétacées, *Géol. Vjesnik*, Zagreb, v. 18, p. 490–509.

Rampnoux, J. P., 1973, Essai de reconstitution géotectonique des Dinarides internes yougoslaves (Serbie) au Jurassique et au Crétacé, *Réunion Annu. Sci. de la Terre*, Paris, p. 353.

Richter, M., 1966, Über Zusammenhänge der Gebirge im östlichen Mittelmeer, *N. Jahrb. Geol. Paleont. Mh.*, v. 2, p. 73–87.

Richter, M., 1974, Die geologische Stellung des Mte Gargano (Apulien), *N. Jahrb. Geol. Paleont. Mh.*, p. 600–608.

Roksandic, M., 1966, Structures profondes et superficielles des Dinarides externes et de l'Adriatique, *Vesnik Z. za geol. i geof. istrav.*, Ser. C, Belgrade, v. 7, p. 101–161.

Ryan, W. B. F., 1969, *The Floor of the Mediterranean Sea*, Ph. D. thesis, Columbia Univ., New York, 421 p.

Ryan, W. B. F., Stanley, D. J., Hersey, J. P., Fahlquist, D. A., and Allan, Th. D. 1971, The tectonics and geology of the Mediterranean Sea, in: *The Sea*, Part II, Maxwell, A. E., ed., New York: Wiley–Interscience, p. 387–492.

Scheffer, V., 1963, Questioni regionali geofisiche riguardanti la geologia dell'Appennino, *Bol. Soc. Geol. Ital.*, v. 82, p. 445–464.

Sestini, G., 1970, Development of the Northern Apennines Geosyncline, *Sed. Geol.*, v. 4, 647 p.

Sikosek, B., and Maksimovic, B., 1971, Regional geotectonic pattern of the Adriatic sea belt, *Nafta*, Zagreb, v. 4–5 (XXII), p. 298–304.

Simpozij Zadar, 1971, Istrazivanju lezista nafte i plina na jadranu i u zoni vanjskih Dinarida, *Nafta*, Zagreb, v. 4–5 (XXII), p. 189–500.

Smith, A. G., 1971, Alpine deformation and the oceanic areas of the Tethys, Mediterranean and Atlantic, *Bull. Geol. Soc. Amer.*, v. 82, p. 2039–2070.

Subbotin, S., Sollogub, V., Prosen, D., Dragesevic, T., Mituch, E., and Posgay, R., 1969, Junction of deep structures of the Carpatho-Balkan region with those of the Black and Adriatic seas, *Can. J. Earth. Sci.*, v. 5, p. 1027–1035.

Turk, M., 1971, Structure of the Tertiary basin in the Northeastern Adriatic, *Nafta*, Zagreb, v. 4–5 (XXII), p. 275–282.

Van Straaten, L. M., 1970, Holocene and late Pleistocene sedimentation in the Adriatic sea, *Geol. Rundsch.*, v. 60, p. 106–131.

Vercellino, J., 1970, Here's what's known about the geology of the Italian Adriatic, *Oil and Gas Int.*, v. 10, p. 70–78.

Chapter 5B

THE DINARIC AND AEGEAN ARCS: GREECE AND THE AEGEAN SEA

J. K. Melentis

Department of Geology and Paleontology
University of Thessaloniki
Thessaloniki, Greece

I. INTRODUCTION

The present high relief of Greece is primarily the result of Alpine folding. For most of the Mesozoic and Cenozoic, Greece formed part of the Tethys. Paleozoic rocks are known in a number of regions, both in the Aegean Islands and on the mainland. These rocks were affected by Caledonian and Hercynian tectonic movements to form a Hercynian landmass over which the Tethyan seas transgressed in the Triassic. The Caledonian and Hercynian craton is here referred to as the Rhodope Massif. Around it are disposed three arcs—the innermost, Pelagonic Arc, which is composed mainly of crystalline schists; a second, median arc, formed by extinct and active volcanoes; and the outermost, Dinaro-Tauric Arc. The Pindos Mountains form the backbone of the Dinaro-Tauric Arc and are considered as an extension of the Dinaric Mountains of Yugoslavia and Albania. In southern Crete the arc bends eastward and then northeastward to form the Taurus Mountains of Asia Minor.

Except for some parts of the Rhodope Massif, Greece was covered by the Tethys and its present form is essentially the result of sedimentary and tectonic evolution during the Mesozoic and Cenozoic. It is this evolutionary history which is the particular concern of this article. In common with other areas subjected to Alpine tectogenesis, there is a close relationship here between tectonic and sedimentary history, and the elucidation of the one is closely related to understanding of the other. Since the region is structurally and tectonically a continuation of the Dinarides, it is not surprising that the stratigraphy and tectonic development of Yugoslavia, Albania, and Greece are similar.

II. OUTLINE OF STRATIGRAPHY

No rocks of Precambrian age have been positively identified, although they may be represented in the most highly altered parts of the Rhodope Massif. In common with the other Balkan countries and Italy there was a Middle Cambrian transgression followed by an Upper Cambrian regression. For much of the rest of the Paleozoic, mainland Greece, particularly in the eastern and central regions, was a marine domain. Lower Paleozoic rocks with a characteristic shelly fauna (Ordovician bryozoa and brachiopods on Kos; Silurian graptolites, lamellibranchs, and trilobites on Chios) are found on certain of the Aegean Islands. These conditions continued into the Upper Paleozoic as may be seen from the Devonian fauna of Chios and the widely developed Carboniferous and Permian marine faunas (Kos, Chios, the Orthys Range, Evia, Attica and Salonika regions of the mainland, Crete, etc.). Graywacke shales of Upper Devonian age in Chios constitute one of the few signs of contemporary tectonic activity in the area. During the Variscan orogeny, Greece and the Balkans were folded and a new geosyncline was created. The sediments laid down in this geosyncline were destined to form the Balkan Mountains after the Alpine orogeny.

The Tethys transgressed over the old Hercynian landmass during the Triassic, covering parts of Chios, Hydra, eastern Argolis, western Parnis, and other regions until by Upper Triassic time the entire Hellenic geosynclinal area was flooded. This condition persisted through the Jurassic, except in the Parnassus–Giona–sub-Pelagonic–Axios zones where a (lower) bauxite horizon is found, suggesting long emergence. This emergence is attributed to the brief Agassiz phase of movement. A second phase of emergence at the end of the Jurassic neo-Kimmeridgian movements formed another (middle) bauxite horizon found in the region of Kilkis, over much of the sub-Pelagonic zone and over parts of Parnassus–Giona. This region was submerged by Middle Cretaceous time.

During the Cretaceous, tectonic unrest in the Hellenic geosyncline became pronounced and Austrian phase-folding resulted in emergence of the Pelagonic Ridge and part of the Axios and sub-Pelagonic zones. Sub-Hercynian folding at the beginning of the Upper Cretaceous influenced the Parnassus–Giona and sub-Pelagonic zones. Still another (upper) bauxite horizon developed on the exposed parts of the Parnassus zone. However, by the end of the Cretaceous the Tethys again covered all of the Hellenic geosyncline. These mainly vertical movements affected only the internal zones while the external zones remained tranquil.

In summary then, the paleogeography of the Mesozoic was dominated by the Pindus Trough or Furrow together with the Ionian Trough formed in Middle Liassic time. Deposition in the troughs consisted of limestones with jasper dolomites and in the axial "starved" regions of the troughs radiolarites formed. Along the trough margins in the Upper Liassic section the red, nodular limestone facies of the "ammonitico rosso" is found. This zone becomes a site of limestone microbreccia deposition at the end of the Jurassic and during the Senonian. The main change in depositional region was the introduction of flysch as a result of orogenic movements by Barremian time in the Pindus Trough, and the development of bauxites over emergent surfaces.

The beginning of the Tertiary was marked by withdrawal of the sea from the Rhodope Massif, which marked the first emergence of mountain ranges following Laramide folding. The Pelagonic Ridge and sub-Pelagonic zone emerged and thereafter remained land areas, as compression of the Axios Trough separated them from the Rhodope Massif. Thus, in the Lower Tertiary eastern and central Sterea, Thessaly, central Macedonia, and eastern Peloponnesus were land areas while the western zones were unaffected by this movement. The Olonos–Pindus zone appeared after the Helvetian fold movements, the Gavrovon–Tripolitza followed the Savenian folding, and the Adriatic–Ionian and the Paxi zones, with the older and younger Tyrolian phases, respectively, appeared in Miocene time. Following the end of orogenic activity Greece began to take on its present shape. Gradual subsidence along normal faults active from the Lower Pliocene into the Quaternary began to affect the continuous land area (Aegis) which stretched from the Hellenides to the Rhodope Massif, Asia Minor, and Cyprus. This subsidence resulted in the formation of the Aegean Sea with only the mountain tops and young volcanoes emerging to form the Aegean Islands of today.

III. TECTONIC EVOLUTION OF GREECE

A. Geotectonic Zones

In Greece as in other regions affected by Alpine folding, it is difficult to divorce stratigraphic from tectonic evolution (see Celet, Chapter 5A, this volume, and Caire, Chapter 4A, Volume 4B). It is intended, therefore, to present here a framework of tectonic zones, the result of work by many authors (Aubouin, 1958, 1959; Aubouin *et al.*, 1963; Brunn, 1956; Celet, 1961; Dercourt, 1962; Renz, 1940, 1955; and others), as the basis for stratigraphic discussion.

A sequence of nine narrow but enlongate, subparallel zones which follow the trend of the Hellenic ranges are recognized from the general geologic and tectonic structure of Greece (Fig. 1). These nine zones, five external and four

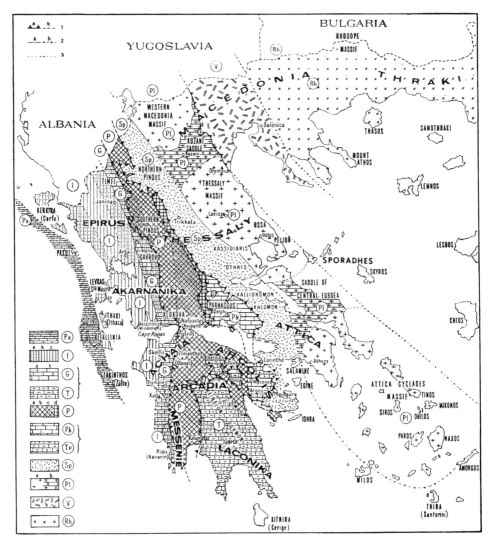

Fig. 1. Sketch-map showing the isopic zones of the Hellenides and their structural relationship (after Aubouin *et al.*, 1963): Pa = pre-Apulian zone; I = Ionian zone; G-T = Gavrovon zone and Tripolitza zone; P = Pindus zone; Pk-Tp = Parnassus zone and Trapezona zone; Sp = sub-Pelagonian zone; Pl = Pelagonian zone, a = basement, b = sedimentary cover; V = Vardar zone; Rh = Rhodope zone.

internal zones, essentially wrap around the nucleus formed by the Rhodope Massif which itself extends northward into central Bulgaria. Only the southern margin of the massif was flooded by the Tethys and now is covered by the Aegean Sea. The northern Aegean Islands of Lesbos, Samothrace, Limnos, and Aghios Evstratios are regarded as parts of the Rhodope Massif.

Surrounding the Rhodope Massif is a system of trenches and ridges. The most internal, the Axios (Vardar) eugeogsynclinal trough, is separated from the Olonos–Pindus Trench by the Pelagonian–sub-Pelagonic–Parnassus zones. The Pindus Trench was in turn separated from the Adriatic–Ionian Trough by the Gavrovon–Tripolitza Ridge. The most external zone, the Paxi zone, represents the transition from the deep water of the Ionian Trough to the shallow-water conditions along the margins of the Apulian Massif (or African promontory, see Caire, Volume 4B). Brief descriptions of the individual zones are given in Table I.

B. Stratigraphy of the Isopic Zones

1. *Paxi or Pro-Alpine Zone*

Within this zone lie the islands of Paxoi and Antipaxoi, the western part of Zankinthos (Zante), and Cephalonia where the Enos Range is important. The oldest beds exposed are Upper Cretaceous rudist limestones and dolomites with radiolaria. These are unconformably overlain by a series of Paleocene, Eocene, and Oligocene foraminiferal limestones. No flysch is found.

2. *Adriatic–Ionian Zone*

This zone includes southwestern Albania, Epiros, the Akarnika Mountains, the Ionian Islands, and western Peloponnesus. The zone then appears to follow an arcuate course and to be represented in Crete, Rhodes, and Cyprus. The characteristic stratigraphic sequence consists of 1500 m of basal gypsum followed by a variety of limestones, the oldest of which is dated Upper Triassic. During Upper Liassic time, along the margins of this deep trough, the "ammonitico rosso" facies of red, nodular limestones occurs while the central part is occupied by normal *Posidonomya* limestones. The limestones often contain chert. During the Lower–Middle Eocene, limestone microbreccias occur along the trough slopes.

3. *Gavrovon–Tripolitza Zone*

This zone can be considered in three segments, the northern or continental Greece, the Peloponnesus, and the Kythira–Crete–Rhodes segments, with the zone at its widest extent (100 km) in the central Peloponnesus segment.

TABLE I

The Isopic Zones of the Hellenides (see also Fig. 1)

Isopic zone	General description (see Fig. 1)
a. The internal zones	
Rhodope (Rh)	The "nuclear" zone, extending from central Bulgaria south into the Aegean Sea, affected by Hercynian (possibly also Caledonian) folding. Alpine and pre-Alpine granite batholiths, with basic and ultrabasic rocks, are found. Metamorphosed Mesozoic rocks also occur.
Vardar (Axios) (V)	Eugeosynclinal trough now 40 km wide, extending from the Greek–Yugoslavian border to Chios. Basement of trough formed of crystalline rocks affected by Hercynian movements. Western margin thrust over the Pelagonic zone.
Pelagonic (including the Altic–Cycladic–Lydocaric and Menderes extension) (Pl)	A ridge extending from the Greek–Yugoslavian border southeastward to include the Cyclades Islands, swinging east and northeast to the Menderes Massif. Affected by Hercynian (possibly also Caledonian) folding, it reacted as a rigid body to Alpine movements, deforming by fractures and sliding and some folding. Consists of Paleozoic and Mesozoic metamorphic rocks (gneisses, schists, and marble), large granite masses, and Mesozoic flysch.
Sub-Pelagonic (eastern Greece) (Sp)	Essentially the western slope of the Pelagonic Ridge over which the Tethys Sea transgressed. A zone of schists and ophiolites. The top of the succession is formed by flysch.
b. The external zones	
Parnassus–Giona	An irregularly shaped ridge zone inserted between the sub-Pelagonic and Pindus zones in central Greece and the northeastern Peloponnesus. A characteristic feature of the zone is a transverse fault which creates a tectonic trough (Corinthiakis).
Olonos–Pindus	A deep eugeosynclinal furrow, pre-Alpine basement unknown, extending from the Albanian border in an arc to Rhodes and Asia Minor. Characterized by overthrusting with thrust slices riding over the Gavrovon–Tripolitza zone.
Gavrovon–Tripolitza	The ridge of Paleozoic metamorphic rocks (phyllites, marble) extending from the Albanian border to the Dodecanese, which separates the Olonos–Pindus zone from the Adriatic–Ionian zone. It is characterized by large normal fault displacements (e.g., Camel fault 4000 m) and major anticlines (e.g., Gavrovon).

TABLE I (*continued*)

Isopic zone	General description (See Fig. 1)
Adriatic–Ionian	A deep trough stretching from the Albanian border toward Cyprus. The pre-Alpine basement is unknown. The base of the Alpine sedimentary sequence, 1500 m of gypsum, has moved diapirically. A zone of complex tectonics with moderate scale upthrusting in the west, and major faulting with the formation of basins (Ionian Sea) in the central part of the basin.
Paxi (pro-Alpine)	Most external, westernmost zone; a transition from deep to shallow water; pre-Alpine basement not known. Tectonically complicated with considerable faulting.

The sequence in the northern segment ranges from Upper Jurassic to the lowest horizons of the Upper Eocene and consists primarily of limestones which contain foraminifera, rudists, nummulites, alveolines, etc. Beginning in the Upper Eocene and extending into the Oligocene, a flysch wedge in excess of 2000 m developed. The same lithological succession is known in the Peloponnesus with thick flysch once again developing in the Upper Eocene and continuing into Oligocene time. Despite dolomitization, sufficient fossils remain to date the onset of calcareous sedimentation as Upper Triassic–Lower Jurassic.

4. Olonos–Pindus Zone

This former deep trough is characterized by reduced thickness or condensed sequences of cherty, pelagic limestones and radiolarites, beginning with the Upper Triassic (150 m), Jurassic, and Cretaceous (upper part of the Lower Cretaceous, 50 m). The Upper Cretaceous is the thickest sequence and consists of 500 m of limestones with rudist fragments and Maestrichtian–Oligocene flysch.

5. Parnassus–Giona Zone

Paleogeographically, this zone formed a ridge, which according to tectonic activity was sometimes emergent, and sometimes submerged. From the Upper Triassic until the Eocene, submergent intervals were characterized by the deposition of limestones and dolomites with gastropods, lamellibranchs, algae, rudists, and foraminifera. Emergent intervals are recognized by the formation of bauxites through weathering of the limestones. The succession terminated with an Eocene flysch sequence of alternating marls and sandstones.

6. Sub-Pelagonic (Eastern Greece) Zone

The characteristic feature of this zone, essentially the western edge of the Pelagonic Ridge, is the transgression of Upper Cretaceous rocks over Jurassic ophiolites and radiolarites. The zone may be considered as a narrow region on either side of the zone of ophiolite emplacement.

The Middle Triassic alternations of dolomites and dolomitic limestones are many hundreds of meters thick. They are followed by thinner Jurassic limestones with trocholines and clypeines. Formation of the ophiolites was a terminal Jurassic event. The Upper Cretaceous transgression varies somewhat in age from place to place. However, there is usually a basal conglomerate followed by limestones and the sequence is typically completed by flysch deposits which may be dated from the included *Globotruncana* and orbitoid fauna.

7. Pelagonic Zone

The continuity of basement outcrop on this former geanticlinal ridge is broken by two zones where Mesozoic sedimentary cover is preserved (Fig. 1, Kozani and Euobia Saddles). The age of the Attica–Cyclades basement is open to debate, for some Mesozoic beds are incorporated in the Pelagonic mass. The crystalline schists must, however, be considered to be older than Triassic. The lowest horizons, granites and gneisses, may be considerably older.

In the two saddles, thin neritic Triassic and Jurassic limestones are found. They rest unconformably upon the metamorphic rocks of the basement and are in turn unconformably overlain by Cretaceous beds. There is a basal conglomerate followed by sandy and microbrecciated limestones of Turonian age and capped by about 300 m of Campanian rudist limestone. Toward the top of the latter, horizons of pelagic limestone with *Globotruncana* and flysch occur.

8. Vardar (Axios) Zone

The eugeosynclinal trough lying between the Rhodope Massif and the Pelagonic zone was a major basin of sedimentation. The Vardar zone may itself be broken up into three parts (subzones) disposed from east to west: (1) The eastern part (formerly known as the Peonia zone) consists of a Triassic–Jurassic sequence of limestones, shales, and sandstones, ending with ophiolites and unconformably overlain by an Upper Jurassic–Lower Cretaceous clastic sedimentary sequence with conglomerates, shales, and limestones. (2) The central part (formerly known as the zone of Mount Paikon) was a geanticlinal ridge consisting of schists, marbles, and crystalline limestones with intercalations of rhyolites, spilites, and keratophyres of the same age as the eastern part; this sequence is overlain by the Upper Cretaceous limestones. (3) The western part (Almopia zone) consists of ophiolites and their cover of Upper Cretaceous limestones and flysch.

There was a marked difference in the development of the three areas, with the folding at the end of Jurassic, while in the Cretaceous two independent troughs developed, one in the east (Peonia Trough) and another in the west (Almopia Trough), separated by the Paikon Ridge.

9. Rhodope Massif Zone

The stratigraphy of the zone is poorly known, for fossils are not common in these metamorphosed rocks. The Rhodope Massif, from the Middle Carboniferous onward, appears to have formed an independent mass dividing the Dinaric from the Balkan Arc within the Alpine geosyncline. Its existence in Triassic time is clearly established, but it appears to have been submerged during the Jurassic. The massif was perhaps partly emergent at times during the Eocene, for there is a terrestrial sequence of conglomerates with alternations of sandstones, sands, and clays, but with intercalated horizons of nummulitic limestones with corals.

The metamorphic basement of Crete and the Peloponnesus should be considered separately for they are not continuous with the Rhodope Massif. Here are recrystallized limestones of presumed Carboniferous age, a series of phyllites, and schists and micaschists with dolomite and gypsum in the lower part. Permian fossils have been found in the dolomite and there is an upper phyllite and micaschist sequence in which a Triassic fauna has been discovered. The sequence in the Peloponnesus is similar.

IV. THE VOLCANOES OF GREECE

The volcanoes of Greece have formed subsequent to Alpine folding and are located along major fault zones, a clear indication of tectonic control. Three zones can be defined: the volcanoes of the south Aegan Sea (Fig. 2); the volcanoes between the Lihades Islands and the Dardanelles; and the volcanoes paralleling the coast of Asia Minor.

They can be divided according to their volcanic products as Pacific (calcalkalic), Atlantic (alkalic), or Mediterranean (potassic) types (see Table II).

Volcanic activity has extended from Neogene to Recent times (Fig. 3). The most famous volcanic event was the catastrophic eruption of Santorini in

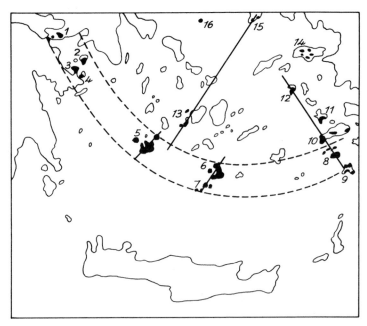

Fig. 2. The volcanoes of the southern Aegean Sea: 1. Aghi Theodori, 2. Aegina, 3. Methana, 4. Poros, 5. Melos, 6. Santorini, 7. Christiana, 8. Nissyros, 9. Telos, 10. Kos, 11. Kalymnos, 12. Patmos, 13. Antiparos, 14. Samos, 15. Emporio of Chios, 16. Kalogeri.

1500 B.C., with the formation of the Ghira Caldera. This caldera is only one of many known. A further evidence of thermal disequilibrium is the common occurrence of hot springs and fumaroles on some of the Greek islands (e.g., Nissyros, Methana, Sousaki, Melos, Lesbos, etc.).

TABLE II

The Volcanoes of the Aegean, According to Composition

Petrological character	Volcanoes
"Pacific" calcalkalic type	Aegina, Methana, Melos, Santorini, Kalymnos, Limnos, Western Thrace, Halkidiki, North Sporades, Lihades Islands, Lokris
"Atlantic" alkalic type	Antiparos, Kos, Samos, Mikrothive and Aghios Evstratios Islands
"Mediterranean" potassic type	Patmos, Nissiros, Ghyali (near Rhodes), Samothraki, Mytilini, Imvros

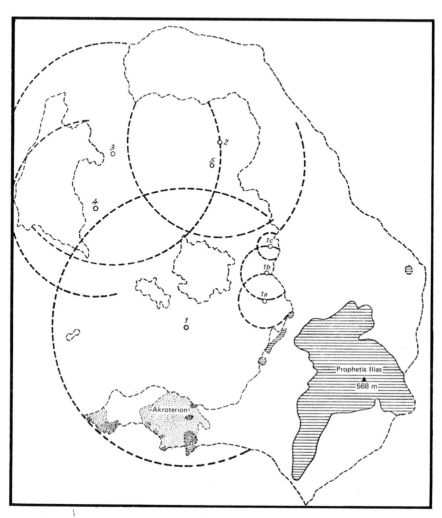

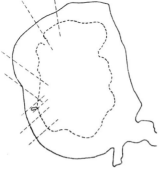

Fig. 3. Volcano of Santorini. Above: The unique prehistorical volcanic island "Strongili," before the year 1500 B. C., with the areas of volcanic activity of the prehistorical volcanoes. Scale 1:96,000. After the lava flows of Akroterion, the succeeding series 1–5: 1. Volcano of Thira, 2. Volcano of Peristeria, 3. Volcano of Semandiri, 4. Volcano of Thirasia, 5. Volcano of Skaros. Below: The site of the caldera's demolition and the radiating sinkings which formed the channels between Thira, Thirasia, and Aspronesi. Scale 1 : 80,000. – – – – – – = Faults. (After Georgalas, 1971.)

V. TECTONIC INTERPRETATION

Some attempts have been made to consider the tectonic evolution of the Hellenides in terms of plate tectonics. The present situation according to McKenzie (1970) is one in which there are two microplates, the Aegean and the Turkish, whose movements reflect the stresses involved in the collision of the African and Eurasian plates. The movement of these microplates is responsible for the high seismicity of the eastern, as compared to the western Mediterranean.

The northern boundary of the Aegean plate is an extension of the Anatolian (or Pamphlagonian) fault which passes through the Sea of Marmara, the northern Aegean cutting the Greek peninsula north of Euobia. The fault has an average horizontal displacement of 11 cm a year, and along this zone a trough 1500 m deep has been formed. The southern margin of the Aegean plate is moving southwestward with the African plate passing below it. McKenzie (1970) has assembled evidence to support this view—the existence of the volcanic arc of the Aegean, the negative gravity anomalies south of Crete, and the existence of the Hellenic Trough. Papazachos (1973) and Papazachos and Comninakis (1971) have identified the descending plate dipping under the Aegean at an angle of 30° from a consideration of the seismic data. The plate, 90 km thick, is said to be moving at a rate of 2.5 cm/yr.

The plate tectonic interpretation of formation of the Hellenic Alpine chains is more speculative. According to some, the Tethys opened, separating the Rhodope plate from the Apulian plate (or African promontory of Caire, this volume) with the generation of the Ionian Trough. Spreading was terminated by movements of the plates bordering the Atlantic in Upper Jurassic times. Thereafter, the approach of the Eurasian and African plates tended to close this opening with the resulting formation of the Dinaric–Hellenide chain.

REFERENCES

Aubouin, J., 1958, Essai sur l'évolution paléogéographique et le développement tecto-orogénique d'un système géosynclinal: le secteur grec des Dinarides, *Bull. Soc. Géol. France*, v. 8 (6), p. 731–748.

Aubouin, J., 1959, Contribution à l'étude géologique de la Grèce septentrionale: les confins de l'Epire et de la Thessalie (le thèse, Paris, 1958), *Ann. Géol. Pays. Helléniques*, v. 10, p. 1–483.

Aubouin, J., 1965, *Geosynclines, Developments in Geotectonics I*, Amsterdam: Elsevier Publ. Co., p. 1–335.

Aubouin, J., Brunn, J. H., Celet, P., Dercourt, J., Godfriaux, I., and Mercier, J., 1963, Esquisse de la géologie de la Grèce, *Livre jubilaire P. Fallot*, *Mém. Soc. Géol. France*, Hors-sér., v. 2, p. 583–610.

Barbier, R., Bloch, J. P., Debelmas, J., Ellenberger, F., Fabre, J., Feys, R., Gidon, M.,

Goguel, J., Gubler, Y., Lanteaume, M., Latreille, M., and Lemoine, M., 1961, Problèmes paléogeographiques et structuraux dans les zones internes des Alpes occidentales entre Savoie et Méditerranée, *Livre jubilaire P. Fallot, Mém. Soc. Géol. France, Hors-sér.*, v. 2, p. 331–378.

Brunn, J. H., 1956, Etude géologique du Pinde septentrional et de la Macédoine occidentale, *Ann. Géol. Pays Helléniques*, v. 7, p. 1–358.

Celet, P., 1961, Contribution à l'étude géologique du Parnasse–Kiona et d'une partie des régions méridionales de la Grèce continentale, *Ann. Géol. Pays Helléniques*, v. 13, 446 p.

Dercourt, J., 1962, Contribution à l'étude géologique du Péloponnèse; terminaison paléogéographique du haut-fond du Parnasse, *Bull. Soc. Géol. France*, v. 4 (7), p. 304–356.

Godfriaux, I., 1962, L'Olympe: une fenêtre tectonique dans les Hellénides internes, *Compt. Rend.*, v. 255, p. 1761.

Hsü, K., 1971, Origin of the Alps and the Western Mediterranean, *Nature*, v. 233, p. 44–48.

McKenzie, D., 1970, Plate tectonics of the Mediterranean region, *Nature*, v. 226, p. 239–243.

Maratos, G., 1972, Geologie von Griechenlands, Publ. Geotechnic Inst., Athens, v. 1, p. 1–189.

Melentis, J., 1972–1973, Die geologischen und hydrogeologischen Verhältnisse des Gebietes des Gallikos-Flusses, 1: Krithea; 2: Fanarion; 3: Kolchis, *Bull. Geol. Soc. Greece*, v. 9, p. 452–481; *Sci. Ann. Fac. Phys. Math. Univ. Thessaloniki*, v. 13, p. 67–82, 197–213.

Melentis, J., 1973, Die Geologie der Insel Skiros, *Bull. Geol. Soc. Greece*, v. 10 (2), p. 298–322.

Papazachos, B. C., 1973, Distribution of seismic foci in the Mediterranean and surrounding area and its tectonic implication, *Geophys. J.*, v. 33, p. 421–430.

Papazachos, B. C., and Comninakis, P. E., 1971, Geophysical and tectonic features of the Aegean arc, *J. Geophys. Res.*, v. 76, p. 8517–8533.

Renz, C., 1940, Tektonik der griechischen Gebirge, *Mém. Acad. Athènes*, v. 8, 171 p.

Renz, C., 1955, Die vorneogene Stratigraphie der normalsedimentaren Formationen Griechenlands, *Inst. Géol. Subs. Res. Athènes*, 1955, 637 + 54 p.

Renz, D., and Reichel, M., 1946, Beiträge zur Stratigraphie und Paläontologie des ostmediterranen Jungpaläozoikums und dessen Einordnung im griechischen Gebirgsystem, *Eclogae Geol. Helv.*, v. 38 (2), p. 211–315.

Chapter 6

OUTLINES OF THE STRATIGRAPHY AND TECTONICS OF TURKEY, WITH NOTES ON THE GEOLOGY OF CYPRUS

Nuriye Pinar-Erdem

Department of Geology
State Academy of Engineering and Architecture
Yildiz-Istanbul, Turkey

and

Emin Ilhan

Konur Sok 44/11
Yenisehir
Ankara, Turkey

I. GENERAL OUTLINE

Turkey occupies an area of about 760,000 km², and is divided by the Dardanelles–Marmara–Bosphorus depression into a small European part in the west (Turkish Thrace) and a larger Asiatic part to the east (Anatolia). Anatolia, which corresponds roughly to the peninsula of Asia Minor, is bounded by the Dardanelles–Bosphorus depression to the north, the Aegean and Mediterranean Seas on the west and southwest, Syria and Iraq lying to the southeast, and Iran and Russia to the east. Most of the geologically interesting features of the country lie in Anatolia, so this name is frequently used as a synonym for Turkey by geologists.

Turkey lies entirely within the Mediterranean sector of the Alpine orogenic belt. This belt is divided into two east–west-extending segments: the essentially linear northern Anatolian fold belt (with its extension in Thrace, the Istranca Mountains) and the southern Anatolian fold belt which consists of several large arcs (Fig. 1). These major fold belts are topographically important mountain ranges of "Alpine" dimensions. In the central and western parts of Anatolia and in Thrace, these belts are separated by crystalline massifs and secondary folds forming interior highlands, large depressions, and secondary mountain groups. Post-Alpine, epeirogenic movement has created complicated patterns of faults, grabens, and "intra-Alpine" basins that intersect with the folds. Portions of the country are covered by post-Alpine, Tertiary sedimentary series, and Neogene–Pleistocene volcanics.

The highest peaks of Turkey, the Buyuk Agri Dagi (the Biblical Mt. Ararat, 5120 m) and Suphan Dagi (4400 m), are extinct volcanoes lying in northeastern Anatolia. Kackar Dagi (3940 m) and Cilo Dagi (4170 m) are the highest elevations of the northern and southern folded chains, respectively. A number of summits exceeding 3000 m exist in the Tertiary–Pleistocene volcanic areas as well as in the fold belts.

The earliest papers on geological investigations in Turkey were published in 1838 and 1839, but systematic geological studies have been undertaken in the country only since 1923 when the Republic of Turkey was formed and the existing scientific institutions were reorganized. The Maden Tetkik ve Arama Enstitusu (Mineral Research Institute, "M.T.A.") was formed; its geological branch acts as the Geological Survey of Turkey.

Today, about 1500 published papers* and many unpublished documents are available concerning various aspects of the geology of Turkey, and it may be said that, in general, the main stratigraphic and tectonic features of the country are known. There still remain many problems to be solved, for instance, the evaluation of isolated stratigraphic and tectonic observations and the establishment of regional correlations. This lack of knowledge is due primarily to the practical demands placed on Turkish geologists who have been and are still largely occupied with projects involving engineering and mining geology. Consequently, until recently, they have had little time for the solution of purely scientific problems. Visiting scientists often studied limited areas without having occasion to become familiar with larger parts of the country, and consequently sometimes reached erroneous conclusions. The authors of the present paper have more than thirty years' experience in Turkey and will try to describe, as impartially as possible, the major geological features of the country and consider the various interpretations of the stratigraphy and tectonics.

---

* Furon (1953) published a detailed bibliography up to 1952; a list of more recent papers has not been prepared.

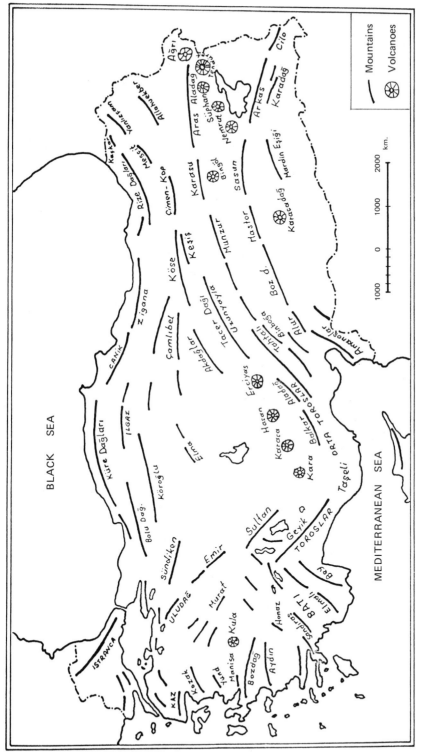

Fig. 1. An orographic map of Turkey naming the mountain chains. The division into a northern Anatolian fold belt and the arcuate, southern Anatolian fold belt is apparent.

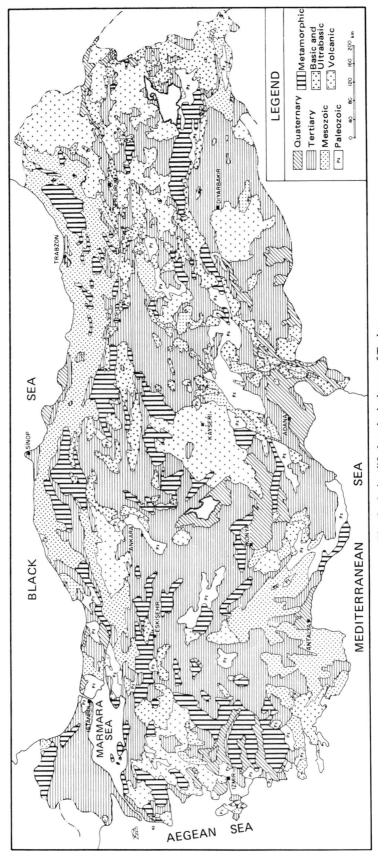

Fig. 2. A simplified geological map of Turkey.

The geology of Turkey (Fig. 2) is controlled by the Alpine orogenic cycle. The Mesozoic and Lower Tertiary rocks and sediments covering most of the country were deposited in Alpine geosynclinal troughs or were formed by syntectonic or posttectonic processes during and after the Alpine orogeny. The tectonic features, as well as the topography of the country, reflect the effects of the Alpine orogeny and the succeeding epeirogenic movements.

II. STRATIGRAPHY

A. Precambrian

Volcanic rocks unconformably overlain by fossiliferous Cambrian strata in the Mardin area (southeast Anatolia) and slightly metamorphosed gray-wackes covered by Cambrian deposits in the Iskenderun region (southern Anatolia) are considered as Precambrian (Erentoz, 1966; Ketin, 1966; Tolun, 1962). Some authors attribute to the Precambrian the metamorphic rocks included in the crystalline series of Anatolia (e.g., Yalcinlar, 1954), though such a classification has not been stratigraphically proven.

B. Paleozoic

The older Paleozoic deposits were affected by Caledonian and Hercynian movements and the entire Paleozoic series was disturbed, partially meta-morphosed, or even eroded during the Alpine orogeny. Therefore, these formations are preserved and exposed in limited areas only and their correlation and the reconstruction of depositional zones is difficult. These rocks reflect orogenic influences that make possible the separation of three sedimentation cycles: (1) a Cambrian–Ordovician–Silurian cycle, (2) a Devonian–Lower Carboniferous cycle, and (3) a Permian cycle (Flügel, 1964). According to Flügel (1964), the first and second cycles are represented in Turkey only by shelf deposits and geosynclinal formations do not exist. (If no other authors are indicated, the data mentioned below are based on papers published by Abdusselamoglu, 1959, 1963; Bilgutay, 1959, 1960; Blumenthal, 1941, 1944, 1947, 1948, 1950a, b; Egemen, 1947; Erentoz, 1956, 1966; Erk, 1942; Erol, 1956; Flügel, 1964; Paeckelmann, 1938; Sayar, 1962; Schmidt, 1964; Tokay, 1953, 1954; Tolun, 1962; Tolun and Ternek, 1952; Unsalaner, 1941; Yalcinlar, 1951.)

1. Cambrian

Successions of quartzite, shale, and carbonate intercalations with trilobites are known from isolated outcrops in northwestern Anatolia (Black Sea coast region) and southern and southeastern Anatolia.

2. *Ordovician–Silurian*

Unmetamorphosed or slightly metamorphosed successions of shales, quartzites, and sandstones with subordinate carbonate intercalations represent the Ordovician–Silurian and occupy larger areas than the Cambrian. The most important occurrences are in the Bosphorus region and northwestern, southern, and southeastern Anatolia. The fossil content—trilobites, graptolites, tabulates, stromatoporiods, and brachiopods—suggests a shelf environment.

3. *Devonian–Lower Carboniferous*

A slight unconformity at the base of the Devonian and frequent changes between clastics and carbonates in the same formation are due to the influence of late-phase Caledonian orogeny. Lithology is variable and ranges from predominantly carbonate to predominantly clastic. These series are well developed in the Bosphorus region, and in northwestern, western and central, and southern and southeastern Anatolia. The fossils most commonly found are octocoralla, brachiopods, some trilobites, and *Orthoceras*.

4. *Upper Carboniferous*

Carbonates with marine Upper Carboniferous fossils are reported from isolated places (Erk, 1942) and most authors agree that a stratigraphic gap, resulting from Hercynian folding, corresponds to the interval between the end of the Lower Carboniferous (locally perhaps the end of the Devonian) and the lowermost Permian (Flügel, 1964; Tolun, 1962). In the Eregli–Zonguldak–Bartin–Cide sector of the Black Sea coast region (northern Anatolia), the Upper Carboniferous is represented by a thick alternation of coal-bearing clastics containing the classic Upper Carboniferous coal flora. There are frequent small marine clastic intercalations which must be explained as short transgressions from the Hercynian geosyncline to the north (Flügel, 1964). These alternations are typical flysch deposits and are frequently formed during and immediately after important orogenies (in this case the Hercynian paroxysm). It is now established that the fusulinids found in Anatolian carbonates are Permian rather than Carboniferous, so the classification of large areas of carbonates as "Permocarboniferous" is no longer valid (Blumenthal, 1944; Bilgutay, 1959, 1960; Flügel, 1964).

5. *Permian*

In the coastal sector of northern Anatolia, continental red beds of probable Permian age cover the productive Upper Carboniferous. Elsewhere, the Permian is represented by carbonates with clastic intercalations (in some cases clastic–carbonate alternations) which are widely developed over much of Anatolia. Some brachiopods and cephalopods, as well as numerous fusulinids and algae (e.g., *Mizzia*) are found in these marine deposits. According to Flügel, the Permian is transgressive over various older formations folded during the Hercynian orogeny. This third sedimentation cycle of the Paleozoic continues into the Mesozoic with the formation of the Alpine geosynclinal troughs. In the Ankara region (inner sector of the northern Anatolian folds) and in the sector of the southern Anatolian folds, which is situated between Adana and Malatya, carbonate sedimentation continued from the Permian into the Lower Triassic and is separated by a hiatus from the Upper Triassic or Liassic (Erol, 1956; Ketin, 1963, 1966).

C. Mesozoic

Mesozoic series are well developed and well preserved in Anatolia. Their lithology reflects the development of geosynclinal troughs in the Lower and Middle Mesozoic as well as the beginning and progress of the Alpine orogeny in the Upper Mesozoic. The gradual formation of the Alpine geosyncline, the differentiation of these troughs into deeper and shallower zones (i.e., the transformation of a former continental or shelf area into open and deep ocean basins and their differentiation into mio- and eugeosynclinal zones) in Lower Mesozoic times, and their deformation and the formation of fold patterns during the Alpine orogeny from the Middle Cretaceous onward have produced a great variety of facies. Carbonates and flysch series typify the miogeosynclinal phases and areas. Neritic to bathyal sediments accompanied by ultrabasic rocks, in part tectonized and affected by regional dynamometamorphism (epi-grade), are characteristic of the eugeosynclinal phases (Ketin, 1966; Ilhan, 1971; Pinar and Ilhan, 1954).

Regional differentiation is impossible because of the poor preservation of the Paleozoic series. In the Mesozoic, however, the formation of geosynclinal troughs permits recognition of a geosynclinal–orogenic facies in the trough areas and the shelf facies of the foredeep (including the border of the stable foreland) in southeastern Anatolia.

1. *Triassic*

Though isolated observations long ago showed the presence of marine Triassic deposits in the Alpine belt of Turkey (e.g., Philippson, 1918; Blumenthal, 1947, 1948; Grancy, 1939), it has often been assumed (e.g., Frech, 1916; Erentoz, 1956) that marine Triassic deposits were not developed and that this sector of the geosyncline had only a Cretaceous or Jurassic history. Recent microfaunal and nannofossil studies have shown the presence of marine Triassic deposits in the monotonous carbonate successions which failed to yield larger fossils. These rocks are developed in many parts of the country, but had previously been mapped as "Permocarboniferous," "Cretaceous," or "Undivided Mesozoic" (unpublished documents, Petroleum Administration, Ankara).

Nevertheless, it must be noted that there are gaps in the Triassic sequence and that a terrestrial influence exists at some horizons, possibly due to the progressive development of the geosyncline or to the effect of early Kimmerian movements that affected some of the Anatolian fold belt (Ketin, 1966; Ilhan, 1971). In some sectors of the Anatolian folds, the Lower Triassic carbonates are concordant on the Permian and are separated by a gap from the Upper Triassic or Jurassic. In other parts of the country, the Triassic succession begins with variegated clastics followed by carbonates, which are sometimes sandy or marly. Algae, corals, some mollusks, brachiopods, and ammonites are the most common fossils. Terrestrial influences in the form of variegated clastics of Upper Triassic age are reported from some parts of the southern Anatolian folds (north of Antalya and southeast of Kayseri) (for further details see: Brunn et al., 1971; Egeran and Ilhan, 1948; Erol, 1956; Erguvanly, 1948; Neumayer, 1887; Philippson, 1918; Paeckelmann, 1938). Among the Triassic carbonates are dense, red, neritic limestones with ammonites, similar to the neritic limestones of the eastern Alps. These sediments, found in the northern and southern Anatolian ranges (Blumenthal, 1947, 1948), represent eugeosynclinal deposits and cannot have been formed during ephemeral transgressions (Kober, 1942). A "condensed" series, only 23 m thick, but extending from the Upper Triassic to the Cretaceous (Gutnic and Monod, 1970) suggests a similar environment. Marly limestones, dolomites, anhydrites, and light-colored clastics represent the Triassic in the shelf area of the southeastern Anatolian foredeep (Dunkee, 1961; Tolun, 1962). Brinkmann (1971) has published a paleogeographic reconstruction (Fig. 3), but his interpretation covers only western Anatolia.

2. Jurassic

In addition to dark slates and shales, flyschlike clastics sometimes with coals, pyroclastics, and lavas (effect of the Kimmerian movements) formed during the Lias. Red and gray dense limestones with ammonites, brachiopods, and crinoids also formed in some areas. These red, neritic carbonates, also common in the eastern Alps, are thought to be typical eugeosynclinal sediments. Generally light-colored carbonates with shale and marl intercalations characterize the Middle and Upper Jurassic. They contain ammonites, belemnites, crinoids, and brachiopods (Otkun, 1942; Stchepinsky, 1946). The "condensed" series of carbonates, radiolarites, and clastics noted above includes Jurassic horizons. This series must be classified as bathyal and proves, together with the

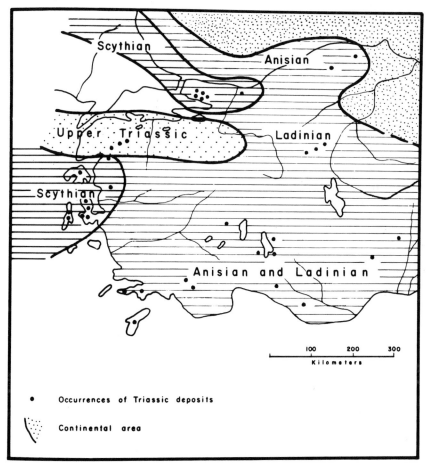

Fig. 3. Paleogeography of the Triassic in western Anatolia, showing the transgressive overlap of the stages of the Triassic. (From Brinkmann, 1971; reproduced with permission of the author and Petrol. Explor. Soc. Libya.)

frequent ultrabasic rocks (supposed to have been erupted during the Jurassic, see below), the existence of typical eugeosynclinal conditions in Anatolia at this time (Fig. 4). Clastics, dolomites, and evaporites are developed in the southeastern Anatolian shelf zone (Dunkee, 1961; Tolun, 1962; Turkse Shell, 1963).

3. Lower Cretaceous (Valanginian to Albian)

The predominately carbonate sedimentation of the Upper Jurassic continued during the Lower Cretaceous, but minor flysch deposits indicate early and local movements of the Cretaceous paroxysm. Ammonites, belemnites, and other large mollusks are frequent fossils (Erentoz, 1966). The southeastern Anatolian shelf area was covered by a thick carbonate succession in shelf

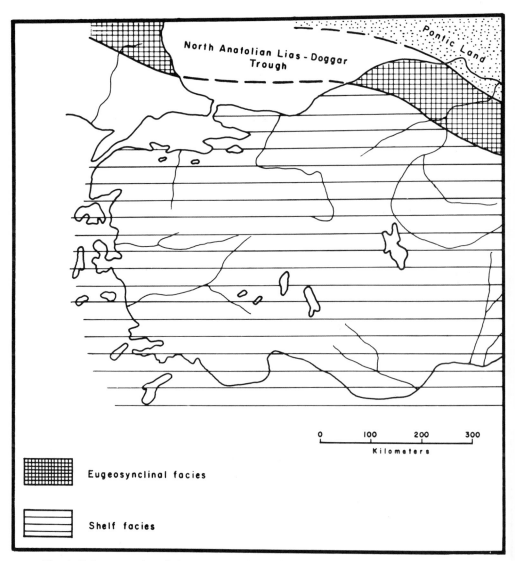

Fig. 4. Paleogeography of the Lower and Middle Jurassic in western Anatolia. (From Brinkmann, 1971; reproduced with permission of the author and Petrol. Explor. Soc. Libya.)

facies (Altinli, 1963*a*; Dunkee, 1961). An interpretation of the paleogeography of western Anatolia was figured by Brinkmann (1971) (Fig. 5).

4. *Upper Cretaceous* (*Cenomanian to Danian*)

Upper Cretaceous deposits, here as elsewhere in the Alpine belt, show the effects of the first big Alpine paroxysm: A thick, flysch series (i.e., alternations of marls, fine- to coarse-grained sandstones, shales, and siltstones), together with frequently variegated or dark-colored successions of conglomerates, breccias, and reefal limestones (formed by corals and rudists) similar to the "Gosau Formation" of the eastern Alps and the Carpathian Ranges, reflect the continuously changing conditions of erosion, transport, and sedimentation during the paroxysm. Generally light-colored, well-bedded to massive carbon-

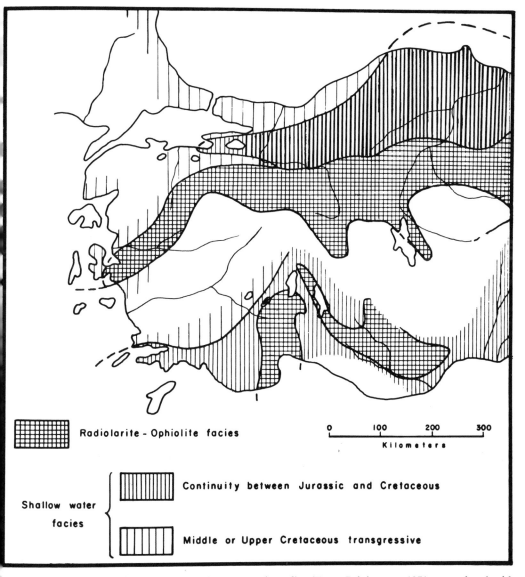

Radiolarite - Ophiolite facies

0   100   200   300
Kilometers

Shallow water facies {
Continuity between Jurassic and Cretaceous

Middle or Upper Cretaceous transgressive
}

Fig. 5. Paleogeography of the Cretaceous in western Anatolia. (From Brinkmann, 1971; reproduced with permission of the author and Petrol. Explor. Soc. Libya.)

ates were formed in the calmer (external) parts of the geosyncline where marine conditions continued without extreme disturbance. Flysch deposits, chalky limestones, and marls with interstratified basalts, tuffs, and agglomerates cut by intrusions are frequent in the eastern half of the northern Anatolian ranges and in some eastern sectors of the southern Anatolian folds.

In the shelf zone of the southeastern Anatolian foredeep, the Lower Cretaceous shelf carbonates were replaced during the Upper Cretaceous by basin shales reflecting the deepening of the foredeep trough as Alpine folds advanced. There are reef carbonates at several horizons (these carbonates are the reservoir rocks and the shales, potential source rocks, of the southeastern Anatolian oilfields; they correspond to the "deeper" productive horizons of Iraq and southern Iran). Fossils are more frequent in Upper Cretaceous deposits than in older formations and include microfossils, solitary corals, bivalves, thick-shelled gastropods, some belemnites, and late ammonites (Egeran and Ilhan, 1948; Erentoz, 1956; Erentoz, 1966; Stchepinsky, 1946; Tolun, 1962).

D. Tertiary–Quaternary

1. *Paleocene (Montian–Sparnacian) and Eocene (Ypresian–Priabonian)*

Sedimentation continued from the uppermost Cretaceous into the Paleocene or even the Eocene in some parts of the Anatolian belt (Fig. 6). If the few characteristic marine fossils of the Paleocene are absent, it becomes difficult to separate these series. Paleocene, Lower Eocene, and Middle Eocene were relatively calm periods with only local tectonic movements. Therefore, compared to the preceding periods, Paleocene–Eocene deposits are relatively uniform. Nevertheless, the Cretaceous paroxysm caused morphological changes which are more important in the inner than in the external zone of the orogenic belt. In consequence, erosion, deposition, and therefore the stratigraphic development during the Lower Tertiary in the inner belt differs from that in the external sector.

In the *inner belt sectors*, flysch sedimentation prevailed in the Paleocene and carbonates are rare; local gaps exist. In the Eocene until Priabonian time, flysch series and carbonates are more or less equally developed. After a major Priabonian regression, an effect of the second important Alpine paroxysm, marine conditions in this sector ended and were gradually replaced by lagoonal, freshwater, and/or continental beds from Upper Eocene to Oligocene and probably Miocene time (Erentoz, 1956; Erentoz, 1966; Egeran and Ilhan, 1948).

In the *external sector* of the southern Anatolian folds, carbonate deposition continued from the Upper Cretaceous into the Eocene (Tietze, 1885). The external zone of the northern Anatolian folds lies below the Black Sea, and thus no comparison can be made with the internal zone. In the eastern part of the northern Anatolian folds, the Eocene is represented by a volcanic facies; flysch and subordinated marls and limestones alternate with or are cut by andesitic lavas (Gattinger, 1963). Basin shales (Paleocene), red clastics sometimes associated with evaporites (Lower Eocene), and marine carbonates (Middle Eocene) are developed in the southeastern Anatolian foredeep. Toward

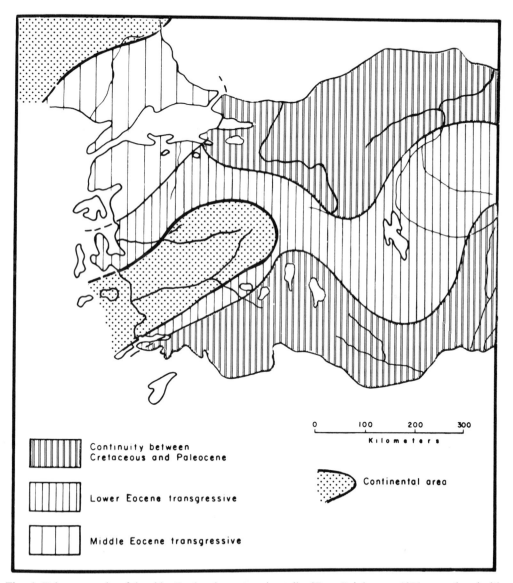

Fig. 6. Paleogeography of the older Tertiary in western Anatolia. (From Brinkmann, 1971; reproduced with permission of the author and Petrol. Explor. Soc. Libya.)

the Arabian shield, these sediments pass into chalky and marly material (Tolun, 1962; Dunkee, 1961).

2. Oligocene

With the Upper Eocene paroxysm, the geosynclinal (oceanic) stage in the *inner parts* of the Turkish Alpine belt ended. From that time on, sediments were no longer deposited in a large oceanic basin, but in larger or smaller depressions formed during epeirogenic movements which succeeded the orogenic events. The Upper Eocene regression represents the gradual passage of the marine environment to lagoonal and finally freshwater conditions. The thickness and lithology of these postorogenic sediments varies depending upon local epeirogenic depressions. In addition, the passage from marine to freshwater deposits occurred progressively and not simultaneously over the entire country. Consequently, the lower and upper time limits of the marine, lagoonal, and lacustrine series are not everywhere the same, though common characteristics can be established. In some of the more important depressions, these facies extend upward into the Miocene and can be subdivided into a lower part consisting of variegated conglomerates, red shales, and siltstones with rock salt, and limited gypsum deposits, and an upper part formed predominantly of light-colored marls and shales and occasional lacustrine limestones, including important gypsum deposits (Stchepinsky, 1942; Alagoz, 1967; Kuttman, 1961; Erentoz, 1966; Ilhan, 1950).

Marine to brackish-water deposition (flyschlike clastics) continued along the *external* (southern) border of the southern Anatolian folds (Tietze, 1885). As the corresponding external (northern) border of the northern Anatolian folds is situated in the Black Sea off the Anatolian coast, it is impossible to establish if marine Oligocene deposits exist there or not (but in the eastern extension of this submarine border zone, marine Oligocene flysch is reported from Transcaucasia) (Moscow, 1937).

3. Miocene

A major Lower Miocene transgression, beginning with the Burdigalian (locally with the Aquitanian) and extending upward to the Helvetian or Tortonian, invaded the Istanbul region and important sectors of southern and eastern Anatolia. This transgression evidently originated from the northern foreland of the belt (Istanbul area) and the Mediterranean basin, respectively. In the larger marine Miocene basins, a littoral facies of bioclastic limestones (algal deposits, but no coral reefs) and coarse clastics can be separated from a basinal shale or marl facies. In the smaller basins bioclastic limestone lenses are intercalated with marls and clastics. Evaporites occasionally mark the beginning and the end of the marine episodes. The rich marine molluscan faunas found in

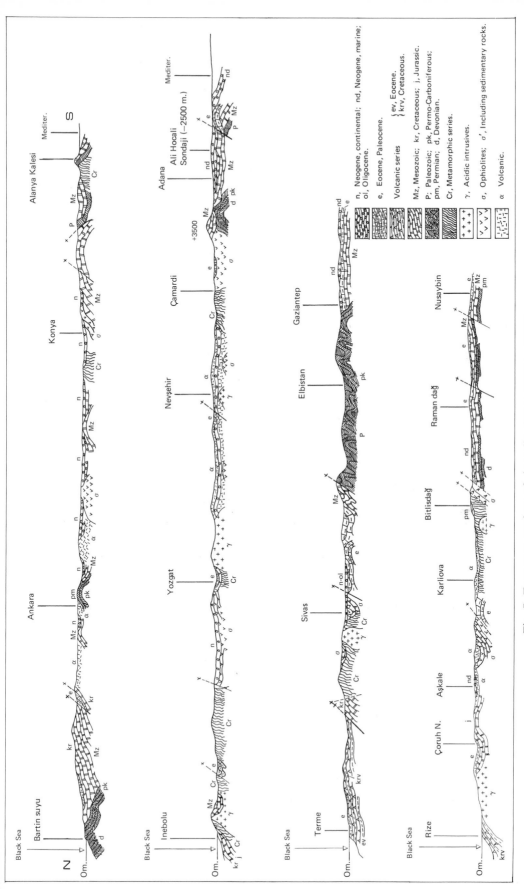

Fig. 7. Transverse sections of Anatolia. (From Erentoz, 1966.)

n, Neogene, continental; nd, Neogene, marine;
ol, Oligocene.

e, Eocene, Paleocene.

Volcanic series { ev, Eocene.
{ krv, Cretaceous.

Mz, Mesozoic; kr, Cretaceous; j, Jurassic.

P, Paleozoic; pk, Permo-Carboniferous;
pm, Permian; d, Devonian.

Cr, Metamorphic series.

γ, Acidic intrusives.

σ, Ophiolites; σ', Including sedimentary rocks.

α, Volcanic.

the Miocene deposits have striking affinities with the present Mediterranean fauna (Erentoz, 1956).

Freshwater deposits are developed in the areas not reached by the marine transgressions. In the larger depressions, lower flyschlike clastics can be separated from a higher marl–limestone series. Similar deposits cover the marine series. Important lignite deposits exist in the freshwater deposits. The freshwater mollusks frequently found in these formation are identical with recent species and cannot be used for stratigraphic time determinations; mammalian remains suggest an Upper Miocene ("Pontian") age (e.g., Yalcinlar, 1952). The freshwater series may extend into the Lower Pliocene (Altinli, 1944; Erentoz, 1956; Erentoz, 1966; Ilhan, 1946; Kurtman, 1961; Schmidt, 1961; Stchepinsky, 1942).

4. *Pliocene*

Marine Pliocene exists only in localities in the neighborhood of the Mediterranean coast, as in the Iskenderun and Datca regions (sandy marls with marine mollusks; Dubertret, 1953; Erentoz, 1956; Tietze, 1885; Turkey, Geol. Map of, 1950–1965). The Pliocene sea did not invade the continental sectors of Turkey; freshwater and continental deposits occur.

5. *Pleistocene*

The Pleistocene sediments can be classifed as follows: (a) Marine terrace deposits existing at various levels along the Mediterranean, Aegean, and Marmara coasts (English, 1904; Yucel, 1953; Pinar, 1943a); (b) lake sediments (sands and marls) with *Dreissensia*, etc., found around some lakes as remains of higher lake levels or in the dry basins of Quaternary lakes (e.g., "Lake Region" of the Toros Mountains and central Anatolian basins) (Chaput, 1936; Erol, 1963; Ilhan, 1946); (c) river terraces and gravels are developed all over the country; (d) glacial deposits exist only in mountains above 3000 m, as in the Toros Ranges in southwestern Anatolia, the Hakkari Mountains in southeastern Anatolia, and the Kackar Daglari in northeastern Anatolia. The only other glaciated area is the region above 2500 m south of the Marmara Sea (Blumenthal, 1938, 1947; Erinc, 1949; Izbirak, 1951; Philippson, 1918).

E. Metamorphic Rocks

Low-grade metamorphic rocks, common in the Anatolian and Thracian folds, generally have been considered as "old" (i.e., related to the Paleozoic orogenies or even older ones) by many authors (e.g., Yalcinlar, 1954). However, Ketin (1966) has shown that some metamorphic complexes grade laterally into unaltered strata of Mesozoic age, and Jurassic fossils recently have been found in similar rocks in Thrace (verbal information). Metamorphosed Cretaceous sediments were reported from the Izmir (Arni, 1939) and the Van–Hakkari areas (Ilhan, 1954). On the other hand, in the Bolu region (northern Anatolia), metamorphic rocks are transgressively covered by Devonian strata (Flügel, 1964) and Silurian fossils were recognized in slates in the Istranca folds in Thrace (Ketin, 1966). Such observations show that metamorphic rocks of various ages exist in Turkey.

F. Magmatic Rocks

1. *Ultrabasic Rocks*

Ultrabasic rocks ("ophiolites," "greenstones," often wrongly called "serpentines") related to Mesozoic formations are more abundant and wide-spread in the Anatolian fold zones than in their European counterparts. These rocks are frequently serpentinized and chloritized and they are generally associated with radiolarites, slates, flyschlike sediments, and various types of limestones. Though some authors consider these rocks as batholiths (e.g., Blumenthal, 1947, 1948), others like Steinmann (1927) and geologists who have worked in Yugoslavia, Greece (e.g., Aubouin, 1960, 1965; Brunn, 1961), and southern Anatolia (Brunn *et al.*, 1971; Dubertret, 1953) consider the "greenstones" as submarine extrusives overlying older and covered by younger sedimentary series. They originate in response to important tectonic events within the eugeosyncline. The fact that the main greenstone zones of Turkey are roughly parallel to the main earthquake zones of the country suggests a relationship to deep-seated, longitudinal faults (cf. Figs. 8 and 9) (Pinar-Erdem, 1974).

Age estimations of these rocks vary from Precambrian to Upper Cretaceous–Eocene (Altinli, 1963a; Arni, 1942; Blumenthal, 1945; Hiessleithner, 1955; Peyve, 1969; Wijkerslooth, 1942; Pinar-Erdem, 1974). It is not possible to confirm the Paleozoic age of the greenstones, although in some limited areas such rocks are associated with Paleozoic formations (see Turkey, Geol. Map of, 1950–1965). Geologists working in Yugoslavia and Greece, the western extension of the southern Anatolian folds, insist on a Jurassic (or even Upper Triassic) age for these rocks (e.g., Aubouin, 1960; Brunn, 1961). In Turkey, the greenstones unconformably overlie Upper Jurassic–Lower Cretaceous carbonates in some places (Pinar-Erdem, 1974), but very frequently they are associated with Upper Cretaceous sedimentary series (e.g., Ketin, 1963, 1966; Norman, 1973). Arni (1942) supposed a possible Jurassic age for greenstones found in the southern Anatolian folds in the Konya region and in the northern Anatolian folds in the Ankara area.

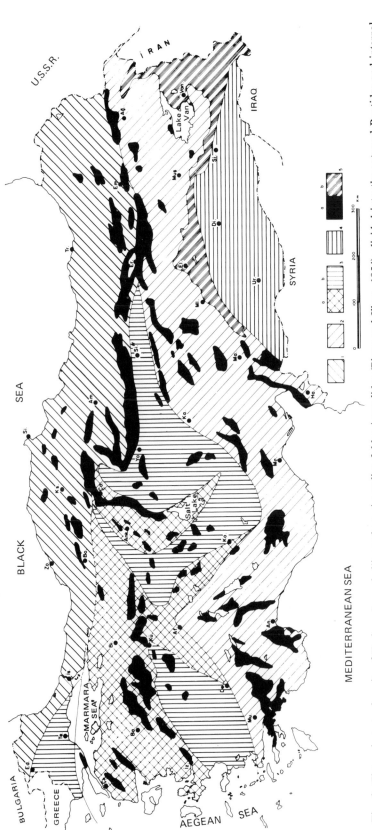

MEDITERRANEAN SEA

Fig. 8. The main tectonic units of Turkey. *Legend:* (1) northern Anatolian folds: Anatolides (Pinar and Ilhan, 1954), divided into the external Pontides and internal Antolides by Arni (1939) and Ketin (1966); (2) southern Anatolian folds: Torides (Pinar and Ilhan and Ketin), divided into the external Iranides and internal Torides (Taurides) by Arni; (3a) intermediate folds and (3b) intermediate massifs: These intermediate units are considered as parts of the Pontides and Anatolides by Ketin, and as a part of the Anatolides by Arni; (4) southern foredeep and foreland of the Alpine belt; (5a) important ophiolite zones; (5b) zone with abundant ophiolites (southern Anatolian folds). *Localities:* Ag = Agri; Am = Amasya; Ank = Ankara; An = Antalya; Ba = Balikesir; Bi = Bilecik; Bo = Bolu; Ck = Canakkale Co = Corum; De = Denizli; Di = Diyarbakir; Ed = Edirne; El = Elazig; Em = Erzincan; Er = Erzurum; Ha = Hatay (Antakya); Is = Istanbul; Iz = Izmir; Ka = Kayseri; Ko = Konya; Ks = Kastamonu; Ku = Kutahya; Ma = Maras; Me = Mersin; Ml = Malatya; Mu = Mugla; Si = Siirt; Si' = Sinop; Si'' = Sivas; Te = Tekirdag; Tr = Trabzon; Ur = Urfa; Yo = Yozgat; Zo = Zonguldak. *Mountains mentioned in the text:* Buyuk Agri Dagi (Ararat), 90 km east of Agri; Cilo Dagi, 125 km east-southeast of Siirt; Erciyes Dagi, 10 km east of Kayseri; Kackar Dagi, 110 km east of Trabzon; Nemrut Dagi, 60 km east of Mus; Suphan Dagi, 70 km northwest of Van.

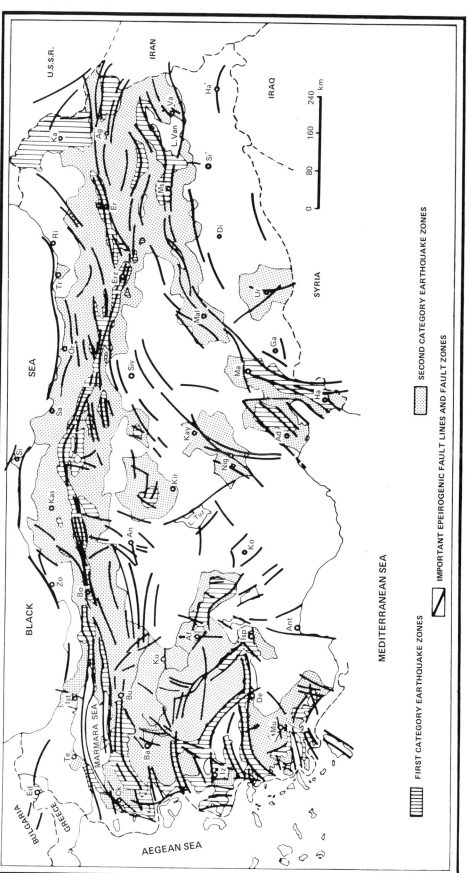

Fig. 9. Epeirogenic fault patterns and seismic zones of Turkey. Explanation of symbols: Ad = Adana; Af = Afyonkarahisar; Ag = Agri; An = Ankara; Ant = Antalya; Ba = Balikesir; Bo = Bolu; Bu = Bursa; Ck = Canakkale; De = Denizli; Di = Diyarbakir; Ed = Edirne; Erz = Erzincan; Er = Erzurum; Es = Eskisehir; Ga = Gaziantep; Ha = Hatay (Antakya): Ha' = Hakkari; Isp = Isparta; Ist = Istanbul; Iz = Izmir; Ka = Kars; Kas = Kastamonu; Kay = Kayseri; Kir = Kirsehir; Ko = Konya; Ku = Kutahya; Mal = Malatya; Ma = Maras; Ms = Mus; Mu = Mugla; Nig = Nidge; Or = Ordu; Ri = Rize; Sa = Samsun; Si = Sinop; Si' = Siirt; Siv = Sivas; Te = Tekirdag; Tr = Trabzon; Ur = Urfa; Va = Van; Zo = Zonguldak.

In order to solve the question of the age of the ultrabasic rocks, two different points must be considered, namely the time of the eruption of these rocks and the time of their emplacement in their present position (Peyve, 1969). Most of these rocks, together with the associated sediments, may have been transported into their present position by tectonic movements, submarine slides (as Rigo and Cortesini, 1964, supposed), thrusts, or combined thrust–slide movements. This may have occurred for instance during the Upper Cretaceous paroxysm. So, the position of the greenstones between, over, or below Upper Cretaceous sediments and the fact that the greenstones are frequently associated in complicated tectonic "melanges" with other formations (Bailey and McCallien, 1953) can easily be explained (Pinar-Erdem, 1974). The present position of the greenstones need not be identical with the place of their eruption and the age of the eruptions need not be identical with the age of the sediments with which the ophiolites are now associated. (It should be noted that the unstable texture of the ophiolites, due to serpentinization and chloritization, facilitates their weathering so that they can form a kind of lubricant for moving masses; many landslides are observed in Turkey in the regions formed by the greenstone series).

2. Other Volcanic Rocks

Volcanic rocks are found in units ranging in age from Precambrian to Pleistocene. Those of pre-Cretaceous age are morphologically and geologically unimportant and consequently will not be discussed.

Basalts are intercalated in the Upper Cretaceous and andesitic rocks in the Eocene series of several parts of the country. They are common, for instance, in the eastern part of the northern Anatolian folds, where the sedimentary succession may be reduced to unimportant intercalations and the volcanics dominate the landscape (Gattinger, 1963). It should be noted that in the neighboring sectors of Russia (Transcaucasia and Caucasus Ranges) volcanic rocks are common in all formations from the Liassic to the Eocene. In addition, post-Alpine (post-Cretaceous) plutonic rocks are widely developed on both sides of the Turkish–Russian frontier. Special conditions, the nature of which are unknown, must have facilitated the ascent of the magmatic masses (Turkey, Geol. Map of, 1950–1965; Moscow, 1937).

The "young" volcanic activity was initiated during the Oligocene, and was widespread in the Miocene and Pliocene, ending in Recent times. In regions with a long history of volcanic activity, very often the sequence begins with an initial phase of tuffs, followed by rhyolites, trachytes, andesites, and basalts, terminating with obsidians. In some areas basalts preceded the acidic and intermediate types. Pyroclastics and lacustrine deposits alternate with the lavas. An older phase of fissure eruptions can be distinguished from a younger period of vent activity. A close relationship exists between epeirogenic fault patterns and the distribution of Tertiary–Quaternary volcanism.

In some parts of the country, lavas cover recent alluvial deposits, recent plains, and thalwegs. Eruptions in historical time are reported from Erciyes Dagi in central Anatolia (Kayseri region), where they were evidently related to a secondary cone, as the main cone and crater are deeply eroded and show traces of the Pleistocene glaciation (Blumenthal, 1938). In the 15th century, Nemrut Dagi (southeastern Anatolia) erupted from a parasitic cone on the northern flank of the mountain (Izbirak, 1951; Ilhan, 1945; Mercier, 1949).

Pleistocene–Recent volcanic rocks are scattered over the country, but most extensive "young" volcanic covers are along the borders of the central Anatolian intermediate zone, including the volcanic areas of Kayseri and Ankara, measuring about 10,000 and 5000 km², respectively; in the Aegean region; in northeastern Anatolia, where various volcanic series extend over a total of about 40,000 km²; and along the southeastern Anatolian fault zone. Minor areas exist in southern Thrace and in the southeastern Anatolian fore-deep (Turkey, Geol. Map of, 1950–1965; Altinli, 1963a, 1964; Ketin, 1962; Gattinger, 1963; Ilhan, 1951; Mercier, 1949; Philippson, 1918).

3. Plutonic Rocks

Plutonic rocks, mostly granites, granodiorites, and diorites, are common in and between the metamorphic series in Thrace and Anatolia. They also are found along the tectonic line separating the northern and southern Anatolian folds in the eastern part of the country. The most important plutonic masses occur in northern Thrace, south of the Marmara Sea, in the inner (eastern) Aegean area, in central Anatolia, and in the northeastern sector of the northern Anatolian folds.

Recent studies indicate that the plutonic bodies are mostly synorogenic or postorogenic with respect to the Alpine orogeny (i.e., Late Cretaceous or post-Cretaceous). However, some of these masses may be of Late Carboniferous age and thus could be related to the Hercynian orogeny (Ketin, 1963, 1966; Gattinger, 1963).

III. TECTONIC FEATURES

A. Pre-Alpine Patterns

For geologists who, like Yalcinlar (1954), consider the metamorphic rocks as remains of Paleozoic orogenies, remains of pre-Alpine features exist in nearly all parts of the country. According to recent studies, however (e.g., Ketin, 1966), such relicts are preserved only in restricted areas, for instance, in the northern Anatolian folds, in Thrace, and in the northwestern sector of the Aegean folds. Flügel (1964) considers these features as the result of epeirogenic (or, following the "Alpine" classification—e.g., Kober, 1942—cratogenic) movements. He interprets the Alpine area, which is actually occupied by the Turkish Alpine belt, as a shelf zone situated to the south of the Caledonian and Hercynian orogenic belts.

Angular discordances within units belonging to the earlier Paleozoic, and the lithological differences that can be observed between Silurian and Devonian strata, indicate the presence of the Caledonian orogeny. More pronounced breaks in the Carboniferous section and, in some places, between the Paleozoic and the Mesozoic deposits are due to disturbances which occurred during the Hercynian diastrophism. As the existing structures were largely obliterated during the Alpine orogenies, the preservation of Caledonian and Hercynian structures on a large scale cannot be expected. The most important of such remains are reported from the Bartin area (northeast of Zonguldak on the Black Sea coast), where they are slightly divergent with respect to the Alpine axis, and from the Bosphorus region (Tokay, 1954). It must be noted that in the latter area, the divergence between Hercynian and Alpine axes appears much less important on more recent maps (Yalcinlar, 1951; Turkey, Geol. Map of, 1950–1965, sheet Istanbul) than on the map of Paeckelmann (1938), where the old features are shown as nearly perpendicular to the WNW–ESE-striking Alpine axes.

B. Alpine Features

It has been mentioned before that the present features of the country are controlled by the Alpine orogeny. The most important stratigraphic series were deposited in the Alpine geosynclinal troughs. The actual structures were formed by the various phases of the Alpine orogeny and the related epeirogenic movements.

All authors agree that the main fold zones of Turkey, the northern and southern Anatolian folds, correspond to the two main branches of the Eurasian orogenic belt. The *northern Anatolian folds* (together with their extension in Thrace, the Istranca folds) form part of the "Alpine" branch of the belt which consists of the northern Alps, the Carpathians, the northern Balkan ranges (west of Turkey), and the Transcaucasian, northern Iran (Elburz), and northern Afghanistan folds (east of Turkey). The *southern Anatolian folds*, together with the southern Alps, the Apennines, the folds of Yugoslavia, Albania, western and southwestern Greece (the west), southern Iran, and the Oman region (the east) form the southern or "Dinaride" branch of the belt (cf. Kober, 1942). Crystalline masses and secondary folds separate the Alpine trends in central and western Turkey.

The *northern foredeep* of the Alpine belt (together with the northern border zone of the northern Anatolian folds) lies in the Black Sea, north of the Anatolian coast, but extensions can be studied in eastern Bulgaria and in western Transcaucasia (Russia). The *southern foredeep* of the Alpine belt extends to Turkey in the southeastern part of the country. The shelf zone of the *Arabian shield*, which limits it in the south, enters Turkey along the Turkish–Syrian frontier (Arni, 1939; Egeran and Ilhan, 1948; Pinar and Ilhan, 1954; Ketin, 1959, 1966).

1. *Age of Folding*

Discontinuities can be observed at several localities between Permian and Upper Triassic or Liassic beds and between various Triassic and Jurassic sequences. The presence of flyschlike clastics, freshwater intercalations, coal horizons, and volcanic rocks in the Liassic of the northern Anatolian folds, sandy intercalations in the Jurassic carbonates of the southern Anatolian folds, and occasional flysch deposits in the Lower Cretaceous carbonates are evidences of *"early" Alpine movements* in Triassic, Jurassic, and Lower Cretaceous times (early and late Kimmerian movements). In general, the effect of these disturbances appears to be more pronounced in the northern than in the southern Anatolian fold belt (Abdusselamoglu, 1959; Erol, 1968; Ketin, 1966; Otkun, 1942).

Evidence of the first big Alpine paroxysm during the *Late Cretaceous* is found over all sectors of the Anatolian folds. This includes angular unconformities between pre-Upper Cretaceous and Upper Cretaceous deposits and between the latter and post-Cretaceous formations and distinctive vertical facies changes, such as that from Middle Cretaceous carbonates to Upper Cretaceous clastic series with reef intercalations. Important erosional gaps exist between pre-Upper Cretaceous and younger formations. The unconformities developed during the Late Cretaceous are more pronounced in the inner (central and axial) zones of the two Anatolian fold belts than in their external sectors, suggesting that the Late Cretaceous orogeny was most intense in the inner zones (Arni, 1939; Ketin, 1966; Egeran and Ilhan, 1948; Pinar and Ilhan, 1954).

Folding ceased by the Late Eocene in the axial parts of the orogenic belt, and lagoonal formations of uppermost Eocene and Oligocene age overlie the eroded folds. In the external zone of the southern Anatolian folds, in the Burdur and Mus regions (southwest and southeast Anatolia, respectively), orogenic movement continued and folded Oligocene clastics are included in the Alpine folds and overlain by Lower Miocene deposits (Altinli, 1944, 1963*b*; Neumayer, 1887; Ketin, 1966).

Along the external border of the southern Anatolian folds, there is evidence of *Late Miocene to post-Miocene* horizontal deformation. In southeastern Anatolia, the large frontal thrust sheets of the southern Anatolian folds rest on the Upper Miocene (Altinli, 1963a; Ozkaya, 1973; Tolun, 1962). Brunn et al. (1971) have described thrusting over the Lower Miocene in the Antalya region, in the external zone of the southern Anatolian folds in southwestern Anatolia. Post-Eocene deposits are absent in the external zones of the northern Anatolian folds, and nothing can be said about possible post-Eocene orogenic movements in this sector of the belt. Folded Oligocene flysch is reported from the eastern extension of this fold zone, in the Transcaucasian "Adjarides" (Moscow, 1937).

2. *Direction and Intensity of Folding*

Contradictory ideas exist concerning the direction and intensity of the Alpine fold movement in Turkey. Many observations show thrusting and folding from south to north in the northern Anatolian folds and from north to south in the southern Anatolian folds (Arni, 1939; Egeran and Ilhan,1948; Pinar and Ilhan, 1954, etc.; see also the sections in the papers of Altinli, 1963b, 1964; Blumenthal, 1944, 1947, 1948, 1949, 1950a; Tolun, 1962). The opposite situation was reported by Blumenthal (1952), who concluded that no distinct sense of movement can be established in the Anatolian belt sector. Observations made in the areas affected by such opposing movements (i.e., from north to south in the northern and from south to north in the southern Anatolian folds) suggest that these folds occurred in zones where younger epeirogenesis has resulted in "back-folding" of orogenic structures in the direction of younger depressions which formed less resistant zones. Such opposing trends are frequent along the border between Alpine folds and the central Anatolian massifs, as well as along epeirogenic fault zones as, for example, the borders of the northern Anatolian fault zone and the Lake Van Basin in southeastern Anatolia (Ilhan, 1971; Pinar and Ilhan, 1952; Ternek, 1953; Tokay, 1973). Data published on these folds suggest that in many cases the inverted movements, which are generally less intense than the "normal" orogenic folding, are younger than the orogenic deformation. In the northern Anatolian fault zone they affected Miocene (maybe even Pliocene) deposits, which are unfolded and unconformably overlie eroded folded rocks.

Some general evidence, like the distribution of greenstones in the Alpine belt and the dynamometamorphism of Mesozoic–Eocene rocks with the consequent formation of slates and marbles, leads one to suppose that the orogenic movements must have been very intense and of "Alpine" dimensions. Examples of this dynamometamorphism are reported from the Izmir (Arni, 1939), Dardanelles, Malatya, Hakkari regions (Ilhan, 1954), northern Anatolia (Ketin, 1966), and the Istranca Mountains in Thrace (verbal information).

Though such evidence is common, the intensity of the orogenic folding processes in Turkey has been the subject of discussion. Borchert (1958), for example, concluded from observations made in southwestern Anatolia that no true orogenic features exist in Anatolia and classified this belt as "para-orogenic." On the other hand, Brunn *et al.* (1971) recently reported complicated piles of thrust sheets from near the same region (Antalya sector of the southern Anatolian folds). Blumenthal (1949) was the first to describe thrust sheets with a width of more than 50 km, which moved from north to south in the area situated east and northeast of Antalya. A thrust front over 400 km long occurs along the southern border of the southern Anatolian folds in southeastern Anatolia and continues eastward into the frontal thrust of the Iranian Zagros folds. In some areas of Anatolia, this front consists of two or three superposed thrust sheets with an established minimum width of 15 to 30 km (Ozkaya, 1973).

Evidence of important horizontal movements is less clear in the northern Anatolian folds, but here too numerous structural details, as well as the distribution of greenstones and the presence of metamorphosed Mesozoic series, suggest important horizontal displacements. Bailey and McCallien (1953) proposed the existence of thrusts with a displacement of more than 200 km in order to explain phenomena related to the ultrabasic rocks in the northern Anatolian folds. Though no evidence for such exaggerated thrusts exists, this attempt illustrates the "need" of important horizontal dislocations felt by geologists who have tried to explain the structures of this belt. Ksiazkiewicz (1930) distinguished several south–north-directed thrust sheets in the metamorphic rocks of the Istranca Mountains in Thrace, the westward extension of the northern Anatolian folds. He classified these structures as Hercynian, but as some of the rocks affected by the movement are of Mesozoic age the thrust features must be "Alpine."

3. *Submarine Slides*

Overthrusts cannot account for all known horizontal displacements in the Anatolian fold belt. Submarine sliding seems to have occurred on a large scale during the Late Cretaceous orogeny and to a lesser degree during the younger paroxysms (Ilhan, 1967, has observed slides of Late Miocene age along the southeastern Anatolian thrust front). Such phenomena were reported at first by Arni (1931) from the Cretaceous "wildflysch" area along the Black Sea coast (see also Tokay, 1954). Submarine slides are also common along the northern border of the southeastern Anatolian foredeep ("exotic masses" in the Upper Cretaceous sediments) and were described by Rigo and Cortesini (1964), who tried to explain the horizontal displacement as being due to "olistostromes." Surely, thrust masses with a length of several hundred kilometers and a width of more than 50 km cannot be explained by simple submarine sliding. This phenomenon may be more important and frequent in the Anatolian fold belt than is generally supposed. Many minor tectonic folds may be related to olistostromes, which probably occur frequently along progressing thrust fronts, as is suggested by tectonic details observed along the southeastern Anatolian thrust front (Ilhan, 1971).

4. *Subdivision and Classification of the Turkish Alpine Zone*

Both paleogeographic and linear tectonic zones can be distinguished in the two main branches of the Alpine belt in Turkey and, beginning with Arni (1939), many attempts have been made to subdivide these folds. A summary of these attempts based on the papers of Arni (1931, 1939, 1942), Egeran and Ilhan (1948), Ketin (1966), and Pinar and Ilhan (1954) is given below:

a. *The Northern Anatolian Folds.* These are divided into external Pontides and internal Anatolides. The *Pontides* are characterized by miogeosynclinal sediments of the Mesozoic. Volcanic rocks of Upper Cretaceous and Eocene age are widespread, but ultrabasic rocks are extremely rare and those that exist may actually be olistostromes originating from the Anatolides. Upper Carboniferous strata consist mainly of freshwater deposits, while the Permian is continental. Middle and late Alpine folding is more pronounced than early Alpine deformation. Oligocene deposits are incorporated in the folds eastward beyond the Turkish–Russian border. The *Anatolides* are dominated by eugeosynclinal conditions during the Mesozoic, and ultrabasic rocks and associated sediments are common. Carboniferous and Permian here are represented by marine facies. Middle Alpine metamorphism is widespread. Oligocene beds unconformably overlie the Alpine folds.

b. *The Internal Torides and the External Iranides.* These are distinguished in the *southern Anatolian folds.* In the *Torides,* carbonates prevail in the Middle and Upper Paleozoic. The Triassic through Upper Cretaceous, and locally the Eocene, is characterized by continuous stratigraphic sequences in which carbonates dominate. Ultrabasic rocks and associated sediments are frequent. Early Alpine metamorphism is developed. Oligocene lagoonal deposits cover unconformably the Alpine folds. In the *Iranides* (not separated from the Torides by Ketin, 1966), eugeosynclinal and pre-Cretaceous Mesozoic deposits are rare, but ultrabasic rocks are widespread. The Eocene is characterized by a miogeosynclinal carbonate facies in the western part of the zone. The Oligocene is included in the Alpine folding.

Though clear evidence of linear tectonic subzones exists in both branches of the Alpine belt, it is still impossible to draw satisfying limits between these secondary units. The proposed lines cut structural units, like the limit between Pontides and Anatolides north of Ankara (Ketin, 1966) in a very artificial manner, or local thrusts and thrust faults are taken as boundaries (limit between Torides and Iranides in southeastern Anatolia; Altinli, 1963a, b, 1964). It seems that either the available data are insufficient for this delineation, or that these subzones gradually pass into one another without linear limits. For these reasons, the present authors (1954) proposed deferring this delineation and suggested applying the following provisional subdivision: northern Anatolian folds subdivided into external and internal Anatolides; southern Anatolian folds; external and internal Torides.

c. *The Question of "Intermediate" Units.* In eastern Anatolia the northern and southern fold belts are separated by a narrow, strongly disturbed zone characterized by slightly metamorphosed sediments, greenstones, and younger plutons (corresponding to a tectonic scar or root zone). In central and western Anatolia, intermediate features are inserted between the two main belts. These intermediate structural units consist of rigid masses in which pre-Tertiary metamorphic or plutonic rocks predominate. They remained partially or entirely emergent during the Late Cretaceous–Eocene (as the Kirsehir massif in central Anatolia and the Menderes massif in western Anatolia) and folds are distributed irregularly between the massifs and between them and the borders of the main fold zones. These folds (Ankara "fan" in central Anatolia and Aegean folds in the Marmara region and western Anatolia) are distinguished from the main fold belts by their divergent linear exposures and less complete stratigraphic successions (Erentoz, 1966; Ilhan, 1971; Pinar and Ilhan, 1954). Consequently, these zones are both tectonically and stratigraphically distinct from the Alpine main folds. It is a matter of controversy whether these intermediate areas should be included in the Pontides and Anatolides (Arni, 1939; Ketin, 1966) or be considered, at least provisionally, as a zone of intermediate character situated between the northern and southern Anatolian folds, as has been proposed by the authors.

5. Southeastern Anatolian Foredeep and Border Folds

A foredeep, including an autochthonous fold zone, separates the orogenic belt of Turkey from the shelf of the Arabian shield. This is the northwestern extension of the Middle East foredeep belt (including most of the Middle East oilfields), which becomes more shallow and narrow in a westerly direction. The estimated total thickness of the foredeep sediments diminishes from about 16,000 m in the southeastern part of the Persian Gulf area to about 6000 m in the Turkish sector (Ilhan, 1967). To the north, the foredeep is separated from the orogenic belt by a significant thrust zone, the southern front of the southern Anatolian folds. The foredeep has no clearly defined limit to the south and southwest, where its typical facies and features pass into those of the shelf zone of the Arabian shield.

The autochthonous border folds, so well developed south of the thrust front in Iran and Iraq, lose their character as an independent tectonic unit east of Siirt in southeastern Anatolia. Westward from this area, they pass into groups of anticlines which are sometimes asymmetrical, with reversed southern flanks and longitudinal faults on one or both of the flanks. Pre-Alpine highs (surrounded by Mesozoic strata of a littoral facies), as well as features formed before and after the deposition of the Upper Cretaceous reefs, can be distinguished among these structures. Their surface expression suggests that some of them are the reflection of deep fault blocks rather than compressional folds (Bolgi and Onem, 1966; Ilhan, 1967). The foredeep series, noted in the discussion of stratigraphy, are covered by Miocene marine–brackish to Pliocene lacustrine to continental formations.

C. Epeirogenic (Post-Alpine) Features

1. General Character

The orogenic features of the Turkish sector of the Alpine belt are intersected by a number of structural features, including individual faults, fault zones, and tectonic depressions ranging from narrow grabens to large tectonic basins. All these intersecting trends are related to important fault zones and are the result of widespread post-Alpine epeirogenic movement. Frequent strong earthquakes and about 3000 hot and cold mineral springs are related to the epeirogenic fault patterns, according to Pinar (1948).

The "intra-Alpine" epeirogenic depressions are occupied by marine, lagoonal, freshwater, and continental Tertiary deposits and have been briefly described in the section on stratigraphy. These series can attain thicknesses of several thousand meters.

While the smaller depressions are limited by faults, some of the large ones were originally unfaulted, subsiding basins which fractured later as deepening movements became more pronounced. In these cases, the basal deposits unconformably overlie the Alpine basement, while the younger Tertiary series are separated from the Alpine margins and the older Tertiary strata by important fracture systems.

2. Age of the Epeirogenic Movements

The important thickness of the Eocene deposits in Thrace and in the central Anatolian Tuz Golu Basin (shown by wells) indicates that epeirogenic subsidence began during the Eocene. In several places, Oligocene gypsiferous series and Miocene marine series occupy depressions which must have been formed before the deposition of these sediments. This is shown by the development of littoral facies along the basin borders. Such depressions exist in northeastern Anatolia (Pinar and Ilhan, 1952, 1954).

Epeirogenic movements continued until very recent times. Vertical disturbances affecting Miocene and Pliocene deposits are frequent. In the Tuz Golu Basin in central Anatolia and in the "Lake Region" of the southern Anatolian Toros Mountains (west of Konya), fault-bounded younger depressions, occupied by lakes or Pleistocene lacustrine deposits, intersect older basins containing Upper Miocene and Pliocene deposits (Ilhan, 1945).

The physiographic expression of mountain ranges, river valleys, and the broad features of the country reflect very recent epeirogenic movements only slightly modified by erosion (Ilhan, 1948, 1968; Erol, 1968). The depressions occupied by the Aegean Sea and the Marmara Sea, including the Dardanelles and the Bosphorus, which are former river valleys (Penck, 1919) probably following fault zones, reached their present condition in the early Pleistocene (English, 1904; Pinar, 1948).

Deformed river terraces are reported from several regions, and tectonic deformation of the coast and coastal valleys ("suspended valleys") can be seen along the Black Sea and the Gulf of Iskenderun (Ilhan, 1948). Marine terraces containing Pleistocene fossils have been found as much as 180 m above sea level (English, 1904; Yucel, 1953).

3. Intensity of Epeirogenic Movements

Vertical displacement along the epeirogenic faults, and regional deformation of large basins affecting the bottom of these depressions, ranges from a few meters to several thousand meters. In the Oligocene and Miocene lagoonal and lacustrine deposits of Anatolia, fault displacements of several hundred meters are common. In the grabens of the Aegean region, displacements of 800–1000 m have been measured, and along the large fault system of southern Thrace and northern Anatolia, movements of 1000–2000 m occur (Pinar and Ilhan, 1952, 1954). In eastern Anatolia, the elevation of the base of the Lower Miocene series varies between 900 m above sea level in the Kelkit Valley (region lying north of Sivas) and 2300 m in the Van region. In the southern part of the Adana Basin, the base of the formation is about 5000 m below sea level (result of drilling), while in the northern part of the same basin it is between 2000 and 2300 m above sea level (Turkey, Geol. Map of 1950–1965, documents of the Petroleum Administration).

4. Classification and Distribution of Epeirogenic Features

Among the epeirogenic features of Thrace and Anatolia, several patterns of faulting can be distinguished: single faults; complicated patterns of faults (both normal and reversed faults, local thrust faults) and narrow grabens; and finally, large subsiding basins with partially fractured margins ("intra-Alpine" or "intra-Cordilleran" basins).

a. *Important Epeirogenic Basins*. The following important intra-Alpine basins and basin patterns can be distinguished in the country:

The *Thrace Basin* covering about 5600 km² is filled with shales, sandstones, and marls which range in age from Eocene to Miocene (in the Eocene series, a littoral reef facies is developed) and show sorting from coarse to fine material from the basin borders to the center of the depression. The thickness of these marine, lagoonal, and freshwater deposits exceeds 4000 m in the central part of the basin (Chaput, 1936, and unpublished documents of the Petroleum Administration).

The basins of *Antalya* and *Adana–Iskenderun* (surface area of the continental sectors of these basins: about 4000–12,500 km², respectively) are occupied by shales, marls, sandstones, and littoral reef limestones (algal deposits, no coral reefs). In the latter depression, the maximum thickness of the Tertiary section must be more than 5000 m according to the well drilled offshore from Mersin (documents of the Petroleum Administration; Erentoz, 1956; Schmidt, 1961). Extending southward, below the Mediterranean Sea, these basins are probably connected with the Miocene depressions of northern Cyprus.

The *Tuz Golu depression* (about 8000 km²) and *Konya Basin* (approximately 6000 km²) in central Anatolia, as well as the *depressions of western Anatolia* (estimated total surface area 3500 km²) are occupied by Miocene and Pliocene lacustrine marls and limestones with subordinate sandstones and shales. In the larger western Anatolian basins, this series is underlain by a flyschlike clastic lacustrine sequence. Pleistocene freshwater and brackish water deposits (the latter in closed basins) sometimes overlie the Tertiary (Chaput, 1936; Philippson, 1918; Ilhan, 1946; Turkish Gulf Oil, 1961).

In addition to the Miocene and Pliocene lacustrine deposits, the "gypsiferous formation," a succession of lagoonal, freshwater, and continental clastics and marls that range in age from the Late Eocene to the Miocene, and Lower Miocene marine deposits (clastics and littoral algal reefs) occupy the depressions situated in the eastern part of central Anatolia and in northeastern and eastern Anatolia. Facies changes in these various strata make it impossible to establish a common stratigraphic column for all these depressions.

The *Kizilirmak–Delice Irmak Basin* (east of Ankara), which covers about 3800 km² and is occupied by the gypsiferous formation, and the *Sivas Basin* (between Kayseri and Erzincan), extending over about 14,000 km² and containing gypsiferous deposits together with marine Miocene and lacustrine Miocene–Pliocene strata, are the most important Tertiary depressions of eastern central Anatolia. The *Yukari Murat Basin* can be considered as the biggest depression of this kind in eastern Anatolia; it is situated between Erzurum and Mus, and covers an area of about 5000 km², including its continuation below younger lavas. The Tertiary sequence consists of marine, lacustrine, and lagoonal Miocene deposits (Alagoz, 1967; Baykal, 1966; Kurtman, 1961; Erentoz, 1966; Ilhan, 1949; Stchepinsky, 1939, 1942). Numerous minor depressions, scattered over the eastern part of central Anatolia and over eastern Anatolia, have similar stratigraphic sequences.

b. *Important Epeirogenic Fault Patterns.* The innumerable faults which intersect with the Alpine folds, intermediate massifs, and intermediate folds in Anatolia and Thrace can be grouped in patterns. The more important of these patterns are listed below (following summaries published by Pinar, 1948; Pinar and Ilhan, 1952, 1954; Ilhan, 1971):

Marmara–Aegean Region. In this area a series of east–west-extending grabens 60 and 250 km in length intersect with minor north–south striking features. The formation of these patterns may be related to the down-faulting movements in the Aegean Basin.

Central Anatolia. Fault systems intersect with the central Anatolian intermediate area and separate the latter from the surrounding Alpine folds; the longest single fault (125 km) of this zone forms the eastern border of the central Anatolian Tuz Golu Basin.

Southern Thrace and Northern Anatolia. A major zone of epeirogenic disturbances, consisting of an elongated series of faults, grabens, and basins which succeed one another "en echelon," extends roughly west–east through southern Thrace, the northern Marmara Sea, and the northern Anatolian folds. The pattern continues into Russia in the east and to Greece and Yugoslavia in the west. The length of the Turkish sector is about 1500 km. This pattern corresponds to the "northern Anatolian earthquake zone" and evidently reflects an important deep-seated tectonic structure, probably a series of fractures or the root zone of a big thrust sheet. This relation between surface faults and deep-seated structures is supported by the fact that the major manifestations of ultrabasic rocks in the northern Anatolian folds form an alignment that is roughly parallel to the fault zone.

Southeastern Anatolian Fault Zone. This zone is the northeastern extension of the big rift system, which includes the Dead Sea graben and extends northward to the Maras region in southeastern Turkey. Here it reaches the southern Anatolian fold belt and turns sharply eastward, following the Alpine folds to Iran. The length of the Turkish part of this zone is about 700 km and corresponds to the "southeastern Anatolian earthquake zone." Tectonic considerations concerning the nature of the northern Anatolian fault area (see above) are also valid for the southeastern Anatolian fault zone.

Black Sea and Aegean Coasts. The Turkish sector of these marginal areas probably corresponds to a complicated fault pattern. This is suggested by the alignment of seismic epicenters along the coast and the striking parallelism of the coast lines with important fault zones.

The majority of the epeirogenic alignments are oriented more or less parallel to the orogenic fold axes. This suggests that the development of the epeirogenic structures has been influenced to some degree by the orogenic trends. On the other hand, in the Aegean region and in the northern extension of the Dead Sea rift valley (area between Mersin, Hatay, and Maras), there is a pronounced north–south orientation of epeirogenic alignments. Some important transverse trends cross the generally west-extending orogenic and epeirogenic lines (Paréjas, 1940).

All the above-mentioned epeirogenic fault patterns are very active earthquake zones and affect about 40% of the surface area of Turkey. In 1952, the authors of this paper compiled a list of more than 750 major shocks recorded in Turkey since the First Century A.D. More complete catalogues published in 1967 and 1971 by Ergin *et al.* list about 2000 shocks with epicenters in or near Turkey which are related to the above mentioned patterns and have been recorded by seismographs since 1904 (for more data concerning the relations between geology and seismicity in Turkey, see Pinar 1943*a*, 1951, 1953*a*, 1954, 1960).

IV. NOTES ON THE GEOLOGY OF CYPRUS

A. Introduction

Geologically, Cyprus forms the southwestern extension of the external zones of the southern Anatolia folds. West of the Gulf of Iskenderun, these external zones are submerged by the Mediterranean Sea. The island belongs to a major arc formed by the southern Anatolian folds between the Arabian promontory of the African–Arabian shield to the east and a small, western promontory of this shield thought to pass northward below the Gulf of Antalya.

The separation of Cyprus from Anatolia must have occurred in relatively recent geological time, for the Miocene Adana Formation seems to extend southwestward into the Miocene Kytherea Basin in Cyprus (Weiler, 1970).

The geological problems of Cyprus are identical with those of southern Anatolia; for example, the same questions recur as to the ultrabasic rocks and the stratigraphic position of the "comprehensive" Mesozoic carbonate series. The reasons for a separate section on Cyprus are political rather than scientific.

B. Stratigraphy

(If no other authors are indicated, the data mentioned below are based on papers published by Ginty, 1940; Hassan, 1962; Henson, 1950; Reed, 1921; Survey of Groundwater and Mineral Resources of Cyprus, 1970.)

No *pre-Mesozoic* formations are reported from Cyprus. Paleozoic material may perhaps exist among the crystalline carbonate blocks found in the tectonic "melanges" associated with the ultrabasic rocks.

The *Triassic* is represented by red clastics and limestones, radiolarites, and various recrystallized carbonates. *Diplopora* and *Megalodus* are the most frequently mentioned key fossils.

The *Jurassic* is also present in the form of more or less crystalline carbonates and marls; *Ellipsactinia* is a frequently reported fossil from these formations.

Carbonates, including shale and marl horizons, extending from the *Triassic up to the Cretaceous* ("Mamonia complex"), can be compared with the "comprehensive Mesozoic carbonate series" of the southern Anatolian fold area.

The Upper Cretaceous is characterized by limestones with subordinate clastics. The Paleocene–Eocene is represented in its lower part by an alternation of slates, marls, limestones, and siliceous limestones with nummulites and orbitoids, followed by an upper, more uniform flysch succession.

Marine Miocene series, which occupy tectonic depressions of Cyprus, are similar to those developed in the Adana Basin on the Turkish mainland. There is a Lower Miocene marl and sandstone series, Lower Tortonian flyschlike sediments grading toward the basin borders into coarse clastics and bioclastic limestones, and an Upper Tortonian sequence composed of brackish-water marls with some gypsum.

The *Pliocene* consists of continental brackish water and marine deposits.

In the Tertiary basins, where outcrop conditions are poor and the unconformities between the various formations mostly hidden, the lithologic similarity between Miocene and Eocene "flyschlike" series, and between the Miocene, Eocene, and Upper Cretaceous limestones, was the reason for the establishment of a supposedly "comprehensive series" reaching from the Upper Cretaceous to the Miocene, similar to the "Lefkara group" (see Reed, 1921).

C. Tectonic Features

(If no other authors are indicated, the data mentioned below are based on papers published by Ginty, 1940; Reed, 1921; Survey of Groundwater and Mineral Resources of Cyprus, 1970.)

Cyprus is an Alpine country dominated by orogenic features strongly affected by postorogenic vertical movements. The "timetable" of the structural development is the same as that of the external zones of the southern Anatolian folds on the Turkish mainland. The effects of post-Cretaceous orogenies predominate, but orogenic movements continued until very recent times.

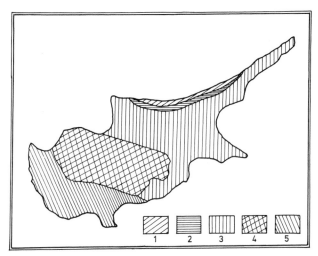

Fig. 10. Principal tectonic units of Cyprus. Explanation of legend: (1) coastal plain strip of Girne (Kyrenia); (2) Girne Range (Alpine folds); (3) Mesaoria (Lefkose) Basin (intra-Alpine tectonic depression); (4) Troodos Mountains area (plutonic, volcanic, and ultrabasic rocks); (5) southern coastal strip of Baf (Paphos)–Limasol–Larnaka (young tectonic depression).

The mountain ranges of the island, the Kyrene (Girne) Range in northern Cyprus and the Troodos Mountains in southwestern Cyprus are Alpine folds and lie along the arc of the southern Anatolian folds, extending from the Gulfs of Iskenderun and Antalya.

As a result of the epeirogenic movements, these folds are broken up into a number of fault blocks separated by intra-Alpine depressions. These are the northern coastal zone of Kyrene (Girne), the central depression of Kytherea (Lefkose), and finally the southern coastal zone of Paphos (Baf)–Limasol (Lemasun)–Larnaca (Iskele). It is the effects of these fault-block features rather than those of the Alpine fold structures that dominate the actual topography.

D. The Main Geological Units of Cyprus

(If no other authors are indicated, the data mentioned below are based on papers published by Ginty, 1940; Reed, 1921; Survey of Groundwater and Mineral Resources of Cyprus, 1970; Weiler, 1970.)

A series of five tectonic units crossing the island with a roughly east–west trend show the effects of the Alpine fold movements in forming the basic structure of the island (Fig. 10). The zones are from north to south:

1. The Kyrene (Girne) Coastal Plain

The plain is composed of Miocene and Pliocene deposits unconformably overlying Eocene and Upper Cretaceous beds. The southern margin of the depression is an important fault zone.

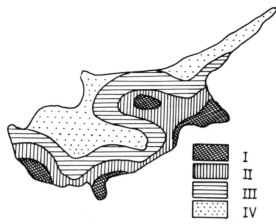

Fig. 11. Earthquake zones of Cyprus: (I) 1st-degree earthquake zone (intensity = IX–X); (II) 2nd-degree earthquake zone (intensity = VIII); (III) 3rd-degree earthquake zone (intensity = VI–VII); (IV) 4th-degree earthquake zone (intensity = IV–V). Degrees of intensity after modified Mercalli scale (Earthquake Research Institute, 1973).

2. The Kyrene (Girne) Ranges and Their Northeastward Extension

The Cape Andreas Range is an uplifted tectonic block formed by Alpine folds. The rocks consist of a Triassic–Jurassic sedimentary series and volcanic beds, as well as Upper Cretaceous–Eocene sediments. The block is overturned to the south and thrust over the Miocene cover of the Kytherea Basin. This movement was the result of the Late Miocene orogenic phase so well developed in the external border zone of the southern Anatolian folds.

3. The Kytherea (Lefkose) or Mesaoria (Mesaya) Basin

The basin occupies the central part of the island. It was formerly an intra-Alpine tectonic depression and is now filled with marine Miocene shales and flysch series. This strongly faulted complex of rocks, more than 3000 m thick, was compared with the fine clastics of the Adana Basin on the mainland by Weiler (1970), who supposed that the latter basin extended southwestward into the Kytherea depression. Thus, the sea floor between Cape Andreas (northeastern promontory of Cyprus) and the Anatolian coast should be formed of submerged Miocene beds.

The Miocene complex is limited in the north by the thrust margin of the Kyrene (Girne) Range. Erosional remains of Miocene strata unconformably covering the northern slopes of the Troodos Mountains in the south suggest that originally the Miocene covered the margins of this tectonic high, but with continuing epeirogenic movements this margin finally became a fault zone.

4. *The Troodos Mountain Area*

This region shows complicated tectonic features because of the widespread development of tectonic "melanges" with ultrabasic rocks, and also because of the intersection of a dense network of epeirogenic faults and Alpine folds. Various interpretations of these features have been published. The classification as "massif" found in some older papers (e.g., Reed, 1921), which does not fit the geological feature, is evidently due to the frequence of ultrabasic rocks considered previously as related to plutonic bodies.

In addition to the Mesozoic–Eocene series in the Kyrene Range, "Mesozoic" or "post-Triassic" ultrabasic rocks with associated radiolarites and red limestones, volcanic rocks (Jurassic?), and pillow lavas (Senonian) are common in this area. Most of the igneous rocks and sediments are intensively folded, contorted, and sometimes recrystallized. It is not clear if the gabbros and granophyres reported from this area are really plutonic material (included in the tectonic mixtures) or simply ultrabasic material showing more granular texture. Descriptions and geologic maps (Ginty, 1940; Survey of Groundwater and Mineral Resources of Cyprus, 1970) suggest that most of these formations are not *in situ*, but form complicated tectonic "melanges" due to submarine slides; however, the possibility of thrusts cannot be excluded.

5. *The Southern Coastal Strip of Paphos (Baf)–Limasol (Lemasun)–Larnaca (Iskele)*

This coastal strip is formed by a young tectonic depression where highly contorted and faulted marine Miocene deposits overlie the Upper Cretaceous–Eocene cover of the Troodos area.

NOTE

No special attempts have been made in Turkey to discuss the factors of "worldwide" range that have produced the orogenic and epeirogenic phenomena. The authors generally agree with the classic theories on lateral compressional stress originating in the movements of the rigid blocks limiting the belt in the south and north (e.g., Argand, 1922; Aubouin, 1962, 1965; Kober, 1942).

New ideas based on such regional geotectonic considerations as "plate tectonics" (e.g., Çoğulu, 1973; Brinkmann, 1973), longitudinal horizontal strike-slip movements (e.g., Brune, 1968; Pavoni, 1961), and "anticlockwise" rotational movements of the Arabian shield (e.g., Gildler, 1966) will be important in tectonic synthesis. The importance of these ideas can be judged only when the entire Mediterranean Alpine belt is taken into consideration and not just a limited part, such as that found in Turkey. Our aim, therefore, has been to summarize for the reader the geology of Turkey so that he may consider for himself how well it fits the various tectonic models that are currently being proposed for the Mediterannean region.

REFERENCES

Abdusselamoglu, S., 1959, Die Geologie des Almacikdagi und der Umgebung von Mudurnu und Goynuk, *Istanbul Univ. Fen Fak. Monogr.*, no. 14.

Abdusselamoglu, S., 1963, Nouvelles observations stratigraphiques et paléontologiques sur les terrains paléozoiques affleurant à l'est du Bosphore, *Bull. Miner. Res. Explor. Inst. Turk.*, v. 60, p. 1–6.

Alagoz, C., 1967, Sivas cevresinde ve dogusunda karst olaylari, *Ankara Univ. D.T.C. Fak. Yaynnlari*, v. 175.

Altinli, E., 1944, Etude stratigraphique de la région d'Antalya, *Rev. Fac. Sci. Univ. Istanbul*, v. 9 (3), p. 227–238.

Altinli, E., 1963a, Explanatory text of the geological map of Turkey, sheet Cizre, *Bull. Miner. Res. Explor. Inst. Turk.*, p. 51–103.

Altinli, E., 1963b, Explanatory text of the geological map of Turkey, sheet Erzurum, *Bull. Miner. Res. Explor. Inst. Turk.*, p. 55–103.

Altinli, E., 1964, Explanatory text of the geological map of Turkey, sheet Van, *Bull. Miner. Res. Explor. Inst. Turk.*, p. 41–90.

Argand, E., 1922, La tectonique de l'Asie, *C. R. Congr. Géol. Int. (Liège)*, v. 1, p. 171–372.

Arni, P., 1931, Zur Stratigraphie und Tektonik der Kreideschichten östlich Eregli an der Schwarzmeerküste, *Eclogae Geol. Helv.*, v. 24, p. 305–355.

Arni, P., 1939, Tektonische Grundzüge Ostanatoliens und benachbarter Gebiete, *Bull. Miner. Res. Explor. Inst. Turk.*, v. 4, p. 53–89.

Arni, P., 1942, Materialien zur Altersfrage der Ophiolite Anatoliens, *Bull. Miner. Res. Explor. Inst. Turk.*, v. 7, p. 481–488.

Aubouin, J., 1960, Essai sur l'ensemble italo-dinarique et ses rapports avec l'arc alpin, *Bull. Soc. Géol. France*, v. 2, p. 487–526.

Aubouin, J., 1962, Propos sur les géosynclinaux, *Bull. Soc. Géol. France*, v. 7, II, p. 487–526.

Aubouin, J., 1965, *Geosynclines*, Amsterdam: Elsevier Publ. Co.

Bailey, E. B., and McCallien, W. J., 1953, Serpentine lavas, the Ankara Melange, and the Anatolian Thrust, *Trans. Roy. Soc. Edinburgh*, v. 23, p. 1.

Baykal, F., 1966, Explanatory text of the geological map of Turkey, sheet Sivas, *Bull. Miner. Res. Explor. Inst. Turk.*, p. 49–116.

Bilgutay, U., 1959, The Permian calcareous algae from Southeastern Anatolia, *Bull. Miner. Res. Explor. Inst. Turk.*, v. 52, p. 48–58.

Bilgutay, U., 1960, Geology of the Hasanoglan region, *Bull. Miner. Res. Explor. Inst. Turk.*, v. 54, p. 44–51.

Blumenthal, M., 1938, Der Erdschias Dagh, *Die Alpen*, Bern, v. 14, p. 82–87.

Blumenthal, M., 1941, Contribution à la connaissance du Permo-Carbonifère du Toros entre Kayseri–Malatya, *Bull. Miner. Res. Explor. Inst. Turk.*, v. 1 (31), p. 118–139.

Blumenthal, M., 1944, Schichtfolge und Bau der Torosketten im Hinterland von Bozkir (vilayet Konya), *Rev. Fac. Sci. Univ. Istanbul*, v. 9, p. 95–125.

Blumenthal, M., 1945, Sind gewisse Ophiolitzonen Nordanatoliens prae-liassisch? *Bull. Miner. Res. Explor. Inst. Turk.*, v. 33, p. 125–132.

Blumenthal, M., 1947, Geologie der Torosketten im Hinterland von Seydisehir und Beysehir, *Publ. Miner. Res. Explor. Inst. Turk.*, Ser. D, no. 2.

Blumenthal, M., 1948, Un aperçu de la géologie des chaînes nord-anatoliennes entre l'ova de Bolu et le Kizilirmak inférieur, *Publ. Miner. Res. Explor. Inst. Turk.*, Ser. B, no. 13.

Blumenthal, M., 1949, Les lambeaux de recouvrement du Toros occidental, *Bull. Geol. Soc. Turk.*, v. 2, p. 31–40.

Blumenthal, M., 1950a, Beiträge zur Geologie der Landschaften am mittleren und unteren Yesil Irmak, *Publ. Miner. Res. Explor. Inst. Turk.*, Ser. D, no. 4.

Blumenthal, M., 1950b, Geologische Forschungen in Hinterland von Alanya, Westlicher Toros, *Publ. Miner. Res. Explor. Inst. Turk.*, Ser. D, no. 5.

Blumenthal, M., 1952, Sur l'inconstance du dejettement tectonique dans la zone orogénique anatolienne, *Rep. 18th Int. Geol. Congr.*, London (1948), v. 13, p. 23–32.

Bolgi, T., and Onem Y., 1966, Regional geological sections in southeastern Anatolia. Petroleum Activities in Turkey, *Petr. Dairesi Nesr.*, v. 11, p. 52–59.

Borchert, H., 1958, Die Chrom- und Kupfererzlagerstaetten des initialen Magmatismus in der Türkei, *Publ. Miner. Res. Explor. Inst. Turk.*, no. 102.

Brinkmann, R., 1971, The Geology of Western Anatolia, in: *Geology and History of Turkey*, Campbell, A. S., ed., Tripoli: Petrol. Explor. Soc. Libya, p. 171–190.

Brinkmann, R., 1973, Über Salzwasser-Thermen im Küstenland von West-Anatolien (Türkei), *Chem. Geol. (Amsterdam)*, v. 12, p. 171–187.

Brune, N., 1968, Seismic moments, seismicity, and rate of slip along major fault zones, *J. Geophys. Res.*, v. 73, p. 777.

Brunn, J. H., 1961, Les sutures ophiolitiques, *Rev. Géogr. Phys. Géol. Dyn.*, v. 2/4, no. 1.

Brunn, J. H., Dumont, J. F., de Graciansky, P. C., Gutnic, M., Juteau, T., Marcoux, J., Monod, O., and Poisson A., 1971, Outline of the Geology of the Western Taurids, in: *Geology and History of Turkey*, Campbell, A. S., ed., Tripoli: Petrol. Explor. Soc. Libya, p. 225–255.

Chaput, E., 1936, Voyages d'études géologiques et géomorphologiques en Turquie, *Mém. Inst. Franc. d'Archéol. Istanbul*, v. 2, 312 p.

Çoğulu, H. E., 1973, New data on the petrology of Kizildag Massif (Hatay, Turkey), in: *Congress of Earth Science on the Occasion of the 50th Anniversary of Turkish Republic*, Abstracts, Miner. Res. Inst. and Geol. Soc. Turkey, eds., Ankara, p. 112.

Dubertret, L., 1953, Notice explicative, feuille d'Antioche, *Rép. Syrienne Miner. Trav. Publ. (Damascus)*, 67 p.

Dunkee, E. F., 1961, Proposed stratigraphic nomenclature, District VI, Southeastern Turkey, *Petroleum Activities in Turkey (Petrol. Administr., Ankara)*, v. 6, p. 38–46.

Earthquake Research Institute, 1973, Bull. no. 1.

Egemen, R., 1947, A preliminary note on fossiliferous Upper Silurian beds near Eregli, *Bull. Geol. Soc. Turk.*, v. 1, p. 56–59.

Egeran, N., and Ilhan, E., 1948, *Turkiye Jeolojisi*, Ankara: published by the authors, 206 p.

English, T., 1904, Eocene and later formations surrounding the Dardanelles, *Q. J. Geol. Soc. London*, v. 60, p. 243–295.

Erentoz, C., 1966, Contribution to the stratigraphy of Turkey, *Bull. Miner. Res. Explor. Inst. Turk.*, v. 58, p. 25–41.

Erentoz, L., 1956, Mollusques du Néogene des bassins de Karaman, Adana et Hatay (Turquie) *Publ. Miner. Res. Explor. Inst. Turk.*, Ser. C, no. 4, 53 p.

Ergin, K., Guçlu, U., and Uz, Z., 1967, A catalogue of earthquakes for Turkey and surrounding area (11 A.D. to 1964), *Istanbul Teknik Univ. Maden Fak.*, paper 24, 146 p.

Ergin, K., Guçlu, U., and Uz, Z., 1971, A catalogue of earthquakes for Turkey and surrounding area (1965–1970), *Istanbul Teknik Univ. Maden Fak.*, paper 28, 75 p.

Erguvanly, K., 1948, *Hereke pudinglari ile Gebze taslarinin insaat bakimindan etudu ve civarinin jeolojisi*, Thesis, Istanbul Teknik Univ. Insaat Fak.

Erinc, S., 1949, Eiszeitliche und gegenwärtige Gletscher in der Kackardaggruppe, *Rev. Fac. Sci. Univ. Istanbul. Ser. B*, v. 14, p. 243–246.

Erk, S., 1942, Etude géologique de la région entre Gemlik et Bursa (Turquie), *Publ. Miner. Res. Explor. Inst. Turk., Ser. B*, no. 9.

Erol, O., 1956, A study of the geology and geomorphology of the region southeast of Ankara in Elma Dagi and its surroundings, *Publ. Miner. Res. Explor. Inst. Turk., Ser. D*, no. 9.

Erol, O., 1963, Geographical researches in the east of Tuz Golu, the Salt Lake, Central Anatolia, *Rev. Turque Géogr.*, XVIII–XIX, v. 22–23, p. 65–78.

Erol, O., 1968, Paleozoic formations of the Problem of Paleozoic–Mesozoic boundary in the Ankara region, *Bull. Geol. Soc. Turk.*, XI, v. 1–2, p. 1–16.

Flügel, H., 1964, Die Entwicklung des Vorderasiatischen Paleozoikums, in: *Geotektonische Forschungen*, no. 18, Stuttgart, 68 p.

Furon, R., 1953, Introduction à la géologie et à l'hydrogéologie de la Turquie, *Mém. Mus. Nat. Hist. Nat., Ser. C*, III/1, p. 1–128. (Contains a complete bibliography until 1952.)

Frech, F., 1916, Geologie Kleinasiens im Bereich der Bagdadbahn, *Zeitschr. Deutsche Geol. Ges. (Stuttgart)*, v. 68, p. 1–325.

Gattinger, H., 1963, Explanatory text of the geological map of Turkey, sheet Trabzon, *Bull. Miner. Res. Explor. Inst. Turk.*, p. 35–75.

Gildler, R. W., 1966, The role of translational and rotational movements in the formation of the Dead Sea. Symposium on the World Rift System, *Geol. Surv. Canada*, Paper 66–14, p. 65–77.

Ginty, J. M., 1940, Geological map of Cyprus, *S.A.E.C., G.H.Q., Middle East Forces (Report)*.

Grancy, S., 1939, Überblick über die bisherigen Aufschlussarbeiten und Ergebnisse im östlichen Anatolischen Steinkohlenbecken, *Bull. Miner. Res. Explor. Inst. Turk.*, v. 4/17, p. 35–65.

Gutnic, M. D., and Monod, O., 1970, Une série mésozoique condensée dans le nappes du Taurus occidental: La série du Boyali Tepe, *C. R. Somm. Soc. Géol. France*, v. 5, p. 166–167.

Hassan, A. H., 1962, Distribution of Triassic and Jurassic formations in east Mediterranean, *Proc. 4th Arab Petrol. Congr. (Beirut)*, Paper 25 (B-3).

Henson, F. S., 1950, Cretaceous and Tertiary reef formations and associated sediments in Middle East, *Bull. Amer. Assoc. Petrol. Geol.*, v. 34 (2), p. 215–338.

Hiessleithner, G., 1955, Neue Beiträge zur Geologie Chromerz führenden Peridotitserpentine des südanatolischen Taurus, *Bull. Miner. Res. Explor. Inst. Turk.*, v. 46/47, p. 17–45.

Ilhan, E., 1945, Le volcanisme néogène et quaternaire en Anatolie, *Turk. Cogr. Dergisi (Ankara)*, v. 3, p. 1–8.

Ilhan, E., 1946, Les dépots pliocènes et quaternaires de la région de Konya–Burdur, *Rev. Fac. Sci. Univ. Istanbul*, v. 11, p. 85–106.

Ilhan, E., 1948, Mouvements tectoniques jeunes en Anatolie, *Bull. Soc. Géol. France*, v. 5/18, p. 521–527.

Ilhan, E., 1949, On the geology of Central Anatolia, *Bull. Geol. Soc. Turk.*, v. 2, p. 116–125.

Ilhan, E., 1950, La formation gypsifère en Anatolie (Asie Mineure), *Bull. Soc. Géol. France*, v. 5/20, p. 451–457.

Ilhan, E., 1951, Note sur la géologie de l'Anatolie Orientale, *Eclogae Geol. Helv.*, v. 44, p. 307–314.

Ilhan, E., 1954, Sur la répartition des terrains métamorphoses mésozoiques dans les plis alpins de l'Anatolie, *Bull. Soc. Géol. France*, v. 6/3, p. 975–980.

Ilhan, E., 1967, Toros–Zagros folding and its relation to the Middle East oil fields, *Bull., Amer. Assoc. Petrol. Geol.*, v. 51 (5), p. 651–667.

Ilhan, E., 1968, Turkiye Tektoniginin Jeomorfolojisi ile iliskisi, *Turk Jeomorfoloji Dergisi* (*Ankara*), v. 1, p. 11–32.

Ilhan, E., 1971, The structural features of Turkey, in: *Geology and History of Turkey*, Campbell, A. S., ed., Tripoli: Petrol. Explor. Soc. Libya, p. 159–170.

Izbirak, R., 1951, Cilo Daglari ve Hakkari ile Van Golu cevresinde cografya arastirmalari, *Ankara Univ. D.T.C. Fak. Yayinlari* (*Ankara*), v. 67, 120 p.

Ketin, I., 1959, Turkiye'nin orojen gelismesi, *Bull. Miner. Res. Explor. Inst. Turk.*, v. 53, p. 78–86.

Ketin, I., 1962, Explanatory text of the geological map of Turkey, sheet Sinop, *Bull. Miner. Res. Explor. Inst. Turk.*, p. 55–106.

Ketin, I., 1963, Explanatory text of the geological map of Turkey, sheet Kayseri, *Bull. Miner. Res. Explor. Inst. Turk.*, p. 35–71.

Ketin, I., 1966, Tectonic units of Anatolia (Asia Minor), *Bull. Miner. Res. Explor. Inst. Turk.*, v. 66, p. 23–34.

Kober, L., 1942, *Tektonische Geologie*, Berlin: Edit. Borntraeger.

Ksiazkiewicz, F., 1930, Sur la géologie de l'Istranca et des terrains voisins, in: *Scientific Results of the Voyage of the "Orbis"*, no. 3, Krakow.

Kuttman, F., 1961, Sivas civarindaki jips serisinin stratigrafik durumu, *Bull. Miner. Res. Explor. Inst. Turk.*, v. 56, p. 39–67.

Mercier, J., 1949, Observations géologiques dans la région de Malazgirt–Bulanik, NW du Lac de Van, Anatolie orientale, *Bull. Geol. Soc. Turk.*, v. 2, p. 127–133.

Moscow, 1937, *International Geological Congress*, 1936, Guidebooks.

Neumayer, M., 1887, Über Trias- und Kohlenkalkversteinerungen aus dem nordwestlichen Kleinasien (Balia Maden), *Anz. Akad. Wiss. Math. Nat. Kl.* (*Vienna*), v. 21, p. 242.

Norman, T., 1973, On the structure of Ankara Melange, in: *Congress of Earth Sciences on the Occasion of the 50th Anniversary of Turkish Republic*, Abstracts, Miner. Res. Inst. Turk. and Geol. Soc. Turk., eds., Ankara, p. 30.

Otkun, G., 1942, Etude paléontologique de quelques gisements du Lias d'Anatolie, *Publ. Miner. Res. Explor. Inst. Turk.*, no. 8,

Ozkaya, I., 1973, Structural Geology of Sason Daglari, in: *Congress of Earth Sciences on the Occasion of the 50th Anniversary of Turkish Republic*, Abstracts, Miner. Res. Inst. Turk. and Geol. Soc. Turk., eds., Ankara, p. 16.

Paeckelmann, W., 1938, Neue Beiträge zur Kenntnis der Geologie, Paläontologie und Petrographie der Umgegend von Konstantinopel, *Abh. Preuss. Geol. Landesanst.* (*Berlin*), N. F. 106, 202 p.

Paréjas, E., 1940, La tectonique transversale de la Turquie, *Rev. Fac. Sci. Univ. Istanbul Ser. B*, fasc. 3/4.

Pavoni, N., 1961, Die Nordanatolische Horizontalverschiebung, *Geol. Rundsch.*, v. 51 (1), p. 122–139.

Penck, W., 1919, Grundzüge der Geologie des Bosphorus, *Veroff. Inst. Meereskunde* (*Berlin*), 71 p.

Petroleum Administration (Petrol Isleri Genel Mudurlugu, formerly Petrol Dairesi), Ankara, Unpublished reports in the files of this administration.

Peyve, A. V., 1969, Oceanic crusts of the past (English translation: "Geotectonics"), *Geotektoniks*, v. 4. (In Russian; copy without bibliographic data.)

Philippson, A., 1918, Kleinasien, in: *Handbuch der Regionalen Geologie*, Heidelberg: Steinmann-Wilckens, 183 p.

Pinar, N., 1943*a*, Géologie et météorologie sismiques du Bassin de la mer de Marmara, *Rev. Fac. Sci. Univ. Istanbul Ser. A*, v. 7, fasc. 3/4.

Pinar, N., 1943b, Etude géologique et météorologique du séisme d'Adapazar du 20 Juin 1943, *Rev. Fac. Sci. Univ. Istanbul Ser. A*, v. 8, fasc. 1.

Pinar, N., 1948, Ege Havzasinin tektonigi, sicaksu ve maden suyu kaynaklari, *Istanbul Univ. Fen Fak. Monogr.* 12. (Lignes tectoniques et sources thermale-minérales de la région égéenne, *Rev. Fac. Sci. Univ. Istanbul.*)

Pinar, N., 1950, Etude géologique et sismologique du tremblement de terre de Karaburun du 23 Juillet 1949, *Rev. Fac. Sci. Univ. Istanbul, Ser. A*, v. 15, fasc. 4.

Pinar, N., 1951, Les régions sismiques de l'Anatolie occidentale, *Publ. Bur. Cent. Seismol. Int., Ser. A*, no. 18.

Pinar, N., 1953a, Etude géologique et macroseismique du tremblement de terre de Kursunlu du 13 Aôut 1951, *U.G.G.I. Bull. Inform. (Paris)*, no. 2.

Pinar, N., 1953b, La géologie du bassin d'Adana et le séisme du 22 Octobre 1952, *Rev. Fac. Sci. Univ. Istanbul Ser. A*, v. 18, fasc. 3.

Pinar, N., 1953c, Relation entre la tectonique et la seismicité de la Turquie, *U.G.G.I. Bull. Inform. (Paris)*.

Pinar, N., 1953d, Preliminary note on the earthquake of Yenice–Gonen, *Bull. Seismol. Soc. Amer. (Berkeley)*, v. 43, p. 307–310.

Pinar, N., 1954, Le seisme de Yenice–Gonen du 18 Mars 1953 en relation avec les éléments tectoniques, *Bur. Cent. Seismol. Int. Ser. A*, fasc. 19.

Pinar, N., 1956a, Note préliminaire sur le séisme d'Eskisehir du 20 Février 1956, *U.G.G.I., Comm. Seismol. (Vienna)*.

Pinar, N., 1956b, Historical and modern earthquake-resistant construction in Turkey, *World Conference on Earthquake Engineering (Berkeley)*, 8 p.

Pinar, N., 1960, La carte seismotectonique de la Turquie, *U.G.G.I. Monogr.*, no. 6.

Pinar, N., and Ilhan, E., 1952, Turkiye Depremleri Izahli Katalogu. *Bayindirlik Bakanligi, Yapi ve I.I. Rsl. Publ. (Ankara)*, v. 6, no. 36, 153 p., 5 maps.

Pinar, N., and Ilhan, E., 1954, Nouvelles considérations sur la tectonique de l'Anatolie (Turquie, Asie Mineure), *Bull. Soc. Géol. France*, v. 6/5, p. 11–34.

Pinar-Erdem, N., 1974, The Turkish green rocks, *Bull. Miner. Res. Explor. Inst. Turk.*, v. 8, p. 119–131.

Reed, C. F., 1921, *Geology of the British Empire*, London.

Renz, C., 1929, Geologische Untersuchungen auf den Inseln Cypern und Rhodos, *Praktika Acad. Athenes*, v. 4, p. 307.

Rigo de Righi, M., and Cortesini, A., 1964, Gravity tectonics in foothill structure belt of southeast Turkey, *Bull. Amer. Assoc. Petrol. Geol.*, v. 48 (12), p. 1911–1937.

Sayar, C., 1962, New observations in the Paleozoic sequence of the Bosphorus and adjoining areas, Istanbul, Turkey, *Symposium Silur/Devon-Grenze (Stuttgart)*, p. 222–223.

Schmidt, G. C., 1961, Stratigraphic nomenclature for the Adana region, Petroleum District VII, *Petroleum Activities in Turkey (Ankara)*, v. 6, p. 47–63.

Schmidt, G. C., 1964, A review of Permian and Mesozoic formations exposed near the Turkish–Iraq border at Harbol, *Bull. Miner. Res. Explor. Inst. Turk.*, v. 62, p. 103–119.

Seidlitz, 1932, *Diskordanz und Orogenese der Gebirge am Mittelmeer*, Berlin.

Stchepinsky, V., 1939, Faune miocène du vilayet de Sivas, *Publ. Min. Res. Explor. Inst. Turk., Ser. C*, no. 1.

Stchepinsky, V., 1942, Stratigraphie comparée des régions situées entre Bursa et Tercan, *Bull. Miner. Res. Explor. Inst. Turk.*, v. 7, (27) p. 307–321.

Stchepinsky, V., 1946, Fossiles charactéristiques de Turquie, *Publ. Miner. Res. Explor. Inst. Turk.*, D. 1, 151 p.

Steinmann, G., 1927, Die ophiolitischen Zonen in den Mediterranean Kettengebirgen, *Rept. Int. Geol. Congr. Madrid*, fasc. 2.

Survey of Groundwater and Mineral Resources of Cyprus, 1970, New York: United Nations Development Program (Report).

Ternek, Z., 1953, Geological study southeastern region of Lake Van, *Bull. Geol. Soc. Turk.*, v. 4, p. 1–22.

Tietze, E., 1885, Beträge zur Geologie von Lykien, *Jahrb. K.K. Geol. Reichsanst.* (*Vienna*), v. 35, p. 283–384.

Tidewater Petroleum Company, 1961, Geology of the Sinop area, *Petroleum Activities in Turkey* (*Ankara*), no. 6, p. 33–35.

Tokay, M., 1953, Contribution à l'étude géologique de la région comprise entre Eregli, Alapli, Kiziltepe et Alacaagzi, *Bull. Miner. Res. Explor. Inst. Turk.*, v. 44/45, p. 35–78.

Tokay, M., 1954, Géologie de la région de Bartin, *Bull. Miner. Res. Explor. Inst. Turk.*, v. 46/47, p. 46–63.

Tokay, M., 1973, Kuzey Anadolu Fay Zonunun Gerede ile Ilgaz arasindaki kisminda jeolojik gozlemler. Kuzey Anadolu Fayi ve Deprem Kusagi Sempozyumu, *Bull. Miner. Res. Explor. Inst. Turk.*, v. 29–31, p. 12–29.

Tolun, N., 1962, Explanatory text of the geological map of Turkey, sheet Diyarbakir, *Bull. Miner. Res. Explor. Inst. Turk.*, p. 33–69.

Tolun, N., and Ternek, Z., 1952, Notes géologiques sur la région de Mardin, *Bull. Geol. Soc. Turk.*, v. 3, p. 15–19.

Toula, F., 1903, Ubersicht über die geologische Literatur der Balkanhalbinsel, etc., *C. R. 9th Congr. Géol. Int.* (*Vienna*), v. 1, p. 185–330.

Turkish Gulf Oil Company, 1961, Regional geology and oil exploration in the Tuz Golu basin, *Petroleum Activities in Turkey* (*Ankara*), v. 6, p. 29–32.

Turkse Shell, 1963, Geological History of the Gaziantep region, District VI, *Petroleum Activities in Turkey* (*Ankara*), v. 8, p. 31–37.

Turkey, Geological Map of, 1950–1965, scale 1:500,000, 18 sheets, *Bull. Miner. Res. Explor. Inst. Turk.* (Turkiye Jeolojik Haritasi).

Unsalaner (Kiragli), C., 1941, A preliminary description of the Carboniferous and Devonian fauna discovered in the western Taurus, *Bull. Miner. Res. Explor. Inst. Turk.*, v. 6, p. 599–605.

Weiler, Y., 1970, Mode of occurrence of pelites in the Kytherea flysch basin (Cyprus), *J. Sed. Pet.*, v. 40, 55 p.

Wijkerslooth, P. de, 1942, Die Chromerzprovinz der Türkei und des Balkans und ihr Verhalten zur Grosstektonik dieser Länder, *Bull. Miner. Res. Explor. Inst. Turk.*, v. 26, p. 54–75.

Yalcinlar, I., 1951, Nouvelles observations sur les terrains paléozoiques des environs d'Istanbul, *Bull. Geol. Soc. Turk.*, v. 3, p. 127–130.

Yalcinlar, I., 1952, Les vertèbres fossiles néogènes de la Turquie occidentale, *Bull. Mus. Nat. Hist. Nat.*, v. 24, p. 423–429.

Yalcinlar, I., 1954, Les lignes structurelles de la Turquie, *C. R. Int. Géol. Congr.*, 19th Session (Algeria, 1952), Sect. XIII, XIV, p. 293–299.

Yucel, T., 1953, Izmir'de eski bir kiyi cizgisi, *Ankara Univ. D.T.C. Fak. Bull.*, v. 2–4, p. 279–286.

Chapter 7A

THE LEVANTINE COUNTRIES: THE GEOLOGY OF SYRIA AND LEBANON (MARITIME REGIONS)

Ziad R. Beydoun

Department of Geology
American University of Beirut
Beirut, Lebanon

I. INTRODUCTION

Syria and Lebanon occupy the northern part of the Levant, bordering the eastern side of the Mediterranean Sea. Syria has an area of 184,920 km² and extends, at its widest in the north, from the Mediterranean Sea eastward to longitude 42°24′E, a distance of about 600 km. Lebanon has an area of 10,400 km² and lies immediately south of maritime Syria, separating southern Syria from the Mediterranean. At its widest point the east–west distance across Lebanon is 90 km (Fig. 1). The areas of Syria and Lebanon under discussion here are limited to the maritime regions of the two countries and comprise the continental shelf, coastal region, and bordering mountain ranges. The eastern limit of the area is taken at the Ghab and Bekaa structural depressions, bordering the coastal mountain ranges inland and constituting a northward extension of the Jordan–Dead Sea–Aqaba rift system.

The coastline of Syria extends from the Turkish border, formed by the Iskenderun salient at Jabal Akra (latitude 35°56′N, elevation 1728 m), southward to the mouth of the river Nahr el Kabir South (latitude 34°38′N), forming the coastal border with Lebanon, a total coastline distance of about 170 km

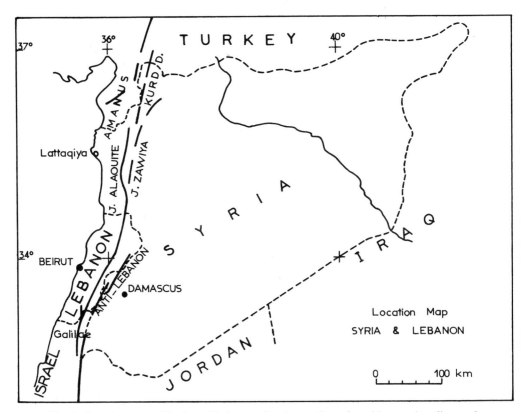

Fig. 1. Location map of Syria and Lebanon, showing outline of maritime region discussed.

(Fig. 2). Maritime Lebanon has a coastline extending from Nahr el Kabir southward to Ras el Nakoura (latitude 33°05'N), a coastline distance of about 210 km (Fig. 3). The area discussed here extends from the coast for a distance ranging from 30 to 60 km inland.

To understand the geology of northwestern Syria, it is necessary to include in the discussion some considerations of that part of Turkey lying between Jabal Akra and the head of the Gulf of Iskenderun along the coast and extending inland across the Amanus Range (Kizil Dagh and Giaour Dagh) to the Kara Su Valley (Fig. 2).

The coastal mountain ranges, the bounding interior fault zones, and to a considerable degree the coastline itself, trend NNE–SSW in the Amanus sector of Turkey as far as the Bassit region of northwestern Syria. The Alaouite (Nussairiyah) Range of maritime Syria extends north–south. In the Lebanon Range to the south the trend shifts again to the NNE–SSW, but in the Galilee Range of hills straddling the southern border, the north–south trend is resumed. Viewed as a whole the region is one of structural blocks in which the more

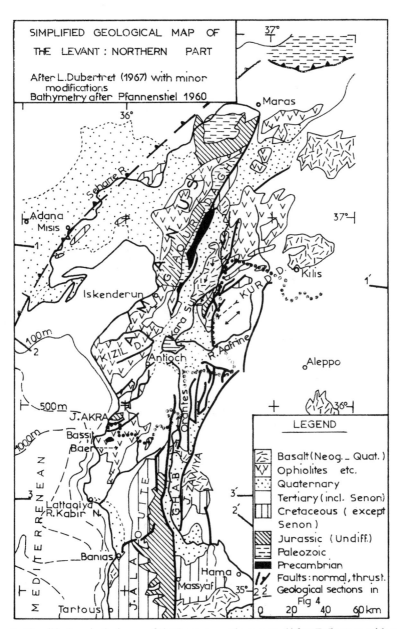

Fig. 2. Simplified geological map of the Levant, northern part. (After Dubertret, with minor additions and modifications.)

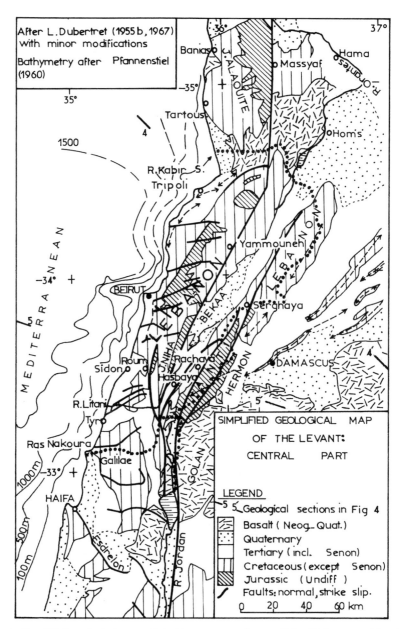

Fig. 3. Simplified geological map of the Levant, central part. (After Dubertret, with minor additions and modifications.)

southerly are progressively offset to the west. The general region is quite mountainous, with elevations ranging up to 3088 m in northern Lebanon.

The continental shelf of this region is generally narrow, less than 5 km to the 200 m isobath. The shelf widens somewhat off southern Lebanon (8–10 km) and reaches a maximum width of about 20 km off the coast of northern Lebanon and southern Syria. The continental slope is generally steep.

The first modern geological work on Lebanon, which included a geological map, was by Zumoffen (1926). Subsequently, a number of systematic geologic investigations have been carried out on land in Syria and Lebanon, mostly by a French group associated with Dubertret and a Soviet group under Ponikarov. The former began work in the 1930s and are still active. Their work has concentrated upon surface mapping in Lebanon and parts of maritime Syria on a scale of 1:50,000 and 1:200,000, the establishment of stratigraphic and faunal successions, the investigation of some local and regional structural problems, the initiation of geophysical work, and the compilation of stratigraphic and faunal information in the *Lexicon of Stratigraphy*. The Soviet group, working under contract to the Syrian Government, synthesized published data and unpublished oil company exploration reports, including subsurface and geophysical data. In addition they carried out extensive geological and some field geophysical studies, the results of which were published between 1964 and 1967 as a series of maps (with explanatory texts) using scales of 1:1,000,000; 1:500,000; 1:200,000; and for selected areas of economic interest, a scale of 1:50,000. In addition to these two main groups, a number of individuals and organizations have concentrated on specific geological and geophysical problems. An overall synthesis on Syria and Lebanon was published in German by Wolfaart (1967). The references at the end of this paper provide a selection of the principal works of all groups. The newly established Geological Survey Departments in the two countries have embarked upon more detailed mapping, but results are not yet available.

II. STRATIGRAPHY

A. Precambrian

Quartzites, schists, marbles, and amphibolites, regionally metamorphosed to green-schist facies, outcrop sparingly in the Bassit area of Lattaqiya in northwestern Syria. These rocks underlie the Mesozoic ophiolites of the region and were considered by Chenevoy (1952, 1959) as pre-Silurian or possibly Precambrian. A Precambrian age is generally accepted now, and Ponikarov *et al.* (1967) regard these rocks as of lower to middle Riphean age (Proterozoic) on the basis of the regional geology. About 110 km NNE of the Lattaqiya area, in the Iskenderun salient of Turkey, nearly 1000 m of steeply dipping, well-bedded quartzite, sandstone, and phyllitic shale are found in a NNE striking belt about 35 km long. These rocks are overlain by a thick series of gently dipping clastics and carbonates containing Middle Cambrian trilobites (Dean and Krummenacher, 1961). Thus, it seems clear that the underlying series are older than Middle Cambrian, and they are generally assigned to the Upper Riphean (Upper Proterozoic) because they are also clearly younger than the metamorphic complex of Bassit (Ponikarov *et al.*, 1967). An infra-Cambrian age has also been suggested because of similarities to folded, but little metamorphosed, series elsewhere in the Middle East. A borehole at Baflioun, along the eastern foothills of the Kurd Dagh (a little outside the area of discussion), passed through a peneplaned surface into contorted micaceous shales tentatively assigned to the Ordovician–Silurian Abba group (Daniel, 1963; Dubertret, 1966), but which are probably correlative with the Upper Proterozoic of the Amanus (Ponikarov *et al.*, 1967).

B. Paleozoic

No Paleozoic exposures occur in maritime Syria, but a thick series of 1000–2000 m of clastics crops out on the eastern slopes of the Amanus Range, a little north of the Syrian border, and rests unconformably upon the peneplaned Upper Proterozoic series described above. Dubertret (1936) first thought this succession to be Devonian, resting unconformably on subvertical Ordovician clastics. Dean and Krummenacher (1961), however, established that this thick sequence of sandstones, siltstones, shales, limestones, and dolomitic limestones contain Middle Cambrian trilobites in its upper part. This Cambrian series is unconformably overlain by another clastic succession, ranging in thickness from 85 m in the central Giaour Dagh to 1115 m some 50 km northward and dated as Ordovician–Silurian. No Paleozoic strata younger than Ordovician occur in the central part of the Giaour Dagh (Dubertret 1966) for the succession is overlain by transgressive Mesozoic carbonates (with a basal conglomerate). However, in the northern Giaour Dagh, Dean and Krummenacher (1961) found a Devonian fauna in dolomitic limestones overlying the Ordovician–Silurian clastic sequence.

Paleozoic rocks are not exposed elsewhere in Syria or the Lebanon except for detached blocks of Carboniferous age within Cretaceous sediments in the core of Jabal Abd-el-Aziz in northeast Syria (Fairbridge and Badoux, 1960). Paleozoic sediments, however, have been encountered in a number of deep boreholes in the interior of Syria and these have established the presence of Cambrian, Ordovician, Silurian, Carboniferous, and Permian rocks, often with considerable intervals missing between them. The existing control suggests the presence of similar facies and thicknesses under the Mesozoic cores of the maritime ranges. These subsurface Paleozoic rocks may be on the order of 1700 m thick and are predominantly of clastic facies with some carbonate horizons (e.g., Daniel, in Beydoun, 1972).

C. Mesozoic

1. *Triassic*

No *in situ* rocks of Triassic age are exposed in the maritime region, but scattered large blocks of limestone (olistostromes) with Triassic fossils are found in the upper lavas of the Mesozoic ophiolite series or associated with radiolarites (Dubertret, 1955a) of the Bassit–Baer area in the Lattaqiya region. Similar carbonate blocks of Jurassic and Lower and Middle Cretaceous age also occur. Controversy over the age of the ophiolite–radiolarite rocks in this region is discussed later in this section.

Extrapolation to the maritime region of data from Triassic rocks encountered in the deep boreholes of interior Syria, Jordan, and Israel suggests that up to 1000 m of neritic carbonates, with interbedded marls and probable evaporites towards the top, may be present (Beydoun, 1972).

2. *Jurassic*

The oldest exposed rocks in the maritime region of Syria and Lebanon are Jurassic in age (Table I). They are found in the cores of the coastal uplifts, extending from the eastern part of southern Lebanon northward to Jabal Akra. The core of the Amanus Range further north exposes the Precambrian and early Paleozoic rocks already discussed (Section IIA, B), but on the flanks Mesozoic sediments unconformably overlie these older rocks (Figs. 2 and 4). The base of the Jurassic system is not visible in the maritime region, but the thickest successions and the oldest horizons exposed are found in Lebanon in the gorge and valley of the Nahr Ibrahim, about 30 km northeast of Beirut. The succession here begins with some 500 m of dolomites followed by limestones with marl intercalations which Renouard (1951) tentatively ascribed to the Liassic. The rest of the Jurassic succession, as described by Dubertret (1955b, 1963), ranges up through the Middle Jurassic Bajocian and Bathonian stages into the Upper Jurassic to include Kimmeridgian and probably Portlandian stages. The rocks consist predominantly of well-bedded to massive blue-grey limestones and dolomitic limestones, which are increasingly fossiliferous upwards and which are interrupted locally toward the top by lava flows and pyroclastics, now frequently laterized, especially in the northern part of the country. Laterally these volcanics grade into a series of ocherous marls, detrital fossiliferous limestones, and chocolate shales. This interval is followed

TABLE I

Stratigraphic Summary Table for the Mesozoic Series and Stages in Maritime Syria and Lebanon (compiled from Dubertret 1955b, 1963, 1966; Ponikarov et al., 1967; Renouard, 1955)

Epoch	Series	Age/stage	Lebanon Range (predominant lithology)[a]	Max. thickness (m)	Alaouite Range–Bassit (predominant lithology)[a]	Max. thickness (m)
Cretaceous	U	Senonian				
		Maestrichtian	chalky marly ls, phosph.	400	ch. ls, marly ch.	300
		Campanian	?chalky ls	—	ch. ls, cong. base ophiolites–radiolarites[b]	
		Santonian	absent		absent	?
		Coniacian	absent			
	M	Turonian	ls, reef ls, marl interbeds	200	marly ls, ls	110
		Cenomanian	ls, some marl interbeds	700	ls with marl; local volcanics	400
		Albian	marl, ls	200	ls, argill. ls, marl	100
	L	Aptian	clastics, ls, volcanics	300	sh, ls, rare ss and volcanics	35
		Barremian	not recognized		not recognized	
		Neocomian				
		Hauterivian	ss, siltst, sh, lignite, mudst, volcanics, probable Neocomian, undifferentiated	250	not recognized	—
		Valanginian			ls with Jurassic pebbles	5
		Berriasian			not recognized	—

Jurassic	U	Portlandian	ool. ls, ocherous marl, probable Portlandian	180	absent	—
		Kimmeridgian	ls with chert, volcanics laterally to detr.	220	ls	450
		Oxfordian	ls and marl; base ls		ls, rare clay interbeds	150
	M	Callovian	dol. ls, ls	650	dol. and ls	400
		Bathonian	dol. ls, ls			
		Bajocian	dol. ls	150	dol. and dol. ls	470
	L	Lias				
		Aalenian	unfoss. ls and dol. ls ?probable Lias (undifferentiated)	400	not exposed	?
		Toarcian				
		Pliensbachian				
		Sinemurian				
		Hettangian				
Triassic			not exposed	1000?	exotic blocks of *Halobia* limestone in ophilites–radiolarites	?

[a] ss = sandstone; s = shale; ls = limestone; siltst = siltstone.
[b] Age of ophiolites–radiolarites disputed (see Section IID in text), but field position under Upper Cretaceous (Maestrichtian) sediments is not.

by a return to well-bedded aphanitic limestones with chert and then by terminal deposition of ocherous, neritic, oolitic limestone, marl, and shale. The end of the Jurassic was characterized by emergence, uplift, and in the north particularly, by block movements and differential erosion which in places had stripped off the uppermost units and had peneplaned the landscape by the onset of the Cretaceous.

In maritime Syria the Jurassic is less well exposed and no definite Bajocian is known (Ponikarov *et al.*, 1967), but the monotony of the carbonate succession and the comparative rarity of good index fossils make subdivisions difficult. About 1470 m of rocks, ranging in age from Bathonian to Callovian–Kimmeridgian, are exposed, but they do not contain volcanics here and the sequence is predominantly aphanitic and dolomitic limestones with some marl interbeds. Sedimentation appears to have stopped with the Kimmeridgian and then resumed during the Lower Cretaceous Aptian stage (Dubertret, 1963).

An electric survey of the area with the oldest outcrops in maritime Lebanon indicates a sharp resistivity contrast some 650 m below the deepest exposures, which was interpreted by Renouard (1955) as indicating a marked change in lithology from Jurassic carbonates to probable evaporites or marls, possibly representing the top of the Triassic. If this is correct, then over 2250 m of Jurassic sediments, mainly carbonates, have been deposited in this part of the basin and paleogeographic reconstructions for the region suggest that the basin axis swung from Egypt NNE through Lebanon and then northeast across Syria, with the thickest section present in Lebanon and southern maritime Syria (Hassan, 1963; Beydoun, 1965).

3. Cretaceous

The Cretaceous succession totals about 1900–2050 m in maritime Lebanon and ranges in age from ?Neocomian to Maestrichtian. The lower part of the section is sandy, the middle part calcareous, and the upper part is a chalky marl. The lower unit begins with a continental fluviatile–deltaic–littoral clastic succession resting disconformably on the Upper Jurassic. This basal clastic section reaches its maximum development in the southern and central part of the maritime region (*ca.* 250 m), but pinches out toward the north. The included plant remains do not yield a precise age. Dubertret (1955*b*, 1963) assigned this series to the basal Cretaceous (tentatively Neocomian), though the sands are diachronous upwards. A volcanic phase frequently characterizes the lower part. The succession becomes increasingly marine upwards with alternations of shale and thin limestones, which contain an Aptian fauna, and culminates in a cliff-forming succession of limestone about 50 m in thickness, which forms a widespread marker extending northward into Syria. A short return to clastic deposition, often accompanied by periodically intense local volcanism, marks the end of the Aptian.

The middle calcareous unit commences in the Albian with return to open marine conditions and begins with interbedded marls and shales and detrital fossiliferous limestones. Upwards, the succession becomes increasingly calcareous, so that by Cenomanian times it consists predominantly of thin to well-bedded limestones and dolomitic limestones with thin marl intervals. Saint-Marc (1974) has recently subdivided the monotonous Cenomanian succession into lithological field units. During the Turonian, carbonate deposition continued but was marked by local shoaling and reef development. The total Middle Cretaceous succession is over 1100 m thick.

The chalky–marly, globigerinal facies of the upper unit extends through the Upper Cretaceous and into the Cenozoic. The Upper Cretaceous generally has not been subdivided, and is collectively labeled Senonian, though locally Upper Santonian–Lower Campanian rocks are recognized (Ejel and Dubertret, 1966) and the Maestrichtian is generally present at the top, while older parts of the Senonian are certainly absent. Upper Cretaceous deposits are confined to the coastal region, southern Lebanon, and the Bekaa and do not occur in the main ranges.

In the Alaouite Range of maritime Syria, between 700 and 950 m of Cretaceous rocks are exposed. The tripartite lithologic division seen in most of Lebanon does not continue here. The Lower Cretaceous is represented by only about 30 m of Aptian limestone, clay, marl, and occasional thin sandstones and tuffaceous intervals. However, Ponikarov et al. (1967) report the local presence of a Valanginian fauna in the basal part, and in some places the base of the Aptian includes Jurassic pebbles and glauconite. The Middle Cretaceous succession is well developed and generally calcareous. Dubertret (1963) reports local development of both green-marl and green-sand facies in the Albian, and both he and Ponikarov et al. (1967) report interstratified volcanics within the Cenomanian succession of the southern Alaouite Range. The Turonian is represented by limestone, sometimes in reefoid facies.

The Upper Cretaceous chalky–marly unit is represented on the western slopes of the coastal ranges and in the Lattaqiya area of Syria by thicknesses of over 300 m, but consists there of Maestrichtian resting unconformably on Cenomanian or Turonian. The Coniacian, Santonian, and Campanian stages are absent, as is revealed by detailed microfaunal work by Krasheninnikov (1965) and others.

4. Ophiolite–Radiolarite Complex

Ophiolites (greenstones) and associated radiolarites and other igneous and sedimentary rocks are exposed in the Lattaqiya region of maritime Syria (Bassit–Baer: southern projection of the Amanus), as well as in the Amanus Range proper in Turkey (Kizil and Giaour Daghs) and in the Syrian Kurd Dagh east of the Kara Su Valley (Fig. 2). They have been examined and reported upon by Kober (1915), Dubertret (1933, 1936, 1937, 1939, 1955a, 1963, 1966, 1976), Majer (1962), Kazmin and Kulakov (1965, 1968), Ponikarov *et al.* (1967), Piro (1967), and Vaugnat and Cogulu (1968), among others, and have been reviewed in a regional context by Ricou (1971). The age, genesis, and mode of occurrence of these rocks is controversial. Dubertret (1955a, 1963) regarded the greenstones (ophiolites) as one huge submarine flow of Upper Cretaceous (Maestrichtian) age and of autochthonous nature, differentiated by gravity with gradational transitions from coarse-grained ultramafics at the base to fine-grained basalts at the top. He recognized from the base upwards: peridotites and pyroxenites, gabbros, diabases, and at the top, pillow lavas (sakalavites). The pillow lavas entrain, or are overlain by, confused blocks of contorted radiolarites (cherts, limestones, and mudstones) which are penecontemporaneous, and also entrain, or are overlain by, exotic blocks of Triassic, Jurassic, Lower and Middle Cretaceous limestone, which are often silicified or recrystallized. The latter were brought up from the substratum during emplacement of the greenstones. Dubertret (1937) originally assumed the complex to be thrust, but detailed field work caused him to conclude (1955a, 1963) that the greenstones locally may rest normally on datable Maestrichtian sediments. They also locally rest on Precambrian strata with slight contact metamorphism (Dubertret, 1966). In many places the greenstones are overlain by the Maestrichtian, which contains a basal conglomeratic facies which includes greenstone. The Maestrichtian grades up into chalky marls and limestones, but also shows lateral variations to reef limestones.

Kazmin and Kulakov (1965, 1968) and Ponikarov *et al.* (1967) report the ophiolite series to consist of: (1) a basic effusive and tuff unit incorporating hypabyssal gabbroid inclusions, (2) intrusive ultrabasics, (3) minor intrusions of gabbro, gabbro–pegmatite, and occasional diorite, and (4) an igneous–sedimentary series consisting of basic to intermediate effusives and tuffs, mudstones, radiolarites, limestones, and siliceous rocks. The sequence in which these subdivisions are listed is thought to correspond to their decreasing age. The authorities noted divide the suite into two main groups: an older, ultrabasic and basic phase comprising (1) to (3) above and a younger phase including the volcanics (pillow lavas) and sediment (4). Limestones with *Halobia* form the youngest component of the radiolarites, and thus date the series as Upper Triassic to Jurassic. An Upper Jurassic–Lower Cretaceous absolute age determination on a nepheline–syenite intruding the radiolarites (*ca.* 122 m.y.) gives an upper limit for formation. If this dating scheme is correct, the ultramafics and gabbros are pre-Upper Triassic in age. The sedimentary blocks of Jurassic and Cretaceous age are considered klippe or small-thrust sheets. The ophiolites are not regarded as autochthonous, but as having suffered some thrusting, sliding, and deformation during the Maestrichtian, after vertical uplift of deep-seated fracture zones nearby. One of the presumed fracture zones is the northeast–southwest trending Lattaqiya–Kilis fault, which is postulated to separate the present ophiolite domain of Bassit–Baer–Amanus and Kurd Dagh from the typical Arabian shelf area to the southeast (see Section IIID).

Piro (1967) carried out further detailed mapping of the area and agrees with Dubertret as to age and position of emplacement of the greenstones, but declares that the ophiolite contact with the metamorphic rocks of Bassit–Baer is tectonic. Majer (1962) considered the peridotites and gabbros to be of Paleozoic age, while the basic effusives, diabase, and other associated rocks are Mesozoic. A Cenozoic phase of extrusive volcanism, ranging from trachytes to phonolites and alkali basalts preceded by a possible intrusive phase, has also been postulated. Vaugnat and Cogulu (1968), working in the nearby Kizil Dagh of Turkey, disagree with Dubertret's interpretation and indicate that, at least in part, the diabases are not flows but constitute a sheeted complex of dykes showing chilled edges against gabbro. They believe that the transition zone between the latter and the peridotites is hybrid, the gabbro injecting the peridotite with inclusions of ultramafic xenoliths, more or less digested. The peridotite is considered as part of the deep simatic layer (upper Mantle?) which underwent partial remelting during Upper Cretaceous tectonism with the resulting basic magma crystallizing as the gabbro of the complex, followed by emplacement of the sheeted complex. Continued activity resulted in fragmentation of the peridotite and serpentinization of ultramafic wedges which penetrated the upper levels. The analogy between the interpretation of this region and the interpretation of the Troodos complex in Cyprus (e.g., Gass and Masson-Smith, 1963) is stressed by these authors.

It is evident from the above that differing opinions on the greenstones of the Bassit–Amanus Range exist, but a number of facts do emerge, which together with consideration of regional data from adjacent areas, enable some conclusions to be drawn. The ophiolites, radiolarites, and associated sedimentary blocks are everywhere overlain by Upper Cretaceous sediments, or contain pre-Upper Cretaceous fossils within their entrained sedimentary blocks. In many places, and in particular in the Amanus, the evidence is quite strong for Upper Cretaceous emplacement despite the conclusions reached by the Soviet teams concerning a long, pre-Maestrichtian Mesozoic history, or those by Majer (1962) invoking Paleozoic age. The ophiolite–radiolarite complexes, often showing strong contortions, rest on a generally undeformed substratum; if this latter is contorted, the deformation can be demonstrated to have occurred after emplacement of the complexes during Late Tertiary orogenesis. These facts strongly indicate that gravity tectonics have played a major role in transportation of the ophiolite complexes here, as has been very clearly demonstrated in the adjacent Taurus foothill belt of Turkey by Rigo de Righi and Cortesini (1954). The ophiolites are, in fact, to be found (though subjected to deformation with the associated radiolarites) between autochthonous shelf sediments of

Upper Cretaceous age, and must have arrived there by gravity slides rather than in tectonic thrusts in view of the generally reported absence of substratum tectonization (Ricou, 1971). How far these rocks have traveled from their place of origin along the Arabian shelf periphery is a matter of dispute, but there is little doubt that they are to be found in a nonorogenic shelf or platform environment. This relationship can only be explained by deep-seated upwelling along extension fractures within the platform, or by gravity tectonics during the Late Cretaceous along the peripheral northern edge of the Arabian shelf (see Sections IIIC, D, and IV).

D. Cenozoic

1. *Paleogene*

Paleogene deposits have been well studied in the Syrian maritime region (Krasheninnikov, 1965; Ponikarov *et al.*, 1967), and it is now possible to subdivide them from Upper Cretaceous deposits of similar facies and to map Paleocene, Eocene, and Oligocene rocks and subdivide them into zones. Paleocene rocks are confined to patches on the lower slopes of the coastal side of the northern Alaouite Range in the region north of Lattaqiya and to the river Kabir North and northern Ghab Valleys (Table II). They consist of argillaceous and chalky facies indistinguishable from the Upper Cretaceous. The Danian stage is recognized, however, on the basis of its microfauna and has been subdivided. Part of the Lower Eocene is absent and the nummulitic to marly carbonates of the Middle Eocene rest on Paleocene in places. In turn, the upper part of the latter is often missing and it is transgressively overlain by the Neogene. The Upper Eocene is absent except for the eastern slopes of the Kurd Dagh (outside the area discussed here) where it is in a shallow-water, often sandy limestone facies. The Oligocene is also absent except for the northern Ghab and the Aafrine Basin (again outside the area). The total Paleogene thickness in the Lattaqiya region reaches 1050 m.

Work in progress on the Upper Cretaceous–Paleocene boundary in the Lebanese maritime region reveals a substantial development of Paleocene sediment in the Tripoli region (and also in parts of southern Lebanon). The facies of the Upper Cretaceous (Senonian) and Paleocene–Lower Eocene deposits in these regions is very similar (chalky–marly–globigerinal lithologies), and only by microfaunal work is subdivision and correlation with similar zones in the Syrian region possible. The Lower Eocene appears to contain frequent chert bands. In southern Lebanon, Dubertret (1963) reports the Paleogene to be discordant on Senonian chalks, but some of this chalky succession is now thought to be Paleocene overlain by cherty, marly, chalky limestones of the Lower Eocene, in turn overlain by chalky Middle Eocene marls. The latter

Stratigraphic Summary Table for the Cenozoic Series and Stages in Maritime Syria and Lebanon (compiled from Dubertret, 1955b, 1963, 1966; Ponikarov et al., 1967; Renouard, 1955)

Epoch	Series	Age/stage	Lebanon Coast and Bekaa Depression (absent in mountains) (predominant lithology)	Max. thickness (m)	Syria Coast, Kabir, and Ghab Depressions (absent in mountains) (predominant lithology)	Max. thickness (m)
Holocene			marine and river terr.; alluv.	—	marine and river terr.; alluv.	—
Pleistocene		Tyrrhenian Sicilian Calabrian	7 marine terraces Calabr.–Tyrrhen., 200-m–6-m levels; Paleolith. tools; river terraces, congs., dunes, alluv., soils	—	6 marine terraces Calabr.–Tyrrhen., 215-m–10-m levels; Paleolith. tools; river terraces, congs., alluv., soils	—
Neogene Pliocene		Astian Plaisancian	ls, marl, sand; marine, laterally continent.; volcanics	360	Clay, ss, ls interbeds; local cong. volcanics; continent. in S. Ghab	470
Miocene	U	Messinian (Pontian)	coarse clastics, lacustr., gyp. marl	800	ls, clay marl, gyp.; volcanics S. Coast	40
	M	Vindobonian Tortonian Helvetian	ls, reefoid, basal clastics, littoral top	320	cl. marl, ls, ss; continent. in S. Ghab / coarse–fine clastics, calc. clay ls	225 / 170
	L	Burdigalian Aquitanian	marly ls / not recognized	80 / —	ls, argill. and ss interbeds / ls, clay marl interb. locally	225 / 150
Paleogene Oligocene		Chattian Stampian Sannoisian	absent	—	ls. and argill. ls in N. Ghab only, absent in Coast	120
Eocene	U	Priabonian	absent	—	absent	—
	M	Lutetian	ls, ch. ls, reefoid ls, S. Leb. and Bekaa only; max. S. Bekaa	800	ls, argill. ls	185
	L	Ypresian	ls, marly ls, ch., marls, chert	370	ls, marl, chert	65
Paleocene		Landenian Montian Danian	ch. marl, ch. ls / not recognized	360 / —	marl, clay marl, chalky marl / marly chalk	650 / 30

become subreefal eastward. No Upper Eocene or Oligocene deposits are known from Lebanon. The Paleogene thickness may be in the order of 800 m or more in the coastal region, and even greater in the Bekaa (Table II).

2. *Neogene*

Recent detailed studies of Cenozoic deposits in Syria by Soviet teams (Krasheninnikov, 1965, 1971; Ponikarov *et al.*, 1967) have extended the work of Dubertret (1955*a*) in northwestern Syria and provided more information. The marine Miocene is present in all its stages and is well developed in the Lattaqiya region, in the Nahr el Kabir North river valley and in the Ghab depression, as well as in patches along the coast between Lattaqiya and the Lebanese border (Table II). The Aquitanian stage is represented by carbonates with lenses and bands of sandstones, conglomerates, and some lignite. The Burdigalian has similar facies, without the clastics but with reefal development. The Lower Miocene is unconformable on the Paleogene, the Cretaceous, or on the ophiolites and may reach a thickness of 375 m. The Middle Miocene is represented by transgressive deposits of Helvetian age, generally lying unconformably on the Burdigalian and often having a basal clastic unit followed by fine clastics and limestones. In the Ghab area, these rocks are present in a limestone facies. The Tortonian has a similar lithology, but is regressive, becoming continental southward in the Ghab depression. Middle Miocene deposits reach 350–400 m in thickness.

The Upper Miocene, Messinian, or Pontian marks a time of regression and restriction of deposition. In the Nahr el Kabir North Valley, the section consists of limestone conformable on the Tortonian, but passes upwards into a facies of about 40 m of gypsum with limestones and marl bands. Pontian (Messinian) movements caused a retreat of the sea from northwestern Syria, but beginning in the Pliocene accentuation of local relief differences allowed marine ingressions to penetrate up the Nahr el Kabir North Valley and into the Ghab. The Plaisancian is represented by clays and marls with levels of sandstone and conglomerate, which become continental during the Astian and are replaced in the Ghab depression by thick conglomerates and lacustrine deposits. Basalt extrusions occurred during the Upper Miocene near the Lebanese border and again in the Pliocene over a larger region extending inland to Homs, as well as in the Banias region along the coast.

In Lebanon, the Neogene is best developed in the Tripoli region, but it is also found in the Nahr el Kalb area, 10 km north of Beirut, in Beirut itself, and south of Sidon (Table II). The Aquitanian is not recognized while the Burdigalian is only reported from the Sidon region, where it is in a limestone–marly limestone facies (Keller, 1933; Dubertret, 1963). In the Tripoli region Vindobonian (Middle Miocene) deposits transgressed over older sediments and

extended to the foot of the already established coastal mountain range. Limestone deposition gave way to red marls and conglomerates by the Pontian
(Messinian), and thickness and facies variations result from local uplift and
downwarping. The only evidence for Messinian deposition under restricted
conditions was found at a depth of 225 m in a water well at Sammakieh near
the Syrian border, close to the coast, where evaporites (gypsum) and associated
chalky limestones with an impoverished, dwarfed microfauna were identified
by Edgell (1968) as Messinian. Nowhere else in Lebanon are there indications
of similar evaporites. In the Nahr el Kalb section the Vindobonian is transgressive with sharp angularity on steeply dipping Upper and Middle Cretaceous
sediments and its terraced shoreline is marked by basal conglomerates and
sandstones which pass seaward to reef limestones (Dubertret, 1966). The sea re-
turned in a final short-lived incursion to the foot of the Lebanon uplift in the
Tripoli region, where marine Plaisancian limestones unconformably overlie
continental Pontian (Messinian) beds, but conditions rapidly turned to brackish
water and continental deposition, both laterally and vertically, with an inter-
fingering of sands, conglomerates, and volcanics toward the Tripoli–Homs
depression. Further movement in the Pliocene led to accentuation of established
structures and in places uplift of Pliocene marine deposits to over 400 m. Inland
in the Bekaa depression, the Miocene–Pliocene deposits are entirely lacustrine,
fluviatile, and continental.

3. *Quaternary*

Recent work by the Soviet teams in the Syrian maritime region, amplifying
earlier work by Dubertret and others (see Dubertret, 1963; de Vaumas, 1953)
has shown the existence of six raised beaches or marine terraces (Ponikarov
et al., 1967). The highest, of Calabrian age, is between 180–215 m above sea
level, while the next lower is between 90–120 m, and the next between 60–80 m.
Correlation with the Lebanese terraces suggests that the last two may be of
late Calabrian and Sicilian age, respectively (late Lower Pleistocene and early
Middle Pleistocene), although the Soviet geologists group them within the
Lower Pleistocene and have found Abbevillian tools associated with the 60–80-
m terrace. The next lower terrace at 30–40 m yields Acheulean tools (Middle
Pleistocene). The 10–20-m terrace belongs to the Upper Pleistocene (Tyrrhenian) and contains Lavalloisian and Mousterian tools of the Upper Paleolithic.
The lowest terrace, 2–4 m high, belongs to the Holocene. These terraces are
often eroded and their original morphology is difficult to define, hence the
variations in elevation from place to place.

In Lebanon, Dubertret (1946, 1963), Wetzel and Haller (1945–48), Fleisch
(1962), and Sanlaville (1969, 1970) have recognized seven marine terraces
which correspond quite closely to those of Syria. An equivalent of the highest

level (180–215 m) was not recognized until recently (Guerre and Sanlaville, 1970), but one occurs undeformed in several localities at an average elevation of 200 m. More precise identification of levels, with the recognition of the four pluvials and a marked regression between the Pleistocene and the Holocene, make a more precise classification in Lebanon than in Syria (e.g., a 6-m and 15-m level as opposed to the 10–20-m level of coastal Syria), and identification of variations of levels during the Holocene has recently been made by Sanlaville (1970).

Variations during the Pleistocene were due to eustatic changes in sea level, continuing into the Holocene and accompanied by continental uplift of the land and subsidence of the adjacent sea. Red, brown, and black earth, beach-rock and fossil and moving dunes, reflecting pluvials and interpluvials, and river terraces away from the coast constitute the other Quaternary deposits of the maritime region.

III. STRUCTURAL FRAMEWORK

A. Land Region

1. *Structure*

The structural style of the Lebanon, Alaouite, and Amanus coastal mountain ranges and the flanking depressions (the Bekaa, Ghab, and Kara Su inland on the east, and the Mediterranean Sea on the west) reflects predominantly vertical tectonics. The ranges are generally arched block uplifts with west dipping, locally strongly flexured, seaward flanks and truncated eastward flanks bordering the rift faults. The magnitude of subsidence of young rocks on the east varies in places and is considerable (e.g., 3000 m or more in southern Lebanon between the Jabal Barouk-Niha and the Bekaa (Dubertret, 1955b; Tiberghien, 1974), but in other places it may consist of only a few hundred meters (e.g., north Lebanon–Bekaa, where displacement is within the Cenomanian strata and the down-dropped part shows strong easterly dip). Similar variations, but with lesser throw, occur in places between the Alaouite (Nusairiyah) Range and the Ghab, and in the Amanus–Kara Su area between Precambrian–Cambrian sediments and nearby Cambrian and ?Mesozoic sediments (Fig. 4). Thus, vertical movements of basement blocks with draping, flexuring, and faulting of the overlying sediments is the structural style in the region rather than pronounced compressional folding. Local compressional structures are present, but they are associated with small-scale strike-slip faults, or with local jostling of blocks and squeezing of the sediments, and they may involve gravity sliding and decollement of the upper cover. Movement, or

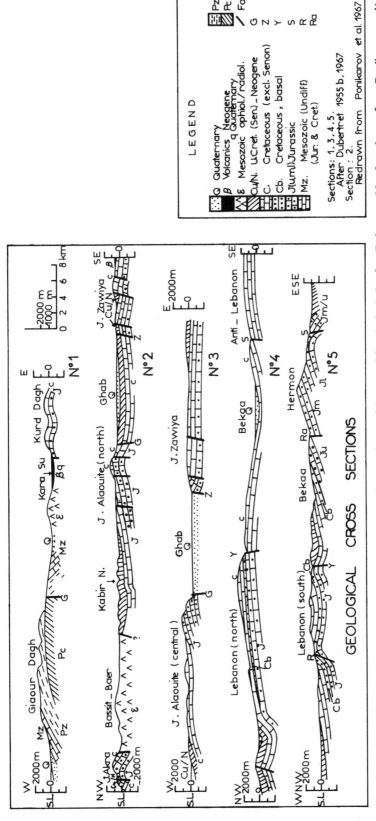

Fig. 4. Geological cross sections across the northern and central Levant. (Nos. 1, 3, 4, 5, after Dubertret; No. 2, redrawn from Ponikarov et al.)

LEGEND

Q	Quaternary
β	Volcanics Neogene
ε	q Quaternary
	Mesozoic ophiol./radiol.
Cu/N.	U.Cret. (Sen.) – Neogene
C.	Cretaceous (excl. Senon)
Cb.	Cretaceous , basal
J(uml)Jurassic	J(uml) Jurassic
Mz.	Mesozoic (Undiff.) (Jur. & Cret.)
Pz.	Paleozoic
Pt.	Pre-cambrian

Faults:
G   Ghab
Z   Zawiya
Y   Yammouneh
S   Senghaya
R   Roum
Ra Rachaya

Sections: 1, 3, 4, 5.
  After Dubertret 1955 b. 1967
Section : 2.
  Redrawn from Ponikarov et al. 1967

GEOLOGICAL CROSS SECTIONS

rejuvenation of movement, along one or more of the main fault trends or within portions of the principal uplifts results in a diversity of local structures, including horsts, bending, and gravity sliding, but vertical tectonics dominate with normal steep to vertical and locally overturned faults and subsidiary transcurrent movements. There is no indication of any important reverse or thrust faulting on the surface or at depth, while the evidence for vertical tectonics is clear.

The structure of the Lebanon Range is both more complex and more diverse than that of the Alaouite Range. The northern part of the Lebanon Range is a large horst-anticline with its east flank truncated by the Yammouneh fault and the west flank exhibiting steep dips (Fig. 4). The range plunges north into the Tripoli–Homs depression and in the core, subsidiary horst-anticline structures are superimposed on the main uplift. A parallel, but topographically lower structural trend separates the main uplift from the coast and is cut into offset segments by small east–west strike-slip faults. The southern part of the main uplift divides into parallel asymmetric structures separated by a tight syncline wedged into a "V," formed by the north-northeast-trending Yammouneh fault (the main boundary fault of the Bekaa Rift) and the north–south-trending Roum fault extending from the western Jordan boundary fault. The southern part of the Lebanon Range is a monocline with gentle west dip (Figs. 3 and 4). The Alaouite uplift is a sharply asymmetrical horst-anticline, with the steep eastern flank truncated by the Ghab fault and showing a vertical displacement of up to 1000 m, while the western flank is gentle but flexured near the coast. This uplift plunges to the north and south beneath Neogene deposits (Ponikarov *et al.*, 1967) (Figs. 2 and 4).

Precambrian orogenic effects are locally revealed in the metamorphics of the Bassit area near Lattaqiya, and steeply tilted, Late Precambrian beds, peneplaned prior to the Cambrian transgression, are seen in the eastern Amanus Range. Subsequent structural evolution is poorly known, but was predominantly epeirogenic with periodic taphrogenic phases. Several unconformities and disconformities in the known stratigraphic record reveal regional and local vertical movement which caused broad emergence and submergence, local uplift and depression, and erosion or nondeposition on the uplifts.

Apart from post-Silurian emergence of the Amanus, little is known directly of the pre-Jurassic history of the maritime area. Regional consideration based on subsurface information from interior Syria and neighboring areas shows that it was an integral part of the "Unstable Shelf" or platform. This shelf was characterized by marine, generally neritic, sedimentation and a regional pattern of basins and swells (Henson, 1951) locally developing and changing through time. The first recorded Mesozoic structural developments are seen in Late Jurassic (Kimmeridgian) fracturing, volcanic eruption, and then emergence of much of north Lebanon with accompanying differential erosion. A slow return to marine conditions, accompanied by two further episodes of volcanicity in Lebanon during the Lower Cretaceous and locally in Syria in Middle Cretaceous time, was followed by local shoaling in the Turonian and nondeposition in the early Senonian. Gentle movements, begun at that time, diversified local marine bottom conditions on either side of the present coastal ranges, and culminated with ophiolite emplacement in the Amanus and other peripheral areas of the shelf (in Turkey, Iraq, Iran, Oman) prior to the end of the Cretaceous. Paleogene movements, at first on a local scale, increased in areal extent and culminated in the emergence of the coastal mountain ranges as vertical uplifts with a complete retreat of the sea by the Late Eocene. Miocene and Pliocene movements increased the uplift and gave it its block-faulted character, while simultaneously further depressing the adjacent low areas of the Bekaa–Ghab–Kara Su on the one side and the Mediterranean on the other.

The presence of interbedded Mesozoic volcanics within some Mesozoic sedimentary formations in various parts of the coastal ranges (and the emplacement of ophiolites in the Amanus, if the autochthonous nature of their occurrence is correct), as well as the presence of locally widespread Neogene lava flows at or near the interior fracture zones, suggest that the latter are ancient sutures and involve the basement at depth. Although the age of these geosutures is probably considerable, the present structure of the maritime region is young, dating initially from the Late Paleogene and taking true form in the Neogene.

2. Geophysical Data

Gravity maps have been published for all of Lebanon (Stahl and Plassard, 1959; Tiberghien, 1974) and for much of Syria inland from the eastern edge of the Alaouite Range (Ponikarov et al., 1967), but not for the Alaouite uplift or for the Amanus. Lejay (1938), Bourgoin (1945–1948), and de Cizancourt (1945–1948) have earlier expressed opinions on the deep structure of Syria and Lebanon, based on the gravity surveys of the late 1930's with generalized maps. From these works and from compilations from eastern Mediterranean marine surveys (compiled among others by Pfannensteil, 1960; Badrawi, 1963; Tiberghien, 1974), certain facts emerged which Tiberghien (1974) has summarized within the context of his own recent gravity work over the whole of Lebanon. The conclusions for the Lebanon maritime region apply for most of the Syrian maritime region as well.

Positive gravity values, which occur along the western part of Lebanon (similar values occur along the Syrian coast but the survey is discontinuous), suggest that it is a continuation of the crustal structure of the eastern Mediterranean (Fig. 5). The Lebanon (and Alaouite) uplift is not isostatically compensated and has no "roots." "Roots" start further inland, but where they begin to appear, they are not regarded as sufficient to support the parallel

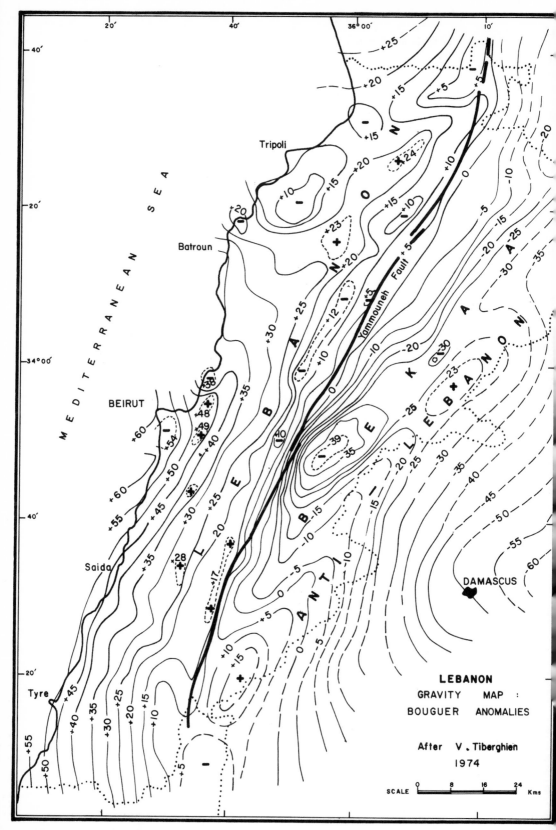

Fig. 5. Lebanon: gravity map, Bouguer anomalies. (After Tiberghien, with minor addition.)

Anti-Lebanon uplift. Based on model calculations for a two-layer case, the Lebanon uplift is regarded as due to the elevation of the second layer whose configuration corresponds to the structural configuration of the top of the Jurassic system. The Bekaa depression (Fig. 5), and to the north the Ghab depression of Syria, show negative gravity values indicative of a deeper crust below them. Calculations for the depth of the Moho give values of 27 km below sea level along the coast from Beirut south, but over 28 km from Beirut north and 30–31 km in the Bekaa, with crustal thickening increasing eastward into Syria (Tiberghien, 1974). The top of the Jurassic is calculated to be 1000 m below sea level in the northern Bekaa and some 2500 m below sea level in the central and southern Bekaa, indicating a very large throw between uplift and graben along the Yammouneh fault. The Lebanon (and Alaouite) uplift mark young Neogene blocks rising between foundering blocks to east and west. The Bekaa and the Ghab form a graben system resulting from distension in the first 15 km of crust within the twin uplifts of Lebanon and Anti-Lebanon northward to the Amanus–Kurd Dagh. The main eastern foundering area, however, is further inland in interior Syria. Foundering of the western block, consisting of the eastern Mediterranean, disrupted isostatic equilibrium and started an upward movement of crustal material under the maritime region (cf. Dubertret, 1967). Calculations by Goedicke (personal communication, 1974), using the gravity values indicated by Stahl and Plassard (1959), suggest that the base of the sedimentary succession (top of "Basement") is about 6 km below the Bekaa floor and perhaps 3 km below sea level under the Lebanon uplift (where the top of the Jurassic is at over 2200-m elevation above sea level). These figures agree fairly closely with calculations of thickness for the pre-Jurassic sedimentary succession based on extrapolation from borehole information in interior Syria (see Section IIB and C). The "Basement" depth does not, however, correspond to that of the top of the lower layer of Tiberghien (1974) (the basaltic or intermediate layer) which is still deeper. According to him, this horizon occurs 8–10 km below sea level under the uplifts and 11–15 km under the Bekaa graben (Tiberghien, 1974). Justification for the relatively shallow depth of that layer is provided by the common presence of extrusive intercalations in the exposed sedimentary section, suggestive of proximity of the basaltic layer.

An aeromagnetic map for most of Syria has been published (e.g., Ponikarov et al., 1967), but no magnetic surveys, other than on a purely local scale and for specific purposes, have been conducted in Lebanon. Pronounced, smooth, positive magnetic anomalies over the Alaouite Range (Fig. 6) suggest that the interpretation by Tiberghien (1974) for Lebanon, of a relatively shallow depth for the intermediate (basaltic) layer, also applies here. Surprisingly, lower intensity positive values (but showing a sharply differentiated and disturbed magnetic field) are mapped over the southern portion of the Bassit

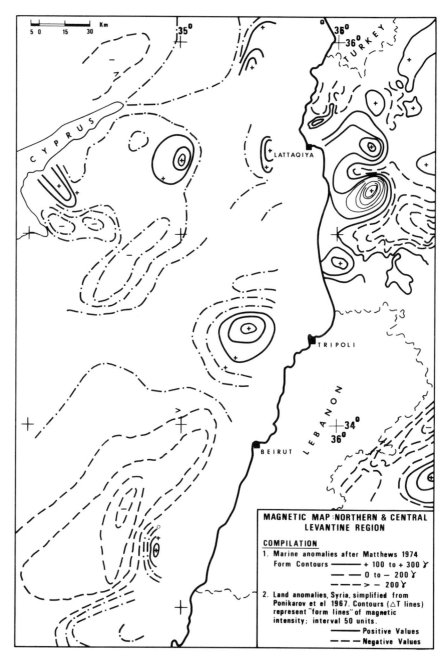

Fig. 6. Magnetic map, northern and central Levantine region. (After Matthews for Levantine Sea and simplified from Ponikarov *et al.* for Syria.)

ophiolite outcrop and possibly are indicative of their ?allochthonous character. Insufficient coverage northward does not permit the extension of this conclusion.

Some paleomagnetic determinations on basalts from Lower Cretaceous and Upper Jurassic sediments in the Lebanon uplift and on Pliocene basalts from coastal Syria are available by Van Dongen *et al.* (1967). Cretaceous results appear to be conformable with African data, but show rotations from African directions which are at times considerable, while the Upper Jurassic results seem quite different. Gregor and Nairn (1971) independently confirmed these results, and it would appear that no satisfactory explanation of the differences will be possible until further data from Africa and Europe are available for comparison. Then, perhaps, it can be determined if local tectonics have played a major role (Zijderveld and Van der Voo, 1973).

No other regional geophysical surveys have been conducted in the land portion of the maritime region under discussion, though limited reflection seismic surveys in connection with oil exploration have been conducted over small portions of the northern Bekaa and over the southern Lebanon coast. These studies failed to reveal deep structure. Lebanon is located in a region of medium (average) seismicity, according to records kept for many years by the observatory at Ksara in the Bekaa.

B. Offshore Region

1. *Bathymetry*

The subbottom structure of the offshore region of Syria and Lebanon is poorly known, but bathymetric soundings summarized by Pfannenstiel (1960) and Emery *et al.* (1966) reveal a generally narrow continental shelf (3–5 km) with a marked shelf-break at average depths of a little over 100 m, suggesting a recent sea level lowering. The average depth of the flat part of the shelf is between 20 and 40 m (Goedicke, 1972). Beyond the shelf-break, which roughly parallels the coast, the slope drops sharply to depths of between 1100 and 1400 m over the relatively short distance of 25–30 km before leveling out in the abyssal plain. Several exceptions are found, such as the east–west-trending "horst" in the vicinity of Beirut where the top is at a depth of 366 m, some 25 km from the shore (Pfennanstiel, 1960), causing a sharp westward projection of the continental slope. In the area extending from just south of Tripoli to south of Tartous in Syria in the north, the shelf is at its widest, reaching some 20 km (flanked inland by a wide coastal plain), and the break in slope is nearer the 200-m isobath. The continental slope itself is gentler in this region. Another noticeable projection in the shelf and slope occurs in a west-northwest-trending promontory linking Tripoli with a series of islands. Along the southern part of Lebanon a slight widening of the shelf (8–10 km) also makes a gentler continental slope with leveling out at a depth of 1100 meters (Fig. 3).

Undulations in the bathymetric contours on the Lebanese shelf, especially between latitude 33°32′N near Sidon and a little south of Tripoli in the area of the narrowest shelf (about 3 km or less), have been shown to be linked to submarine valleys (Goedicke, 1972), most, but not all, of which are seaward extensions of important land valleys. Probably these valleys were carved subaerially during Pleistocene sea level changes or possibly during a Late Neogene phase, after the main mountain uplift and once drainage patterns had been established. Other valleys appear to form seaward continuations of onshore faults, but the main rivers to the north and south of the area indicated (coinciding with a wider shelf area) do not appear to have submarine continuations (except perhaps off the Litani river at latitude 33°22′N), probably because sediments have filled their channels. None of these valleys has as yet been followed and mapped over the edge of the continental slope.

Emery and George (1963) have shown that overall sediment movement by longshore currents is from south to north and that the dominant waves are from the southwest, though local reversals of sediment movement have been noted. Most sediments are derived locally; however, a small fraction of the fines may have been transported from the Nile Delta. A high percentage of calcium carbonate in the sand is indicative of local origin, for the greater portion of the exposed strata in the country are carbonates. Work on estimation of the sediment load delivered to the shelf or bypassed to the continental slope and rise is in progress. Graded bedding and slumped, unconsolidated sediments have recently been cored from one of the submarine valleys (Goedicke and Sagabiel, 1976), suggesting that the valleys form important avenues for sediment delivery from the shelf to the abyssal plain where thicknesses of about 1 km of unconsolidated sediments have been indicated (Wong and Zarudski, 1969). A very rich sub-Recent foraminiferal fauna is present among these sediments of the shelf (Kafescioglu, 1976). No data are available on the Syrian part of the shelf.

2. Geophysical Data

Bourgoin (1945–48), in discussing the positive gravity anomalies along the coast of Lebanon, concluded that the eastern side of the Mediterranean represented the deeply down-buckled and broken edge of a continental platform. Thick and widespread Upper Miocene evaporites (Messinian salt and gypsum–anhydrite) have recently been discovered in most of the eastern Mediterranean where they sometimes form diapiric structures (e.g., Finetti and Morelli, 1973; Mulder, 1973; IFP-CNEXO, 1974). Late Cenozoic–Quaternary basin fill overlies the evaporites in places to quite considerable thickness (e.g., about 1 km in the Levantine Basin). The evaporites may be traced along seismic profiles to the eastern edge of the Mediterranean and into the continental slope (D. H. Matthews and J. M. Woodside, personal communication, 1974). Similar Upper Miocene evaporite basins occur onshore in Cyprus in the Mesaoria plain (Henson *et al.*, 1949; IPF-CNEXO, 1974), in coastal Syria (Poni-karov *et al.*, 1967), and in the Sammakieh water well in north Lebanon (Section IID), as well as in the Adana Basin of southeast Turkey, both onshore and offshore (Mulder, 1973), and in interior Syria (Daniel, 1963; Dubertret, 1966; Ponikarov *et al.*, 1967), strongly suggesting Upper Miocene dessication of the northwestern part of the Arabian shelf and the eastern Mediterranean. This large region was divided into a number of smaller evaporite basins, but all were part of the same climatic event. The evidence favors the shallow-basin model for the Mediterranean desiccation, the basin subsequently foundering during Pliocene and Pleistocene times. Continued relative uplift of the adjacent land in Pliocene and Pleistocene times is suggested by marine Pliocene deposits several hundred meters above sea level in the Tripoli and Lattaqiya regions and by the 200-m elevation of the older Pleistocene terrace (see Section IID).

Unfortunately, marine seismic profiles over the eastern Mediterranean stop some 20–25 km short of the Levant coast, and it is not possible to trace the now deeply foundered evaporites in the region between the abyssal plain and the continental slope and shelf. In the offshore marine seismic lines, sur-veyed in connection with oil exploration over the continental shelf of the Tripoli region, a thick Cenozoic–Mesozoic–pre-Mesozoic sedimentary section is revealed as being very closely correlative with the land geology and adjacent deep boreholes. The lines do not go far enough over the continental slope to pick up recognizable Messinian evaporites, but a Middle Miocene reflector shows the Miocene section thinning from the inland basin in which Sammakieh is located over a local nearshore structure and then extending offshore part way down the slope. Identity of the reflector is lost because of the thick de-velopment seaward of unconsolidated turbidites (Beydoun, 1972).

An offshore negative magnetic anomaly trending approximately northeast–southwest has recently been mapped (Matthews, 1974), and preliminary con-touring shows it to extend 25 km offshore to Beirut, southwestward for at least 160 km, and to merge into other negative features off the Israeli coast (Fig. 6). This anomaly also runs parallel to the trend of the Eratosthenes Seamount (positive anomaly) south of Cyprus. A small positive anomaly begins to border the larger negative one landward, some 30 km west of Tyre, while another larger positive anomaly some 900 km^2 in area is centered 35 km west of Tripoli, and a smaller north–south positive anomaly lies 22 km southwest of Lattaqiya. These marine anomalies line up with those in the Alaouite Range of maritime Syria on a north-northeast trend (Fig. 6). Matthews (1974) observes that these and other positive and negative anomalies in the Levantine sea to the north and west do not appear to be sea-floor spreading anomalies, but are consistent with a continental crust under the Levantine portion of the eastern Mediterranean (with its 10-km thickness of sediments). The evidence suggests that there is no oceanic crust south of the Hellenic Trench.

C. The Eastern Rift Margin

The eastern structural margin of maritime Syria and Lebanon is taken at the Rift zone, marked by the structural depressions of the Bekaa in Lebanon, the Ghab (and Massyaf) in Syria, the Kara Su Valley in Turkey, and the Jordan–Dead Sea depression to the south of Syria–Lebanon. These depressions, however, separate twin uplifts, the Lebanon and Anti-Lebanon and the Alaouite and Jabal Zawiya to the north (the latter, however, having a much less striking orographic expression by comparison to the Alaouite). Further north the Kara Su depression separates the Amanus and Kurd Dagh uplifts, and in the south the Jordan–Dead Sea–Aqaba depression separates the Trans- and Cis-Jordanian uplifted blocks. The trend of this zone of major fracturing and subsidence runs north–south along the Jordan, NNE–SSW in the Bekaa, north–south along the Massyaf–Ghab depression, and then again NNE–SSW in the Kara Su depression, until terminating against the Poturge and Bitlis Massifs lying north of the thrust front of Siirt–Maras in southeastern Turkey (Fig. 2).

Controversy exists as to the nature of the inland boundary fault system. In terms of plate tectonics, this zone is regarded as a sliding plate margin along which a postulated 106 km of left lateral movement has taken place in the Neogene to Pleistocene interval, with the Arabian block moving northward relative to the Levantine block. Dubertret (1932), following the ideas of Lartet (1869), developed this hypothesis, but later rejected it for lack of supporting evidence from detailed field mapping in Lebanon (the reasons for this are outlined in some of his recent works, e.g., Dubertret 1967, 1970).

Quennell (1958) and later Freund and Raab (1969) and Freund *et al.* (1970) developed the hypothesis of left lateral movement, citing stratigraphic, geomorphologic, paleontologic, and structural evidence from the Jordan and Palestine (Israel and West Bank) blocks and offering in support even the offset of the ophiolites of the Amanus area on both sides of the Kara Su depression of northwestern Syria and southeastern Turkey. Bender (1968, and in Freund *et al.*, 1970) disagrees with this interpretation and cites the Aqaba–Dead Sea–Jordan fracture as an old geosuture along which movement had taken place in Precambrian as well as in later times, as it played the role of a hinge zone along which facies changes occurred (see also Wetzel and Morton, 1959; Picard, 1967). Disagreement exists over the age of this geosuture as well as over the amount and type of movement that has occurred along it. Evidence is cited for Precambrian, Upper Jurassic, post-Turonian (e.g., see de Sitter, 1962), Upper Cretaceous, Miocene, and Plio-Pleistocene movement, the last two phases of movement being responsible for a total of 106 km of northward shift of Arabian block (Quennell, 1958; Freund *et al.*, 1970) after the opening of the Gulf of Aden and the Red Sea in the Miocene and the separation of Arabia from Africa. The idea of dissociation of the complex Lebanon–Bekaa–Anti-Lebanon structure (the Bekaa being neither graben nor syncline), necessitated by having the Anti-Lebanon (especially Mount Hermon) start developing 106 km south of its present position and then assume its present form 45 km south of where it is located today, is unacceptable to Dubertret (1976) (Fig. 4).

Presently, there is insufficient stratigraphic and structural evidence from Lebanon for horizontal movements large enough to accomodate all the postulated displacement without leaving a crustal gap where the trend of the plate margin changes from north–south to NNE–SSW. There is ample evidence, however, for considerable vertical displacement. Freund *et al.* (1970) indicate that accomodation of most or all of such a crustal gap is possible by a complex combination of sinistral, dextral, and vertical movements in the Lebanon, Anti-Lebanon, and the Bekaa, as well as in Galilee to the south, over a wide zone within and outside the rift. Recent work in the Palmyride region of Syria by geologists of the French Technical Cooperation Mission and the Syrian Geological Research Directorate (M. Khoury, 1974, personal communication) suggests that some such accomodation to horizontal movement could have taken place by compressional folding and dextral northeast and east–northeast faulting in the south of the Palmyrides and tensional movements in the north. Tiberghien (1974), on the other hand, after concluding detailed gravity work over the whole Lebanese territory, finds it difficult to accept other than small scale horizontal movements, not exceeding the order of ten or so kilometers in the central Lebanon. Clearly, scope for additional work exists, especially as the postulated northward movement of the Arabian block relative to the Levant block (which should form part of the African plate) does not appear to take into account the implications of the underthrusting of Cyprus by a portion of that plate from the south.

D. The Northern Shelf Margin

This region falls outside the area under review in this chapter, lying principally in southern Turkey, northern Iraq, and southwestern Iran. Nevertheless, it is pertinent to an understanding of the evolution of this part of the Levantine region, within the broader context of the evolution of the eastern Mediterranean.

The preceding discussion of the stratigraphy and structure of the region, from the Amanus Range in the north, through the Alaouite Range, southward to southern Lebanon and the bordering structural depressions to the east, shows that they belong to the Arabian shelf, or more precisely, the mobile or "Unstable Shelf" (Henson, 1951). The basement generally lies well below this unstable shelf, except for the anomalous outcrops of the Bassit–Amanus Range (Dubertret and André, 1969). The ophiolites in the northwestern part of the region appear to belong to the "eugeosyncline," but occur within the "miogeosynclinal" part of the shelf, and this has led to differing explanations as to how they arrived where they are found. Blumenthal (1938) suggested that the northwestern limit of the shelf lay along the Kara Su and lower Orontes Valley south of Kizil Dagh while admitting that ophiolites are not restricted to the orogenic zone but can occur on the platform. Stchepinsky (1947) proposed the southeastern boundary of the ophiolites, from the Kurd Dagh southwest to Lattaquiya, as the limit of the shelf. Soviet workers in Syria subsequently adopted this trend as marking a fracture which became known as the Lattaqiya–Kilis fault (Ponikarov et al., 1967), separating the marginal part of the Tauride Alpine geosyncline from the platform or shelf. This boundary places the Amanus Range within the Alpine geosyncline and it is stated that the Amanus has been, as a result, affected by Neogene compressional movements. Dubertret (1976) rightly points out that the Lattaqiya–Kilis line is in reality an orographic and not a structural boundary, and that behind it typical shelf deposits and platform-type vertical tectonics are undeniably present in the Jabal Akra north of the Bassit and in the facies and gentle structures of the Amanus (Giaour Dagh) and Kurd Dagh (Fig. 4).

Work over the last thirty years in Iran, Iraq, and Turkey by many individuals and organizations has now demonstrated that Arabian shelf-type deposits extend well into Iran and Turkey, but that the present boundary of the shelf is clearly at the 1500-km-long northwest–southeast-trending Zagros thrust zone of Iran and northeastern Iraq and at the more sinuous, broken and arching Siirt–Maras thrust zone of the Taurides of Turkey (Ten Dam, 1965). Maras lies about 100 km northeast of the head of the Gulf of Iskenderun. Recent seismic profiling in the Adana–Iskenderun basin of Turkey, which extended well offshore, revealed that the Siirt–Maras thrust extends below Quaternary cover and Miocene and Cretaceous sediments on land, offshore to the northeast tip of the Kyrenia range of Cyprus (Misis–Kyrenia zone), and shows folding and outward thrusting to the southeast. This zone now divides the formerly more extensive Adana Basin into its Iskenderun and Adana components (Mulder, 1973; and others). Data on the Middle East region, including the evolution of the peripheral part of the Arabian Shelf and the bordering "orthogeosynclines," has recently been summarized (Beydoun and Dunnington, 1975).

IV. CONCLUDING REMARKS

Placing the northern limit of the Arabian Shelf as indicated above clearly includes much of Cyprus and the eastern Levantine Sea within the Arabian Shelf. Recent geophysical data on the crust of the Levantine Sea and on the considerable thickness of overlying sediment lends support to this view. It makes of the Jordan–Bekaa–Ghab–Kara Su rift zone a fracture within the platform and brings into question the rate of relative northward movement of the two components (including the Levantine Sea and Cyprus), since both parts appear to be pushing against the edge of Anatolia and central Iran to create thrusts that are directed southward.

The ophiolites of northwestern Syria and the Amanus region are clearly located within this shelf, as are the ophiolites of Oman at the southeastern end of Arabia (e.g., Ricou, 1971).

Questions are posed as to their environment of occurrence and the age and duration of their emplacement, as well as to their allochthonous or autochthonous nature. The ophiolites of nearby Cyprus are thought to represent buoyed-up mantle material, underthrust and kept in place by lighter crust, but it is difficult to accept a similar explanation for the Amanus–Kurd Dagh–Bassit occurrences, and it may well be that the latter are associated with the old geosuture now marked by the meridional rift from Aqaba at the head of the Red Sea to Maras in Turkey. If so, they are autochthonous or para-autochthonous in nature with latitude for gravity tectonics to have played a part in their distribution. Ophiolites in the fold region of the Taurus Mountains of Turkey, not too far to the east-northeast, are clearly allochthonous and emplaced by gravity sliding (Rigo de Righi and Cortesini, 1954). Elsewhere in Iraq? and Iran, the ophiolites are allochthonous, but are adjacent to the Zagros thrust zone (Ricou, 1971). In Oman they are also allochthonous, as demonstrated by Wilson (see Ricou, 1971), but here any deep extension fracture, oceanic ridge, or thrust zone lies offshore and has not been located. Emplacement of the Levant ophiolites is different from the rest (except Oman?) in not being close to a conveniently oriented thrust zone and in lying at the northern extremity of a meridional geosuture thought to be of considerable antiquity, so that extension and upwelling in place within the shelf seems a plausible explanation.

Clearly, considerable work remains to be carried out to solve the problems posed. The age and mode of emplacement of the ophiolites can be determined only by very detailed field studies and absolute age determinations. Marine geophysical work along the offshore portion of Syria and Lebanon is necessary to tie eastern Mediterranean seismic, gravity, and magnetic surveys to surveys on land. Careful determination of the rate of northward movement of the Arabian and Levantine parts of the shelf is required since both appear to be moving northward. Some of this work is already in progress, but a long road lies ahead.

ACKNOWLEDGMENTS

Appreciation is expressed to Dr. L. Dubertret for permission to reproduce some of his figures, making available manuscript before publication, and for critical discussions and suggestions; and to my colleagues in the Department of Geology, Dr. T. R. Goedicke and Dr. I. A. Kafescioglu, for encouragement, suggestions, and availability of research results. Acknowledgment is also expressed to Dr. D. H. Matthews, Dr. V. Tiberghien, and the Ministry of Industry, Syrian Arab Republic, for the incorporation of (simplified) geophysical results in two of the figures. Thanks are also extended to Mr. Ali Karaki for drawing the figures.

REFERENCES

Badrawi, R., 1963, The geological significance of the regional gravity data, eastern Mediterranean region, *Proc. Fourth Arab Petrol. Cong.*, Beirut, Paper 25 (B-2), 6 p.

Bender, F., 1968, Über das Alter und die Entstehungsgeschichte des Jordangrabens am Beispiel seines Südabschnittes (Wadi Araba, Jordanien), *Geol. Jahrb.*, v. 86, p. 177–196.

Beydoun, Z. R., 1965, A review of the oil prospects of Lebanon, *Proc. Fifth Arab Petrol. Congr.*, Cairo, Paper 29 (B-3), 20 p.

Beydoun, Z. R., 1972, A new evaluation of the petroleum prospects of Lebanon with special reference to the pre-Jurassic, *Proc. Eighth Arab Petrol. Congr.*, Algiers, Paper 80 (B-3), 15 p.

Beydoun, Z. R., and Dunnington, H. V., 1975, *The Petroleum Geology and Resources of the Middle East*, England: Beaconsfield, Scientific Press, 99 p.

Blumenthal, M., 1938, Die Grenzzone zwischen syrischer Tafel und Tauriden in der Gegend des Amanos, *Eclogae Geol. Helv.*, v. 31 (2), p. 381–383.

Bourgoin, A., 1945–1948, Sur les anomalies de la pesanteur en Syrie et au Liban. Discussion et interprétation géologique des observations faites par le R.P.P. Lejay, *Notes Mém. Syrie Liban*, v. 4, p. 59–90.

Chenevoy, M., 1952, Sur la présence d'une série métamorphique au Nord de Lattaquie (Syrie), *C. R. Acad. Sci. France*, v. 234, p. 2087–2088.

Chenevoy, M., 1959, Le substratum métamorphique des roches vertes dans le Baer et le Bassit (Syrie septentrionale), *Notes Mém. Moyen-Orient*, v. 7, p. 1–19.

Cizancourt, H. de, 1945–1948, La tectonique profonde de la Syrie et du Liban. Essai d'interprétation géologique des mesures gravimétrique, *Notes Mém. Syrie Liban*, v. 4, p. 157–191.

Daniel, E. J., 1963, Syrie intérieure, in: *Lexique Stratigraphique Internationale*, v. 3, *Asie*, Dubertret, L., ed., Paris: C.N.R.S., fasc. 10 cl, p. 151–289.

Dean, W. T., and Krummenacher, R., 1961, Cambrian trilobites from the Amanos mountains, Turkey, *Paleontology*, v. 4 (1), p. 71–81.

Dubertret, L., 1932, Les formes structurales de la Syrie et de la Palestine, *C. R. Acad. Sci. France*, v. 195, p. 66.

Dubertret, L., 1933, La tectonique de la Syrie septentrionale à la fin du Crétacé et au debut du Tertiare, *Notes Mém. Syrie Liban*, v. 1, p. 13–28.

Dubertret, L., 1936, Stratigraphie des régions recouvertes par les roches vertes du Nord-Ouest de la Syrie, *C. R. Acad. Sci. France*, v. 203, p. 1173.

Dubertret, L., 1937, Sur la constitution et la genèse des roches vertes syriennes, *C. R. Acad. Sci. France*, v. 204, p. 1663.

Dubertret, L., 1939, Sur la genèse et l'âge des roches vertes syriennes, *C. R. Acad. Sci. France*, v. 209, p. 763.

Dubertret, L., 1946, Sur le Quaternaire côtier libanais et les oscillations de niveau de la mer au Quaternaire, *C. R. Acad. Sci. France*, v. 223, p. 431.

Dubertret, L., 1947, Sur la limite nord du plateau Syrien, *C. R. Soc. Géol. France*, p. 107–108.

Dubertret, L., 1955*a*, Géologie des roches vertes du Nord-Ouest de la Syrie et du Hatay (Turquie), *Notes Mém. Moyen-Orient*, v. 6, p. 5–224.

Dubertret, L., 1955*b*, *Carte Géologique du Liban au 1:200,000ᵉ, avec Notice Explicative*, Beirut: Republique Libanaise, Ministère du Travaux Publics, 108 p.

Dubertret, L., 1963, Liban, Syrie: chaîne des grands massifs contiers et confins à l'Est, in: *Lexique Stratigraphique Internationale*, v. 3, *Asie*, Dubertret, L., ed., Paris: C.N.R.S., fasc. 10 cl, p. 9–153.

Dubertret, L., 1966, Liban, Syrie, et bordure des pays voisins, *Notes Mém. Moyen-Orient*, v. 8, p. 251–358.

Dubertret, L., 1967, Remarques sur le fossé de la Mer Morte et ses prolongements au Nord jusqu'au Taurus, *Rev. Géogr. Phys. Géol. Dyn.* (2), v. 9 (1), p. 3–16.

Dubertret, L., 1970, Revue of the structural geology of the Red Sea and the surrounding areas, *Phil. Trans. Roy. Soc. London Ser. A*, v. 267, p. 9–20.

Dubertret, L., 1976, La péninsule Arabique: explanatory text to sheet 16, Tectonic map of Europe 1:2,500,000, Chapter 6, Section 4, *I.G.C. Sub.Commission of Tectonic Maps* (in press).

Dubertret, L., and André, C., 1969, Péninsule Arabique: carte orographie–hydrographie— mers et cartes structurales avec texte explicative, *Notes Mém. Moyen-Orient*, v. 10, p. 285–318.

Edgell, H. S., 1968, Paleontological report on the Sammakieh water well, Ministry of Water and Electric Resources, Beirut, open file report.

Ejel, F., and Dubertret, L., 1966, Sur l'âge précis du gisement de poissons et du Crustacés de Sahel Alma, Liban, *C. R. Soc. Géol. France*, v. 9, p. 353.

Emery, K. O., and George, C. J., 1963, The shores of Lebanon, *Contrib. Woods Hole Oceanographic Institution, 1385*, and American University of Beirut, *Misc. Papers Nat. Sci.*, No. 1, 10 p.

Emery, K. O., Heezen, B. C., and Allan, T. D., 1966, Bathymetry of the eastern Mediterranean Sea, *Deep Sea Res.*, v. 13, p. 173–192.

Fairbridge, R. W., and Badoux, H., 1960, Slump blocks in the Cretaceous of northern Syria, *Proc. Geol. Soc. London*, pt. 1581, p. 113–117.

Finetti, I., and Morelli, C., 1973, Geophysical exploration of the Mediterranean Sea, *Bull. Geofis. Teor. Appl.*, v. 15 (60), p. 263–341.

Fleisch, H., 1962, La côte libanaise au Pleistocène ancien et moyen, *Quaternaria*, v. 6, p. 497–524.

Freund, R., and Raab, M., 1969, Lower Turonian ammonites from Israel, *Tectonophysics*, v. 2, p. 457–474.

Freund, R., Garfunkel, Z., Zak, I., Goldberg, M., Weissbrod, T., and Derin, B., 1970, The Shear along the Dead Sea Rift, *Phil. Trans. Roy. Soc. London Ser. A*, v. 267, p. 107–130.

Gass, I. G., and Masson-Smith, D., 1963, The geology and gravity anomalies of the Troodos massif, Cyprus, *Phil. Trans. Roy. Soc. London Ser. A*, v. 255, p. 417–467.

Goedicke, T. R., 1972, Submarine canyons on the central continental shelf of Lebanon, in: *The Mediterranean Sea*, Stanley, D. J., ed., Stroudsburg: Dowden, Hutchinson and Ross, p. 655–670.

Goedicke, T. R., and Sagabiel, S., 1976, Sediment movement in Lebanese submarine canyons, in: *Symposium on the Eastern Mediterranean*, Hulings, N., ed., *Acta Adriatica*, Split, v. 18, p. 117–128.

Gregor, C. B., and Nairn, A. E. M., 1971, Palaeomagnetic results from Lebanon (abstract), *Trans. Amer. Geophys. Union*, v. 52 (4), p. 188.

Guerre, A., and Sanlaville, P., 1970, Sur les hautes niveaux marins quaternaires du Liban, *Hannon, Rev. Libanaise Géogr.*, v. 5, p. 21–26.

Hassan, A. A., 1963, The distribution of Jurassic and Triassic formations in the East Mediterranean, *Proc. Fourth Arab Petrol. Congr.*, Beirut, Paper 25 (B-3), 10 p.

Henson, F. R. S., 1951, Observations on the geology and petroleum occurrences of the Middle East, *Proc. Third World Petrol. Congr.*, The Hague, v. 1, p. 118–140.

Henson, F. R. S., Browne, R. V., and McGinty, J., 1949, A synopsis of the stratigraphy and geological history of Cyprus, *Q. J. Geol. Soc. London*, v. 105, p. 1–41.

IFP-CNEXO (Institut Français du Pétrole and Centre National pour l'Exploration des Océans), 1974, Carte géologique et structurale des bassins Tertiares du domaine Méditerranéene, échelle 1:2,500,000, Paris: Editions Technip.

Kafescioglu, I. A., 1976, Preliminary results of quantitative distribution of foraminifera on the central continental shelf of Lebanon, in: *Symposium on the Eastern Mediterranean*, Hulings, N., ed., *Acta Adriatica*, Split, v. 18.

Kazmin, V. G., and Kulakov, V. V., 1965, Formation of ophiolites in the Northwest of Syria, *Izvest. Vysz. Ucheb. Zaved. Geol. y Razv.*, v. 2, p. 3–14 (In Russian).

Kazmin, V. G., and Kulakov, V. V., 1968, *The Geological Map of Syria, Scale 1:500,000: Explanatory Notes* [for Part of the Lattaqiah Region], Damascus: Syrian Arab Republic, Ministry of Petroleum, Electricity and Execution of Industrial Projects, 124 p.

Keller, A., 1933, Le Miocène au Liban avec liste des faunes néogènes connu en Syrie, au Liban, à Chypre, et en Cilicie, *Notes Mém. Syrie Liban*, v. 1, p. 155–182.

Kober, L., 1915, Geologische Forschungen in Vorderasien zur Tectonik des Libanon, Das Taurus Gebirge, *Denkschr. Akad. Wiss. Wien*, v. 91, p. 379–428.

Krasheninnikov, V. A., 1965, Zonal Paleogene stratigraphy of the eastern Mediterranean, *Acad. Sci. U.S.S.R. Geol. Inst. Trans.*, v. 133, p. 5–75 (In Russian).

Krasheninnikov, V. A., 1971, Stratigraphy of the Miocene deposits of the Mediterranean on the basis of foraminifera, *Acad. Sci. U.S.S.R. Geol. Inst. Trans.*, v. 220, p. 5–238 (In Russian).

Lartet, L., 1869, *Essai sur la Géologie de la Palestine*, Thesis, Fac. Sci. Paris, v. 316, Paris: Masson, 292 p.

Lejay, P., 1938, *Exploration Gravimétrique des Etats du Levant sous Mandat Français*, Paris: Comité National Française de Géodésie et Géophysique, 54 p.

Majer, V., 1962, Magmatische Gesteine im Gebiet von Bassit zwischen Latakia und Kessab im Nordwestlichen Syrien, *N. Jahrb. Min.*, v. 98 (2), p. 250–282.

Matthews, D. H., 1974, Bathymetry and magnetic anomalies of the eastern Mediterranean (abstract and preliminary maps), *Committee for Marine Geology and Geophysics, XXIV Congrès–Assemblée Plenière de la C.I.E.S.M.*, Monaco.

Mulder, C. J., 1973, Tectonic framework and distribution of Miocene evaporites in the Mediterranean, in: *Messinian Events in the Mediterranean*, Drooger, C. W., ed., Amsterdam: North-Holland Publ. Co., p. 44–59.

Pfannenstiel, M., 1960, Erläuterungen zu den bathymetrischen Karten des ostlichen Mittelmeeres. *Bull. Inst. Océanogr. Monaco*, v. 59 (1192), p. 1–60.

Picard, L., 1967, Thoughts on the graben system in the Levant, *Geol. Surv. Canada*, Prof. Paper 66-14, p. 22–32.

Piro, Y., 1967, *Contribution à l'Etude des Roches Vertes du Nord-Ouest de la Syrie*, Thèse 3e cycle, Montpellier.

Ponikarov, V. P., Kazmin, V. G., Mikhailov, I. A., Razvaliayev, A. V., Krasheninnikov, V. A., Kozlov, V. V., Soulidi-Kandratiyev, E. D., Mikhailov, K. Y., Kulakov, V. V., Faradzhev, V. A., and Mirzayev, K. M., 1967, *The Geology of Syria: Explanatory Notes on the Geological Map of Syria, Scale 1:500,000, Part I, Stratigraphy, Igneous Rocks and Tectonics*, Damascus: Syrian Arab Republic, Ministry of Industry, 229 p.

Quennell, A. M., 1958, The structural and geomorphic evolution of the Dead Sea Rift, *Q. J. Geol. Soc. London*, v. 114, p. 1–24.

Renouard, G., 1951, Sur la découverte du Jurassique inférieur (?) et du Jurassique moyen au Liban, *C. R. Acad. Sci. France*, v. 232, p. 992–994.

Renouard, G., 1955, Oil prospects of Lebanon, *Bull. Amer. Assoc. Petrol. Geol.*, v. 39, p. 2125–2169.

Ricou, L. E., 1971, Le croissant ophiolitique péri-arabe, une ceinture de nappes mises en place au Crétacé supérieur, *Rev. Géogr. Phys. Géol. Dyn.* (2), v. 13, p. 327–349.

Rigo de Righi, M., and Cortesini, A., 1954, Gravity tectonics in foothill structure belt of southeast Turkey, *Bull. Amer. Assoc. Petrol. Geol.*, v. 48, p. 1911–1937.

Saint-Marc, P., 1974, Etude stratigraphique et micropaléontologique de l'Albien, du Cénomanien et du Turonien du Liban, *Notes Mém. Moyen-Orient*, v. 13, p. 9–298.

Sanlaville, P., 1969, Les bas niveaux marins Pleistocènes du Liban, *Méditerranée* (3), p. 257–292.

Sanlaville, P., 1970, Les variations Holocènes du niveau de la mer au Liban, *Rev. Géogr. Lyon*, v. 45 (3), p. 279–304.

Sitter, L. U. de, 1962, Structural development of the Arabian shield in Palestine, *Geol. en Mijnb.*, v. 41, p. 116–124.

Stahl, P., and Plassard, J., 1959, *Carte Gravimétrique du Liban au 1:200,000ᵉ avec Notice Explicative*, Beirut: Republique Libanaise, Ministère des Travaux Publics, 23 p.

Stchepinsky, V., 1947, Sur la limite septentrionale du plateau Syrien, *Bull. Soc. Géol. France* (5), v. 17, p. 33–39.

Ten Dam, A., 1965, La bordure nord de la platforme arabique, *C. R. Soc. Géol. France*, p. 153–156.

Tiberghien, V., 1974, Le champ de la pesanteur au Liban et ses interpretations, *Publications Techniques et Scientifiques de l'Ecole Supérieure d'Ingénieurs de Beyrouth*, no. 26, 193 p.

Van Dongen, P. G., Van der Voo, R., and Raven, T., 1967, Paleomagnetic research in the central Lebanon mountains and in the Tartous area (Syria), *Tectonophysics*, v. 4 (1), p. 35–53.

Vaugnat, M., and Cogulu, E., 1968, Quelques réflexions sur le massif basique–ultrabasique du Kizil Dagh, Hatay, Turquie, *C. R. Soc. Phys. Hist. Nat. Genève*, 2 (3), p. 210–216.

Vaumas, E. de, 1953, Sur les terraces d'abrasion marine de la région de Lattaquie (Syrie), *C. R. Acad. Sci. France*, v. 237, no. 20.

Wetzel, R., and Haller, J., 1945–48, Le Quaternaire côtier de la région de Tripoli (Liban), *Notes Mém. Syrie Liban*, v. 4, p. 1–48.

Wetzel R., and Morton, D. M., 1959, Contribution à la géologie de la Transjordanie, *Notes Mém. Moyen-Orient*, v. 7, p. 95–191.

Wolfaart, R., 1967, Geologie von Syrien und dem Libanon, in: *Beiträge zur Regionalen Geologie der Erde*, Martini, H. J., ed., Berlin: Gebrüder Bornraeger, v. 6, 326 p.

Wong, H. K., and Zarudski, E. F. K., 1969. Thickness of unconsolidated sediments in the eastern Mediterranean Sea, *Bull. Geol. Soc. Amer.*, v. 80, p. 2611–2614.

Zijderveld, J. D. A., and Van der Voo, R., 1973, Palaeomagnetism in the Mediterranean area, in: *Implications of Continental Drift on the Earth Sciences*, Tarling, D. H., and Runcorn, S. K., eds., London: Academic Press, v. 1, p. 133–161.

Zumoffen, G., 1926, *Géologie du Liban*, Paris: Henry Barrère, 165 p.

Chapter 7B

THE LEVANTINE COUNTRIES: THE ISRAELI COASTAL REGION*

David Neev

Israel Geological Survey
Jerusalem, Israel

and

Zvi Ben-Avraham

Israel Oceanographic and Limnological Research
Haifa, Israel

I. INTRODUCTION

The shore of Israel forms a rather smooth, concave arc which trends east in the south (northern Sinai), and then farther north gradually turns to the north-northeast. In some areas the coastline formed depositionally, in some erosionally, but for the most part it is structurally controlled. This paper describes for the coastal plain, the continental shelf, and the continental slope: (1) the basic geological results obtained in the last few years, (2) the structure, and (3) the tectonic setting in relation to the structural elements known on land and in the sea.

* Contribution No. 81 of the Israel Oceanographic and Limnological Research Ltd.

II. PHYSIOGRAPHY

A. Coastline and Coastal Plain

The Mediterranean coastline of Israel is relatively smooth, the most prominent exceptions being the projection of Mount Carmel and the associated indentation of Haifa Bay. In northern Sinai, the Tineh Embayment and the adjoining Bardawil Lagoon disturb the smoothness of the coastal arc. The coast of Lebanon to the north is notably more indented (Fig. 1).

The coastline of Israel was described by Emery and Neev (1960), Schattner (1967), and Neev et al. (1973a, b). It can be divided into a southern and a northern segment. The southern segment, from Rafah to the southern plunge of the Mount Carmel block, is marked by a continuous, low, linear escarpment (10 to 50 m), which is interrupted by several gaps formed by Late Pleistocene rivers. These old estuaries have been filled by beach, swamp, and alluvial deposits. The coastal cliff is composed of carbonate-cemented quartz sandstone (kurkar), originally deposited as NE-facing barchan dunes, interbedded with reddish-brown loams (hamra). The cliff reaches its greatest height in the central part of this segment, between Tel Aviv and some 10 km north of Netanya, where the beaches are narrowest (as little as 10 m). A rocky (kurkar), sandy strip extends along this segment for several hundred meters offshore, forming rimmed terraces at sea level as well as submerged knolls. This strip terminates seaward with a low step (0 to 3 m), which forms a patchy straight line parallel to the coastline (Fig. 2). The coastal plain of this segment narrows from about 50 km in the south to about 20 km in the north, where it approaches Mount Carmel. Three longitudinal kurkar ridges, trending subparallel to the coastline, occur in this segment.

The northern segment extends northward along Mount Carmel and the Lebanese coast. Along this segment, the inland hills, composed mostly of Cretaceous chalk, limestone, and dolomite, occasionally approach the coastline, leaving a narrow coastal plain or none at all. Mount Carmel appears to be a large, tilted block, plunging to the SSW and terminating northward at a steep NW-trending fault escarpment. Along the Lebanese coast, high, steep escarpments are formed where the mountains reach the shore. In places, wide, flat coastal plains breach the mountainous coastline, as at the Qishon Valley and Zevulun Plain north of Haifa, and at the plains north of Tripoli and south of Lataqia (Fig. 1).

B. Continental Margin (Shelf and Slope)

The continental shelf off Israel, which reaches about 100-m depth, is about 25 km wide off Rafah and narrows toward the north to 10 km off Mount Carmel. Mount Carmel itself extends some 12 km offshore to form a notable topographic hump across the shelf called the Haifa Nose (Figs. 1 and 7). From Mount Carmel northward, the shape of the shelf, and especially that of the slope, becomes irregular. Two pronounced canyons, the Carmel Canyon, which is an extension of the Qishon Graben (Nir, 1973), and the Akhziv Canyon (Fig. 3) (Nir, 1965), cross the shelf as the extensions of onshore river valleys. Between them, the shelf broadens to about 15 km, forming the Akko (Acre) Nose, a

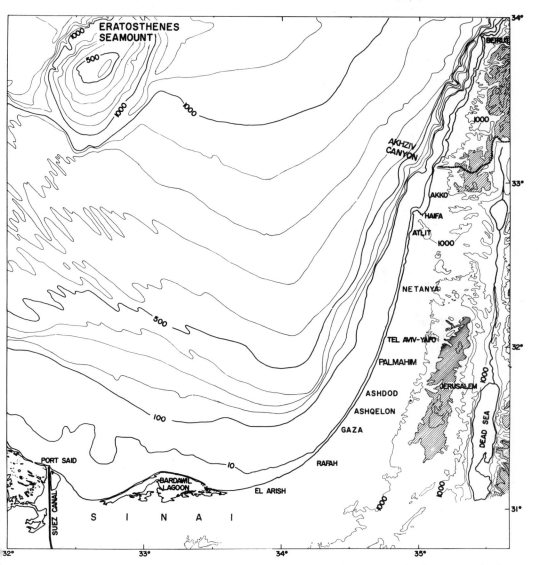

Fig. 1. Bathymetry and topography of Israel and adjacent eastern Mediterranean Basin. Contours on land are in feet at 1000-ft intervals. Contours at sea are in fathoms at 100-fm intervals. The 10-fm contour is also shown. Based on Defense Mapping Agency Hydrographic Center Map No. 54030.

Fig. 2. Air photo of the coastline (central part), showing the coastal cliff and the associated abrasion and erosion effects. Note the straight line formed by the sharp termination to the west of the abraded terrace (rocky strip composed of cemented eolianite–"kurkar").

feature similar to the Haifa Nose. North of the Akhziv Canyon the shelf narrows to 3 to 0.5 km and is dissected by numerous structures and canyons (Boulos, 1962; Carlisle, 1965; Goedicke, 1972).

Emery and Bentor (1960) divided the shelf off Israel into two main topographic belts. The first, a near-shore belt, extends from the shoreline to the 30-m isobath, about 3 km offshore. The slope of this belt is relatively steep (0.5° to 1°). The second belt forms a wide and flat (10′ to 20′ slope) area extendding approximately to the 80-m isobath. The transition zone between these two belts, approximately between 30- and 50-m isobaths, is characterized along most of the shelf by a patchy, reeflike, rocky ridge, subparallel to the coastline, and protruding sharply through the sedimentary cover (Emery and Bentor, 1960; Neev *et al.*, 1966, 1973*b*).

Between Rafah and the southern plunge of Mount Carmel, the continental slope drops gently down from 80 to about 1000 m, with slopes increasing from 2° in the south to 8° in the north. However, certain topographic irregularities were noted, such as asymmetric hillocks on the upper slope, niches on its

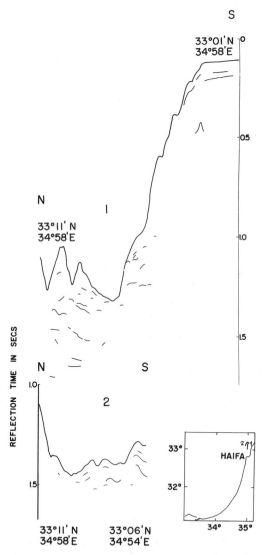

Fig. 3. Line drawings of seismic profiler traverses across the Akhziv Canyon. Based on 1000-J sparkarray profile.

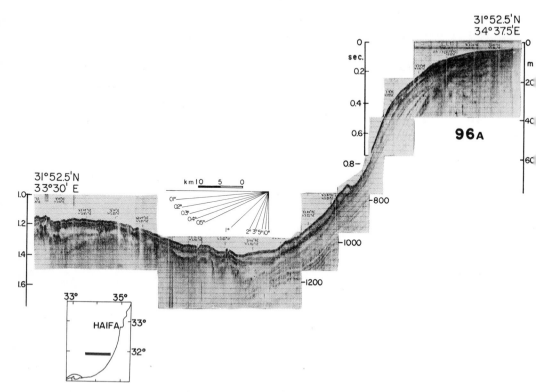

Fig. 4. East–west 1000-J sparkarray profile off southern Israel. The scar at the midslope was formed by downslope gravitational gliding of the uppermost sedimentary layer. (From Neev *et al.*, 1973*b*.)

middle and lower parts, and a rough bottom, which is acoustically fuzzy below the niches. The niches were formed by the frequent occurrences of downslope sliding of the uppermost sedimentary unit (Holocene) (Fig. 4). The main feature interrupting this segment of the slope is the 10-km-wide NW-trending depression of Palmahim (southwest of Tel Aviv), the Palmahim Graben. The continental slope, from the southern plunge of Mount Carmel northward, steepens to an average of about 10°.

III. SEDIMENTS

A. Coastal Plain

A fairly detailed description of the sediments in the beaches of Israel is given by Emery and Neev (1960). They state that the bulk of the beach sand comes to Israel from the southwest by longshore transportation, provided by both wave-induced and general offshore Mediterranean currents. It is believed

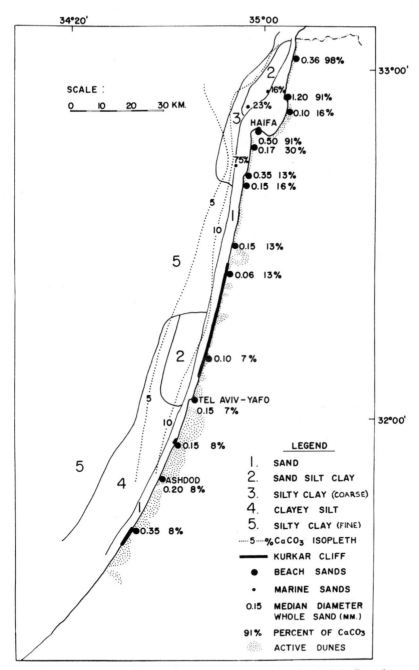

Fig. 5. Sediment types and calcium carbonate content in the beaches and the littoral zone of Israel. (From Nir, 1973; Emery and Neev, 1960.)

that the Nile is the chief source of the beach sands of Israel, although there is probably some supplementary contribution from sea cliffs and the seasonal streams of Sinai. Appreciable contributions of calcareous organic debris were noticed along the beaches off Mount Carmel. The calcareous component becomes more important north of Akko, where the bulk of the sand consists of broken shells. Emery and Neev (1960) also found that similar sand distribution patterns existed during the Pleistocene epoch, as is illustrated by the position and nature of eolianites (Fig. 5). Recently, Bakler *et al.* (1972) studied the environment of deposition of sandy units in the coastal plain of Israel, using statistical analysis of grain size data. These authors found that the coastal Quaternary sands of Israel can be divided into three main groups: sediments deposited in a shallow marine environment, in an eolian environment, and in a surf zone.

B. Continental Margin

The nature of the Recent and sub-Recent sediments of the Israel Mediterranean margin is described by Nir (1973). His main conclusions are: that the grain size decreases with distance from shore and with water depth; that there is a strip of sand extending from the shore 4 to 5 km westward, where it changes to a region of silty clays; and that the main type of sediment derived from the Nile is clayey silts and silty clays. The main clay minerals found in the Nile lutite are montmorillonite (60 to 80%), illite (0 to 20%), and kaolinite. The heavy mineral suite that characterizes the shelf and slope is hornblende–augite with occasional mica assemblages. The calcium carbonate content along the continental shelf and slope is variable, but generally increases from south to north, due to the intensive growth of organic banks and the decreasing influence of Nile-derived sediments (Nir, 1973). Calcium carbonate content averages about 9 to 10% in the shelf and 5% in the slope sediments, as compared with 30 to 50% in other parts of the Levantine Basin, which are not influenced by the Nile. The rate of deposition in the areas influenced by the Nile sediments (i.e., the shelf and slope north of Sinai and off southern and central Israel) is very high, reaching 650 mm/1000 years off the Nile Delta (Nir, 1973).

IV. STRATIGRAPHY

The crystalline basement of the Arabo-Nubian Massif crops out in central and southern Sinai. The entire sedimentary column thickens and changes from a continental to a marine facies toward northern Sinai and the southern Negev. However, the facies and thickness patterns of the sedimentary sequence, at least from the Jurassic upward, in Israel, northern Sinai, and their offshore

Stratigraphic Correlation

Chronostratigraphy	Offshore Basin	Lithostratigraphy — Coastal Hinge Line	Eastern Platform
Quaternary / Holocene / Pleistocene	Kurkar Group — carbonate cemented eolianites interbedded with marine clay	Kurkar Group — carbonate cemented eolianites interbedded with loams and swamp deposits	Nondeposition, erosion, inland alluvium
Tertiary — Pliocene: Upper, Lower	Upper Yaffo Formation—marl		Some alkali-olivine basalts
Miocene: Upper (Messinian)	Lower Yaffo Fm—marl		
Miocene: Middle / Lower	Mavqi'im Fm—evaporites; Ziqim Fm—marl	Extensive alkali-olivine basalts	Extensive alkali-olivine basalts
Oligocene: Upper / Middle / Lower	Beit Guvrin Formation—marl		Inland deposits
Eocene: Upper / Middle / Lower	Shephela Group chalk (Campanian and L. Eocene flint)	Thinning to nondeposition	Shephela Group chalk (Campanian and L. Eocene flint)
Mesozoic — Cretaceous: Upper (Senonian)	Talmei Yaffe Fm marl, chalk, and flint	Judea Group — carbonates and reefs	Judea Group — dolomite and marl
Cretaceous: Middle (Turonian, Cenomanian)			
Cretaceous: Lower (Albian; Aptian to Berriasian)	Gevar'am Fm shale and marl	Helez Fm shale with some sand and limestone	Kurnub Group sand ("Nubian") with marl and some limestone; extensive alkali-olivine basalts
Jurassic: Upper	Undivided basinal facies, some alkali-olivine basalts	reefs with carbonate and shale	Arad Group and younger Jurassic sediments limestone and shale with some sand

∧∧∧∧∧ unconformity; ∧∧∧∧∧ major unconformity associated with rejuvenation of the coastal faulting system; – – – vaguely defined contact; ——— clear contact.

areas are influenced by two main structural elements: (1) the NNE-trending coastal hinge line and its associated fault system, and (2) the NE-trending fold system. In northern Israel and its offshore area, the influence of a third element is superimposed: (3) the NW-trending tensional system of Mount Carmel and the Haifa Nose (Section V on Structural Framework describes these elements in detail). Table I is an attempt to correlate the different lithostratigraphic units which are found in three zones: (a) the Offshore Basin, which stretches between the present coastline and Offshore Structure No. 1, located on the lower part of the continental slope (see Section V on Structural Framework); (b) the Coastal Hinge Line, which extends along the present coastline and east of it to a distance of 10 to 15 km; (c) the Eastern Platform, which extends further on to the east to about the Jordan–Dead Sea Graben.

The deepest stratigraphic unit, studied by means of drilling and geophysical information in the coastal plain and shelf of Israel, is of Jurassic age. The dominant shale, limestone, and sand of the Arad Group in the Eastern Platform changes at the Coastal Hinge Line to a more reefoidal facies, associated with remarkable thinning and truncation of the Upper Jurassic formations. Offshore, rapid thickening and facies change into basinal dark shale, marl, light limestone, and turbidites occurs. The NW-trending Mount Carmel Ridge, with its offshore Haifa Nose, is dominated by reef facies. Extrusive rocks, composed of alkali-olivine basalts, are interbedded in the uppermost Jurassic units off central Israel, in the Carmel Ridge, and in the northeastern Negev.

A similar facies transition is recognized during the Lower Cretaceous: the Kurnub Group (much of the so-called "Nubian Sandstone" with some limestone) in the east changes into the Helez Formation (shale with some sand and limestone) at the Coastal Hinge Line, and into the shale and marl of the Gevar'am Formation in the Offshore Basin.

The Middle Cretaceous (Albian to Turonian) Judean Group is dominated by reef facies along the Coastal Hinge Line. Backreef dolomites and marl characterize the Eastern Platform to the east. Abrupt change of facies, associated with appreciable thickening, is recognized in the Talmei Yaffe Formation (marl and tight chalk and flint) to the west. The Lower and Middle Cretaceous Formations in the Carmel Ridge are dominated by reef facies, occasionally interbedded with basic to ultrabasic eruptives. This volcanic activity has been intermittently maintained in the Carmel Ridge through the Upper Cretaceous and into the Neogene or Pleistocene.

The chalky and marly formations of the Upper Cretaceous to Lower Tertiary Shephela Group (Senonian to Middle Eocene) demonstrate appreciable thinning and truncation along the Coastal Hinge Line and the Carmel Ridge. The interbedded Campanian and Lower Eocene flints, which characterize the basinal facies, are found both on the Eastern Platform and in the Offshore Basin.

A detailed description of the Late Cenozoic stratigraphy of the coastal plain of Israel is given by Gvirtzman (1970). It is believed that a major regression took place at the end of the Oligocene, resulting in erosion and channelling, and an extreme restriction of marine sediments younger than Eocene from the area east of the coastal plain. Miocene, Pliocene, and Pleistocene marine transgressions filled the erosional channels with continental, lacustrine, fluviatile, and lagoonal sediments, while subsequent regressions caused re-deepening of the system.

The Neogene geological history of the Coastal Hinge Line and the Offshore Basin, which is associated with the Miocene evaporites (Messinian), appears to be instructive also from a broader, regional point of view. During the Upper Miocene, damming by a sill between the Mediterranean Sea and the Atlantic Ocean resulted in the gradual desiccation of the former, according to Hsü (1972, 1973). This process caused precipitation of an evaporitic layer, expressed as Reflector M on continuous seismic profiles (Ryan, 1973). Data recovered from closely-spaced drilling on land in the coastal plain of Israel indicate that the gypsum deposits were precipitated while the Mediterranean water body in this belt was still relatively deep (at least 200 m). The restriction of halite deposits to paleotopographic lows, such as the Palmahim Graben, corroborates the above conclusion. The association of potash deposits in the Sicilian evaporites (Decima and Wezel, 1973) may indicate that this area was situated in one of the deepest parts of the desiccated Mediterranean Basin. Decima and Wezel (1973) estimate, however, that the evaporites in this site were deposited on a sea floor 200 to 500 m below the Late Miocene Atlantic sea level. This conclusion is in agreement with the concept of a general subsidence phase, which has been taking place since the Early Miocene (Neev, 1960; Neev et al., 1973b) and thereby explains the great depths at which the evaporite deposits are found today. The evaporites are missing from the sedimentary sequence of vast areas in the shelf off Israel and its coastal plain (Gvirtzman, 1970), as well as in the northern limits of the Nile Delta (Said, 1973). The Reflector M is still recognized, however, in the seismic profiles in these places. The high reflectivity of this horizon is interpreted as being associated with the Mio-Pliocene unconformity. Nondeposition and exposure of this surface have occurred due to the joint effect of the local epeirogenic fluctuations and the eustatic drop of sea level caused by the desiccation process. Exceptionally intensive subsidence has occurred in several areas of a negative structural nature since the Upper Miocene, such as in the Palmahim Graben and the Nile Delta. Only gypsum deposits were found in the northern delta embayment; the association and range of thickness of these deposits are similar to those found in the coastal plain of Israel.

On the shelf and upper slope, the uppermost (Holocene) unit overlies a highly irregular, probably erosional surface (Neev et al., 1966). This erosional

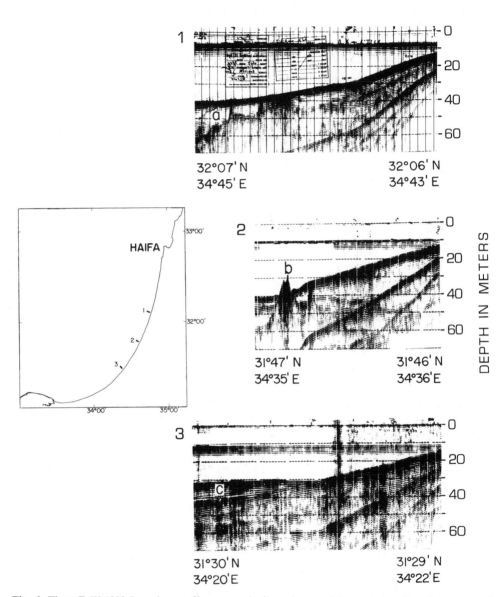

Fig. 6. Three E–W 1000-J sparker profiles across the littoral zone of the central and southern parts of the coastline. Note the uppermost erosional surface (Würm) (a) and the associated easternmost kurkar ridge (a fossil coastal dune). This ridge (b) protrudes, in Profile 2, through the younger sedimentary cover which thickens both to the east and west. The westward tectonic tilt is more emphasized in Profiles 1 and 3. An additional (youngest) sedimentary unit (c) is seen in Profile 3 (closest to the Nile Delta).

surface includes several low ridges, separated by broad, sediment-filled depressions, which extend subparallel to the present coastline (Fig. 6). Seismic reflections within and below the ridges indicate that these are sedimentary features and not of tectonic origin. Analogy with similar onshore Pleistocene ridges suggests that they may be eolianite (Emery and Bentor, 1960). They probably formed as coastal dunes, accumulated and stabilized during various Late Pleistocene phases. The erosional surface was probably formed during the Würm low-sea-level stage. The easternmost ridge, located along the 30- to 40-m isobaths, protrudes, in many cases, through the cover of the youngest sedimentary unit (Fig. 6). High-resolution seismic profiles (using 200-J boomer) show that the Holocene sediments east of the ridge thicken eastward toward the coastline.

V. STRUCTURAL FRAMEWORK

A. General Structure

The geophysical properties of the eastern Mediterranean Basin are basically different from those of the adjacent land. The former is dominated by positive Bouguer anomalies, while negative anomalies dominate the Levant, Sinai, and northern Egypt; the zero line conforms roughly to the coastline (De Bruyn, 1955; Woodside and Bowin, 1970). A significant difference in crustal thickness exists between the Nile Delta and the area south of it (Gergawi and Khashab, 1968); the crust under the former is 24 km thinner than at Helwan. Lort et al. (1974) concluded that the crust–mantle boundary at the northern foot of the Nile cone is found at a depth of 27 km. Assuming a combined thickness of 15 km for the sea water and sedimentary column at that place, the crust in the eastern Mediterranean may be considered as semi-continental to oceanic. Thus, the eastern Mediterranean Basin differs from the adjacent land areas to the east and south, which are characterized by a continental crust. Neev et al. (1973b) suggest that the contact between the geological provinces of the eastern Mediterranean Basin and the Levant–Sinai is marked by a NNE–SSW-trending compressional zone, which is stretched along the base of the continental slope of the Levant, some 60 to 80 km off the coast of Israel. These authors named it the Pelusium Line, since it extends southwestward to join the Pelusiac arm (now abandoned) of the Nile Delta. This arm formed the northeastern limit of the delta, at least since the Paleonile stage (Said, 1973). The northern extension of the Pelusium Line probably joins the northeastern Mediterranean coastline at Lataqiya (Fig. 7). Differences also exist between other geological properties of the provinces on each side of the Pelusium Line.

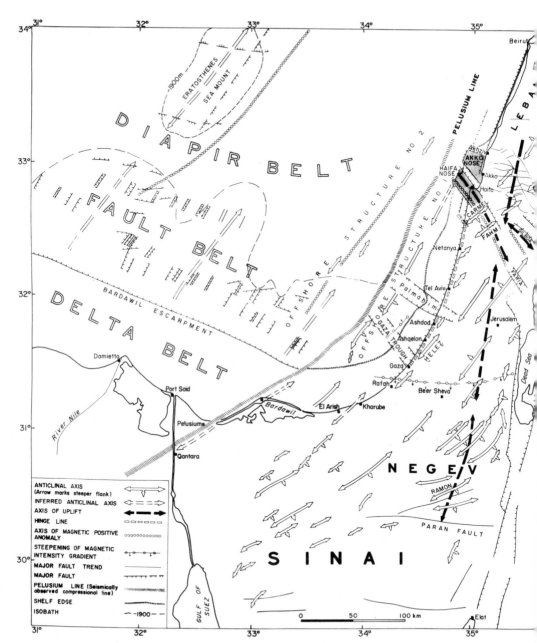

Fig. 7. Sketch map of major tectonic elements in Israel and adjacent waters. The regional tectonic trends show
the map are: NE-trending folding system; the NW tensional elements; and the N–S Dead Sea Rift and the struc
backbone of Israel. (From Neev et al., 1973b.)

B. The Coastal Fault System

The coastal fault system runs NNE (Gvirtzman and Klang, 1972), slightly diagonal to the NE-trending fold system (see below). A hinge line, which acted as a positive structural element, was formed just to the east of this fault system. Faults of this system are of the normal gravity type, and no indications were found for horizontal movement along them. The earliest proved activity along this system occurred during the uppermost Jurassic, when the continental (eastern) side moved upward, thereby inducing extensive erosion of the exposed land areas. However, the appreciable thickening of the Jurassic and older formations on the western side (Gvirtzman and Klang, 1972) probably indicates earlier phases of faulting activity. The last full-scale activity along this system took place during the Late Oligocene or earliest Miocene (Neev, 1960; Gvirtzman, 1970). The first phase of this event involved a differential faulting, which caused the continental side to be uplifted regionally to about 1500 m above the Oligo-Miocene sea level. Long, deep canyons were entrenched during this exposure. The next phase, which extended until the Pleistocene, involved regional subsidence. Abundant extrusions of olivine basalts were associated with the initial stages of this phase (Miocene). The subsidence, however, was similar for both the land and offshore blocks. The broad and elongated negative magnetic anomaly found between the Coastal Hinge Line and the Pelusium Line may have resulted from multiple subsidence events, which involve an increasing crustal thickness. This anomaly is cut off diagonally to the north by the NW extension of the Haifa Nose.

We believe that the present-day coastline is the surface expression of a fault belonging to the coastal fault system, which has been rejuvenated during the Holocene (Neev et al., 1973a, b) (Fig. 8). This conclusion is based on the following evidence:

1. The physiography of the coastline is linear and bordered to the east by an almost continuous line of low, steep cliffs (up to 50 m). Young erosional features, including hanging valleys, are abundantly found along these cliffs (Fig. 2).

2. Naturally deposited shell beds (beach deposits) occur on top of the coastal cliffs. Where found, these beds overlie the Upper Pleistocene–Holocene sediments, which are either the friable-cemented eolianites (kurkar) or red loams (hamra). In several localities, however, they overlie ancient sites of Roman to Crusader times. Moreover, pottery shards, part of them rounded to pebble shapes due to wave activity, are imbricated in the shell beds. The ages of these shards range from Roman to Medieval times.

3. The hamra beds (red loam), which unconformably overlie the Upper Pleistocene kurkar, were dated by flint implements belonging to the epipaleo-

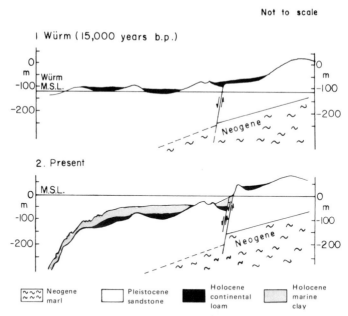

Fig. 8. Evolution of the coastal fault and shelf off Israel during the Holocene. Westward downwarping of the present-day slope occurred or intensified since the Holocene. (From Neev *et al.*, 1973*b*.)

lithic stage (15,000 y. b. p.). The beds are composed in places of alternating graded, bedded sand (wind blown) settled into a red clay matrix, which has been deposited in aquatic mud flats. The 4 to 5° eastward dips, which are measured in these beds along the coastal cliff, indicate a Recent or sub-Recent eastward tectonic tilt.

4. A layer of cemented coastal dune sands, a few meters thick, composed mostly of organic $CaCO_3$ particles, overlies the hamra beds along the coast. This layer, dated by ^{14}C as 7000 to 10,000 years old, is located at elevations as high as 50 m above m.s.l. (i.e., at Netanya). It is unlikely that these Holocene beach sediments, which were deposited close to the base level of erosion, could be found today at such a high elevation without a tectonic adjustment, even if an appreciable regression of the cliff due to abrasion is considered.

5. Northeastward-facing barchan dunes, now stabilized, cover Byzantine and early Moslem settlements, which were originally built on clean and stable hamra (red loam) plains. In the central coastal plain, some of these dunes are located today on top of the coastal cliffs at elevations of up to 50 m above m.s.l. In these locales, the beaches are as narrow as 10 to 20 m. There is no doubt that these dunes were derived and fed from the beaches, similar to the active processes observed today at the broad and shallow outlets of the rivers

which form gaps in the coastal cliff. It appears, therefore, that the above-mentioned "hanging" and fossil dunes drifted shoreward and accumulated on land when the coastal cliffs were at a much lower elevation above m.s.l. This process probably occurred only a few hundred years ago.

6. Data from submarine archaeological and geophysical studies indicate that the Herodian harbor of Caesarea (north of Netanya) has subsided by about 10 m since its construction. Abundant Roman quarries and fresh-water wells are found at the present-day sea level and below it along the central and northern parts of the coastline, as well as in small islands which are located 1 to 2 km off the coast of the western Galilee.

7. Eastward thickening of the Holocene sedimentary sequence was noticed in continuous seismic profiles, which were carried out at the littoral zone between Tel Aviv–Yafo and Gaza. This sequence is abruptly terminated along the patchy strip of kurkar rocks (Upper Pleistocene age) which parallels the coastline (Figs. 2 and 6).

8. Preliminary examination of C.D.P. seismic profiles, carried out recently across the coastline by the Institute for Petroleum Research and Geophysics of Israel, indicates some disturbances close to the coastline at the Caesarea (north of Netanya), Ashdod, and Gaza segments.

On the basis of the above indications, it is suggested that the entire coastal zone of Israel was downwarped and submerged under the Mediterranean waters some time in the post-Crusader period. After this, the offshore side of the coastline remained submerged, whereas the area east of it was uplifted to its present-day position. It is probable that the present elevation of the upthrown block is rather close to the prefaulting one (Crusaders). The coastal fault line runs at least from Rafah in the south to the Haifa Nose, where it probably alternates in an "en echelon" way with the fault along the coast of the western Galilee. The maximal throw along the fault line is estimated to be about 50 m.

C. Northeast-Trending Fold System

A regional NE–SW-trending fold system dominates Israel and northern Sinai (the "Syrian Arc" of Krenkel, 1924). At sea, a NE–SW-trending anticline was found along the lower continental slope, some 30 to 40 km off the coast of Israel, and named by Neev et al. (1973b) "Offshore Structure (OSS) No. 1." Seismic reflection data indicate that OSS No. 1 has developed since the Jurassic, with only one interruption (base Miocene to mid-Pliocene). An isopach map of interval "M" (mid-Neogene reflector) to the present day sea bottom indicates that OSS No. 1 is a bead in a chain of structures which extends subparallel to the present coastline of Israel, starting from the shelf off Sinai and continuing to the Haifa (Carmel) Nose. This chain of structures appears to be part of the regional fold system. The Helez Structure, which is located about 10 km east of the coastline, is the closest member of this system to OSS No. 1 (Fig. 7). Druckman (1974) deduced from stratigraphic data that the activity of the regional fold system in the northern Negev was initiated by Triassic time. According to similar information presented by Weissbrod (1969), this folding activity could have started even earlier, in the Upper Paleozoic. Second-order folded structures of shorter wavelength and perhaps different amplitude have also been developed during this very long period of fold activity. West of the Pelusium Line, NE-trending features are also recognized. They are expressed mainly as strong positive magnetic anomalies, but also as physiographic features. The most prominent of them is associated with the Eratosthenes Seamount (Fig. 1). We believe that the NE-trending features west and east of the Pelusium Line were generated due to the same compressional folding process, although there is a marked difference in the wave lengths of the fold system on each side of the Line.

D. The Present-Day Shelf Break

The present-day shelf break off the coast of Israel is an entirely new hinge line along which the eastern Mediterranean Basin has recently been downwarped and subsided. This new hinge line is located along the trough of the offshore basin which separates the OSS No. 1 and the coastal fault system. Seismic profiles indicate that the OSS No. 1, although presently located along the lower part of the continental slope, has never acted as a dam for sediments derived from the east. This holds true also for the Upper Pleistocene, when rejuvenated uplifting and folding along the axis of the OSS No. 1 are recognized. Thickening of sediments from the OSS No. 1 toward the trough stretching along the Pelusium Line is also recognized. However, a westward creep of Holocene-aged sediments was noticed at the shelf break and on the upper part of the continental slope. There are indications that these sediments were originally deposited as nearly horizontal layers, and they are, therefore, interpreted as having been tilted to the west in later times. The westward creep is limited to the Holocene sequence, for similar features were not observed in the underlying Pleistocene layers. This creep phenomenon is recognized in almost all profiles which cross the shelf edge. Moreover, other features which indicate westward gravitational sliding and crumpling of the entire uppermost Holocene sedimentary layer are abundantly recognized along the lower half of the continental slope off southern Israel (Fig. 4). These slides have produced big scars, 1 to 5 km wide, with steep sidewalls and eastward scarps about 50 to 70 m high. This sliding of the uppermost layers did not affect the underlying sediments. The downslope sliding effects are choked and terminated at the trough along the Pelusium Line.

A similar history of very young subsidence of the floor along the western border of the eastern Mediterranean basin was recorded by Finetti and Morelli (1972), on the basis of similar structural–stratigraphic evidence.

E. Northwestern Structural Elements

The NW-trending Haifa Nose (Fig. 9) extends from the coastline more than 50 km northwestward as a physiographic and structural element, which is expressed also as positive magnetic and gravity anomalies (Neev et al., 1973b). A geophysical survey conducted in order to study the nature and extent of the Carmel structure into the eastern Mediterranean (Ben-Avraham and Hall, 1977) shows that the structure is very well defined across the shelf. This structure disappears at the continental slope, which was found to be very much broken in this area. Beyond the continental slope, a structure with no clear magnetic signature was detected. It has the same strike as the Carmel structure. A small structure with the same strike as the Haifa Nose was detected off Atlit and may be a part of an "en echelon" structure, which typifies the Haifa Nose. The Haifa Nose is not recognized west of the Pelusium Line, but it does extend inland toward the SE up to the Jordan Rift. The thinning of the Cretaceous and Lower Tertiary formations toward this ridge, and the dominance of reefoidal facies there, indicate its continuing development as a positive structural element, at least since the Jurassic. The Haifa Nose is associated with deep tensional fissures, which probably affected the deeper part of the crust. This is indicated by the repetitive interbedding of extrusives composed of alkali-olivine basalts and ultrabasic xenoliths (Sass, 1957) in the sedimentary sequence of this element, and by the normal fault system which separates it from the graben of Jezreel to the northeast. It is assumed that the pressures released by this fissuring are responsible for the uplift of the ridge. The Haifa Nose tensional element was active contemporaneously with the activity of the NE-trending compressional fold system, but strikes almost perpendicular to it.

The Palmahim Graben (Fig. 7) and the offshore extension of the Gaza Canyon have also been developed as NW-trending negative structural elements, at least since the Miocene (Neev et al., 1973b). It seems, however, that at present, activity along the former is of an appreciably greater intensity in the outer shelf and in the continental slope than in the inner shelf and coastal plain.

VI. A MODEL EXPLAINING THE TECTONIC ORIGIN OF THE COASTAL REGION

In terms of plate tectonics, it is possible to explain the rather complicated set of events described above. We suggest that the major factor controlling the tectonic style in the coastal region is the northward drift of the African

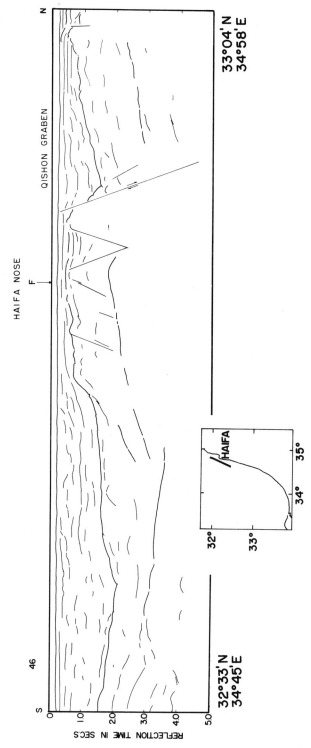

Fig. 9. Line drawing of seismic profile traverse across the Haifa Nose on the shelf off Israel. Based on part of the Belpetco Israel Seismic Profile 46 (a 12-fold C.D.P. profile, shot with a 900-cu in-air-gun). F marks the location of Belpetco Israel Foxtrot wildcat drillhole.

plate and its underthrusting beneath the Alpine system at Cyprus and Crete. This drift was rejuvenated after the Upper Pliocene. It creates a stress-couple which increases in intensity with the decreasing distance between the African continent and the Alpine front range (i.e., with time). As a result, the floor of the eastern Mediterranean Basin has been subsiding through time. This subsidence became catastrophic during the Holocene.

The Pelusium Line may be the western boundary of the Arabian plate along which differential movement occurs. The region east of the Pelusium Line, including the coastal region, shelf and slope of Israel, is influenced by the subsidence of the eastern Mediterranean, which caused the slope to be dragged down during the relative motion of the two plates. Thus we assume that the continental slope and shelf off Israel were dragged down and arched to the west, thereby causing the submergence of the coastline. This line forms the apex of the arch which extends between the Pelusium Line and the Judean upwarp (axis of uplift in Fig. 7). The tension created as a result of this arching was probably released by uplift of the eastern side of the fault along the present-day coastline. The coastline also coincides with the coastal fault system, hence the faulting is considered as a rejuvenation of this system. Compared with the catastrophic nature and magnitude of the earth movements in the eastern Mediterranean mentioned above, the movements along the coastal fault line must be considered as mild. Rejuvenation of activity along this fault since the Oligocene–Miocene phases occurred only recently, and the magnitude of displacement has been just a few tens of meters. Additional movements along this fault can be expected only if additional large-scale subsidence occurs in the eastern Mediterranean, west of the Pelusium Line.

ACKNOWLEDGMENTS

The contributions to this paper of our colleagues in the Marine Geology Division of the Geological Survey of Israel, G. Almagor, A. Arad, D. Argas, N. Bakler, J. K. Hall, J. Levi, U. Levi, R. Madmon, and Y. Nir, are invaluable. We appreciate their help with the data acquisition at sea, data processing, and interpretation. The present paper depends heavily on the results of the joint studies which were summarized by Neev et al. (1973a, b) and by Nir (1973). We thank N. Bakler for his critical help in preparing the manuscript.

REFERENCES

Bakler, N., Denekamp, S., and Rohrlich, V., 1972, Sandy units in the coastal plain of Israel: environmental interpretation using statistical analysis of grain size data, *Israel J. Earth Sci.*, v. 21, p. 155–178.

Ben-Avraham, Z., and Hall, J. K., 1977, Geophysical survey of Mount Carmel structure and its extension into the Eastern Mediterranean, *J. Geophys. Res.*, v. 82, p. 793–802.

Boulos, I., 1962, *Carte du Reconnaissance des Côtes du Liban, 1:150,000*, Beirut: Bassile Frères, 2nd Ed.

Carlisle, D. H., 1965, *A Continuous Seismic Profiling Survey off the Coast of Lebanon*, M. Sc. Thesis, M.I.T., Cambridge, Mass., 138 p.

De Bruyn, J. W., 1955, Isogam maps of Europe and North Africa, *Geophys. Prosp.*, v. 3, p. 1–14.

Decima, A., and Wezel, F., 1973, Late Miocene evaporites of the central Sicilian Basin, Italy, *Initial Reports of the Deep Sea Drilling Project Nat. Sci. Found.*, 13 (2), Ch. 44-1, p. 1234–1240.

Druckman, Y., 1974, The stratigraphy of the Triassic sequence in southern Israel, *Bull. Geol. Surv. Israel*, v. 64, 94 p.

Emery, K. O., and Bentor, Y. K., 1960, The continental shelf of Israel, *Bull. Geol. Surv. Israel*, v. 26, p. 25–41.

Emery, K. O., and Neev, D., 1960, Mediterranean beaches of Israel, *Bull. Geol. Surv. Israel*, v. 26, p. 1–24.

Finetti, I., and Morelli, C., 1972, Wide-scale digital seismic exploration of the Mediterranean Sea, *Boll. Geofis. Teor. Appl.*, v. 14 (56), p. 291–342.

Gergawi, A., and Khashab, H., 1968, Seismicity of the U.A.R., *Bull. Helwan Observ.*, v. 76, 27 p.

Goedicke, T. R., 1972, Submarine canyons on the central continental shelf of Lebanon, in: *The Mediterranean Sea*, Stanley, D., ed., Stroudsburg: Dowden, Hutchinson and Ross, Inc., p. 655–670.

Gvirtzman, G., 1970, The Saqiye Group (Late Eocene to Early Pleistocene) in the coastal plain and Hashephela regions, Israel, *Rep. Geol. Surv. Israel*, no. OD/5/67, and *Rep. Inst. Petrol. Res. Geophys.*, no. 1022, 170 p. (In Hebrew, English abstract.)

Gvirtzman, G., and Klang, A., 1972, A structural and depositional hinge-line along the coastal plain of Israel, evidence of magneto-tellurics, *Bull. Geol. Surv. Israel*, v. 55, 18 p.

Hsü, K. J., 1972, When the Mediterranean dried up, *Sci. Amer.*, v. 227 (6), p. 27–36.

Hsü, K. J., 1973, The origin of Mediterranean evaporites, *Initial Reports of the Deep Sea Drilling Project Nat. Sci. Found.*, 12 (2), Ch. 43, p. 1203–1235.

Krenkel, E., 1924, Der Syrische Bogen, *Zentralbl. Mineral.*, v. 9, p. 274–281; v. 10, p. 301–313.

Lort, J. M., Limond, W. Q., and Gray, F., 1974, Preliminary seismic studies in the eastern Mediterranean, *Earth Planet. Sci. Lett.*, v. 21, p. 355–366.

Neev, D., 1960, A pre-Neogene erosion channel in the southern coastal plain of Israel, *Bull. Geol. Surv. Israel*, v. 25, 20 p.

Neev, D., Edgerton, H. E., Almagor, G., and Bakler, N., 1966, Preliminary results of some continuous seismic profiles in the Mediterranean shelf of Israel, *Israel J. Earth Sci.*, v. 15 (4), p. 170–178.

Neev, D., Bakler, N., Moshkovitz, S., Kaufman, A., Magaritz, M., and Gofna, R., 1973a, Recent faulting along the Mediterranean coast of Israel, *Nature*, v. 245, p. 254–256.

Neev, D., Almagor, G., Arad, A., Ginzburg, A., and Hall, J. K., 1973b, The geology of the southeastern Mediterranean Sea, *Rep. Geol. Surv. Israel*, no. MG/73/5, 43 p.

Nir, Y., 1965, Bottom topography, in: *Submarine Geological Studies in the Continental Shelf and Slope Off the Mediterranean Coast of Israel*, Neev, D., ed., *Rep. Geol. Surv. Israel*, no. QRG/1/65, p. 4–6.

Nir, Y., 1973, Geological history of the recent and subrecent sediments of the Israel Mediterranean shelf and slope, *Rep. Geol. Surv. Israel*, no. MG/73/2, 179 p.

Ryan, W. B. F., 1973, Geodynamic implications of the Messinian salinity crisis, in: *Messinian Events in the Mediterranean*, Drooger, C. W., ed., Amsterdam: North Holland Publ. Co., p. 26–28.

Said, R., 1973, The geological evolution of the river Nile, 3 parts, *Int. Conf. NE African and Levantine Pleistocene Prehistory*, Texas, 128 p.

Sass, E., 1957, *Volcanological Phenomena of Mt. Carmel*, M.Sc. Thesis, Hebrew University, Jerusalem, 106 p. (In Hebrew.)

Schattner, I., 1967, Geomorphology of the northern coast of Israel, *Geogr. Ann.*, v. A49, p. 310–320.

Weissbrod, T., 1969, The Paleozoic of Israel and adjacent countries. Part 1. The subsurface stratigraphy of southern Israel, *Bull. Geol. Surv. Israel*, v. 47, p. 1–35.

Woodside, J., and Bowin, C., 1970, Gravity anomalies and inferred crustal structure in the eastern Mediterranean Sea, *Bull. Geol. Soc. Amer.*, v. 81, p. 1107–1125.

Chapter 8

THE GEOLOGY OF THE EGYPTIAN REGION

E. M. El Shazly

Egyptian Atomic Energy Establishment
Academy of Scientific Research and Technology
Cairo, Egypt

I. INTRODUCTION

Egypt covers a roughly square area of almost 1,000,000 km² at the crossroads of Africa and Asia. It is bounded to the north by the Mediterranean Sea and to the east by the Red Sea. It may be divided into seven main geographic parts: (1) the valley and delta of the Nile, (2) El Fayum, (3) the Suez Canal, (4) the Western Desert, (5) the Eastern Desert, (6) the peninsula of Sinai, and (7) the islands in the Red Sea (Ball, 1939). To these may be added the marine territorial waters in the Mediterranean Sea, Red Sea, Gulf of Suez, and Gulf of Aqaba.

The oldest known rocks are geosynclinal sediments, outcropping at the Abu Swayel area in the southern part of the Eastern Desert (El Shazly *et al.*, 1973), which gave isotopic ages ranging from 1150 to 1300 m.y. The Precambrian in Egypt is considered as belonging to Precambrian I or the Upper Proterozoic (El Shazly, 1964), contrary to previously held opinion assigning to it ages as old as Archean (Hume, 1934, 1935, 1937). The igneous and metamorphic, generally Precambrian, Basement is the northeastern extension of the African Shield, and is separated from the Arabian Shield by the Red Sea.

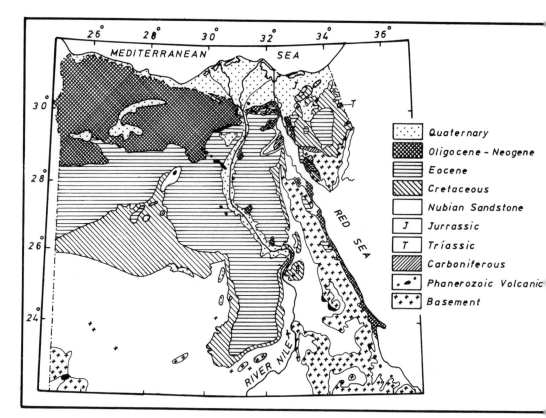

Fig. 1. Sketch map showing major geologic units of Egypt (Survey Dept., Cairo, 1951).

The sedimentary cover embraces the Phanerozoic, with all geologic periods represented. On the simplified geologic map of Egypt (Fig. 1), it can be seen that the Basement occupies about 10% of the surface area, and the Phanerozoic sediments and volcanics about 90%. However, most of the older sedimentary rocks, especially those belonging to the Paleozoic, are hidden in relatively deep basins under younger rocks. Subsurface exploration has added considerably to our knowledge of the geologic history of Egypt. Phanerozoic volcanics are well distributed, both in space and time, invading the sedimentary cover as well as the Basement. The volcanics are associated with crustal disturbances, but Egypt has behaved as a craton throughout the Phanerozoic.

For the purpose of the present work, the geology of Egypt is treated under the following four main headings; the Basement, the Paleozoic, the Mesozoic, and the Cenozoic. Greater emphasis is laid on the sedimentary cover in general, and the way in which its geologic history is related to the Mediterranean Sea and its forerunner, the Tethys.

II. THE BASEMENT

Surface exposure of the Basement in Egypt is governed by tectonism. Outcrops occur in southern Sinai, in the eastern part of the Eastern Desert south of latitude 29°N, in the Nile Valley at Aswan, again south of latitude 24°N, in the southeastern part of the Western Desert from Gebel Abu Bayan to south of El Kharga Oasis, and in the environs of Gebel El Oweinat in the southwestern corner of the Western Desert at the Egyptian–Libyan–Sudanese border.

About 90% of the Basement is hidden under the Phanerozoic sedimentary cover (Fig. 1). El Gezeery *et al.* (1975) give Basement structural contours for northern Egypt, especially as they are revealed from deep wells (e.g., Fig. 2). Due to limited control, this figure shows only general features of the Basement configuration, although geophysical measurements have also been utilized where available. There is a general increase in the Basement depth northward toward the Mediterranean Sea. The Gulf of Suez is a conspicuous Basement low between the structurally high Sinai and Eastern Desert regions, and to the northwestward is separated from the Nile Delta low by the Agrud high located to the southwest of the Bitter Lakes. Another very important Basement low extends NNW–SSE from the northeastern part of the Western Desert toward the Nile Valley in Upper Egypt. This trend incorporates several significant lows,

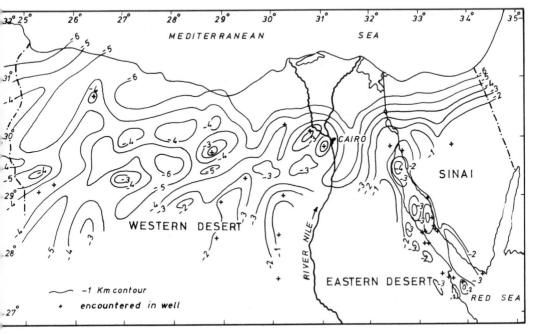

Fig. 2. Northern Egypt depth map of top of Basement (El Gezeery *et al.*, 1975).

e.g., Abu Gharadiq, and highs, e.g., El Nashfa. In the western part of the Western Desert the Siwa Oasis high is notable.

The major rock units in the Basement are Precambrian to earliest Paleozoic, and consist of sedimentary, plutonic, and volcanic sequences, which are briefly outlined in the following paragraphs.

A. First Basement Sediments or Geosynclinal Sediments

These rocks are the oldest known in Egypt, and are thick flysch sediments deposited in eugeosynclines in association with thick volcanics. Both the sediments and volcanics have been regionally metamorphosed, folded, and locally migmatized, and are widely distributed in the Basement. Stratigraphically, they are divided into sediments deposited before the development of the First Basement Plutonites and thinner sediments deposited after these plutonites and containing pebbles derived from them. The earliest sediments, the Wadi Abu Swayel Group, occur in the southern part of the Eastern Desert. This group includes the 5-km calcareous and calcpelitic Wadi Haimur Formation representing the deeper water facies of deposition, the 1–3-km pelitic Wadi Mereikha Formation, and the 1-km psammitic and psammopelitic Wadi Nagib Formation deposited in a shallower water environment (El Shazly et al., 1965, 1975e). Primary structures such as bedding, graded bedding, and occasional wave ripple marks indicate deposition of some of the Wadi Abu Swayel sediments in a shallow marine environment. Migmatized successions occur at several Egyptian localities, including Wadi Hafafit in the southern Eastern Desert (El Ramly and El Akaad, 1960; Basta and Zaki, 1961; El Shazly and Hassan, 1972) and Wadi Feiran in southern Sinai (Schurmann, 1966; El Akaad et al., 1967). Early workers considered these rocks to be "Fundamental gneisses" (Hume, 1934, 1935, 1937). The youngest First Basement Sediments are represented by the 2.7-km Wadi El Muweilih Formation in the central Eastern Desert, which is mainly constituted of conglomerate with tectonized pebbles, including those from the First Basement Plutonites (El Shazly et al., 1971b). This conglomerate has also been encountered in the vicinity of Gebel Atud, about 100 km to the southeast (Amin et al., 1953; El Shazly and Hamada, 1954), and appears to be a shallow-water deposit, possibly formed near the margin of the geosyncline. The Rb/Sr age for metamorphism of schists in the Abu Swayel Group is ±1195 m.y. (El Shazly et al., 1973).

B. First Basement Volcanics or Main Geosynclinal Volcanics

These rocks are generally thick, sheetlike bodies interbedded with the First Basement Sediments, although relations are sometimes cross cutting. Compositionally, the volcanics range from ultrabasic dunites through basic basalts and intermediate andesites, to acidic dacites or even rhyolites, and may be island arc volcanics. The field relations of the ultrabasic rocks are controversial. They have usually been considered as intrusions (Hume, 1934, 1935, 1937), but Rittmann (1958) has advocated the idea that they are geosynclinal submarine lava flows and El Shazly (1964) classified them with the Main Geosynclinal Volcanics.

The First Basement Volcanics are conveniently divided into two associations (El Shazly, 1964): an ultrabasic–basic association in the "Barramiya Group" which attains a thickness of a few kilometers, and the intermediate–acidic "Sheikh Shadli Group" which reaches a thickness of 6 km at the type locality (Mansour, M. S., personal communication). An Rb/Sr isochron based on five basic volcanic samples and one intermediate–acidic sample from the Abu Swayel area gives an age of $\pm 856?$ m.y. (El Shazly *et al.*, 1973). This age is younger than that obtained for synorogenic plutonites occurring with these volcanics. The disparity in the analyses has been attributed to loss of radiogenic Sr by weathering or low Rb/Sr ratios. A galena sample from the Um Samiuki polymetalliferous deposit in the First Basement Volcanics has been given a Pb/Pb age of about 1000 m.y. (Wampler, J. M., personal communication).

C. First Basement Plutonites or Early Orogenic Plutonites

These are infrequent medium-sized elongated plutonic bodies. The composition of these rocks is variable but they are typically represented by gneissic plagioclase granite with bluish quartz grains. The type locality is at Wadi Shait in the southern part of the Eastern Desert (Hume, 1934, 1935, 1937; Schurmann, 1966; Moustafa and El Akaad, 1962), but others occur at Wadi Nagib further south (El Shazly *et al.*, 1975e). An Rb/Sr isochron age of $\pm 876?$ m.y. has been reported for the plagioclase granite of Wadi Shait (Hashad *et al.*, 1972), while an Rb/Sr isotopic age of $\pm 890?$ m.y. has been assigned to a plagioclase granite from Wadi Nagib (El Shazly *et al.*, 1973). Both results should be considered tentative in view of the unfavorable Rb/Sr ratio of these granites.

D. Second Basement Plutonites or Synorogenic Plutonites

Hume (1934, 1935, 1937) distinguished these plutonites by their gray color, but their complex nature was not realized until systematic mapping was undertaken in the central Eastern Desert (Amin *et al.*, 1954; Moustafa and Abdallah, 1954; Mansour and Bassyuni, 1954; El Ramly and Al Far, 1955; etc.). El Shazly and Hamada (1954) applied the term "Granodiorite–Diorite–Grey Granite Complex" to the outcrop of these plutonites in Wadi Garf, where they form great oval-shaped batholiths usually possessing a transitional contact zone with enclosing metamorphosed sediments and volcanics. Moreover, they show gneissic texture and the presence of abundant xenoliths pointing to their anatectic nature. Their chemical composition is comparable to the average composition of the earth's crust. The Second Basement Plutonites are widely distributed in the Basement of Sinai, the Eastern Desert, and the southwestern corner of the Western Desert (El Ramly, 1972; El Shazly et al., 1974a, 1975h). An Rb/Sr isochron age of ±980? m.y. has been obtained for quartz diorite samples from the Um Kroosh plutonic body in the southern part of the Eastern Desert (El Shazly et al., 1973). White granites in the vicinity of Gebel Darhib and Wadi El Gemal in the southern Eastern Desert are differentiated phases of the Second Basement Plutonites.

E. Second Basement Volcanics or Emerging Geosynclinal Volcanics

The type exposure of these rocks ("the Imperial Porphyry") at Gebel El Dokhan in the northern Eastern Desert has been exploited as an ornamental stone since Roman times (Barthoux, 1922; Hume, 1934, 1935, 1937; Andrew, 1938; Schurmann, 1966; Ghobrial and Lotfi, 1967). Compositionally, these volcanics range from intermediate andesites to acidic rhyolites and alkaline trachytes. These rocks have been encountered as far south as Gebel Abu Swayel and vicinity in the southern Eastern Desert (El Shazly et al., 1965); however, they do not form large exposures nor are they as common as the First Basement Volcanics. El Shazly (1964, 1966a) postulated that these volcanics formed along fracture lines developed during emergence from the geosynclinal stage. The Second Basement Volcanics have not been subjected to regional metamorphism, although contact metamorphism by later granites may have influenced them. Rb/Sr isotopic ages for two rhyolite (graphic microgranite) samples from the Gebel Abu Swayel area are ±654 and ±665 m.y. (El Shazly et al., 1973).

F. Second Basement Sediments or Postgeosynclinal Sediments

These sediments are well known in Wadi El Hammamat in the central Eastern Desert near latitude 26°N, where an ornamental stone termed "Breccia Verde Antico" has been exploited since ancient times. Other exposures of these rocks are known in the Basement of the Eastern Desert and Sinai (Hume, 1934, 1935, 1937; Schurmann, 1966; El Akaad and Noweir, 1969; El Shazly et al., 1975f). The Second Basement Sediments were deposited in intracratonic basins caused by the initiation of mountain building. They are preserved only in down-faulted blocks or in topographic lows. Stratigraphically these sediments, the "Wadi El Hammamat Supergroup," are divided into an earlier Wadi Kareem Group (El Shazly and Ghanem, 1975) deposited before erosional exposure of the Late Orogenic Granites, and a later Wadi El Mahdaf Formation containing pebbles of these granites. Sediments of the Wadi Kareem Group are more than 5 km thick at the type locality, where they consist of a 983-m-thick tectonized basal conglomerate (Wadi Melgi Formation), followed unconformably upward by 2186-m-thick graywackes with mudstones and siltstones (Wadi El Tarfawy Formation), overlain by 2700 m of conglomerates (Wadi Abu Ghadir Formation). The topmost 2700-m-thick Wadi El Mahdaf Formation is made up of conglomerates containing, among others, abundant granite pebbles considered to have been derived from the earliest types of Late Orogenic Granites (El Shazly *et al.*, 1971*b*). The Second Basement Sediments show molasse characteristics, and current ripple marks have been noted on the fine-grained detrital rocks.

G. Third Basement Plutonites or Late Orogenic Plutonites

Pink granites form large batholiths, stocks, etc., in the Basement in Sinai, the Eastern Desert, and the southwestern corner of the Western Desert (Ball, 1912; Barthoux, 1922; Hume 1934, 1935, 1937; Shukri and Lotfi, 1954; Amin, 1955; Sabet, 1961; Schurmann, 1966; El Ramly, 1972; El Shazly *et al.*, 1974*a*, 1975*h*). These bodies are normally discordant with relatively sharp contact zones; however, sometimes they are concordant or have gradational contacts. These granites are calcalkalic to alkalic in composition, while in texture they are generally equigranular. The porphyritic, pink granites seem to be the oldest and most tectonized among the granites discussed, but they are not the most common. Considerable isotopic age data are available for the Third Basement Plutonites, and they have yielded a wide range of ages (± 656? to ± 480? m.y.) (Gheith *et al.*, 1959; Afanassiev, 1960; El Ramly, 1963; Schurmann, 1964, 1966; Hashad *et al.*, 1972; El Shazly *et al.*, 1973). The younger ages are contradictory to field data in the southern Sinai because the pink granites are unconformably overlain by Early Cambrian sandstones. There is no doubt, however, that there have been successive intrusions due to differentiation or other phenomena.

H. Fourth Basement Plutonites or Postorogenic Plutonites

These rocks are mainly represented by the granites of Aswan and other less common rocks, including the monumental granodiorites. These granites are alkaline, porphyritic rocks with rapakivi texture. They contain abundant, oriented xenoliths and their contacts are gradational, suggesting a metasomatic origin (Ball, 1907; Rittmann, 1953; El Shazly, 1954; Higazy and Wasfy, 1956; Gindy, 1956). These plutonites occupy an area about 400 km wide and extend NNW–SSE in surface exposures and subsurface (e.g., Nashfa Well No. 1 and Umbarka Well No. 1) in the northern Western Desert. A whole-rock Rb/Sr

isochron age for the porphyritic granites at Aswan is ±590? m.y. according
to Hashad *et al.* (1972); however, Schurmann (1966) obtained K/Ar ages of
±570 m.y. and ±470 m.y. for biotite and potash feldspar, respectively. A
K/Ar age for porphyritic granite obtained from the Basement in the Nashfa
Well No. 1 is 585 ± 20 m.y. (Pan American U.A.R. Oil Company, personal
communication).

I. Major Precambrian Tectonics

Investigations of the Basement rocks in Egypt show a major Precambrian
orogeny ±1000 m.y. ago, which corresponds to the Karagwe–Ankolean cycle
of Holmes and Cahen (1959) in Africa. The termination of the Precambrian
tectonics in Egypt ±600 m.y. ago corresponds roughly to the end of the
Precambrian itself, and is equivalent to the Katanga cycle in Africa (Holmes
and Cahen, 1959). Finally, the transition between the Precambrian and the

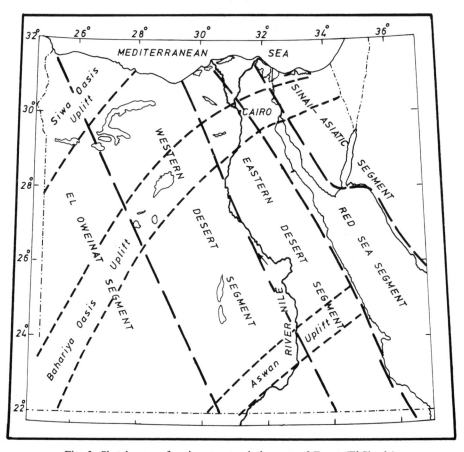

Fig. 3. Sketch map of major structural elements of Egypt (El Shazly).

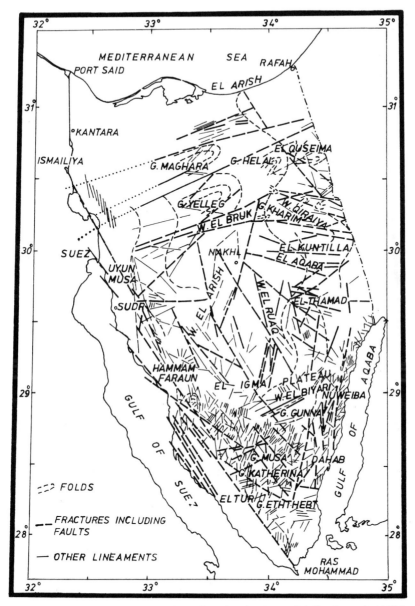

Fig. 4. Structural lineation map of Sinai Peninsula, Egypt (El Shazly *et al.*, 1974*a*).

Early Phanerozoic is characterized by Aswan Plutonism ± 580 m.y., corresponding to the Eocambrian "Miami Plutonism" in Africa (Holmes and Cahen, 1959). El Shazly (1966*a*, 1970, 1972) has suggested the development of rigid crustal plates or segments in Egypt and surrounding countries (Fig. 3) along major NNW–SSE faults. These segments, which may be called old

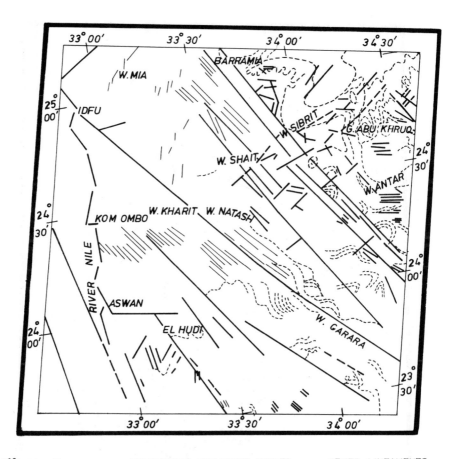

Fig. 5. Structural lineation map of East Aswan area, Egypt (El Shazly *et al.*, 1974*b*).

plates, are shown in Fig. 3, and they are, from east to west: (1) Sinai–Asiatic Segment, (2) Red Sea Segment, (3) Eastern Desert Segment, (4) Western Desert Segment, and (5) Gebel El Oweinat Segment. Segments 1, 3, and 5 act as highs, while Segments 2 and 4 act as lows. The highs are characterized by a thick crust (40 to 50 km) in the Eastern Desert and Gebel El Oweinat Segments, while the lows show a thinner crust, normally 30 to 40 km in the Western Desert (Babaev, 1968) and 20 to 40 km in the Red Sea (Phillips *et al.*, 1969). The highs are noted for the occurrence of orogenic granites, while the lows are characterized by the presence of the alkaline, metasomatic granites of the Fourth Basement Plutonites. In the Red Sea, granite has not been reached, but alkali metasomatism has been demonstrated on St. John's Island at the periphery of the axial trough of the Red Sea (El Shazly and Saleeb, 1972).

Recent structural analyses, especially those using LANDSAT images (Figs. 4 and 5), have led to a partial understanding of the Precambrian fold pattern acting on the Egyptian Basement (El Shazly *et al.*, 1974*b*, 1975*d*, 1975*f*). Early folding of metamorphic rocks along E–W to ESE–WNW trends has been largely destroyed by later movements. The earliest folding susceptible to structural analysis has axes ranging from WNW–ESE to NW–SE and involves rocks from the Precambrian up to the basal formation of the Second Basement Sediments. Later folding acting on the Precambrian rocks up to the porphyritic, pink granites has a NNE–SSW axial trace or sometimes a NE–SW trend, with the force usually exerted from the WNW and directed toward the ESE.

III. THE PALEOZOIC

A. Paleozoic Sediments

The three major Paleozoic exposures shown on the geological map of Egypt (Fig. 1) initially led to the belief that Paleozoic sediments were not common in Egypt and that the younger Nubian Sandstone covered much of the Basement. Drilling in northern Egypt has revealed widespread subsurface Paleozoic sediments. They may be still more widely distributed where sandy facies make them difficult to separate from the Nubian Sandstone.

Paleozoic basins are indicated on maps (Figs. 6–8) prepared by El Gezeery *et al.* (1975) and based on data from some 250 wells and surface outcrops in the northern part of Egypt. The isopach and facies maps show that the Paleozoic basins trend generally NNW–SSE and almost parallel the present Red Sea, which suggests again that NNW–SSE structural lines have been outstanding since the end of Precambrian time (El Shazly, 1966*a*, 1972). Ghorab (1960) and Amin (1961) elucidated important problems regarding the sedimentary basins of northern Egypt, recently classified by El Gezeery *et al.* (1972).

1. *Gulf of Suez Basin*

Two Paleozoic exposures are found at Um Bogma and at Wadi Araba (Fig. 1) on the eastern and western sides of the present Gulf of Suez, respectively. In the subsurface the top of the Paleozoic is encountered at depths ranging from 1 to 3 km in the Gulf of Suez, and becomes increasingly deep northward. The thickness of the Paleozoic sediments increases northward from about 0.5 km to an estimated 2 km toward the present Nile Delta. Sandy facies occur in the south and shaly facies in the north, with the sand/shale ratio passing from 8/1 to 1/8. The Carboniferous exposures of Um Bogma and

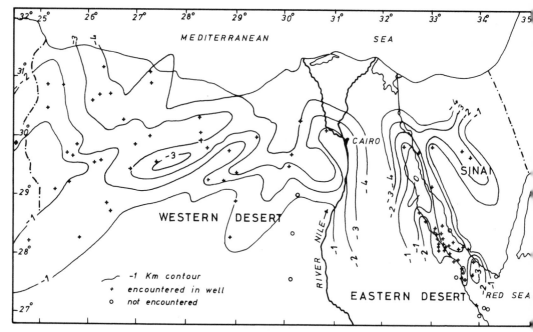

Fig. 6. Northern Egypt depth map of top of Paleozoic (El Gezeery *et al.*, 1975).

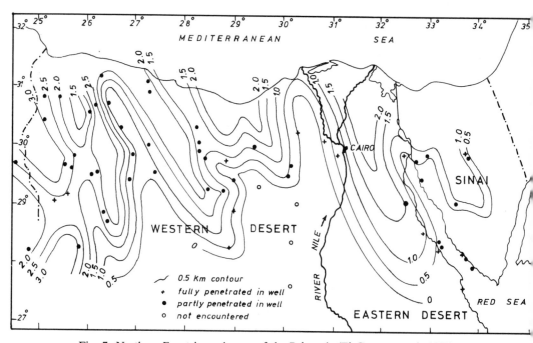

Fig. 7. Northern Egypt isopach map of the Paleozoic (El Gezeery *et al.*, 1975).

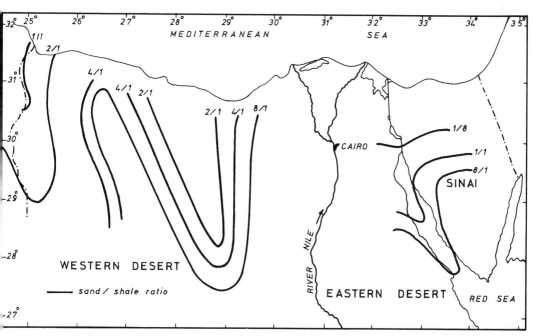

Fig. 8. Northern Egypt facies map of the Paleozoic (El Gezeery *et al.*, 1975).

Wadi Araba are partly in carbonate facies. The Gulf of Suez Paleozoic basin appears to extend toward the Nile Delta; however, drilling has not yet reached Paleozoic sediments there.

2. *Northeastern Western Desert Basin*

This basin extends from south of Bahariya Oasis (not shown on Figs. 6, 7, or 8) toward the present Mediterranean Sea. The top of the Paleozoic sediments is 0.5 km below the surface at Bahariya Oasis and descends northward to an estimated 4 km near the Mediterranean. Paleozoic sediments are 0.5 km thick at Bahariya Oasis and decrease southward, but toward the north they increase to an estimated 2 km near the Mediterranean. Near the basin margins the rocks are sandy with a sand/shale ratio near 8/1, but toward the center the ratio drops to 2/1.

3. *Northwestern Western Desert Basin*

A complex Paleozoic basin exists in the northwestern part of the Western Desert and extends into Libya. The basin is known principally from subsurface data, but Paleozoic outcrops occur near the Egyptian–Libyan–Sudanese borders. North of Siwa Oasis Paleozoic sediments are 2 to 3 km thick. They are sandy near the basin margins but toward the center show a sand/shale ratio reaching 1/1. South of Siwa Oasis the thickness continues in the range of 2 to 3 km.

Contrast in the conditions of Paleozoic deposition can be shown by comparison of the westernmost and easternmost Paleozoic basins. The western basin is well represented in the subsurface of the Siwa Oasis area and a composite log for this area has been prepared by N. A. Khalil (personal communication). The Paleozoic section overlies the igneous–metamorphic Basement unconformably, and is represented by the 207-m-thick Cambro-Ordovician Zeitun Formation, followed upward by the 406-m Silurian Bahrein Formation, which is separated from 1032 m of the Devonian Desouky and Basur Formations by an 11-m-thick basalt sheet. The Devonian is overlain by 1035 m of Carboniferous rocks placed in the Siwa and Kohla Formations, and the whole succession is separated from Mesozoic and Cenozoic sediments by a notable unconformity.

The lithology and oil potential of the Paleozoic sediments at Siwa Oasis area have been considered by El Hashemi (1972) and Vladimirskaja (1973). According to the former, the Cambro-Ordovician is a varicolored arenite with thin shales; the Silurian is constituted of interbedded arenite and thin micaceous shales; the Devonian is a micaceous and glauconitic, sandy shale, followed upward by a thick, partly micaceous sandstone and a thin micaceous shale; the Carboniferous is sandstone with shale followed upward by arenite with thin shales, and higher by sandy shale with limestone near the top; and finally, the Permo-Carboniferous is shale with thin sandstones and a limestone bed near the top. The Paleozoic sandy facies have been divided into three paleo-environmental units, which include a basal arkose, a middle quartz sandstone, and an upper arkose (El Hashemi, 1972). The basal unit extends from the sandy facies of the Cambro-Ordovician to the lower third of the Devonian, where immature sediments indicate high relief, short transportation, and rapid burial. Deposition conditions were oxidizing initially and reducing later. The probable prevalence of an arid climate throughout is indicated by the fresh feldspars, authigenic anhydrite, and dolomite. The quartz sandstone unit occupies the upper two-thirds of the Devonian and the lower half of the Carboniferous. The sediments are mature, pointing to conditions of low relief, long transport, and reworking of preexisting sediments and rocks under reducing conditions. The upper arkose unit, incorporating the rest of the Paleozoic section, suggests more than one source area. The feldspars of this unit are generally decayed, suggesting a humid climate. Energy input is inferred from textural and mineralogical immaturity of the sediments to have been low, and the environment was a reducing one. El Sweify (1975) stressed the prevalence of a shallow marine environment during the deposition of Devonian and Carboniferous sediments in the Siwa–Faghur area, while Andrawis (1972) concluded that the Devonian sediments encountered in Gibb Afia Well No. 2 are offshore marine deposits from turbid water.

By comparison, the Paleozoic sediments in the southwestern corner of the Western Desert at the Egyptian–Libyan–Sudanese borders are more continental, more sandy, less well developed, and contain abundant volcanics (Sandford, 1935; Burollet, 1963; Vittemberga and Cardello, 1963; Conant and Goudarzi, 1967; Klitsch, 1968; Klerkx, 1969; Goudarzi, 1970; El Shazly et al., 1970a, 1975h; Issawi, 1975). There exists the possibility that the Paleozoic sedimentary basin of the Siwa Oasis area is connected with or crosses the basin just described.

Paleozoic sediments have long been recognized at Um Bogma, Sinai, and at Wadi Araba, Eastern Desert, east and west of the Gulf of Suez, respectively. The marine Carboniferous rocks in Sinai contain well-known manganese iron deposits (Walthur, 1890; Barron, 1907; Ball, 1916). The Paleozoic sequence in the environs of Um Bogma lies unconformably on the Basement and contains Early Cambrian sandstones with siltstones and shales, separated by an unconformity from an Early Carboniferous sequence with a massive lower sandstone containing iron and manganese oxides, overlain by a dolomite–limestone section with shale intercalations and with manganese iron ores, which is overlain by pinkish sandstones, carbonaceous shales, coaly seams, and sandstone intercalations, and finally by whitish sandstones (Synelinkov and Kollerov, 1959; Hassan, 1967; Omara, 1967; Weissbrod, 1969). The Early Carboniferous and Early Cretaceous sediments are separated by basalt sheets of Jurassic age.

According to Soliman and El Fetouh (1969), the lower sandstones at Um Bogma were deposited on a peneplaned Basement under transitional, then oscillating transitional, then dominantly transitional conditions, which were followed by a widespread marine transgression leading to the development of the dolomite–limestone sequence with shale intercalations. Regression led to deposition of the upper sandstones under a transitional, fluviatile, and then a deltaic environment in which carbonaceous shales and coal seams were deposited, and finally to fluviatile and sand dune deposits.

The exposed Late Paleozoic succession at Wadi Araba on the eastern side of the Gulf of Suez has been described by Abdallah and El Adindani (1963). The Paleozoic rocks, which are about 350 m thick, are assigned principally to the Late Carboniferous with the Early Permian found occasionally in the higher part of the succession. The probable presence of Permian sediments had been postulated by Kerdany and Abdel Salam (1970) on the basis of palynological analysis of wells drilled in the Gulf of Suez region. The Paleozoic succession exposed at Wadi Araba consists of shales and sandstones followed upward by crinoidal, dolomitic limestones and marls with sandstones, overlain in turn by sandstones and shales with occasional marls and limestones. Underlying black shales reported from wells drilled in the Gulf of Suez region have been assigned to the Early Carboniferous (Babaev, 1968).

Exposures on the two sides of the Gulf of Suez seem to represent a single Paleozoic basin. The facies maps for the Carboniferous sediments here (Kostandi, 1959) seem consistent with the subsurface maps of El Gezeery et al. (1975), and with the outline of the Paleozoic geology of the Gulf by Khalil (1975). For the Carboniferous the maps show increasing shale and carbonate at the expense of sand as the basin center (near the center of the Gulf of Suez) is approached. There is a similar increase of clay and carbonate toward the north. It is clear that the Carboniferous marine transgression came from a NNW direction. The separation of Paleozoic exposures on the two sides of the Gulf of Suez has been visualized by El Shazly and Abdallah (1966) as due to transcurrent faulting leading to horizontal displacement of about 70 km along the Gulf of Suez. The major fault believed to be responsible for this phenomenon has been identified by El Shazly et al. (1974a) on the structural lineation map of Sinai based on LANDSAT-1 satellite images (Fig. 4). This NW–SE-striking fault is from southern Sinai across the Gulf of Suez, with obvious horizontal right-lateral displacement accompanied by chevron folding with a generally north–south trend. This faulting and folding probably started in the Miocene and continued into post-Miocene time.

B. Paleozoic Volcanicity

Volcanism reflects Paleozoic tectonics and is more widespread than previously realized. Cambro-Ordovician and Ordovician ages have been determined for two andesite dikes cutting the Basement at Um Kroosh (El Shazly et al., 1975g) in the southern part of the Eastern Desert, suggesting that the andesites are the oldest dike rocks cutting the Basement. A basalt sheet at the Silurian–Devonian boundary (Khalil, N. A., personal communication) has been encountered in the subsurface successions at Siwa Oasis. Two determinations on altered basalt sheets in Paleozoic sediments at Sharib Well No. 1 (McKenzie, 1971) in the northern Western Desert suggest Silurian–Devonian and Devonian ages.

Numerous age determinations (Table I) suggest volcanic episodes in the Late Carboniferous (around 300 m.y.) and Permian (around 270 m.y. and 245 m.y.). These rocks are mainly trachytes, bostonites, and nepheline syenites, but include some camptonites and olivine basalts. They are widely intruded as ring complexes, dikes, sheets, sills, flows, etc., into the Basement of the Eastern Desert and the sedimentary column of the Western Desert. Considering that most isotopic ages are based on K/Ar and Rb/Sr techniques, the data are remarkably consistent (El Shazly et al., 1975g; El Ramly, 1963; Anonymous).

TABLE I

Isotopic Determinations Giving Paleozoic Ages[a]

Rock type	Locality	Region	Isotopic method[b]	Apparent isotopic age (m.y.)	Reference
Trachyte	Nasb El Qash	C.E.D.	Rb/Sr isochron	245 ± 15	
Trachyte	Um Shaghir	C.E.D.	Rb/Sr isochron	273 ± 15	El Shazly et al. (1975c)
Bostonite	El Atshan	C.E.D.	Rb/Sr	273 ± 20	
Bostonite	G. Um Kibash	S.E.D.	K/Ar	290	El Ramly (1963)
Olivine basalt	Rabat Well No. 1, 4999–5009 ft	N.W.D.	K/Ar	293 ± 12	Anonymous
Camptonite	Hafafit Mine	S.E.D.	K/Ar	300	El Ramly (1963)
Bostonite	W. Kareem Old generation	C.E.D.	Rb/Sr	300 ± 15	
Trachyte	Um El Khors	C.E.D.	Rb/Sr isochron	302 ± 15	El Shazly et al. (1975c)
Nepheline syenite	G. Mishbeh	S.E.D.	Rb/Sr	305 ± 20	
Altered basalt	Sharib Well No. 1, 8183 ft	N.W.D.	K/Ar	347 ± 14	McKenzie (1971)
Altered basalt	Sharib Well No. 1, 8205 ft	N.W.D.	K/Ar	395 ± 16	
Andesite	Um Kroosh	S.E.D.	Rb/Sr	480 ± 25	El Shazly et al. (1975c)
Andesite	Um Kroosh	S.E.D.	Rb/Sr	500 ± 30	

[a] Notation used: G. = Gebel; W. = Wadi; E.D. = Eastern Desert; W.D. = Western Desert; N. = Northern; C. = Central; S. = Southern.
[b] Whole rock, unless otherwise specified.

C. Major Paleozoic Tectonics

There is clear evidence of transgression from the Tethys to the north into Egypt, with the greatest transgression occurring in the Carboniferous. The Paleozoic basins trend generally north-northwest–south-southeast and are largely controlled by the major faults developed at the Late Precambrian–Early Paleozoic transition. The configuration of the basin in the southwestern corner of the Western Desert is not yet well defined.

The facies changes and the Late Carboniferous–Permian volcanicity are reflections of land emergence and diastrophism occurring between the Paleozoic and the Mesozoic. The Paleozoic sediments are more strongly folded than the overlying Mesozoic sediments from which they are separated by a major unconformity, but details of the Paleozoic folding are not yet well known. Unconformities exist within the Paleozoic succession, demonstrating continuing instability.

IV. THE MESOZOIC

A. Triassic Sediments

The best known Triassic succession in Egypt is a Middle Triassic section in an anticlinal structure at Araif El Naga in the northern Sinai, which was discovered by Awad (1946). The succession is 200 m thick and consisted of cross-bedded sandstones with fossil plants at the base, overlain by sandstones and shales, overlain in turn by limestones and marls with Middle Triassic ammonites (the 45-m-thick *Ceratites* beds), while at the top lithographic limestones occur.

Triassic sediments occur in the subsurface of the Sinai and in the Gulf of Suez region. The sections range in thickness from 71 m at Gebel Ataqa to 238 m at Nekhl (Eicher, 1946; Kostandi, 1959; Hassan, 1963). In the south these rocks are in clastic facies with sandstones and shales, but northward they grade into limestones, suggesting transgression from the north.

Recent evidence suggests that the Triassic extends southward into the Gulf of Suez, where it forms a basin. The Wadi Qiseib Formation, probably of Permo-Triassic age, occurs on the eastern flanks of the El Galala El Bahariya Plateau on the western side of the Gulf of Suez, where it reaches a thickness of 80 m (Abdallah *et al.*, 1963). This formation overlies the marine Late Carboniferous sediments and consists of ferruginous sandstones, siltstones, and shales, with minor gypsum and rocks salt, and an upper member of marine limestones and marls. Most of the Permo-Triassic succession is continental, but a Middle Triassic marine transgression may have extended south from the Gulf of Suez to embrace what is now the eastern flank of El Galala El Bahariya Plateau.

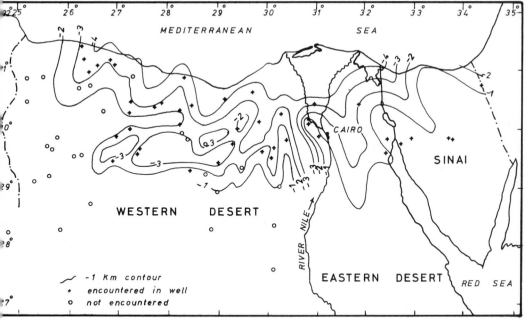

Fig. 9. Northern Egypt depth map of top of Masajid Formation (Jurassic) (El Gezeery *et al.*, 1975).

B. Jurassic Sediments

Marine Jurassic sediments are exposed in northeastern Egypt, and they crop out in northern Sinai, especially in the Maghara structure and on the western side of the Gulf of Suez in the foothills of El Galala El Bahariya Plateau. Jurassic continental rocks have been discovered in the southwestern part of the Western Desert at El Gilf El Kebir Plateau. In the subsurface, Jurassic sediments are well developed in the northern part of the country (Fig. 9) and they have been found as far south as the El Kharga Oasis. Due to the continental nature of Jurassic sediments in southern Egypt, they may have escaped recognition in exposures or in the subsurface.

Jurassic sedimentation is dominated by a widespread invasion of the Tethys, more extensive in the east than in the west, leading to the development of a major basin in the north, with smaller southward-branching basins with N–S to NNW–SSE trends separated by tectonic highs. From east to west the smaller basins are: Sinai, northern Gulf of Suez (extending into the Nile Delta), northeastern Western Desert, and six successive basins further west in the Western Desert.

The Jurassic isopach map of northern Egypt (Fig. 10) shows that the Jurassic sediments increase from about 0.5 km in thickness near latitude

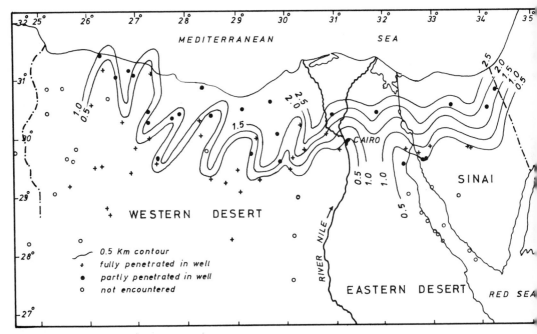

Fig. 10. Northern Egypt isopach map of the Jurassic (El Gezeery *et al.*, 1975).

29°30'N to more than 2 km in the northeast near the Mediterranean Sea (El Gezeery *et al.*, 1975). In western Egypt subsurface Jurassic sediments range in thickness from 0.5 km in the south to 1 km in the north. The facies map (Fig. 11) shows a tendency toward more marine sediments in the northeastern part of the country, with sand/shale ratios changing from 1/1 in the south and in the outer parts of the basins to 1/8 northward and in the basin interior. The clastic ratio changes from 8/1 to 1/4 northward and toward the inner parts of the basins. In western Egypt sand increases relative to shale, and clastics dominate at the expense of carbonates. The sand/shale ratio in this region changes from 8/1 to 1/8 from south to north and from the outer to the inner parts of the basins, while a parallel change in the clastic ratio from 8/1 to 1/1 occurs. It is possible that the dominantly continental Jurassic basins may extend from the northern Western Desert to its southern part in the subsurface. Shale intercalations in sandstones in the subsurface of El Kharga Oasis have yielded Jurassic fossils (Helal, 1966). Continental Jurassic sandstones have been identified by El Shazly *et al.* (1975*h*) at the El Gilf El Kebir Plateau in the southwestern Western Desert, while Issawi (1975) has assigned a Jurassic–Cretaceous age to these sandstones.

The sedimentary succession of Gebel Maghara in northern Sinai best illustrates the north Egyptian Jurassic basins (Moon and Sadek, 1921; Shata, 1951; Farag, 1959*b*; Al Far *et al.*, 1965; Al Far, 1966). The Jurassic section at

Gebel Maghara is as follows (Al Far, 1966). At the base lies the 272-m-thick Shusha Formation, probably of Pliensbachian age, and represented by sandstones and clays sometimes carrying coaly material and deposited in a fluvio-marine environment. The overlying Bir Maghara Formation of Late Liassic to Late Bajocian age consists of 444 m of massive marine limestones and clays, which are in turn followed upward by the 215-m-thick Safa Formation of Bathonian age, which consists of cross-bedded sandstones and clays with coaly seams of fluvio-marine character and intercalated marine shales. At the top of the section is the 576-m-thick El Masajid Formation, incorporating marine limestones of Bathonian–Kimmeridgian age. Middle Jurassic outcrops have been reported in northern Sinai at El Minsherha (Farag and Shata, 1954).

Faulted outcrops on the slopes and foothills of El Galala El Bahariya Plateau (Barthoux and Douvillé, 1913; Arkell, 1956; Farag, 1959a) on the western side of the Gulf of Suez represent the Gulf of Suez Basin. Bathonian strata at Khashm El Galala are about 170 m thick and consist of limestones, marls, and sandstones. Middle Callovian, Oxfordian, and Kimmeridgian sandstones, shales, and limestones up to 30 m thick have been described from Abu Darag and Wadi Qiseib (Abdallah *et al.*, 1963).

In the Western Desert Jurassic sediments are extensively developed toward the north. The continental facies is equivalent to the Permo-Jurassic Eghei Group constituted of shales, siltstones, and sands. The marine facies is repre-

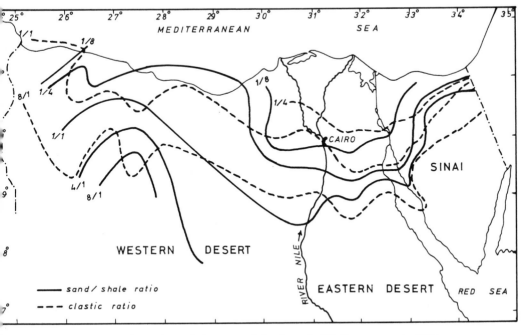

Fig. 11. Northern Egypt facies map of the Jurassic (El Gezeery *et al.*, 1975).

sented by the Wadi El Natrun, El Khatatba, and El Masajid Formations. Transgression was initiated during deposition of the early Middle Jurassic Wadi El Natrun Formation, which consists of lagoonal to marginal marine clastics, carbonates, and evaporites. The Middle Jurassic El Khatatba Formation is represented by shales and sandstones with limestones of shallow marine to estuarine character. A maximum Jurassic transgression represented by the El Masajid Formation is suggested by marine limestones and shales (Aadland and Hassan, 1972; Barakat, 1970; Khalid, 1975). Relations both laterally and vertically between the continental and marine facies are controlled by position in the basins of deposition.

C. Cretaceous Sediments

Following a regression in the latest Jurassic–earliest Cretaceous, a marine invasion from the north led to development of a North Egyptian Basin. Smaller basins formed trending north–south or nearly so, almost perpendicular to the main basin. These small basins from east to west are: east Sinai, west Sinai, east Nile Delta, west Nile Delta, and four basins in the Western Desert. Epicontinental basins, still partly connected northward, are characteristic features of the Early Cretaceous, especially in the Western Desert. The west Kattaniya-1, Abu Gharadiq, Betty-1, and Shoushan-1 are typical.

The Early Cretaceous isopach map (El Gezeery et al., 1975) shows a thickness of 2 km to be normal, while the section thickens to more than 3 km at Mersa Matruh (Fig. 12). Sediments in the subsidiary basins range in thickness from 0.5 to 2 km along a northward traverse in the Western Desert. In the inner epicontinental basins thicknesses of 1.5 km or more occur in the central regions, with thinning outward to less than 0.5 km. Early Cretaceous sandstones and clays continue south of latitude 29°N.

The Early Cretaceous sediments are predominantly clastics. In the northern part of the North Egyptian Basin in the Western Desert, the clastic ratio is 8/1 and the sand/shale ratio is 1/1 (Fig. 13). Southward, the carbonates decrease drastically and the sand/shale ratio reaches 8/1. In the inner epicontinental basins sand/shale ratios range from 1/1 in the north to 8/1 in the south. Continental to shallow marine Cretaceous sediments are represented by the Nubian Sandstone, which is widely distributed in southern and central Egypt and occurs over much of northern Africa and southwestern Asia. The Nubian Sandstone ends with fossiliferous Senonian sediments, including the Qoseir variegated shales and the El Sibaiya phosphate formation (Ghorab, 1956; Youssef, 1957; El Akkad and Dardir, 1966b; El Naggar, 1970). Marine fossils of Cenomanian and Turonian age (Hume, 1911) and of Early Cretaceous age (Attia and Murray, 1952) have been found intercalated in the Nubian Sandstone at Wadi Qena.

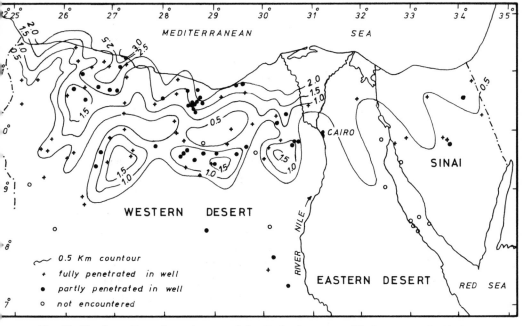

Fig. 12. Northern Egypt isopach map of the Early Cretaceous (El Gezeery *et al.*, 1975).

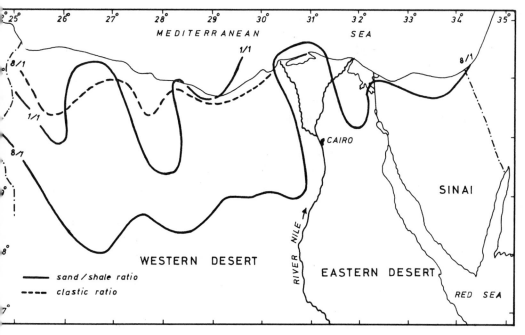

Fig. 13. Northern Egypt facies map of the Early Cretaceous (El Gezeery *et al.*, 1975).

The Nubian Sandstone, especially in southern Egypt, has been described and discussed by numerous authors, including Russegger (1837), Shukri (1945), Attia (1955), and Pomeyrol (1968). East of Aswan it is normally divided into three clastic units reaching 160 m in thickness, with oolitic iron ore deposits in the middle (Attia, 1955). At El Kharga Oasis in the southern Western Desert the subsurface section is about 1573 m thick, consisting of a succession of Nubian Sandstone and sediments of comparable facies; it has been studied by El Shazly and Shata (1960).

On the basis of paleomagnetic and paleogeographic studies, El Shazly and Krs (1972a, 1973) concluded that the Nubian Sandstone was deposited under tropical to subtropical conditions, as the paleoequator at the time passed through central Egypt.

Use of the term "Nubian Sandstone" has tended to obscure the Early Cretaceous facies which occur in Egypt. Fortunately, in recent usage the term is becoming more restricted and accordingly more exact. The Early Cretaceous Malha Formation (Abdallah et al., 1963), exposed on the southeastern flanks of El Galala El Bahariya Plateau, is a 50–100-m-thick continental to shallow marine, snow white, violet, and pinkish sandstone with clay intercalations and kaolin deposits, including marine interbeds identified by H. G. Awad as being of Urgo-Aptian age. The same unit is present in wells drilled in the Gulf of Suez region (Khalil, 1975) and in Sinai. El Shazly et al. (1974a), in their map of the Sinai Peninsula, recognize Early Cretaceous sediments previously termed Nubian Sandstone as belonging to the Malha Formation. Early Cretaceous sediments in the northern Western Desert are essentially sandy, but oil exploration has disclosed the presence of widespread Aptian marine carbonates. These carbonates have been divided into a lower dolomite and limestone unit, and an upper dolomitic unit including the El Alamein Formation, which is the oil reservoir in El Alamein and neighboring oil fields (Abdine and Deibis, 1972; Hamed, 1972). The depth to the top Alamein Dolomite is shown in Fig. 14. According to these authors, transgression from the sea to the north started with the lower carbonate unit and reached a climax during deposition of the upper carbonate unit, after which a significant regression occurred. The Aptian transgression is pronounced in the north and diminishes southward.

Late Cretaceous marine sediments are distributed south to about 23°N (Fig. 1), revealing another major transgression in the north. In the North Egyptian Basin, the Late Cretaceous sediments attain a thickness of 1–1.5 km in the north, and thin to 0.5 km southward (Fig. 15). These rocks are carbonate-rich with a clastic ratio of 1/4 (see Fig. 17) (El Gezeery et al., 1975).

South of the major North Egyptian Basin, subsidiary basins assume two trends, NNW–SSE, represented in particular by the Gulf of Suez Basin, and NE–SW, exemplified by the Western Desert basins.

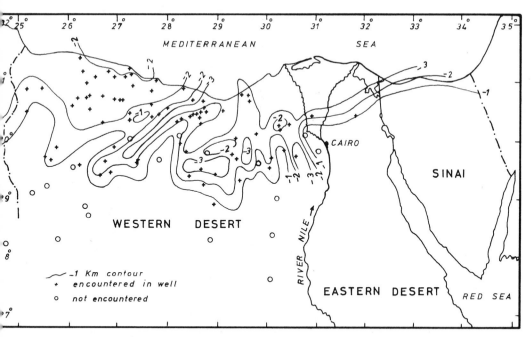

Fig. 14. Northern Egypt depth map of top of Alamein Dolomite (Early Cretaceous) (El Gezeery *et al.*, 1975).

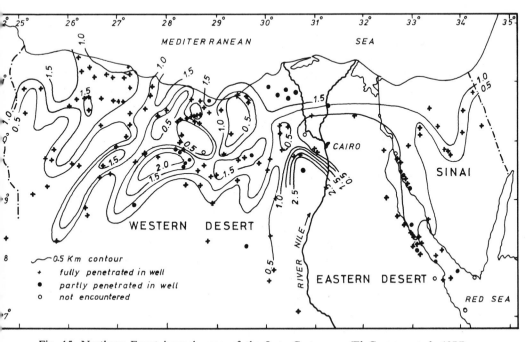

Fig. 15. Northern Egypt isopach map of the Late Cretaceous (El Gezeery *et al.*, 1975).

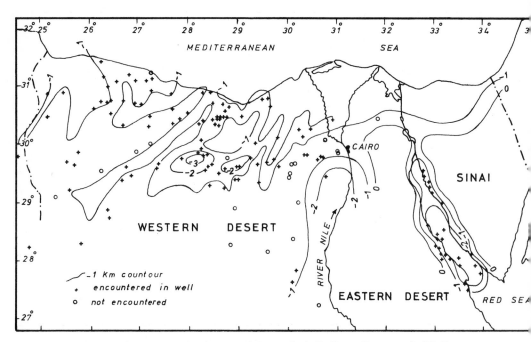

Fig. 16. Northern Egypt depth map of base of chalk (Late Cretaceous) (El Gezeery *et al.*, 1975).

The Gulf of Suez Basin closely parallels the presént Gulf. A structure map on the base of the chalk (Fig. 16) shows a section of more than 2 km at the center of the Gulf and an outward decrease. The average thickness of Late Cretaceous sediments is in the range of 0.5 km. Sandstones are predominant in the south where the clastic ratio is 1/1, but northward limestones increase as the ratio falls to 1/4 (Fig. 17). A Late Cretaceous basin parallels the Eastern Desert high and coincides roughly with the Nile Valley, especially between 29° and 28°N. It is not clear whether the orientation of this basin is dominated by the folding. The more westerly basins include Abu Gharadiq and Shou-shan-1 Basins, which are separated by the intervening Bahariya Oasis and Siwa Oasis highs. The thickness of Late Cretaceous sediments in these basins may exceed 2.5 km in the center, but decreases outward to about 0.5 km. The sand/shale ratio is 1/4 in the center and changes outwards to 1/1, while the clastic ratio of 1/4 in the basin centers changes peripherally to 8/1.

The marine Late Cretaceous sediments in Egypt are represented by three major facies: a phosphatic offshore facies, a basins facies, and a facies found on intervening structural highs. The phosphatic facies occur between 25° to 27°N, extending from El Qoseir–Safaga on the Red Sea to Qena–Idfu in the Nile Valley, and to the El Kharga El Dakhla Oases in the Western Desert (Hume *et al.*, 1920; Ghorab, 1956; Youssef, 1957; El Nakkady, 1958; El Akkad and

Dardir, 1966a; El Naggar, 1970; Issawi et al., 1971; Abd El Razik, 1972). The principal phosphatic facies are Campanian–Maestrichtian in age, but some phosphatics occur in the Senonian. During the Campanian the facies sequence from south to north is as follows: offshore marine and continental sands south of about latitude 24°N; offshore marine sand and clay facies which contain most of the exploitable phosphate deposits; essentially marine clayey facies between latitudes 27° to 29°N where some phosphate deposits are present; and finally, predominantly marine carbonates north of 29°N (Ghanem et al., 1970).

The facies characteristic of the highs is exemplified by the Late Cretaceous sediments of El Bahariya Oasis. From base to top the section consists of the El Bahariya Formation (45 to 303 m), El Heiz Formation (0.5 to 30 m), El Hafhuf Formation (1.6 to 123 m), and the Chalk Formation (17 to 24 m) covered unconformably by the iron-ore-bearing Eocene sediments (Ball and Beadnell, 1903; Faris et al., 1956; El Shazly, 1962; El Akkad and Issawi, 1963; Said and Issawi, 1964; El Bassyouny, 1975). The El Bahariya Formation of Early Cenomanian age is carbonaceous sandstone and shale with bone fragments, and is of fluvio-marine, essentially pro-delta origin. The El Heiz Formation of Late Cenomanian age is made up of dolomitic sandstones and biomicrites deposited in an open marine environment. The El Hafhuf Formation of Campanian age is formed of dolostones with basal and intraformational

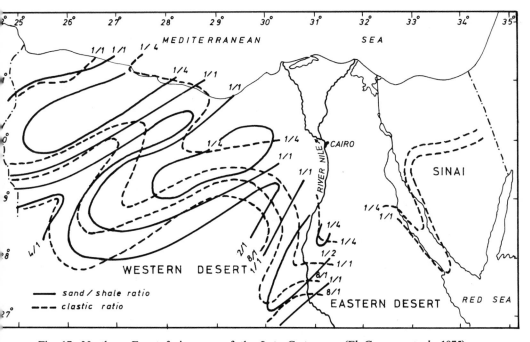

Fig. 17. Northern Egypt facies map of the Late Cretaceous (El Gezeery et al., 1975).

TABLE II

Isotopic Determinations Giving Mesozoic Ages[a]

Rock type	Locality	Region	Isotopic method[b]	Apparent isotopic age (m.y.)[c]	Reference
Hornblende basalt	G. Abu Khruq	S.E.D.	K/Ar	85	El Ramly (1963)
Granosyenite	G. Abu Khruq	S.E.D.	K/Ar	75–80	
Bostonite	W. Kareem New generation	C.E.D.	Rb/Sr	74 ± 2.0	El Shazly et al. (1975c)
Basanite basalt	El Gilf El Kebir	S.W.D.	K/Ar	75 ± 3.2	Greenwood (1969)
Olivine basalt	Darb El Arbain?	S.W.D.	K/Ar	76 ± 2 77.5 ± 3	Meneisy and Kreuzer (1974a)
Olivine basalt	Darb El Arbain?	S.W.D.	K/Ar	77.5 ± 2 78.5 ± 2.5	
Trachyte	G. El Nuhud North	S.E.D.	K/Ar	78	El Ramly (1963)
Quartz syenite	G. Zargaat El Naam	S.E.D.	Rb/Sr	80 ± 10	El Shazly et al. (1975c)
Granite porphyry	Um Shilman	S.E.D.	Rb/Sr	80	El Shazly et al. (1973)
Nepheline syenite	G. El Naga	S.E.D.	K/Ar	84 ± 3* 82 ± 3†	
Nepheline syenite	G. Abu Khruq	S.E.D.	K/Ar (Biotite)	86 ± 1.5* 83 ± 3† 93.5 ± 1.5	Meneisy and Kreuzer (1974b)

Rock type	Locality		Method	Age	Reference
Analcite basanite basalt	W. Um Hokban	S.E.D.	K/Ar	85.6 ± 4.3	Greenwood (1969)
Alkali olivine basalt	Nubia	S.W.D.	K/Ar	88.9 ± 4.4	
Granite	G. Zargaat El Naam	S.E.D.	Rb/Sr	90 ± 10	El Shazly et al. (1975c)
Yellow muscovite granite	Um Shilman	S.E.D.	Rb/Sr	90	
Red muscovite granite	Um Shilman	S.E.D.	Rb/Sr	93	El Shazly et al. (1973)
Quartz syenite	El Mansoury	S.E.D.	Rb/Sr	95 ± 10	El Shazly et al. (1975c)
Alkali olivine basalt	W. Natash	S.E.D.	K/Ar	100	
Olivine basalt	W. Abu Darag	N.E.D.	K/Ar	117 ± 3 / 115 ± 3	Meneisy and Kreuzer (1974a)
Olivine nephelinite	W. Araba	N.E.D.	K/Ar	126 ± 4 / 125 ± 4	
Basalt	Um Bogma	SW Sinai	K/Ar	178	Weissbrod (1969)
Olivine basalt	Kattaniya Well No. 1, 13,291.5–13,293 ft	N.W.D.	K/Ar	191 ± 19	Anonymous
Granosyenite	Bir Um Hebal	S.E.D.	Rb/Sr	200 ± 20	El Shazly et al. (1975c)
Nepheline syenite	G. El Naga	S.E.D.	Rb/Sr	220 ± 20	

[a] Notation used: G. = Gebel; W. = Wadi; E.D. = Eastern Desert; W.D. = Western Desert; N. = Northern; C. = Central; S. = Southern.
[b] Whole rock, unless otherwise specified.
[c] *: Sieve fractions 1.0–0.5 mm; †: Sieve fractions 0.5–0.16 mm.

conglomerates, while the Chalk Formation consists of porous dolostones with calcite veins (Soliman, *et al.*, 1970).

The Late Cretaceous basinal facies are thicker, more typically marine, and richer in carbonates than those of the highs. The best example in the Western Desert is illustrated by the Abu Gharadiq Basin (El Gezeery *et al.*, 1972; Aadland and Hassan, 1972). At this locality the Late Cretaceous environment of deposition was shallow marine in the early Cenomanian (El Bahariya Formation of sandstone, shale, and limestone), followed upward by shallow to open marine conditions in the late Cenomanian–Turonian–early Senonian (Abu Roash Formation made up of limestones, shales, and sandstone), and ends with late Senonian–Maestrichtian shale to clayey limestone in the lower part and chalky limestone in the upper part (Khoman Formation).

D. Mesozoic Volcanicity

Mesozoic volcanic rocks (Table II) are abundant, especially in the Late Cretaceous, and reflect crustal disturbances, especially Late Cretaceous–Early Tertiary Laramide diastrophism. Triassic volcanicity (200 to 220 m.y.) is represented by alkaline ring complexes and volcanic plugs in the southern Eastern Desert (El Shazly *et al.*, 1975g). Alkali olivine basalts (with an apparent age of 190–115 m.y.) occur as sills, sheets, and dikes at the Triassic–Jurassic boundary (Anonymous), and in Jurassic (Weissbrod, 1969) and Early Cretaceous rocks (Meneisy and Kreuzer, 1974a) in the Sinai and the Eastern and Western Deserts. Late Cretaceous volcanics and small intrusions are widespread in the Eastern and Western Deserts, especially in the south, and yield apparent Rb/Sr and K/Ar ages ranging from 100 to 75 m.y. These rocks include small granite bodies and veins, alkaline ring complexes and volcanic plugs, alkali olivine basalt sheets, and the lava flows of Wadi Natash type (Greenwood, 1969; El Ramly, 1963; El Shazly *et al.*, 1973, 1975g; Meneisy and Kreuzer, 1974b). Wadi Natash lavas, ranging from alkaline basalts to trachytes to acidic rhyolites, have been encountered in the southern Eastern Desert interbedded with the basal strata of the Nubian Sandstone (Barthoux, 1922), while ring complexes are notable in the southern Eastern Desert (El Ramly *et al.*, 1971).

E. Major Mesozoic Tectonics

The Mesozoic is notable for invasions of the Tethys into Egypt from the north. These transgressions reached their climax in Middle Triassic, Middle Jurassic, Kimmeridgian, Aptian, and in the Late Cretaceous starting in Cenomanian time. The younger transgressions are more pronounced and cover increasingly wide areas (Awad and Fawzi, 1956). A Mesozoic basin existed in the area of the present Mediterranean Sea. Subsidiary basins extending into the cratons were controlled initially by major fault lines. However, in the Late Cretaceous the basins and intervening highs were controlled to a great extent by Laramide diastrophism with folding along an axial trace trending northeast–southwest. The Late Cretaceous–Early Tertiary diastrophism produced major unconformities between Mesozoic and Tertiary sediments and has been the subject of considerable discussion (Hume, 1911, 1962, 1965; Faris, 1947; El Nakkady, 1958; Farag, 1959a; Said, 1962; and others).

The most obvious folds caused by the Laramide diastrophism are the "Syrian arcs," which have been noted and described in northern Egypt, especially in the northern Sinai and the northern Western Desert (Beadnell, 1902; Blankenhorn, 1921; Shata, 1953, 1956; Knetsch, 1958; Sigaev, 1967; Youssef, 1968). These folds extend across the central Sinai (El Shazly et al., 1974a) and the Suez Canal Zone (El Shazly et al., 1975a), nearly as far south as the Egyptian–Sudanese border, and affect both the sedimentary cover and the Basement (El Shazly et al., 1974b, c, 1975d, f). Open, plunging folds, with their axial trace striking NE–SW to ENE–WSW in response to forces from the NNW–SSE, are well displayed. Another, less clear, folding trend occurs in which axial traces strike NW–SE to NNW–SSE. The latter is believed to have been generated by vertical forces associated with block faulting, and it seems to be somewhat older.

V. THE CENOZOIC

A. Paleocene–Eocene Sediments

Eocene–Paleocene outcrops are widespread from about latitude 30°N to about 22°N near the Egyptian–Sudanese border. Smaller exposures are present in northeastern Egypt between latitudes 30°N and 31°N (Fig. 1). Subsurface maps of the Eocene–Paleocene sediments (Figs. 18–20) by El Gezeery et al. (1975) show sections near the present Mediterranean Sea ranging in thickness from 0.1 to 0.5 km, while to the south thicknesses reach 2 km. Near the Mediterranean Sea the facies is dominantly clastic, whereas carbonates dominate in the south. The thickness distribution suggests that northern Egypt near the Mediterranean Sea was structurally high, and it seems probable that the southern Mediterranean area may have supported a landmass or very shallow sea with the center of the main basin moving progressively southward. In contrast to Late Cretaceous basins, the Eocene–Paleocene basins seem to have been controlled mainly by faults trending NNW–SSE (Gulf of Suez Basin) to north–south (Nile Valley Basin), but relics of NE–SW Syrian arcs are still evident, especially in subsidiary basins to the west of the Nile Delta.

Unlike the shallow basins developed in the Sinai (maximum thickness about 0.2 km), the Gulf of Suez Basin is conspicuous with the sequence 0.2

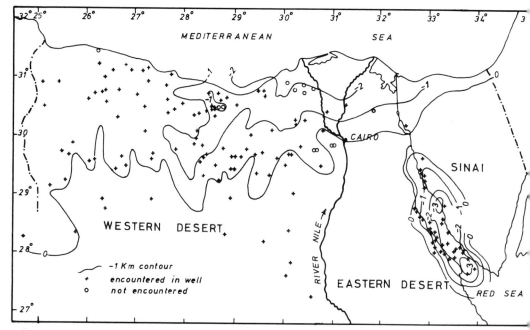

Fig. 18. Northern Egypt depth map of top of Middle Eocene (El Gezeery *et al.*, 1975).

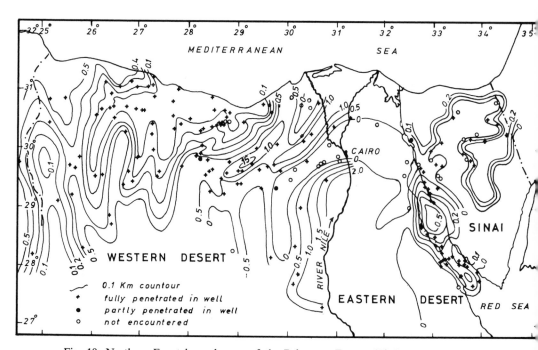

Fig. 19. Northern Egypt isopach map of the Paleocene–Eocene (El Gezeery *et al.*, 1975).

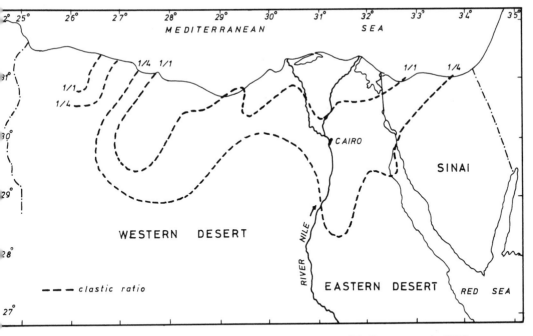

Fig. 20. Northern Egypt facies map of the Paleocene–Eocene (El Gezeery *et al.*, 1975).

to 0.5 km thick in the center of the basin. Toward the south, east, and west the section disappears, but northward it continues into the western Sinai. The clastic ratio in the Gulf of Suez Basin is 1/4 in the south, with clastics increasing northward. The most important Eocene–Paleocene basin is related to the Nile Valley, especially from the latitude of Cairo southward. Sediment thicknesses in the basin center reach about 2 km, but decrease peripherally. In the center of the basin carbonates dominate, but the clastic ratio of 1/4 changes northward to 1/1 (Fig. 20).

The Eocene–Paleocene sediments cover large areas in the Western Desert, but are thinner. In the Abu Gharadiq Basin the maximum thickness in the basin center is about 1 km, while farther west in the center of the NNW–SSE-trending basin passing by Zeitun Well No. 1 in the Siwa Oasis area the maximum is only 0.5 km. The facies in the Western Desert basins are more calcareous southward, with a clastic ratio of 1/4 as compared to the northern strip near Mediterranean Sea where the clastic ratio is 1/1. Eocene sediments in El Bahariya Oasis are characterized by their ferrugination and the development of economically important iron ore bodies.

The Paleocene sediments are generally well represented in Egypt. They are typically developed in the Nile Valley at Gebel Oweina, where El Naggar (1970) described a sedimentary succession unconformable on Maestrichtian shales and marls and overlain by Early Eocene, which is divided into three members: a lower member (37 m) of light-gray conglomeratic shale, chalky on top, passing upward into a middle chalk member (23 m), followed upward by an upper (60 m) light-gray shale members with thin gypsum–anhydrite bands.

The Paleocene succession in the vicinity of Kurkur Oasis in the southeastern part of the Western Desert attains a maximum thickness of 123.5 m (Issawi, 1968, 1972), and it is constituted of alternating chalk and limestone with subsidiary marl, shale, and sandstone, suggesting deposition under shallow marine to littoral conditions alternating with a deeper marine environment. Shallower conditions predominate southward where the Paleocene–Eocene basin terminated. Paleocene sediments are exposed in the Western Desert even beyond Farafra Oasis (Le Roy, 1953).

The Early Eocene is typically divided into lower calcareous shales (56 m) and upper limestones (about 300 m) with flint bands. These rocks are characteristically developed at Luxor and Gebel Serai in the environs of the Nile Valley at about latitude 26°N. Soliman and Korany (1972) interpreted the early Eocene depositional environment in the El Qoseir area, at about 26°N but near the Red Sea coast, as having been shallow marine with comparatively high salinity, which allowed formation of dolomite and the eventual deposition of silica, which has been mobilized to form chert bands and nodules.

The Middle Eocene is developed in central Egypt, the Gulf of Suez, the Sinai, and the Western Desert south of the latitude of Cairo (Cuvillier, 1930, 1953; Ghorab, 1960; Hume, 1962, 1965; Krasheninnikov and Ponikarov, 1964; Farag and Sadek, 1966; Viotti and El Demerdash, 1968; Fahmy et al., 1968; Barakat and Aboul Ela, 1970; Kerdany and Abdel Salam, 1970; Abd El Razik, 1972; Khalil, 1975; etc.).

Ghorab and Ismail (1957) and Ismail and Farag (1957) described the microfacies and depositional environment of Eocene and Pliocene sediments exposed in the vicinity of Helwan south of Cairo. The Middle Eocene at Wadi Hof is 120 m thick, and is constituted of white, hard, poorly fossiliferous, chalky limestone with bands of dolomitic limestone, covered by 80 m of yellowish-white, hard, highly fossiliferous, chalky limestone at the Helwan Observatory. The environment of deposition is interpreted as warm marine with alternating neritic and littoral facies. Ansary (1956) studied the habitat of foraminifera in the Late Eocene sea and concluded that comparatively deep-water initial conditions shoaled toward the top of the section. Comparable conclusions were reached by Ghorab and Ismail (1957) regarding Late Eocene depositional conditions near Helwan, where these rocks attain a thickness of some 174 m and are constituted of chalky, marly, and sandy limestones, marls and sandy marls, and clays and sandy clays. These sediments are considered to have been deposited in a shallow neritic sea merging into shoreline conditions, with periodic access to detrital material brought from land by rivers and intermittent development of clear seas with nummulites. Deltaic and interdeltaic Late Eocene facies have been described by Vondra (1974) at Qasr El Sagha just north of Lake Qarun in the northern Western Desert. These sediments, transported by a river system draining into the Tethys, are some 200 m thick and have been differentiated by the author into four types: an arenaceous, bioclastic, carbonate facies; a gypsiferous and carbonaceous, laminated claystone and siltstone facies; an interbedded, claystone, siltstone, and quartz sandstone facies; and finally, a quartz sandstone facies. These facies represent a southward migrating complex of offshore bars and barrier beaches, back-bar open and restricted lagoons, and a prograding delta with distributary channels. The Late Eocene and Oligocene in the area north of Lake Qarun are noted for the presence of important vertebrate remains.

B. Oligocene Sediments

The Oligocene in Egypt was marked by general uplifting, and accordingly continental sediments are characteristic; however, extensive subsurface marine sediments occur in the northern Western Desert.

In the Gulf of Suez and the Red Sea regions, as well as in the Cairo–Suez district, Oligocene sediments are either continental or entirely missing. On the eastern side of the Gulf of Suez near Abu Zenima, the Oligocene consists of a 49-m-thick basal, red sandstone unit and a 103-m-thick upper, red shale unit (Moon and Sadek, 1923; Ghorab, 1964). In the El Qoseir–Safaga area near the Red Sea coast (26–27°N) the southernmost continental sediments, which are probably of Oligocene age, were deposited in local basins. These sediments are about 149 m thick and consist of conglomerate with sandstone and shale incorporating thin limestone intercalations (El Akkad and Dardir, 1966b; Issawi et al., 1971). The well-known Oligocene continental sediments at Gebel El Ahmar in the vicinity of Cairo and in the Cairo–Suez district are mainly ferruginated sands, sandstones, and quartzites, with cylindrical pipes (Shukri, 1953; Farag and Sadek, 1966).

The depositional environments of Oligocene sediments in the northern Western Desert (Fig. 21) have been studied by Marzouk (1970). Continental conditions prevailed around the latitude of Cairo and to the south. This lithology is interrupted by deltaic deposition (Gebel El Qatrani sediments) west of the River Nile (Beadnell, 1905). El Shazly et al. (1974c) divided the 170-m-thick Gebel El Qatrani sediments, best developed at Widan El Faras (Fig. 22), into lower sandy, middle clayey, and upper sandy members, deposited by a river system originating in what is now the Eastern Desert, and with material coming in the earlier Oligocene from southeast and in the later

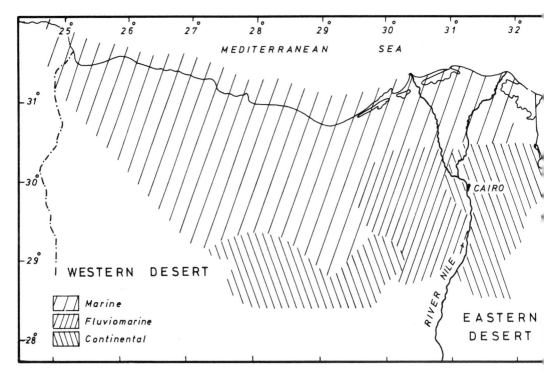

Fig. 21. Environments of deposition of Oligocene sediments in northern Egypt, simplified
(Marzouk, 1970).

Oligocene from the northeast. On the basis of an analysis of the cross-bedding
of Gebel El Qatrani sediments and the observations of Oligocene river channels
in the Eastern Desert southeast of Cairo, the author has concluded that the
Oligocene river system draining into the Tethys should not be considered as a
paleo-Nile but as a separate river system. According to Bowen and Vondra
(1974) the Gebel El Qatrani sediments represent an alluvial complex deposited
in a low-forested deltaic plain by a sinuous to meandering stream.

In the northern Western Desert the deltaic environment gives way to
marine conditions (Fig. 21). The Late Eocene to Late Oligocene marine section
is largely shale and reaches 580 m in thickness (Marzouk, 1970) at the Alam
El Bueib Well No. 1 and about 337 m at the Ghazalat Well No. 1. At the
Qaret Shoushan Hole No. 1, west of longitude 27°30′E, it is overlain by 71 m
of white, milky limestone of latest Oligocene age (Ouda, 1971; Omara and
Ouda, 1972). The latter yields shallow-water, benthonic fossils, indicating a
shoaling of the sea in this part of the Western Desert near the end of the Oligocene.

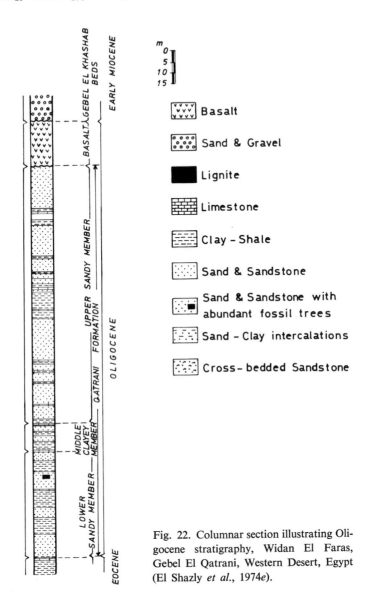

Fig. 22. Columnar section illustrating Oligocene stratigraphy, Widan El Faras, Gebel El Qatrani, Western Desert, Egypt (El Shazly et al., 1974e).

C. Miocene Sediments

The Miocene was a period of great transformation leading to the present-day configuration of Egypt, the Red Sea, the Mediterranean Sea, and the Nile Basin. After Oligocene uplift, an Early and Middle Miocene transgression from the Tethys began, only to be followed by Late Miocene regression related to Alpine orogeny to the north.

Early Miocene sediments are well developed in the Gulf of Suez Basin,

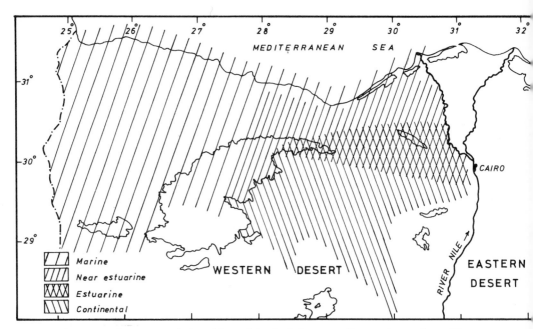

Fig. 23. Environments of deposition of Early Miocene sediments in northern Western
Desert, Egypt (Ouda, 1971).

both in the surface and in the subsurface. These rocks are represented by the
thick, marine Gharardal Group (Moon and Sadek, 1923; Sadek, 1959; Ghorab,
1964; El Gezeery and Marzouk, 1974), representing mainly open marine to
lagoonal shales and marls with limestones and sandstones with anhydrite near
the top, as well as by equivalent nonmarine and coastal facies of reefal limestones and calcareous conglomerates toward the basin periphery. In the Cairo–
Suez district the Early Miocene sediments were deposited in a shallow marine
environment, while in the Sadat area south of Suez, they are represented by
36 m of reefal, white limestone with shale (Abdallah and Abd El Hady, 1966).

Marzouk (1970) and Ouda (1971) studied the subsurface distribution and
depositional environments of Early Miocene sediments in the northern Western
Desert. In the eastern part of the Western Desert, continental conditions
dominate south of 30°N while estuarine facies prevail to the north near Wadi
El Natrun, and farther north near the Mediterranean pass into an intermediate
marine to estuarine environment (Fig. 23). In the northwestern part of the
Western Desert marine facies prevail, which are divided into lower clays and
sandstones and upper shales and limestones (Omara and Ouda, 1972).

The Miocene geology of the northern Western Desert was studied during
the early stages of petroleum exploration by Shata (1955), Ghorab (1960),
Amin (1961), Said (1962), and others. A new geological map gives the distri-

bution and facies patterns of exposed Miocene and younger rocks in this area
(El Shazly et al., 1975b). The Early Miocene is mapped as ferruginous sandstones and sands intercalated with shales and silts, essentially deposited in a
continental environment and exposed in the northeastern part of the Western
Desert as far north as the latitude of Wadi El Natrun.

The Middle Miocene represents the peak of the Miocene transgression,
and its sediments are widespread in Egypt. Middle Miocene sediments, generally
overlying the Basement and less commonly the Cretaceous or younger sediments, are well developed on the coastal strip by the Red Sea. These rocks
have been described by Beadnell (1924), El Shazly and Hassan (1962), El
Akkad and Dardir (1966a), Souaya (1963), and others. El Shazly (1966b)
described a well-developed succession of Middle Miocene and younger sediments at Zug El Bohar, south of El Qoseir, which attain a thickness of 612 m
and consist of a basal, shallow marine limestone passing into sandstone, containing detrital material derived from the Basement, and an upper unit of
anhydrite–gypsum.

In the Gulf of Suez Basin the thick Middle Miocene Ras Malaab Group is
best developed in a subsiding basin, where it is mainly lagoonal but partly
open marine and contains anhydrite and rock salt associated with sands and
shales. Peripheral parts of the basin show reefal to lagoonal conditions (Ghorab,
1964; Heybroek, 1965; Ghorab et al., 1969; Zaghloul and El Sawi, 1971;
Hilmy and Ragab, 1972; El Gezeery and Marzouk, 1974). In the Suez Canal
Zone Middle Miocene sediments consist of sandstones, clays, and limestones
with gypsum (El Shazly et al., 1975a), while in the Cairo–Suez district the rocks
are shallow marine limestones with intercalated sandstones and clays (Shukri
and El Ayouti, 1956; Farag and Sadek, 1966; Abdallah and Abd El Hady,
1966).

The deepest boreholes in the Nile Delta encountered 931 m of the Middle
Miocene Sidi Salem Formation, which consists of shales becoming sandy
southward and enclosing dolomitic limestones. A thick anhydrite body, which
may well be of Late Miocene age, has been encountered on the top of the Sidi
Salem Formation in the northwestern part of the Nile Delta. A marine environment (see Fig. 24) is reported for the Middle Miocene in the Delta (IEOC,
1970; El Sherbini, 1973; El Gezeery and Marzouk, 1974); however, Ouda
(1971) believes that the Middle and Late Miocene sediments underlying the
delta are marine in the north, passing to fluviatile in the south and west. In the
northern Western Desert Middle Miocene sediments occur mainly in a near-
shore marine environment, represented by highly fossiliferous clayey and sandy
limestones and the shallow marine limestones of the Jaghbub Formation best
developed among the Middle Miocene sediments; and a shallow, marine to
lagoonal environment, which is considered either terminating the Middle
Miocene or of Late Miocene age (El Shazly et al., 1975b). Detailed paleonto-

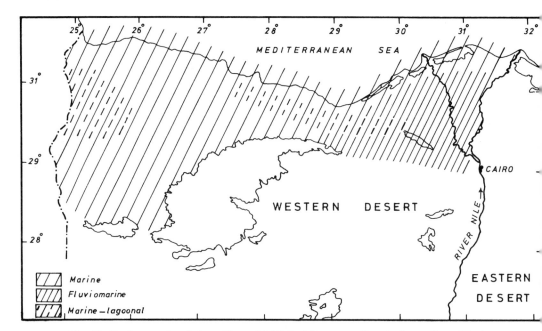

Fig. 24. Environments of deposition of Middle Miocene sediments in northern Western Desert and Nile Delta, Egypt. Some environments probably extend into Late Miocene (Ouda, 1971; El Shazly *et al.*, 1975*b*).

logical investigations are being carried out to resolve this problem. If the second alternative is correct, the latter environment may correspond to the Messinian salinity crisis in the Mediterranean.

The Late Miocene was a period of uplift and folding. Six curved lineaments (El Shazly *et al.*, 1975*b*) between the El Qattara Depression in the northern Western Desert and the Mediterranean littoral represent Late Miocene–Early Pliocene folding with WNW–ESE axial traces. Most of the Late Miocene deposits of northern Egypt are continental; marine late Miocene may exist. Omara and Ouda (1972) consider the Jahbub Formation to range in age from Middle to Upper Miocene, while Desio (1929), Bellini (1969), and Gindy and El Askary (1969) assigned it to the Middle Miocene. At Zug El Bohar on the Red Sea coast, 75 m of sandstones, conglomerates, and limegrits overlying the Middle Miocene anhydrite–gypsum series were considered to be Upper Miocene (Beadnell, 1924), but more recently have been assigned a Pliocene age by Souaya (1963) and others. However, Ross *et al.* (1973) report that the Red Sea is underlain by Late Miocene evaporites, including 53-m-thick layers of gypsum and rock salt correlated with those in the Mediterranean 16 km east of the *Atlantis II* Deep, the latter discussed by Ryan (1973). Late Miocene tectonics in the Tethys and the Red Sea are thought to have formed huge, closed lagoons in which evaporites were deposited and then covered by localized continental and lagoonal deposits (Benson and Ruggieri, 1974; etc.).

D. Pliocene Sediments

The Early Pliocene was a period of uplift which continued Late Miocene movements. As one result, a "Proto-Nile" developed in the Nile River Basin. Several authors have suggested the existence of older versions of the Nile, but in the author's opinion older river systems are not ancestral to the Nile, although the latter here and there may have used parts of these older river systems.

The Pliocene sea transgressed along the fracture system, which is more or less parallel to the Red Sea tectonic lines along most of its course, and reached as far south as Wadi Halfa on the Egyptian–Sudanese border (Fig. 25). Chumakov (1967, 1968) described three series of Late Pliocene–earliest Pleistocene sediments in the vicinity of Aswan. The oldest series is of Plaisancian age, corresponding to the maximum marine transgression. It consists of clay with sand lenses at −172 to −35 m absolute level, and was deposited in a marine to estuarine environment. The middle series belongs to a regressive phase and deposition occurred in a fresh-water basin. This series is of Astian age and consists of sands and sandy loams and gravel interbedded with clay lenses at

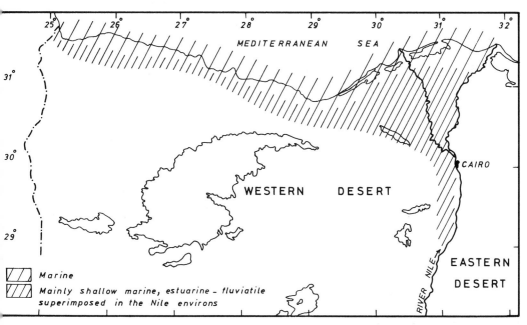

Fig. 25. Environments of deposition of Pliocene sediments in northern Western Desert and Nile Delta, Egypt (Ouda, 1971; El Shazly *et al.*, 1975*b*).

+7 to +116 m absolute level. The upper series of Villafranchian age is made up of alluvial gravels at +160 to +200 m absolute level, pointing to the alluviation of the Nile Valley.

Pliocene sediments have also been described farther north along the Nile Valley and El Fayum Depression (Sandford and Arkell, 1939; Ball, 1939; Arkell and Ucho, 1965; etc.). At Helwan, in the vicinity of Cairo, Ghorab and Ismail (1957) and Ismail and Farag (1957) described 20 m of Late Pliocene Astian impure limestones, sandstones, and conglomerates of shelly, shoreline facies. On the other side of the Nile at Kom El Shellul near the pyramids of Giza, Calabrian (earliest Pleistocene) marine to littoral facies have been recorded. At Wadi El Natrun to the west of the present Nile Delta, Plaisancian–Astian sandstones, clays, and limestones containing vertebrate remains with an estimated age of 7 to 9 m.y. outcrop at Gar El Meluk, where the depositional environment was predominantly fresh water in the lower part of the section and included stream channels, and marine to brackish water in the upper part (Stromer, 1902; Thenius, 1959; James and Slaughter, 1974).

Under the present Nile Delta thick Pliocene strata (Fig. 26), including the Lower Pliocene(?) Abu Madi Formation of sand with shale, were deposited in a shallow marine environment. The Pliocene Kafr El Sheikh Formation, which is essentially marine shale although partly fluvio-marine, is followed upward by the Upper Pliocene(?) El Wastani Formation constituted of estuarine to fluviatile sand and shale (IEOC, 1970; El Sherbini, 1973). At El Qantara in the northern part of the Suez Canal Zone, El Shazly et al. (1975) found Late Pliocene sand and clay, considered as an extension of the Pliocene of the Nile Delta, to be below Quaternary sediments. It may be concluded that the earliest sediments of true Nile origin found in various parts of the Nile Valley and delta and their peripheries are of Late Pliocene origin.

The Pliocene sediments are well represented along the Red Sea coastal strip where they overlie the Miocene. They are typically represented by the Laganum–Clypeaster Series of Beadnell (1924) with an Indo-Pacific fauna indicating the opening of the Red Sea through Bab El Mandab. At Zug El Bohar these sediments attain a thickness of 150 m (El Shazly, 1966b) and contain algae-bearing calcareous sandstone and shelly, algal, sandy biomicrite, indicating deposition in a shallow marine environment not far from land with an abundant supply of detrital material (Ghorab, M. A., and Ismail, M. M., personal communication). A 75-m succession of Upper Miocene or Pliocene age underlies the previously mentioned sediments, and is constituted of microfacies of calcareous sandstone, dolostone, and sandy oobiomicrite, suggesting a depositional environment ranging from marine littoral or neritic zone to deeper marine conditions.

In the Gulf of Suez, Pliocene and younger sediments encountered in the subsurface may reach a thickness of 500 m. The sediments are generally marine

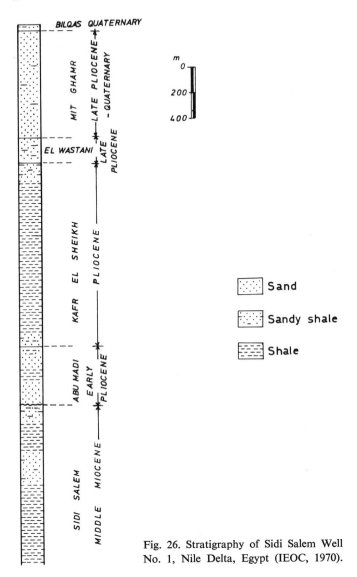

Fig. 26. Stratigraphy of Sidi Salem Well
No. 1, Nile Delta, Egypt (IEOC, 1970).

in the south, and are usually constituted of fluvial or other continental gravels
and sands in the north (Khalil, 1975; etc.). Abdel Salam and El Tablawy (1970)
described three Pliocene diatom zones encountered in the East Bakr and East
Gharib exploratory wells drilled on the western side of the Gulf of Suez. The
conditions of deposition of these zones are given as lagoonal to sublittoral and
suggest an oscillating and unstable environment. These authors maintain that
the top part of the section, usually considered as Middle Miocene, is actually
Pliocene in age. In the Cairo–Suez district, the southern Suez Canal Zone, and
Sinai Peninsula, scattered Pliocene deposits including gravels, sands, and

porcellaneous limestones have been encountered and appear to have been deposited in environments ranging from continental to shallow marine (Shukri and El Ayouti, 1956; Farag and Sadek, 1966; and others).

In the Western Desert, in addition to the exposures at Wadi El Natrun and related ones at Gebel El Hamzi, Pliocene limestone has recently been reported by Shata (1972), who described it as the "pink oolitic limestone," and concluded that its faunal content indicated deposition under shallow and warm marine conditions. The limestone is said to be almost horizontal, either pseudo-oolitic, pseudo-conglomeratic, or detrital and to attain a maximum thickness of 60 m and lie unconformably on Miocene sediments. El Shazly et al. (1975b) mapped this limestone in the northern Western Desert and found it to extend in a zone 5 to 40 km wide from west of the Wadi El Natrun to the Egyptian–Libyan border, reflecting the transgression of the Mediterranean Sea in the northern Western Desert.

The environments of deposition of Pliocene sediments in the northern Western Desert and the Nile Delta, as inferred from surface exposures or from subsurface occurrences, are compiled in Fig. 25. Apart from the previously mentioned Pliocene sediments, the calcareous tufas are of particular interest from the genetic point of view. Tufa deposition began during the Pliocene and continued into the Quaternary. These rocks have been encountered in several depressions and oases in the Western Desert and in the Nile Valley of Upper Egypt (Sandford and Arkell, 1939; Ball, 1939; Caton-Thompson, 1952; Wendorf, 1965; Butzer and Hansen, 1968).

E. Quaternary Sediments

Work on the Quaternary geology of Egypt prior to 1939 has been summarized and commented upon by Ball (1939), especially in relation to the Nile Valley. Ball divided the Quaternary sediments as follows: raised beaches and coral reefs along the coast of the Red Sea; oolitic limestones on the Mediterranean coast; alluvial deposits in the Nile Valley and the delta; lacustrine deposits and Nile mud in El Fayum Depression; alluvial deposits in the drainage channels and depressions of the deserts and on the coastal plains; calcareous tufa in the oases of El Kharga and Kurkur; and dunes and other accumulations of wind-borne sand. The recent surface features of various environments in Egypt have been described by Hume (1925).

Considerable research has been done since 1939 (Sandford and Arkell, 1939; Huzayyin, 1941; Keldani, 1941; Caton-Thompson, 1952; Attia, 1954; Hurst, 1957; El Shazly, 1957; Said, 1957; Shukri, 1960; Fairbridge, 1963; Arkell and Ucho, 1965; Chumakov, 1967, 1968; Wendorf, 1965; Butzer and Hansen, 1968; Shata and El Fayoumy, 1970; Shata, 1971; El Shazly et al., 1970b, 1975b; Hanter, 1975; etc.).

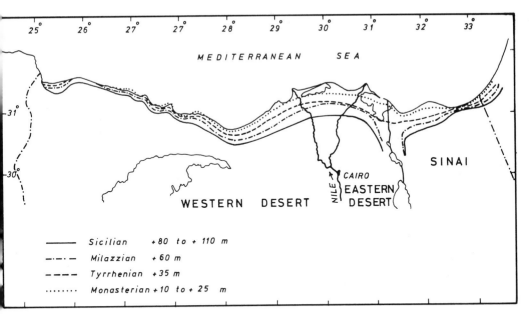

Fig. 27. Pleistocene Mediterranean shorelines, Egypt (compiled by Shata, 1971).

In general, all studies show that there has been a progressive regression of the Mediterranean Sea since it reached its peak during the Pliocene (Fig. 27), though minor transgressions occur in northern Egypt. Deep drilling in the present Nile Delta has contributed greatly to our understanding of the Quaternary geology of this area and is summarized as follows (IEOC, 1970; El Sherbini, 1973): the fluvio-marine Mit Ghamr Formation is Upper Pliocene to Quaternary (Fig. 26), constituted of sand with clay beds, attaining a maximum thickness of 852 m in Sidi Salem Well No. 1, and thinning northward and southward.

The Bilqas Formation is Quaternary and consists of clayey sand, which thickens from 4.4 m at Shibin El Kom in the southern delta to 60.5 m northward in the Rosetta Offshore Well No. 2. El Shazly et al. (1975) studied the Quaternary sedimentary succession at El Qantara in the northern Suez Canal Zone as revealed in shallow boreholes and noted the strong similarity to the Quaternary sediments of the Nile Delta. Recent works on the Quaternary sediments of the Red Sea coastal zone include those of Butzer and Hansen (1968), El Akkad and Dardir (1966a), and El Shazly (1966b). At Zug El Bohar (El Shazly, 1966b) the thickness of the Pleistocene sediments is about 110 m, and they are represented by a shelly, coralline, limegrit facies with sandy, algal biomicrite, indicating deposition under a fairly shallow, reefal, marine environment with access to abundant detrital material from land (Ghorab, M. A., and Ismail, M. M., personal communication).

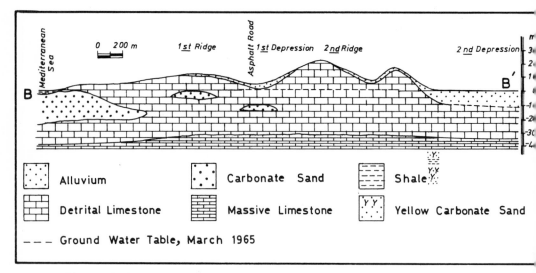

Fig. 28. Geological cross-section B–B' at Sidi Kreir, west of Alexandria; Pleistocene sediments (El Shazly *et al.*, 1970*b*).

The Mediterranean coastal zone is characterized by the presence of Pleistocene carbonate bars normally termed oolitic limestone; however, El Shazly *et al.* (1970*b*) prefer to apply the term detrital limestone (Fig. 28) to suggest conditions of transportation and deposition. The bars, especially those of the Arab Gulf to the west of Alexandria, have been studied by several authors [Zeuner, 1950; Shukri *et al.*, 1956; Butzer, 1960; Shata, 1971 (Fig. 27); etc.]. El Shazly *et al.* (1975*b*) mapped the Pleistocene detrital limestones and associated near-basal gypsum and clay along the coastal zone of the Mediterranean Sea from the northern delta to the Egyptian–Libyan border. There are various opinions regarding the origin of these bars, but recent observations point to the formation of carbonate granules in the Mediterranean Sea near the coast and their transportation inland by the retreat of the sea under general arid, warm conditions. The oolites thus formed in the sea are transported and accumulated on land by the northern wind as dune bars during the very dry periods. In wet periods, however, fine detrital material is brought from the southern tableland by channels draining the land and thus accumulating clay and loam, while in intermediate periods gypsum with clay is accumulated in closed or almost closed basins behind the bars.

F. Cenozoic Volcanicity

The earliest Cenozoic volcanism is of Paleocene age (Table III) and consists of the nepheline syenites from Gebel Abu Ghruq in the southern Eastern Desert (El Ramly, 1963). However, in this case, there is doubt about the age because isotopic determinations on this same ring complex have also yielded older ages (Table II). Isotopic ages corresponding to Late Eocene and the Eocene–Oligocene boundary have been reported for olivine basalts and Hawaiite sheets from the southern Western Desert (Greenwood, 1969; Meneisy and Kreuzer, 1974a).

The most widespread Cenozoic volcanicity in Egypt is represented by basalts of Oligocene to Early Miocene age found as sheets, flows, and dikes in the southern, central, and northern Western Desert, and in the Nile Valley and the Nile Delta (Greenwood, 1969, 1970; Meneisy and Kreuzer, 1974a; El Shazly et al., 1975c). This volcanicity was the first to be recognized in Egypt and it is widely distributed in the northern part of the country, where its paleomagnetism has been studied by El Shazly and Krs (1972b). Oligocene tholeiite basalts have been noted in the Red Sea coastal zone by several authors, including Beadnell (1924) and Sabet (1958), but no isotopic ages are available.

The only definitely known Quaternary volcanics in Egypt are represented by basalt flows and dikes outcropping in St. John's Island (El Shazly et al., 1967) on the periphery of the Red Sea axial trough. These rocks are comparable to basalts encountered on the bottom of the Red Sea (Degens and Ross, 1969; Falcon et al., 1970). The apparent isotopic age of a sample of this basalt as determined by the K/Ar method is ± 1.5 m.y. (Brandt, S., personal communication), which corresponds to the Pliocene–Quaternary boundary.

G. Major Cenozoic Tectonics

During the Cenozoic, Egypt underwent great changes in its relation to the Tethys and the subsequent Mediterranean Sea. The great Paleocene–Eocene transgression was accompanied by southward migration of the axis of deposition of the main sedimentary basin; while northern Egypt became a high, uplift and emergence were conspicuous during the Oligocene, though the tendency was already apparent in the Late Eocene. It resulted in the persistence of marine conditions only in the northern Western Desert. Uplift was accompanied by important fracturing and essentially basic volcanicity through the Late Eocene to Early Miocene. Zittel (1883) considered the Western Desert as an unfolded continental block; however, subsequent mapping in the northern Western Desert has revealed the presence of notable structures (Shata, 1953).

The last major transgression of the Tethys reached its climax in the Middle Miocene and terminated with gypsiferous carbonate deposition of suspected Late Miocene age in the northern Western Desert. Folding and diastrophism during the Late Miocene and Early Pliocene led to major changes, including the transformation of the Tethys into the Mediterranean Sea. Structural analysis of Eocene to Miocene exposures in the Gebel El Qatrani area pointed to the existence of an early N–S major stress axis of Oligocene age, and a younger

TABLE III

Isotopic Determinations Giving Cenozoic Ages[a]

Rock type	Locality	Region	Isotopic method[b]	Apparent isotopic age (m.y.)[c]	Reference
Basalt	St. John's Island	Red Sea	K/Ar	1.5	Brandt, S. (personal communication)
Olivine basalt	Basalt Hill, Bahariya Oasis	N.W.D.	K/Ar	19.5 ± 1.1* 19.6 ± 0.9†	Meneisy and Kreuzer (1974a)
Olivine basalt	SW Mubarak Well No. 1, 12,390–12,430 ft	N.W.D.	K/Ar	19.6 ± 2.0	Greenwood (1970)
Olivine basalt	G. Meyesra, Bahariya Oasis	N.W.D.	K/Ar	19.6 ± 0.8* 20.5 ± 0.7†	
Basalt	Abu Zaabal	Nile Delta	K/Ar	21.7 ± 0.8* 23.2 ± 0.8†	Meneisy and Kreuzer (1974a)
Basalt	Samalut	Near Nile Valley, Upper Egypt	K/Ar	22.9 ± 1.6* 22.5 ± 1.7†	
Basalt	G. Qatrani	N.W.D.	K/Ar	24.7	
Basalt	G. Qatrani	N.W.D.	K/Ar	27 ± 3	El Shazly et al. (1975c)

Rock type	Locality	Region	Method	Age	Reference
Olivine basalt	SW Mubarak Well No. 1, 12,460–12,530 ft	N.W.D.	K/Ar	27.4 ± 2.7	Greenwood (1970)
Basalt	Samalut	Near Nile Valley, Upper Egypt	K/Ar	27.0 ± 2.4* 28.4 ± 1.8†	Meneisy and Kreuzer (1974a)
Basanitic alkali olivine basalt	Near Pottery Hill	S.W.D.	K/Ar	27.7 ± 2.8	Greenwood (1969)
Olivine basalt	SW Mubarak Well No. 1, 12,540–12,600 ft	N.W.D.	K/Ar	31.7 ± 3.2	Greenwood (1970)
Olivine basalt	Darb El Arbain?	S.W.D.	K/Ar	32.4 ± 2.0* 33.9 ± 0.8†	Meneisy and Kreuzer (1974a)
Hawaiite	Oweinat Approach	S.W.D.	K/Ar	37.1 ± 1.9	Greenwood (1969)
Olivine basalt	Darb El Arbain?	S.W.D.	K/Ar	41.0 ± 1.7* 39.9 ± 1.5†	Meneisy and Kreuzer (1974a)
Nepheline syenite	G. Abu Khruq	S.E.D.	K/Ar	40	El Ramly (1963)
Nepheline syenite	G. Abu Khruq	S.E.D.	K/Ar	55	

[a] Notation used: G. = Gebel; E.D. = Eastern Desert; W.D. = Western Desert; N. = Northern; S. = Southern.

[b] Whole rock, unless otherwise specified.

[c] *: Sieve fractions 1.0–0.5 mm; †: Sieve fractions 0.5–0.16 mm.

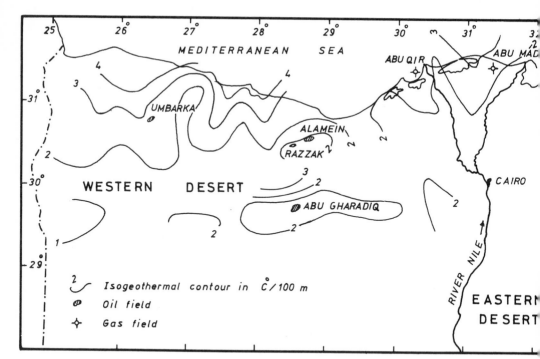

Fig. 29. Geothermal gradient map of northern Western Desert and Nile Delta, Egypt (Khalil and El Demerdash, 1975).

E–W major stress axis (El Shazly *et al.*, 1974*e*). Transgression of the Mediterranean Sea during the Late Pliocene, following the fracture system developed earlier and alluviation by the Nile River from Late Pliocene to Quaternary time, changed this fracture system into the Nile Valley. In Egypt, the fracture system along the Nile has been investigated by Yallouze and Knetsch (1954). The Quaternary has been a period of regression interrupted by minor transgressions. The continuation of Mediterranean tectonics to the present has been inferred from the geothermal gradient map prepared by Khalil and El Demerdash (1975) for the northern Western Desert and the Nile Delta, the detection on infrared thermal imagery of a possible active E–W fault at Gebel El Qatrani area (El Shazly *et al.*, 1974*d*), and by seismicity. The mean geothermal gradient in the northern Western Desert and the Nile Delta has been found to be 2.15°C/100 m (Khalil and El Demerdash, 1975), and the map shows that the gradiet generally increases toward the Mediterranean coast where it ranges from 3 to 4.5°C/100 m (Fig. 29). Farther south there are some local increases, e.g., at Abu Gharadiq Basin and along a major, roughly E–W-trending fault passing through the El Qattara Depression.

Although the Red Sea is believed to have originated in Late Precambrian–Early Paleozoic tectonics (El Shazly, 1966*a*), it acquired its characteristics and independence from the Tethys, due to the Cenozoic tectonics, starting with Late Eocene–Oligocene–Early Miocene uplift and development of a major fault system in the Red Sea region. These faults trend NW–SE and show considerable horizontal displacement along fault planes dipping away from the Red Sea. Oligocene volcanicity accompanied these movements, which are also reflected in the sedimentary column by nondeposition or continental sedimentation.

The Red Sea coastal zone typically shows Middle Miocene to Pliocene strata, which are notably inclined toward the Red Sea and contain numerous diastems. However, the Pliocene sediments are separated from the Quaternary sediments by a conspicuous angular unconformity. The Pliocene–Quaternary contact is also marked by basalts in the Red Sea axial trough and on St. John's Island. The Neogene fault system is well developed in the Red Sea region, usually acquiring a NNW–SSE trend and sometimes showing lateral displacements, with fault planes dipping toward the Red Sea. There is evidence that some of the faults of this system were still active during the Quaternary. Crystals of halite in Middle Miocene evaporites at Salif in Yemen contain oil bubbles (El Shazly, 1967). Furthermore, the carbon isotope ratio of the petroleum in the Gulf of Suez province points to fractionation due to rising temperatures. This suggests that the Red Sea and the Gulf of Suez were thermally active during the Miocene and that evaporite deposition may have occurred without the presence of a lagoonal environment. The geothermal gradient map of the Gulf of Suez, prepared by El Adl and El Naggar (1975), shows some anomalies in the Gulf of Suez (Fig. 30). The mean geothermal gradient is 2.5°C/100 m toward the center. The positive anomalies also coincide with the NW–SE fault lines passing near Sudr Oilfield where thermal waters have been encountered, as well as near El Tor farther south. The metabasalt in St. John's Island, although believed to be of Precambrian age (El Shazly et al., 1967), gave an apparent isotopic age by the K/Ar method of ±12 m.y. This indicates at least notable metasomatic activity in the Red Sea axial trough during the Late Miocene. Girdler and Styles (1974) suggested two sea-floor spreading stages for the Red Sea, one of them occurring about 41 to 34 m.y. and coinciding with Oligocene–Late Eocene tectonism, and the other about 4 to 5 m.y. during the Pliocene.

Alpine tectonism, initiated in the Late Miocene, is notable in Egypt. It is represented by open folds particularly evident in the northern Western Desert, where they show a WNW–ESE trend. Three major fault trends belonging to this tectonic episode are distinguished, namely the NW–SE, the NNE–SSW, and the WNW–ESE sets. The NW–SE faults show right–lateral, horizontal movement as well as vertical displacement. An important fault of this type extends from the southern Sinai to the Gulf of Suez through the Eastern Desert to Wadi El Natrun on the western side of the Nile Delta. In the Sinai this

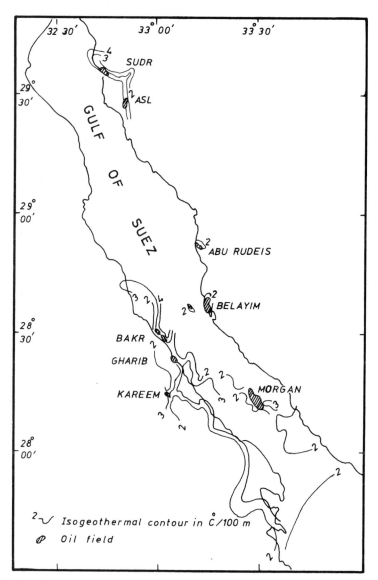

Fig. 30. Geothermal gradient map of Gulf of Suez, Egypt (El Adl and El Naggar, 1975).

fault is accompanied by notable drag folding. The delta collapse is visualized as being due to the combined action of the NW–SE faults which become especially remarkable westward, and the ENE–WSW faults which are particularly notable eastward. The NW–SE faults have also been noted as far as the Egyptian–Libyan border where the Jaghbub Oasis Depression has been moved some 60 km to the northwest relative to the Siwa Oasis Depression.

The NNE–SSW system is well developed in the southern Sinai near the

Gulf of Aqaba, but is still noted as far west as the Egyptian–Libyan border. These are also strike-slip faults, but they belong to the left-lateral type. It is of particular interest that the more or less E–W depressions and uplifts in the northern Western Desert are controlled by the WNW–ESE faults, which possess considerable vertical displacements. The features in question include El Qattara, Siwa Oasis, Maghra Oasis, and Jaghbub Oasis Depressions, as well as El Diffa Plateau (El Shazly *et al.*, 1975*b*).

VI. SUMMARY

The oldest known rocks in Egypt are the Late Precambrian (Upper Proterozoic) igneous and metamorphic basement rocks forming the northern edge of the African Shield. These basement rocks, which may extend into the earliest Phanerozoic, provide evidence of two major sedimentary cycles, two cycles of volcanicity, and four of plutonism. The youngest, postorogenic plutonic cycle contains rocks whose isotopic ages lie close to the Precambrian–Cambrian boundary. The tectonic framework of Egypt has been determined to a considerable extent by the final Precambrian orogeny with the area becoming part of the foreland. As a result, the country can be divided into five major, rigid segments which trend NNW–SSE: the Sinai–Asiatic, Red Sea, Eastern Desert, Western Desert, and Gebel El Oweinat Zones. The second and fourth zones have usually acted as sinks, receiving sediment from the remaining three which acted as highs.

During the Phanerozoic, Egypt was an exorogenic foreland, receiving sedimentation affected by intermittent volcanism. The Paleozoic sediments are exposed at three principal locations, though they are widely spread in subsurface, especially from 27°N to the present Mediterranean Sea. All the divisions of the Paleozoic are represented with volcanicity restricted to the Cambro-Ordovician boundary, Ordovician, Silurian–Devonian boundary, Devonian, Late Carboniferous, and Permian. Major diastrophism accompanied by emergence occurred during the Late Carboniferous and Permian, that is, near the Paleozoic–Mesozoic boundary. Other important movements occurred during the Early Paleozoic, reaching from Cambrian into Devonian time.

The sedimentary basins, the Gulf of Suez Basin which extends toward the present Nile Delta, the northeastern Western Desert Basin, and the northwestern Western Desert Basin have a NNW–SSE trend closely paralleling the Red Sea, and reflect the influence of the Late Precambrian–Early Phanerozoic tectonism on subsequent basin formation. A further basin in the southwestern Western Desert near Gebel El Oweinat is less well known. In general, in the basins the thickness of sediments increases toward the present Mediterranean Sea. All the available evidence points to transgression from a sea, the Tethys, lying to the north. The maximum trangression was during the Carboniferous; the greatest emergence occurred toward the end of the Paleozoic, close to the Mesozoic boundary.

The history of the Mesozoic reflects the effects of tectonic events with transgression and regression of the Tethys. In Egypt the most obvious tectonic event was the formation, during the Laramide orogenic phase, of the Syrian Arcs which trend NE–SW to ENE–WSW. Volcanicity accompanied tectonism and while volcanic rocks of Triassic, Triassic–Jurassic, Jurassic, and Early Cretaceous age are known, they are particularly well developed in the Late Cretaceous. The Tethys transgressed from the north on several occasions in Middle Triassic, Middle Jurassic, Kimmeridgian, Aptian, and Cenomanian times. With the passage of time the transgressions seemed to gain momentum, for while the Middle Triassic basins are restricted to the northern part of the country, during the Late Cretaceous transgressive seas reached as far south as the Sudanese frontier. The later-developed basins also extend farther to the west. The form of the developed basins is controlled by a basement NNW–SSE trend and, in the Late Cretaceous by the ENE–WSW trend of the Syrian Arcs, which dominated the Gulf of Suez and Western Desert Basins in particular.

The Cenozoic witnessed the transformation of the Tethys into the Mediterranean Sea. During the Paleocene–Eocene, deep-sea conditions migrated southward and shallow-water conditions of sedimentation or even a landmass existed to the north in the region which is now the southern Mediterranean.

During the Late Eocene a shallowing of the Tethys became apparent and the Oligocene is marked by emergence. During the period Eocene to Miocene, volcanics developed along the fracture systems associated with these tectonically controlled movements. River systems draining into Tethys near Lake Qarun in the northern Western Desert can be recognized. The headwaters of this Eocene–Oligocene system is in the Eastern Desert; consequently, it cannot be considered ancestral to the Nile. The only well-known marine Oligocene beds are found in the northern Western Desert.

More extensive marine beds are found in the Miocene, with the maximum extent in Middle Miocene time reaching to the Gulf of Suez and beyond the Egyptian border in the Red Sea region. The Red Sea acquired its identity during the Miocene, although the author believes that the region was a zone of weakness or structural depression since the Late Precambrian. Uplift and emergence occurred during Late Miocene time, movements associated with Alpine folding. The presence of Late Miocene sediments in the Gulf of Suez–Red Sea region and in northern Egypt is controversial and requires more detailed investigation. The presence of Late Miocene evaporites in the Red Sea and Mediterranean Sea, and in Middle Miocene and probably Late Miocene rocks in the Gulf of Suez, Red Sea, and northern Egypt has been attributed to deposition in restricted basins.

The Mediterranean Sea acquired, more or less, its present form in Late Miocene time, the Nile Valley assuming its identity as a result of Late Miocene–Early Pliocene tectonics. Late Miocene marine sediments have been traced under the present Nile Delta and at its peripheries, as well as along the Nile Valley to the Egyptian–Sudanese frontier. Late Pliocene Nile sediments indicate the alluviation of the Nile Valley and are the oldest sediments of definite Nile origin found in Egypt. These sediments excluded, transgression from the Mediterranean Sea has been confined to a zone some tens of kilometers wide south of the present shoreline. Pliocene sediments are widely distributed along the Red Sea, and their Indo-Pacific fauna marks the connection of the Red Sea with the Indian Ocean.

The Quaternary has been dominated by regression with only minor transgressions, and excluding the Nile Valley and delta, sediments were formed in a narrow zone bordering the Mediterranean Sea. Uplift and tectonic disturbance marks the Pliocene–Quaternary boundary in the Red Sea region, here indicated by an angular unconformity. Volcanicity occurred in the Red Sea axial trough.

REFERENCES

Aadland, A. J., and Hassan, A. A., 1972, Hydrocarbon potential of the Abu Gharadiq basin in the Western Desert, Arab Republic of Egypt, *Eighth Arab Petrol. Congr.*, Algiers, Paper no. 81 (B-3).

Abdallah, A. M., and Abd El Hady, F. M., 1966, Geology of Sadat area, Gulf of Suez, *J. Geol. U.A.R.*, v. 10 (1), p. 1–24.

Abdallah, A. M., and El Adindani, A., 1963, Stratigraphy of Upper Paleozoic rocks, western side of the Gulf of Suez, *Geol. Surv. Min. Res. Dept.*, Cairo, Paper no. 25.

Abdallah, A. M., El Adindani, A., and Fahmy, N., 1963, Stratigraphy of the Lower Mesozoic rocks, western side of the Gulf of Suez, Egypt, *Geol. Sur. Min. Res. Dept.*, Cairo, Paper no. 27.

Abd El Razik, T. M., 1972, Comparative studies on the Upper Cretaceous–Early Paleogene sediments on the Red Sea Coast, Nile Valley and Western Desert, Egypt, *Eighth Arab Petrol. Congr.*, Algiers, Paper no. 71 (B-3).

Abdel Salam H., and El Tablawy, M., 1970, Pliocene diatom assemblage from East Bakr and East Gharib exploratory wells in the Gulf of Suez, *Seventh Arab Petrol. Congr.*, Kuwait, Paper no. 57 (B-3).

Abdine, A. S., and Deibis, S., 1972, Lower Cretaceous Aptian sediments and their oil prospects in the northern Western Desert, Egypt, *Eighth Arab Petrol. Congr.*, Algiers, Paper no. 74 (B-3).

Afanassiev, G. D., 1960, A few traits in the Magmatism of Egypt, *Izv. Akad. Nauk. USSR, Ser. Geol.*, no. 3, p. 26–40. (In Russian)

Al Far, D. M., 1966, Geology and coal deposits of Gabel El Maghara, northern Sinai, *Geol. Surv. Min. Res. Dept.*, Paper no. 37.

Al Far, D. M., Hagemann, H. W., and Omara, S., 1965, Beiträge zur Geologie des kohleführenden Gebietes von El Maghara, Nord-Sinai, Ägypten, *Geol. Mitt.*, v. 4 (4), p. 397–429.

Amin, M. S., 1955, Geology and mineral resources of Umm Rus Sheet, *Geol. Surv.*, Cairo.

Amin, M. S., 1961, Subsurface features and oil prospects of the Western Desert, Egypt, U.A.R., *Third Arab Petrol. Congr.*, Alexandria.

Amin, M. S., Sabet, A. H., and Mansour, A.O.S., 1953, Geology of Atud district, *Geol. Surv.*, Cairo.

Amin, M. S., Moustafa, G. A., and Zaatout, M. A., 1954, Geology of Abu Diab district, *Geol. Surv.*, Cairo.

Andrawis, S. F., 1972, New biostratigraphic contribution for the upper part of the Paleozoic rocks of Gibb Afia Well No. 2, Western Desert, Egypt, *Eighth Arab Petrol. Congr.*, Algiers, Paper no. 76 (B-3).

Andrew, G., 1938, On the imperial porphyry, *Bull. Inst. Egypte*, v. 20 (1), p. 63–81.

Anonymous, Potassium–argon datings of one core sample each from the Kattaniya No. 1 and Rabat No. 1 Wells drilled by Pan American UAR Oil Company, Robertson Research Company Limited, Project No. 690/182.

Ansary, S. E., 1956, The probable depth of the Upper Eocene sea in Egypt as indicated by Foraminifera, *Bull. Inst. Désert Egypte*, v. 6 (1), p. 193–207.

Arkell, A. J., and Ucho, P. J., 1965, Review of Predynastic development in the Nile Valley, *Curr. Anthropol.*, v. 6, p. 145–166.

Arkell, W. J., 1956, *Jurassic Geology of the World*, London: Oliver and Boyd Ltd.

Attia, M. I., 1954, Deposits in the Nile Valley and the Delta, *Geol. Surv.*, Cairo.

Attia, M. I., 1955, Topography, geology and iron ore deposits of the district east of Aswan, *Geol. Surv.*, Cairo.

Attia, M. J., and Murray, G. W., 1952, Lower Cretaceous ammonites in marine intercalations in the "Nubian Sandstone" of the Eastern Desert of Egypt, *Quart. J. Geol. Soc. London*, v. 107, p. 442–443.

Awad, G. H., 1946, On the occurrence of marine Triassic (Muschelkalk) deposits in Egypt, *Bull. Inst. Egypte*, v. 27, p. 397–429.

Awad, G. H., and Fawzi, I. M. A., 1956, The Cenomanian transgression over Egypt, *Bull. Inst. Désert Egypte*, v. 6 (1), p. 169–184.

Babaev, A. G., 1968, Oil and gas prospects of the U.A.R., Unpublished report, General Petroleum Company, Cairo.

Ball, J., 1907, A description of the First or Aswan Cataract of the Nile, *Surv. Dept.*, Cairo.

Ball, J., 1912, The geography and geology of south-eastern Egypt, *Surv. Dept.*, Cairo.

Ball, J., 1916, The geography and geology of west-central Sinai, *Surv. Dept.*, Cairo.

Ball, J., 1939, Contributions to the geography of Egypt, *Surv. Mines Dept.*, Cairo.

Ball, J., and Beadnell, H.J.L., 1903, Bahariya Oasis, its topography and geology, *Surv. Dept.*, Cairo.

Barakat, M. G., 1970, A stratigraphical review of the Jurassic formations in Egypt and their oil potentialities, *Seventh Arab Petrol. Congr.*, Kuwait, Paper no. 58 (B-3).

Barakat, M. G., and Aboul Ela, N. M., 1970, Microfacies and paleoecology of Middle Eocene and younger sediments in Geneifa area, Cairo–Suez district, *J. Geol. U.A.R.*, v. 14 (1), p. 1–51.

Barron, T., 1907, The topography and geology of the Peninsula of Sinai, western portion, *Surv. Dept.*, Cairo.

Barthoux, J. C., 1922, Chronologie et description des roches ignées du Désert Arabique, *Mém. Inst. Egypte*, v. 5.

Barthoux, J. C., and Douvillé, H., 1913, Le Jurassique dans le désert à l'est de l'Isthme de Suez, *C. R. Acad. Sci. Paris*, v. 157 (5), p. 265–268.

Barthoux, J., and Fritel, P. H., 1925, Flore crétacée du grés de Nubie, *Mém. Inst. Egypte*, v. 7 (2), p. 65–119.

Basta, E. Z., and Zaki, M., 1961, Geology and mineralization of Wadi Sikeit area, south Eastern Desert, *J. Geol. U.A.R.*, v. 5 (1), p. 1–37.

Beadnell, H. J. L., 1902, The Cretaceous region of Abu Roash near the pyramids of Giza, *Surv. Dept.*, Cairo.

Beadnell, H. J. L., 1905, The topography and geology of Fayum Province of Egypt, *Surv. Dept.*, Cairo.

Beadnell, H. J. L., 1924, Report on the region of the Red Sea Coast between Qoseir and Wadi Ranga, *Petrol. Res. Bull.*, Cairo: Government Press, no. 13.

Bellini, E., 1969, Biostratigraphy of the "Al Jaghbub (Giarabub) Formation" in Eastern Cyrenaica, Libya, *Proc. Third African Micropaleontol. Colloq.*, Cairo, p. 165–183.

Benson, R. H., and Ruggieri, G., 1974, The end of the Miocene, a time of crisis in Tethys–Mediterranean history, *Ann. Geol. Surv.*, Cairo, v. 4, p. 237–250.

Blankenhorn, M., 1921, *Handbuch der Regionalen Geologie, Bd. 7, Abt. 9, Heft 23, Aegypten*, Heidelberg: Carl Winters Universitätsbuchhandlung.

Bowen, C. F., and Vondra, C. F., 1974, Paleoenvironmental interpretations of the Oligocene Gabal El Qatrani Formation, Fayum Depression, Egypt, *Ann. Geol. Surv.*, Cairo, v. 4, p. 115–137.

Burollet, P. F., 1963, Reconnaissance géologique dans le sud-est du bassin de Kufra, *Rev. Inst. Franç. Pétrole*, v. 18 (11), p. 1537–1545.

Butzer, K. W., 1960, On the Pleistocene shore lines of Arab's Gulf, Egypt, *J. Geol.*, v. 68 (6), p. 626–637.

Butzer, K. W., and Hansen, C. L., eds., 1968, *Desert and River in Nubia, Geomorphology and Prehistoric Environments at the Aswan Reservoir*, Madison: The University of Winsconsin Press.

Caton-Thompson, G., 1952, *The Kharga Oasis in Prehistory*, London: Athlone Press.

Chumakov, I. S., 1967, Pliozenovaya: Pleistozenovaye otlojeniya dolina Nila v Nubiya: Verkhnem Egipte, *Trudy Akad. Nauk. SSSR*, v. 170.

Chumakov, I. S., 1968, Pliocene ingression in the Nile Valley according to new data, in: *Desert and River in Nubia, Geomorphology and Prehistoric Environments at the Aswan Reservoir*, Butzer, K. A., and Hansen, C. L., eds., Madison: The University of Wisconsin Press, p. 521–522.

Conant, L. C., and Goudarzi, G. H., 1967, Stratigraphic and tectonic framework of Libya, *Bull. Amer. Assoc. Petrol. Geol.*, v. 51 (5), p. 719–730.

Cuvillier, J., 1930, Révision du nummulitique Egyptien, *Mém. Inst. Egypte*, v. 16.

Cuvillier, J., 1953, Paléogéographie de l'Egypte au début des temps tertiaires, *Bull. Soc. Géogr. Egypte*, v. 25, p. 5–7.

Degens, E. T., and Ross, D. A., eds., 1969, *Hot Brines and Recent Heavy Metal Deposits in the Red Sea*, New York: Springer-Verlag.

Desio, A., 1929, Risultati scientifici della Missione alla oasi di Giarabub (1926–1927), Fasc. 2, La Geologia, *R. Soc. Geogr. Ital.*, p. 83–163.

Eicher, D. B., 1946, Conodonts from the Triassic of Sinai, Egypt, *Bull. Amer. Assoc. Petrol. Geol.*, v. 30, p. 613–616.

El Adl, R., and El Naggar, G., 1975, Geothermal gradient map of the Gulf of Suez region, Egypt, Unpublished map, General Petroleum Company, Cairo.

El Akaad, M. K., and Noweir, A. M., 1969, Lithostratigraphy of the Hammamat–Um Seleimat district, Eastern Desert, Egypt, *Nature*, v. 223 (5203), p. 284–285.

El Akaad, M. K., El Gaby, S., and Abbas, A. A., 1967, On the evolution of Feiran migmatites, *J. Geol. U.A.R.*, v. 11 (2), p. 49–58.

El Akkad, S., and Dardir, A., 1966a, Geology of the area between Ras Shagra and Mersa Alam, *Geol. Surv. Min. Res. Dept.*, Cairo, Paper no. 35.

El Akkad, S. L., and Dardir, A. A., 1966b, Geology and phosphate deposits of Wasif, Safaga area, *Geol. Surv.*, Cairo, Paper no. 36.

El Akkad, S., and Issawi, B., 1963, Geology and iron ore deposits of the Bahariya Oasis, *Geol. Surv.*, Cairo, Paper no. 18.

El Bassyouny, A. A., 1975, Geology of the area between Gara El Hamra and Ghard El Moharik, Bahariya Oasis, Western Desert, *Egypt. J. Geol.* (in press).

El Gezeery, M. N., and Marzouk, I. M., eds., 1974, Miocene rock-stratigraphy of Egypt, Stratigraphic Sub-committee of the National Committee of Geological Sciences, *Egypt. J. Geol.*, v. 18 (1), p. 1–59.

El Gezeery, M. N., Mohsen, S. M., and Farid, M., 1972, Sedimentary basins of Egypt and their petroleum prospects, *Eighth Arab Petrol. Congr.*, Algiers, Paper no. 83 (B-3).

El Gezeery, M. N., Farid, M., and Taher, M., 1975, Subsurface geological maps of northern Egypt, Unpublished maps, General Petroleum Company, Cairo.

El Hashemi, M. M., 1972, Sedimentation and oil possibilities of Palaeozoic sediments at Siwa basin, Western Desert, Egypt, *Eighth Arab Petrol. Congr.*, Algiers, Paper no. 72 (B-3).

El Naggar, Z. M., 1970, On a proposed lithostratigraphic subdivision of the Late Cretaceous–Early Paleogene succession in the Nile Valley, Egypt, U.A.R., *Seventh Arab Petrol. Congr.*, Kuwait, Paper no. 64 (B-3).

El Nakkady, S. E., 1958, Stratigraphic and petroleum geology of Egypt, Monograph Series no. 1, University of Assiut, Assiut.

El Ramly, M. F., 1963, The absolute ages of some Basement rocks from Egypt, *Geol. Surv. Min. Res. Dept.*, Cairo, Paper no. 15.

El Ramly, M. F., 1972, A new geological map for the Basement rocks in the Eastern and south Western Deserts of Egypt, scale 1:1,000,000, *Ann. Geol. Surv.*, Cairo, v. 2, p. 1–18.

El Ramly, M. F., and Al Far, D. M., 1955, Geology of El Mueilha–Dunqash district, *Geol. Surv.*, Cairo.

El Ramly, M. F., and El Akaad, M. K., 1960, the Basement Complex of the central Eastern Desert of Egypt between latitudes 24°30′ and 25°40′N, *Geol. Surv. Min. Res. Dept.*, Cairo, Paper no. 8.

El Ramly, M. F., Budanov, V. J., and Hussein, A. A., 1971, The alkaline rocks of south-eastern Egypt, *Geol. Surv.*, Cairo, Paper no. 53.

El Shazly, E. M., 1954, Rocks of Aswan area, *Geol. Surv.*, Cairo.

El Shazly, E. M., 1957, *Review of Egyptian Geology, Part 2*, Cairo: Science Council, p. 51–91.

El Shazly, E. M., 1962, Report on the results of drilling in the iron ore deposit of Gebel Ghorabi, Bahariya Oasis, Western Desert, *Geol. Surv.*, Cairo.

El Shazly, E. M., 1964, On the classification of the Precambrian and other rocks of magmatic affiliation in Egypt, U.A.R., *Twenty-Second Int. Geol. Congr.*, India, Proc. Sect. 10, p. 88–101.

El Shazly, E. M., 1966a, Structural development of Egypt, U.A.R., *Geol. Soc. Egypt, Fourth Annu. Meeting, Programme and Abstracts*, p. 31–38.

El Shazly, E. M., 1966b, Lead deposits in Egypt, *Ann. Mines Géol. Tunisie*, no. 23, p. 245–275.

El Shazly, E. M., 1967, Oil shows in Yemen and oil bubbles in rock salt, *Sixth Arab Petrol. Congr.*, Baghdad, Paper no. 40 (B-3).

El Shazly, E. M., 1970, Evolution of granitic rocks in relation to major tectonics in: *West Commemoration Volume*, Faridabad, India: Today and Tomorrow Printers and Publishers, p. 569–582.

El Shazly, E. M., 1972, On the development of segments in the Basement crust in northern Africa, *Twenty-Fourth Int. Geol. Congr.*, Canada, Abstracts, p. 27.

El Shazly, E. M., and Abdallah, A. M., 1966, Shear faulting along the Gulf of Suez and its implications, *Geol. Soc. Egypt, Fourth Annu. Meeting, Programme and Abstracts*, p. 38–39.

El Shazly, E. M., and Ghanem, M., 1975, Stratigraphy of the late Precambrian molasse sediments in Egypt, *Egypt. J. Geol.* (in press).

El Shazly, E. M., and Hamada, M. M., 1954, Geology of Gebel Mudargag West area, *Geol. Surv.*, Cairo.

El Shazly, E. M., and Hassan, A. K., 1962, Report on the results of drilling at Um Gheig Mine, Eastern Desert, *Geol. Surv. Min. Res. Dept.*, Cairo.

El Shazly, E. M., and Hassan, M. A., 1972, Geology and radioactive mineralization at Wadi Sikait–Wadi El Gemal area, Eastern Desert, *Egypt. J. Geol.*, v. 16 (2), p. 201–234.

El Shazly, E. M., and Krs, M., 1972a, A paleomagnetic study of Cretaceous rocks from Wadi Natash area, Eastern Desert, Egypt, *Trav. Inst. Géophys. Acad. Tchécoslov. Sci. No. 329, Geofys. Sbornik*, v. 18 (1970), p. 323–334.

El Shazly, E. M., and Krs, M., 1972b, Magnetism and paleomagnetism of Oligocene basalts from Abu Zaabal and Qatrani, northern Egypt, *Trav. Inst. Géophys. Acad. Tchécoslov. Sci., No. 356, Geofys. Sbornik*, v. 19 (1971), p. 261–270.

El Shazly, E. M., and Krs, M., 1973, Paleogeography and paleomagnetism of the Nubian Sandstone, Eastern Desert of Egypt, *Geol. Rundschau*, v. 62 (1), p. 212–225.

El Shazly, E. M., and Saleeb, G. S., 1972, Scapolite–cancrinite mineral association in St. John's Island, Egypt, *Twenty-Fourth Int. Geol. Congr.*, Canada, Proc. Sect. 14, p. 192–199.

El Shazly, E. M., Farag, I. A. M., and Bassyuni, F. A., 1965, Contributions to the geology and mineralization at Abu Swayel area, Eastern Desert, Part 1, Geology of Abu Swayel area, *Egypt. J. Geol.*, v. 9 (2), p. 45–67.

El Shazly, E. M., Saleeb, N. S., Saleeb, G. S., and Zaki, N., 1967, A new approach to the geology and mineralization of St. John's Island, Red Sea, *Geol. Soc. Egypt, Fifth Annu. Meeting, Abstracts*, p. 5–8.

El Shazly, E. M., Hermina, M. H., El Tahlawy, M. R., Fahmy, G. T., Assaf, H. S., El Hinnawy, M. E., Habib, M. M., and Abd El Hamid, M. L., 1970a, A general ground-water potential of the southwestern part of the Western Desert, including El Oweinat and El Gilf El Kebir, *Arab Soc. Min. Petrol.*, Cairo, v. 25 (1), p. 125–135. (In Arabic.)

El Shazly, E. M., Shehata, W. M., Sameida, M. M., and El Sayed, M., 1970b, Groundwater studies at Sidi Kreir, Mediterranean coastal zone, *Arab Soc. Min. Petrol.*, Cairo, v. 25 (1), p. 26–39.

El Shazly, E. M., El Sokkary, A. A., and Khalil, S. O., 1971a, Petrography and geochemistry of some pink granite pebbles from Hammamat conglomerates, Eastern Desert, *Egypt. J. Geol.*, v. 15 (1), p. 49–55.

El Shazly, E. M., El Sokkary, A. A., and Khalil, S. O., 1971b, Petrography and geochemistry of some granite pebbles from metamorphosed conglomerates, Qift–Qoseir area, Eastern Desert, *Egypt. J. Geol.*, v. 15 (1), p. 57–64.

El Shazly, E. M., Hashad, A. H., Sayyah, T. A., and Bassyuni, F. A., 1973, Geochronology of Abu Swayel area, south Eastern Desert, *Egypt. J. Geol.*, v. 17 (1), p. 1–18.

El Shazly, E. M., Abdel Hady, M. A., El Ghawaby, M. A., and El Kassas, I. A., 1974a, Geology of Sinai Peninsula from ERTS-1 satellite images, *Remote Sensing Research Project*, Acad. Sci. Res. Technol., Cairo.

El Shazly, E. M., Abdel Hady, M. A., El Ghawaby, M. A., and El Kassas, I. A., 1974b, Geologic interpretation of ERTS-1 satellite images for East Aswan area, Egypt, *Proc. Ninth Int. Symp. on Remote Sensing and Environment*, Environmental Res. Inst. Michigan, Ann Arbor, Michigan, p. 105–117.

El Shazly, E. M., Abdel Hady, M. A., El Ghawaby, M. A., and El Kassas, I. A., 1974c, Geologic interpretation of ERTS-1 images for West Aswan area, Egypt, *Proc. Ninth Int. Symp. on Remote Sensing and Environment*, Environmental Res. Inst. Michigan, Ann Arbor, Michigan, p. 119–131.

El Shazly, E. M., Abdel Hady, M. A., and Morsy, M. A., 1974d, Geologic interpretation of infra red thermal images in East Qatrani area, *Proc. Ninth Int. Symp. on Remote Sensing and Environment*, Environmental Res. Inst. Michigan, Ann Arbor, Michigan, p. 1877–1889.

El Shazly, E. M., El Hazek, N. M. T., Abdel Monem, A. A., Khawik, S. M., Zayed, Z. M., Mostafa, M. E. M., Morsy, M. A., 1974e, Origin of uranium in Oligocene Qatrani sediments, Western Desert, Arab Republic of Egypt, *IAEA-SM-183/1*, Vienna, p. 467–478.

El Shazly, E. M., Abdel Hady, M. A., El Shazly, M. M., El Ghawaby, M. A., El Kassas, I. A., Salman, A. B., and Morsy, M. A., 1975a, Geological and groundwater potential studies of El Ismailiya master plan study area, *Remote Sensing Research Project*, Acad. Sci. Res. Technol., Cairo.

El Shazly, E. M., Abdel Hady, A. M., El Ghawaby, A. M., El Kassas, I. A., Khawasik, S. M., El Shazly, M. M., and Sanad, S., 1975b, Geologic interpretation of Landsat satellite images for west Nile Delta area, Egypt, *Remote Sensing Research Project*, Acad. Sci. Res. Technol., Cairo.

El Shazly, E. M., Hashad, A. H., and El Reedy, M. A., 1975c, Application of carbon isotopic data to the oil genesis of main Egyptian petroleum provinces, *Ninth Arab Petrol. Congr.*, Dubai, Paper no. 50 (B-2).

El Shazly, E. M., Abdel Khalek, M. L., and Bassyuni, F. A., 1975d, Structure of the Greater Abu Swayel area, Eastern Desert, *Bull. Fac. Sci.*, Cairo University, Giza (in press).

El Shazly, E. M., Abdel Khalek, M. L., and Bassyuni, F. A., 1975e, Geology of the Greater Abu Swayel area, Eastern Desert, *Egypt. J. Geol.*, v. 19 (1), p. 1–41.

El Shazly, E. M., El Ghawaby, M. A., and Assaf, H. S., 1975f, Contributions to the geology and structural analysis of the central Eastern Desert of Egypt, *Egypt. J. Geol.* (in press).

El Shazly, E. M., Hashad, A. H., and El Manharawy, M. S., 1975g, Phanerozoic volcanicity in Egypt, *Egypt. J. Geol.* (in press).

El Shazly, E. M., Hermina, M. H., El Tahlway, M. K., Fahmy, G. T., Assaf, H. S., El Hinnawy, M. E., Habib, M. M., and Abd El Hamid, ML., 1975h, Geological reconnaissance of the southwestern part of the Western Desert including El Oweinat and El Gilf El Kebir, *Egypt. J. Geol.* (in press).

El Shazly, M. M., and Shata, A. A., 1960, Contribution to the study of the stratigraphy of El Kharga Oasis, *Bull. Inst. Désert Egypte*, v. 10 (1).

El Shazly, M. M., Sanad, S., and Rofail, N. H., 1975, Hydrogeological and hydrological investigation of the site of proposed tunnel at Kantara, Suez Canal, Engineering and Geological Consulting Office, Cairo.

El Sherbini, M. I., 1973, *Grain Size, Mineralogical and Petrological Studies on the Subsurface Sediments of Sidi Salem Well No. 1, Nile Delta*, M. Sc. thesis, Faculty of Science, Mansoura University, Mansoura, Egypt.

El Sweify, A., 1975, Subsurface Paleozoic stratigraphy of Siwa–Faghur area, Western Desert, Egypt, *Ninth Arab Petrol. Congr.*, Dubai, Paper no. 119 (B-3).

Fahmy, S. E., Mikhailov, I., and Samodurov, V., 1968, Biostratigraphy of the Paleogene deposits in Egypt, *Proc. Third African Micropaleontol. Colloq.*, Cairo, p. 477–500.

Fairbridge, R. W., 1963, Nile sedimentation above Wadi Halfa during the last 20,000 years, *Kush*, v. 11, p. 96–107.

Falcon, N. L., Gass, I. G., Girdler, R. W., and Laughton, A. S., 1970, A discussion on the structure and evolution of the Red Sea and the nature of the Red Sea, Gulf of Aden and Ethiopia Rift Junction, *Phil. Trans. Roy. Soc. London Ser. A*, v. 267.

Farag, I. A. M., 1959a, *Stratigraphy of Egypt*, Giza: Cairo University Press, v. 1 and 2.

Farag, I. A. M., 1959b, Contribution to the study of the Jurassic Formations in the Maghara Massif, Northern Sinai, Egypt, *Egypt. J. Geol.*, v. 3 (2), p. 175–199.

Farag, I. A. M., and Sadek, A., 1966, Stratigraphy of Gebel Homeira area, Cairo–Suez district, *J. Geol. U.A.R.*, v. 10 (2), p. 107–123.

Farag, I. A. M., and Shata, A., 1954, Detailed geological survey of El Minshera area, *Bull. Inst. Désert Egypte*, v. 4 (2), p. 5–82.

Faris, M. I., 1947, Contribution to the stratigraphy of Abu Rauwash and the history of the Upper Cretaceous in Egypt, *Bull. Fac. Sci.*, Cairo University, Giza, no. 27, p. 221–239.

Faris, M. I., Farag, I. A. M., and Gheith, M. A., 1956, Contributions to the geology of Bahariya Oasis, *Geol. Soc. Egypt*, p. 36–38.

Ghanem, M., Zalata, A. A., Abd El Razik, T. M., Abdel Ghani, M. S., Mikhailov, I. A., Razvaliaev, A. V., and Mirtov, Y. V., 1970, Stratigraphy of the phosphate-bearing Cretaceous and Paleogene sediments of the Nile Valley between Idfu and Qena, in: *Studies on some Mineral Deposits in Egypt*, Cairo: Geol. Surv.

Gheith, M. A., and staff, 1959, Program of age measurements in the Middle East, NY03940, *Seventh Ann. Progress Rep.* 1959, Dept. Geol. and Geophys., M.I.T., Boston.

Ghobrial, M. G., and Lotfi, M., 1967, The geology of Gebel Gattar and Dokhan areas, *Geol. Surv.*, Cairo, Paper no. 40.

Ghorab, M. A., 1956, A summary of a proposed rock-stratigraphic classification of the Upper Cretaceous rocks in Egypt, Presented at the Meeting of the Geological Society of Egypt.

Ghorab, M. A., 1960, Geologic observations on the surface and subsurface petroleum indications in the United Arab Republic, *Second Arab Petrol. Congr.*, Beirut, v. 2, p. 25–38.

Ghorab, M. A., 1964, Oligocene and Miocene rock-stratigraphy of the Gulf of Suez region, Unpublished report, Consultative Stratigraphic Committee of the Egyptian General Petroleum Corporation, Cairo, E.R. 275.

Ghorab, M. A., and Ismail, M. M., 1957, A microfacies study of the Eocene and Pliocene east of Helwan, *Egypt. J. Geol.*, v. 1 (2), p. 105–125.

Ghorab, M. A., El Shazly, E. M., Abdel Gawad, A., Morshed, T., Ammar, A. A., and Ibrahim, A. M., 1969, Discovery of potassium salts in the evaporites of some oil wells in the Gulf of Suez region, *Geol. Soc. Egypt, Seventh Annu. Meeting, Abstracts*, p. 3–4.

Gindy, A. R., 1956, The igneous and metamorphic rocks of the Aswan area, with a new geological map, *Bull. Inst. Egypte*, v. 37, p. 83–130.

Gindy, A. R., and El Askary, M. A., 1969, Stratigraphy, structure and origin of Siwa Depression, Western Desert of Egypt, *Bull. Amer. Assoc. Petrol. Geol.*, v. 53 (3), p. 603–625.

Girdler, R. W., and Styles, P., 1974, Two stage Red Sea floor spreading, *Nature*, v. 247, p. 7–11.

Goudarzi, G. H., 1970, Geology and mineral resources of Libya, *U.S. Geol. Surv.*, Prof. paper no. 660.

Greenwood, R. J., 1969, Radiogenic dating of 5 basalt samples submitted by Gulf of Suez Petroleum Company, Oilfields Report No. 297, Robertson Research Company Limited.

Greenwood, R. J., 1970, Potassium/argon dating of three samples from the southwest Mubarak No. 1 Well, Oilfields Report No. 337, Robertson Research Company Limited.

Hamed, A. A., 1972, Environmental interpretations of the Aptian carbonates of the Western Egyptian Desert, *Eighth Arab Petrol. Congr.*, Algiers, Paper no. 79 (B-3).

Hanter, G., 1975, Contribution to the origin of the Nile Delta, *Ninth Arab Petrol. Congr.*, Dubai, Paper no. 117 (B-3).

Hashad, A. H., Sayyah, T. A., El Kholy, S. B., and Youssef, A., 1972, Rb/Sr age determination of some Basement Egyptian granites, *Egypt. J. Geol.* v., 16 (2), p. 269–281.

Hassan, A. A., 1963, The distribution of Triassic and Jurassic formations in East Mediterranean, *Fourth Arab Petrol. Congr.*, Beirut, Paper no. 25 (B-3).

Hassan, A. A., 1967, A new Carboniferous occurrence in Abu Durba, Sinai, Egypt, *Sixth Arab Petrol. Congr.*, Baghdad Paper no. 39 (B-3).

Helal, A. H., 1966, Jurassic plant microfossils from the subsurface of Kharga Oasis, Western Desert, Egypt, *Palaeontographica*, v. 117 (B), p. 83–98.

Heybroek, F., 1965, The Red Sea Miocene evaporite basin, in: *Salt Basins Around Africa*, London: Institute of Petroleum.

Higazy, R. A., and Wasfy, H. M., 1956, Petrogenesis of granitic rocks in the neighbourhood of Aswan, Egypt, *Bull. Inst. Désert Egypte*, v. 6 (1), p. 209–256.

Hilmy, M. E., and Ragab, A. I., 1972, Geology and mineralogy of Wadi Rieina evaporites, Sinai, *Geol. Soc. Egypt, Tenth Annu. Meeting, Abstracts*, p. 1–2.

Holmes, A., and Cahen, L., 1959, Géochronologie Africaine, résultats acquis au 1ᵉʳ Juillet 1956, *Twentieth Int. Geol. Congr.*, Mexico, Assoc. African Geol. Surv., p. 21–26.

Hume, W. F., 1911, The effects of secular oscillations in Egypt during the Cretaceous and Eocene periods, *Quart. J. Geol. Soc. London*, v. 67 (1), p. 118–148.

Hume, W. F., 1925, *Geology of Egypt, Vol. 1, The Surface Features of Egypt; Their Determining Causes and Relation to Geologic Structure*, Cairo: Survey of Egypt.

Hume, W. F., 1934, 1935 and 1937, *Geology of Egypt, Vol. 2, The Fundamental Pre-Cambrian Rocks of Egypt and the Sudan; Their Distribution, Age and Character; Part 1, The Metamorphic Rocks; Part 2, The Later Plutonic and Minor Intrusive Rocks; Part 3, The Minerals of Economic Value*, Cairo: Survey of Egypt, p. 1–900.

Hume, W. F., 1962 and 1965, *Geology of Egypt, Vol. 3, The Stratigraphic History of Egypt; Part 1, From the Close of the Pre-Cambrian Episodes to the End of the Cretaceous Period; Part 2, From the Close of the Cretaceous Period to the End of the Oligocene*, p. 1–712 and p. 1–734, Cairo: Geol. Surv. Min. Res. Dept.

Hume, W. F., Magdwick, T. G., Moon, F. W., and Sadek, H., 1920, Preliminary geological report on the Quseir–Safaga district, particularly the Wadi Mureikha area, *Petrol. Res. Bull.*, Cairo: Government Press, no. 5.

Hurst, H. E., 1957, *The Nile, a General Account of the River and Utilization of its Waters*, London: Constable, Rev. Ed.

Huzayyin, S. A., 1941, The place of Egypt in Prehistory, *Mém. Inst. Egypte*, no. 43.

IEOC, 1970, Stratigraphical reports of IEOC wells drilled in the Nile Delta, Unpublished report, IEOC, Cairo.

Ismail, M. M., and Farag, I. A. M., 1957, Contributions to the stratigraphy of the area east of Helwan, Egypt, *Bull. Inst. Désert Egypte*, v. 7 (1), p. 95–134.

Issawi, B., 1968, The geology of Kurkur–Dungul area, *Geol. Surv.*, Cairo, Paper no. 46.

Issawi, B., 1972, Review of Upper Cretaceous–Lower Tertiary stratigraphy in central and southern Egypt, *Bull. Amer. Assoc. Petrol. Geol.*, v. 56 (8), p. 1448–1463.

Issawi, B., 1975, Contribution to the geology of El Oweinat–El Gilf El Kebir area, Western Desert, Egypt, *Egypt. J. Geol.* (in press).

Issawi, B., Francis, M., El Hinnawy, M., Mehanna, A., and El Deftar, T., 1971, Geology of Safaga–Quseir coastal plain and of Mohamed Rabah area, *Ann. Geol. Surv.*, Cairo, v. 1, p. 1–19.

James, G. T., and Slaughter, B. H., 1974, A primitive new Middle Pliocene Murid from Wadi El Natrun, Egypt, *Ann. Geol. Surv.*, Cairo, v. 4, p. 333–361.

Keldani, E. H., 1941, A bibliography of geology and related sciences concerning Egypt up to the end of 1939, *Geol. Surv.*, Cairo.

Kerdany, M. T., and Abdel Salam, H. A., 1970, Bio- and litho-stratigraphic studies of the pre-Miocene of some off-shore exploration wells in the Gulf of Suez, *Seventh Arab Petrol. Congr.*, Kuwait, Paper no. 56 (B-3).

Khalid, A. D., 1975, Jurassic prospects in the Western Desert, Egypt, *Ninth Arab Petrol. Congr.*, Dubai, Paper no. 100 (B-3).

Khalil, N. A., 1975, Tectonic and stratigraphic aspects in the Gulf of Suez oil province, *Ninth Arab Petrol. Congr.*, Dubai, Paper no. 114 (B-3).

Khalil, N. A., and El Demerdash, A., 1975, Geothermal gradient map of the northern Western Desert and Nile Delta, Egypt, Unpublished map, General Petroleum Company, Cairo.

Klerkx, J., 1969, Expédition scientifique Belge dans le Désert de Libye, Jebel Uweinat 1968–1969, Sect. 3, Géologie, Africa-Tervuren, v. 15 (4), p. 108–110.

Klitsch, E., 1968 Outline of the geology of Libya, in: *Geology and Archeology of Northern Cyrenaica, Libya*, Barr, F. T., ed., Tripoli: Petrol. Expl. Soc. Libya, p. 71–78.

Knetsch, G., 1958, Germanotype Deformationbilder aus Ägypten, *Geologie*, Berlin, v. 7 (3–6), p. 440–447.

Kostandi, A. B., 1959, Facies maps for the study of the Paleozoic and Mesozoic sedimentary basins of the Egyptian region, *First Arab Petrol. Congr.*, Cairo, v. 2, p. 54–62.

Krasheninnikov, V. A., and Ponikarov, V. P., 1964, Zonal stratigraphy of Paleogene in the Nile Valley, *Geol. Surv. Min. Res. Dept.*, Cairo, Paper no. 32.

Le Roy, L. W., 1953, Biostratigraphy of the Maqfi section, Egypt, *Mem. Geol. Soc. Amer.*, no. 54.

Mansour, M. S., and Bassyuni, F. A., 1954, Geology of Wadi Gerf district, *Geol. Surv.*, Cairo.

Marzouk, I., 1970, Rock-stratigraphy and oil potentialities of the Oligocene and Miocene in the Western Desert, U.A.R., *Seventh Arab Petrol. Congr.*, Kuwait, Paper no. 54 (B-3).

McKee, D. E., 1962, Origin of the Nubian and similar sandstones, *Geol. Rundschau*, v. 52, p. 551–587.

McKenzie, A. J., 1971, Potassium–argon age dating of two core samples from the Gharib-IX Well, UAR, Robertson Research Company Limited, Project No. 111B/94.

Meneisy, M. Y., and Kreuzer, H., 1974a, Potassium–argon ages of Egyptian basaltic rocks, *Geol. Jahrb.*, v. 8, p. 21–31.

Meneisy, M. Y., and Kreuzer, H., 1974b, Potassium–argon ages of nepheline syenite ring complexes in Egypt, *Geol. Jahrb.*, v. 9, p. 33–39.

Moon, F. W., and Sadek, H., 1921, Topography and geology of northern Sinai, *Petrol. Res. Bull.*, Cairo: Government Press, no. 10.

Moon, F. W., and Sadek, H., 1923, Preliminary geological report on Wadi Gharandel area, *Petrol. Res. Bull.*, Cairo: Government Press, no. 12.

Moustafa, G. A., and Abdallah, A. M., 1954, Geology of Abu Mireiwa district, *Geol. Surv.*, Cairo.

Moustafa, G. A., and El Akaad, M. K., 1962, Geology of the Hammash–Sufra district, *Geol. Surv.*, Cairo.

Omara, S., 1967, Contribution to the stratigraphy of the Egyptian Carboniferous exposures, *Sixth Arab Petrol. Congr.*, Baghdad, Paper no. 44 (B-3).

Omara, S., and Ouda, K., 1972, A lithostratigraphic revision of the Oligocene–Miocene succession in the northern Western Desert, Egypt, *Eighth Arab Petrol. Congr.*, Algiers, Paper no. 94 (B-3).

Ouda, K., 1971, *Biostratigraphical Studies of Some Surface and Subsurface Miocene Formations from the Western Desert, Egypt*, Ph.D. thesis, Fac. Sci., University of Assiut, Egypt.

Phillips, J. D., Woodside, J., and Bowin, C. O., 1969, Magnetic and gravity anomalies in the central Red Sea, in: *Hot Brines and Recent Heavy Metal Deposits in the Red. Sea*, Degens, E. T., and Ross, D. A., eds., New York: Springer-Verlag, p. 98–113.

Pomeyrol, R., 1968, Nubian Sandstone, *Bull. Amer. Assoc. Petrol. Geol.*, v. 52 (4), p. 589–600.

Rittmann, A., 1953, Some remarks on the geology of Aswan, *Bull. Inst. Désert Egypte*, v. 3 (2), p. 35–64.

Rittmann, A., 1958, Geosynclinal volcanism, ophiolites and Barramiya rocks, *Egypt. J. Geol.*, v. 2 (1), p. 61–65.

Ross, D. A., Whitmarsh, R. B., Ali, S. A., Bourdeaux, J. E., Coleman, R., Fleisher, R. L.,
Girdler, R., Manheim, F., Matter, A., Nigrini, C., Stoffers, P., and Supko, P. R.,1973,
Red Sea Drillings, *Science*, v. 179, p. 377–380.

Russegger, J., 1837, Kreide und Sandstein, Einfluss von Granit auf letzteren Porphyry,
Grunsteine, etc., in Ägypten und Nubien, bis nach Sennar, *Neues Jahrb. Min.*, p. 665–669.

Ryan, W. B. F., 1973, Geodynamic implications of the Messinian crisis of salinity, in: *Messinian Events in the Mediterranean*, Drooger, C. W., ed., Amsterdam: North-Holland
Publ. Co., p. 26.

Sabet, A. H., 1958, Geology of some dolerite flows on the Red Sea coast, south of El Qoseir,
Egypt. J. Geol. v., 2 (1), p. 45–58.

Sabet, A. H., 1961, *Geology and Mineral Resources of Gebel El Sinai Area, Red Sea Hills,
U.A.R.*, Ph.D. thesis, Leiden State University, Leiden.

Sadek, H., 1926, The geography and geology of the district between Gebel Ataqa and El
Galala El Bahariya, Gulf of Suez, *Surv. Dept.*, Cairo.

Sadek, H., 1959, The Miocene in the Gulf of Suez region, Egypt, *Geol. Surv.*, Cairo.

Said, R., 1957, *Review of Egyptian Geology, Part 1*, Cairo: Science Council, p. 1–50.

Said, R., 1962, *The Geology of Egypt*, Amsterdam and New York: Elsevier Publ. Co.

Said, R., and Issawi, B., 1964, Geology of Northern Plateau, Bahariya Oasis, *Geol. Surv.*,
Cairo, Paper no. 29.

Sandford, K. S., 1935, Geological observations on the northwestern frontiers of the Anglo-
Egyptian Sudan and the adjoining parts of the Libyan Desert, *Quart. J. Geol. Soc. London*,
v. 87, p. 193–221.

Sandford, K. S., and Arkell, J. W., 1939, Palaeolithic man and the Nile Valley in Lower
Egypt with some notes upon a part of the Red Sea littoral, Oriental Inst., Chicago,
Publ. 46.

Schurmann, H. M. E., 1964, Rejuvenation of Pre-Cambrian rocks under epirogenetical conditions during old Palaeozoic times in Africa, *Geol. Mijnb.*, v. 43, p. 196–200.

Schurmann, H. M. E., 1966, *The Pre-Cambrian along the Gulf of Suez and the Northern Part
of the Red Sea*, Leiden: E. J. Brill.

Shata, A., 1951, The Jurassic of Egypt, *Bull. Inst. Désert Egypte*, v. 1 (2), p. 68–73.

Shata, A., 1953, New light on the structural developments of the Western Desert of Egypt,
Bull. Inst. Désert Egypte, v. 3 (1), p. 101–106.

Shata, A., 1955, An introductory note on the geology of the northern portion of the Western
Desert of Egypt, *Bull. Inst. Désert Egypte*, v. 5 (3), p. 96–106.

Shata, A., 1956, Structural development of the Sinai Peninsula, Egypt, *Bull. Inst. Désert
Egypte*, v. 6 (2), p. 117–157.

Shata, A., 1971, The geomorphology, pedology and hydrogeology of the Mediterranean
coastal desert of the UAR, *Symp. Geol. of Libya*, Fac. Sci., University of Libya, Tripoli,
p. 431–446.

Shata, A., 1972, The pink pseudo-oolitic limestone, a post-Miocene rock unit in the Mediterranean coastal zone west of the Nile Delta, *Egypt. J. Geol.*, v. 16 (1), p. 1–18.

Shata, A. A., and El Fayoumy, I., 1970, Remarks on the regional geological structure of the
Nile Delta, *Proc. Symp. Hydrology of Deltas*, UNESCO, v. 1, p. 189–197.

Shukri, N. M., 1945, Geology of the Nubian Sandstone, *Nature*, v. 156 (3952), p. 116.

Shukri, N. M., 1953, The geology of the desert east of Cairo, *Bull. Inst. Désert Egypte*, v. 3 (2)
p. 89–105.

Shukri, N. M., 1960, The mineralogy of some Nile sediments, *Quant. J. Geol. Soc. London*, v.
105, p. 511–534.

Shukri, N. M., and El Ayouti, M. K., 1953, The mineralogy of some Nubian Sandstone in
Aswan, *Bull. Inst. Désert Egypte*, v. 3 (2), p. 65–88.

Shukri, N. M., and El Ayouti, M. K., 1956, The geology of Gebel Iweibid–Gebel Gafra area, Cairo–Suez district, *Bull. Soc. Egypte*, v. 29, p. 67–109.

Shukri, N. M., and Lotfi, M., 1954, The geology of the north-western part of Gebel Siwiqat El Arsha area, Eastern Desert of Egypt, *Bull. Fac. Sci.*, Cairo University, no. 32, p. 25–46.

Shukri, N. M., Philip, G., and Said, R., 1956, The geology of the Mediterranean coast between Rosetta and Bardia, 2, Pleistocene sediments, geomorphology and microfacies, *Bull. Inst. Egypte*, v. 37, p. 395–427.

Sigaev, N. A., 1967, The main tectonic map of Egypt, *Geol. Surv.*, Cairo, Paper no. 39.

Soliman, S. M., and El Fetouh, M. A., 1969, Petrology of the Carboniferous sandstones in west-central Sinai, *J. Geol. U.A.R.*, v. 13 (2), p. 49–152.

Soliman, S. M., and Korany, E. M., 1972, Lithostratigraphy and petrology of Paleocene–Lower Middle Eocene, Gebel El Duwi area, Egypt, and evaluation for petroleum prospecting, *Eighth Arab Petrol. Congr.*, Algiers, Paper no. 75 (B-3).

Soliman, S. M., Faris, M. I., and El Badry, O., 1970, Lithostratigraphy of the Cretaceous Formations in the Baharia Oasis, Western Desert, Egypt, U.A.R., *Seventh Arab Petrol. Congr.*, Kuwait, Paper no. 59 (B-3).

Souaya, F. J., 1963, Micropaleontology of four sections south of Quseir, Egypt, *Micropaleontology*, v. 9 (3), p. 233–267.

Stromer, E., 1902, Das Pliocaen des Wadi Natrun, in: *Neue geologisch–stratigraphische Beobachtungen in Aegypten*, Blanckenhorn, M., ed., *Sitz-Ber. Math. Phys. Kl.k.b. Akad. Wiss.* v. 32, p. 419–426.

Synelinkov, A. S., and Kollerov, D. K., 1959, Palinogonic analysis and age of coal samples from Beda–Thora district, west-central Sinai, *Geol. Surv. Min. Res. Dept.*, Cairo, Paper no. 4.

Thenius, E., 1959, *Handbuch der stratigraphischen Geologie, 3*, Tetiaer, Zweiter Teil, Wirbeltierfaunen, Stuttgart.

Viotti, C., and El Demerdash, G., 1968, Studies on Eocene sediments of Wadi Nukhul area, east coast, Gulf of Suez, *Proc. Third African Micropaleont. Colloq.*, Cairo, p. 403–423.

Vittimberga, P., and Cardello, R., 1963, Sédimentologie et Pétrographie du Paléozoique du bassin de Kufra, *First Saharan Symp., Rev. Inst. Franç. Pétrole*, v. 18 (11), p. 1546–1558.

Vladimirskaya, R. A., 1973, Lithologic study of the Cretaceous and Paleozoic sandstone sequence in Siwa-1 and Bahrein-1, Western Desert, *Egypt. J. Geol.*, v. 17 (1), p. 47–55.

Vondra, C. F., 1974, Upper Eocene transitional and near-shore marine Qasr El Sagha Formation, Fayum Depression, Egypt, *Ann. Geol. Surv.*, Cairo, v. 4, p. 79–94.

Walthur, J. K., 1890, Ueber eine Kohlenkalk-Fauna der aegyptisch–arabischen Wüste, *Zeitschr. D. Geol. Ges.*, Berlin, v. 42, p. 419–449.

Weissbrod, J., 1969, The Paleozoic of Israel and adjacent countries, Part II, The Paleozoic outcrops in southwestern Sinai and their correlation with those of southern Israel, *G.S.I. Bull.*, no. 48, p. 1–32.

Wendorf, F., ed., 1965, *Contributions to the Prehistory of Nubia*, Dallas: Southern Methodist Univ. Press.

Yallouze, M., and Knetsch, G., 1954, Linear structures in and around the Nile Basin, *Bull. Soc. Géogr. Egypte*, v. 27, p. 168–207.

Youssef, M. I., 1957, Upper Cretaceous rocks in Kosseir area, *Bull. Inst. Désert Egypte*, v. 7 (2), p. 35–45.

Youssef, M. I., 1968, Structural pattern of Egypt and its interpretation, *Bull. Amer. Assoc. Petrol. Geol.*, v. 52 (4), p. 601–614.

Zaghloul, Z. M., and El Sawi, M. M., 1971, On the subsurface Miocene evaporites in the Gulf of Suez region, *Geol. Soc. Egypt, Ninth Annu. Meeting, Abstracts,* p. 13–15.

Zeuner, F. E., 1950, *Dating the Past,* London: Methuen, rev. ed.

Zittel, A. K., 1883, Beiträge zur Geologie und Palaeontologie der Libyschen Wüste und der angrenzenden Gebiete von Aegypten, *Palaeontographica,* v. 30 (1), p. 1–112.

Chapter 9

THE BLACK SEA AND THE SEA OF AZOV*

David A. Ross

Woods Hole Oceanographic Institution
Woods Hole, Massachusetts

I. INTRODUCTION

The Black Sea is one of the world's largest marginal seas, with an area of 432,000 km² and a volume of 534,000 km³. Most of the basin is deeper than 2000 m (Fig. 1) and the maximum depth reached is 2206 m. It is connected by a narrow passage, the Kerch Strait, to the shallow Sea of Azov which has an area of about 37,500 km² (excluding the Sivash or Putrid Sea) and a volume of about 470 km³. The average depth of the Sea of Azov is close to 12 m (Fig. 2). Sill depth between the two seas at the Kerch Strait is about 5 m. The Black Sea is connected to the Mediterranean Sea via the shallow Bosporus (sill depth about 50 m). The Sivash or Putrid Sea is a shallow area separated from the Sea of Azov by a 110-km-long sand bar called the Arabatskaya Strelka.

A. General Setting

Basically, the Black Sea occupies an oval basin between the folded Alpine belts of the Caucasus and Crimea Mountains to the north and northeast and the Pontic Mountains to the south. The Sea of Azov is situated between the Yaila

---

* Contribution No. 3507 of the Woods Hole Oceanographic Institution.

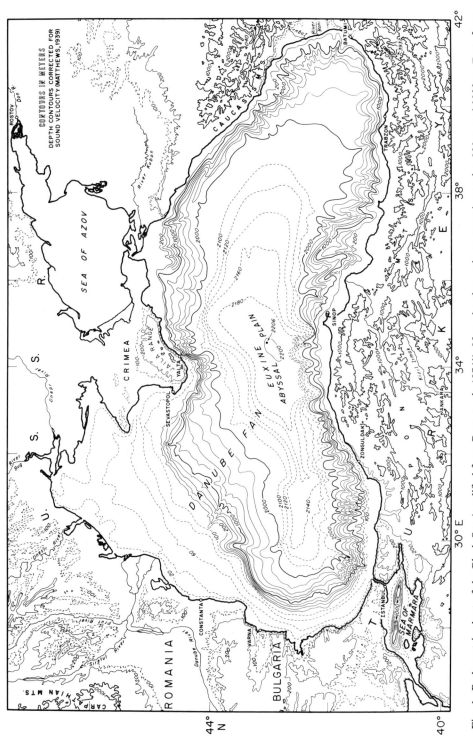

Fig. 1. Bathymetric chart of the Black Sea (modified from Ross *et al.*, 1974*b*). Note the change in contour interval at 200 m and 2000 m. Data from the 1969 *Atlantis II* cruise, a Russian chart supplied by Pavel Kuprin of the University of Moscow, and U.S. Plotting Sheets 108N and 3408N. Land topography is from the Morskoi Atlas, Tom 1, Navigatsionne–Geographicheski Izdanie Morskogogeneralnogo Shtaba. The map was contoured by Elazar Uchupi of the Woods Hole Oceanographic Institution.

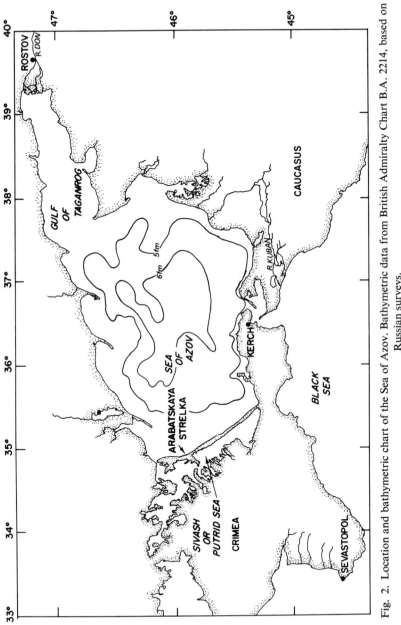

Fig. 2. Location and bathymetric chart of the Sea of Azov. Bathymetric data from British Admiralty Chart B.A. 2214, based on Russian surveys.

and Caucasus Mountains to the south and the Russian Platform to the north. The alignment of the mountain ranges in the area is generally east–west, suggesting that during the Mesozoic and Cenozoic, the Black Sea Basin was an area of north–south compression. However, the more recent development of the Black Sea is the result of subsidence of the platform and coastal folded areas, including some parts of the Caucasus (Balvadze *et al.*, 1966; Tsagareli, 1974; Ross *et al.*, 1974*a*).

One of the most distinctive features of the Black Sea is the thick sediment accumulation of 18 to 24 km in its central portions (Neprochnov *et al.*, 1974). Sediment thickness in the Sea of Azov is not known, but the recent accumulation is high; as much as 30 m of post-Neoeuxinian sediments (post about 7000 years) have accumulated in the central part of that sea.

B. Previous Work

The general outline of the Black Sea was established by early works of Andrusov (1890), Arkhangel'skiy (1928*a*, 1930), Arkhangel'skiy and Strakhov (1932), and Goncharov (1958). The first detailed echo-sounding surveys were made in 1956 (Goncharov and Neprochnov, 1967). Significant sediment studies have been made by Andrusov (1890, 1893), Arkhangel'skiy and Strakhov (1932, 1938), Strakhov (1947, 1954, 1961), Barkovskaya (1961*a*, *b*), and Nevesskiy (1967). More recent studies concerning the structural aspects of the Black Sea have been made by Goncharov *et al.* (1966), Malovitskiy *et al.* (1969*a*, *b*), Neprochnov (1962), Muratov and Neprochnov (1967), and Neprochnov *et al.* (1964); and magnetic data have been obtained by Melikhov *et al.* (1969) and Malovitskiy *et al.* (1969*a*). The bathymetry, water chemistry, sediments, and general structure of the Black Sea were studied during an expedition of the *R/V Atlantis II* of the Woods Hole Oceanographic Institution (Degens and Ross, 1974). Many of the data presented here are extracted from the report of that expedition (which included summary papers by Russian, Bulgarian, and Rumanian scientists) and from a later summary by Ross (1974). Laking (1974) has compiled a very extensive bibliography on the Black Sea, including articles on its biology and chemistry.

Literature on the Sea of Azov is mainly in Russian, but some recent summaries have been translated into English. These include Panov (1965) who wrote on the Holocene development of the Sea of Azov, Aleksina *et al.* (1971) on clay minerals, and Krustalev and Scherbakov (1968) and Panov and Spichak (1961) on the sedimentary budget of the Sea of Azov. Caspers (1957) wrote a general summary of the two basins.

II. BATHYMETRY

A. Black Sea

The Black Sea Basin can be divided into four distinct physiographic provinces (Fig. 3). The provinces and their percentage area (according to Ross *et al.*, 1974*b*) are: shelf (29.9%); basin slope (27.3%); basin apron (30.6%); abyssal plain (12.2%). The shelf is best developed west of the Crimean Peninsula where the Danube, Dnestr, Bug, and Dnepr Rivers enter the Black Sea. In this area the shelf is almost 200 km wide, probably due to infilling by these rivers. Elsewhere, most of the shelf is narrow, rarely exceeding 20 km in width. An exception is found immediately off the Sea of Azov where the shelf is about 65 km wide. The shelf break generally occurs at 100 m, but deepens to about 130 m off Crimea and the Sea of Azov.

Basin slopes (gradient 1:40 or more) are of two types. Relatively smooth slopes are found west of the Crimea and south of the Sea of Azov. These smooth basin slopes are probably due to a large sedimentary blanket, but may represent basic differences between the bordering platform areas and the mountainous regions of the Black Sea which fringe the second type of slope (Ross *et al.*, 1974*b*). The second slope type is fairly steep and highly dissected and is generally found adjacent to mountainous areas. These slopes may reflect a seaward extension of some of the onshore structures further sculptured by submarine canyons and modulated by slumping.

The basin apron province (gradient between 1:40 and 1:1000) is similar to a continental rise. It is the largest physiographic province within the Black Sea and except in the region of the Danube Fan (Fig. 3) is continuous throughout the sea. The Danube Fan, first detected from seismic profiles, is not necessarily related to the present Danube, but the profiles do indicate an area of thick sediments and slumping on the basin apron off the mouths of the Danube and other rivers. During the Pleistocene when sea level was lower, the major rivers deposited considerable amounts of sediment in the deeper parts of the Black Sea.

The deepest physiographic province in the Black Sea has been called the Euxine Abyssal Plain, and it has an extremely gentle gradient of less than 1:1000. Although shallower, it resembles the abyssal plains of deep oceans. Many earlier papers on the Black Sea refer to two distinct topographic depressions—one in the eastern and one in the western part of the Black Sea. Our surveying tracks noted only one depression.

B. Sea of Azov

The shallow Sea of Azov lies entirely within shelf depths. Its bathymetry changes rapidly depending upon wind force and direction. Coastal areas are

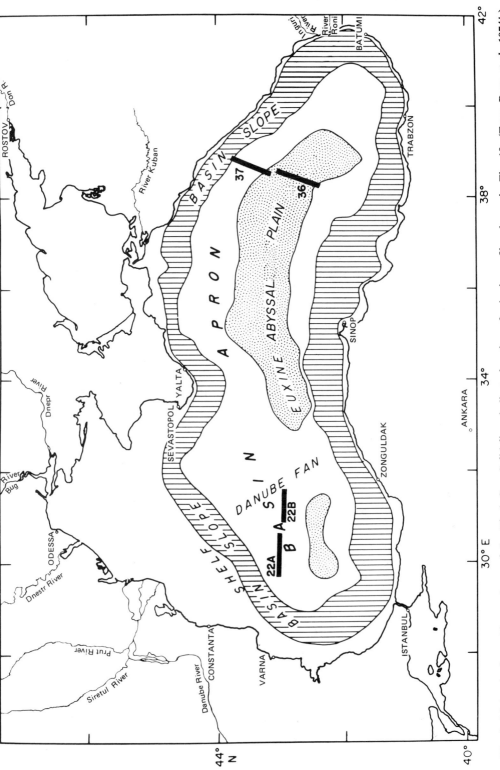

Fig. 3. Main physiographic provinces of the Black Sea. Solid lines indicate location of seismic profiles shown in Fig. 19. (From Ross *et al.*, 1974*b*.)

generally composed of highly eroded Tertiary and Quaternary sedimentary formations (Buachidze, 1974). The low coastal region is characterized by numerous spits, shallow gulfs, and lagoons which are in various stages of silting up. In general, the Sea of Azov is an area of rapid deposition with sediments supplied by the Don and Kuban Rivers. In the Taganrog Gulf, where the Don enters, depths are generally 1 m or less.

III. HYDROGRAPHY

A. Black Sea

The hydrography of the Black Sea is, to a large degree, controlled by exchange with the Mediterranean through the narrow Bosporus. There is a 3000 to 30,000 m^3/sec flow of Black Sea surface water with a salinity of 17.5‰ toward the Mediterranean. There is a corresponding deep-water flow of more dense Mediterranean water (38.5‰) into the Black Sea. The present low salinity of the Black Sea is due to relatively high river inflow and precipitation relative to the Mediterranean. When sea level was lower in the Black Sea during the Pleistocene glaciation, Mediterranean inflow ceased and the Black Sea became almost a fresh water lake (see Section IVA). Manheim and Chan (1974) show that the salinity of interstitial waters in the sediments was as low as 6‰ during the most recent lowering of sea level. When sea level was higher than present levels, increased inflow of Mediterranean water was possible and the Black Sea would have become more saline. Present salinities increase from about 17.5‰ at the surface (Brewer, 1971; Kremling, 1974; Brewer and Spencer, 1974) to about 21.9‰ at 300 m, and then gradually to 22.4‰ at 2000 m.

Deuser (1972, 1974) and Ross et al. (1970) have shown that brackish-water, aerobic conditions existed in the Black Sea during the Late Pleistocene and Early Holocene. Anoxic conditions began about 9000 years ago with the inflow of Mediterranean water due to rising sea level. This inflow resulted in density stratification in the Black Sea. Deuser (1971) has shown that production of organic material in the surface water is sufficient to exhaust oxygen in the bottom waters through decay as it settles out. This situation resulted in the establishment of anaerobic conditions in the bottom waters about 7300 years ago. The interface of oxygen-rich and hydrogen sulfide-rich water has been rising since that time to its present depth of about 200 m.

Water circulation tends to be variable in the Black Sea (Shimkus and Trimonis, 1974). According to Zenkovich (1958) there are two counterclockwise gyres (Fig. 4). Studies made over several years in the early 1950's showed unequal and changing current patterns (Dobrzhanskaya, 1967).

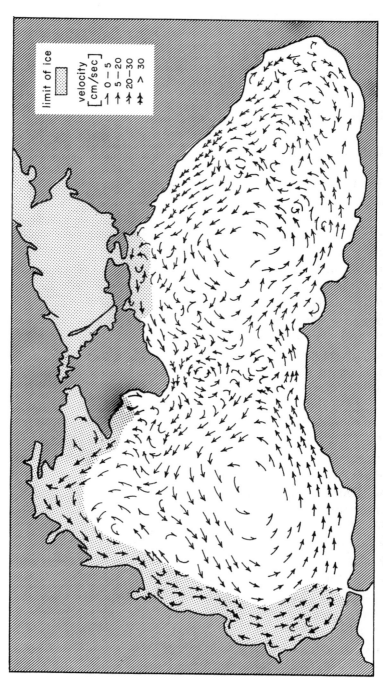

Fig. 4. Current systems in the Black Sea. (After Zenkevich, 1963; Shimkus and Trimonis, 1974.)

B. Sea of Azov

Little has been published on the hydrographic conditions in the Sea of Azov. Salinities are about 11‰, due to dilution from the Don River and lack of exchange across its shallow sill with the more saline water of the Black Sea. Salinities in the Sivash (Fig. 2) are extremely high, reaching as much as 100‰ in this shallow, protected area. Circulation is generally anticlockwise, but is considerably affected by local wind conditions.

IV. SEDIMENTS

A. Black Sea

Strakhov (1954), Raupach (1952), Erünal-Erentöz (1956), and Caspers (1957) have summarized the results of numerous investigations of the surface sediments of the Black Sea. The *Atlantis II* expedition to the Black Sea produced additional studies by Ross *et al.* (1970), Degens (1971, 1974), Ross and Degens (1974), Müller and Stoffers (1974), Shimkus and Trimonis (1974), and Trimonis (1974). Studies of recent turbidite deposition have been made by Jipa (1974), and studies of the mass physical properties of some sediments in the western Black Sea were made by Keller (1974).

The Black Sea has a drainage area of 1,864,000 km² (Müller and Stoffers, 1974). Most drainage comes from the eastern Russian platform, and only 15% comes from the high mountain areas (Müller and Stoffers, 1974). Several large rivers empty into the Black Sea (Table I), including the Danube, Dnestr, Bug, Dnepr, and Don (via the Sea of Azov). These rivers are concentrated in the northern part of the Black Sea. To the east is the Rioni, whereas to the south and west the rivers are small, although they apparently carry large amounts of material into the Black Sea. Müller and Stoffers (1974) have made a detailed study of the mineralogy and petrology of the Black Sea sediments and show that the sediment derived from the north and northwest, especially from the Danube, has a relatively low calcite–dolomite ratio and a relatively high quartz–feldspar ratio. This material differs from that carried in from the Anatolian coast in the east, which has a high calcite–dolomite ratio and a low quartz–feldspar ratio. Illite is the dominant clay mineral from the north, while montmorillonite is the dominant clay mineral from the south.

Texturally, the clay fraction is the major component of the sediments and there is an increasing clay content from the coast to the center of the Black Sea. Other changes in the clay mineralogy can be traced to the different petrologic characteristics of the northern and southern source areas.

TABLE I

Main Rivers Draining into Black Sea: Basin Area, Water and Solid Discharge, and Mechanical Denudation (adapted from Strakhov, 1967; Müller and Stoffers, 1974)

River	Basin area (10^3 km²)	Annual water discharge (km³)	Annual solid discharge (10^6 tons)	Mechanical denudation (tons/km² or g/m²)
Danube	816	201	83	101
Dnepr	503	53	2.02	5
Don	422	28	7.75	18.3
Rioni	13.4	13.5	8.5	633 (2000)
Dnestr	72		2.5	31.5
Kuban				180

In many of the cores that Müller and Stoffers studied, they noted a change in the montmorillonite–illite ratio with depth in the cores. These changes are thought to be related to the differing influences of the two main source areas during recent times. Apparently higher montmorillonite content is indicative of glacial phases and permafrost in the northern areas, when the illite supply was diminished.

Using Russian data, Shimkus and Trimonis (1974) summarized the sediment input into the Black Sea and Sea of Azov by rivers. There is a yearly water discharge of 374 million km³ into the Black Sea and Sea of Azov, which brings in about 150 million tons of solids in suspension, 55% of which comes from the Danube. An additional 15 million tons yearly is supplied by traction load, mainly during the springtime flood season. Total annual contribution of dissolved solids is on the order of about 100 million tons. Shimkus and Trimonis (1974) have also noted the distribution of suspended material during the late summer and early fall (Fig. 5), which is generally a period of low terrigenous influx but rather high productivity from plankton (mainly dinoflagellates and diatoms). The main producers of carbonate material, coccolithophores, are relatively scarce at this time. The largest concentration of suspended material occurs around the edges of the Black Sea and Sea of Azov, while little material is present farther from shore.

The sediments deposited within the last 25,000 years can generally be separated into three main units (Ross et al., 1970). These sedimentary units, which may be recognized over much of the Black Sea, are as follows (Fig. 6):

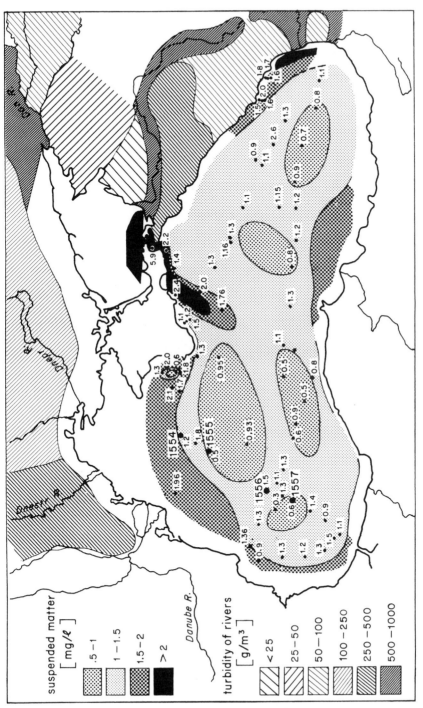

Fig. 5. Concentration of suspended material in surface waters of the Black Sea as measured in September of 1966, and average turbidity of rivers (from Shamov, 1951). (From Shimkus and Trimonis, 1974.)

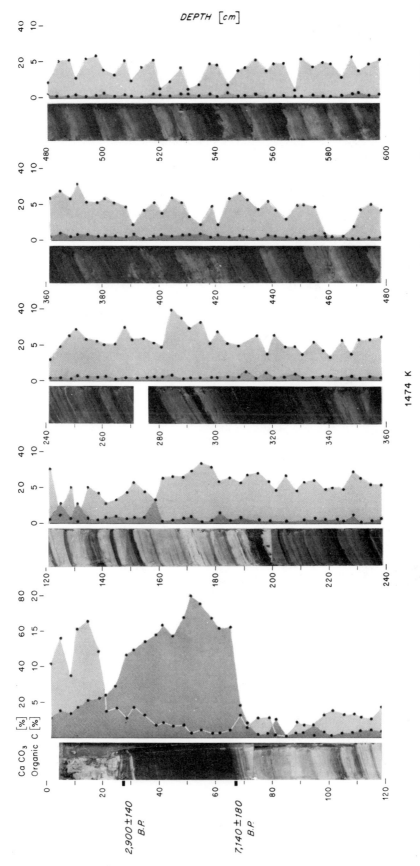

Fig. 6. The three main sedimentary units of the Black Sea and their calcium carbonate and organic carbon content. Shown is the upper part of an 11-m core, Unit 3 extended to the end of core. (From Ross and Degens, 1974.)

1. *Unit 1*

This uppermost unit is generally about 30 cm thick and consists of an alternating sequence of light and dark microlayers, which are essentially composed of the coccolithophore *Emiliania huxleyi* (Fig. 7a–c). The unit may be thicker near coastal areas where terrigenous deposition is higher, or in the deeper basins where it is interrupted by turbidites. Early Russian studies had suggested that the carbonate material was of inorganic origin, but electron microscope studies of the *Atlantis II* material by Müller and Blaschke (1969), Bukry *et al.* (1970), and Bukry (1974) have shown that it is mainly of organic origin. The light and dark layers are apparently due to lesser and greater amounts of organic material, with corresponding change in the amount of carbonate. There are 50 to 100 layers per centimeter within this unit and its radiocarbon age at the base is about 3000 years B. P. Thus, it appears that the layers are due to a yearly phenomenon, probably blooms of *Emiliania huxleyi*. There is an abrupt transition between Unit 1 and Unit 2.

2. *Unit 2*

This middle unit is generally about 40 cm thick, although again its thickness may be increased where terrigenous deposition is high. It is a dark brown, jelly-like sapropel, similar to those collected in the Mediterranean (see, for example, McCoy, 1974). Detailed examination shows that it is also microlaminated, but with a low color contrast between layers. The organic content can be as high as 50% in this unit. Electron microscope studies of this sapropel show it to consist of numerous unidentified membranes (Degens *et al.*, 1970; Degens, 1971; Ross and Degens, 1974). The occurrence of such membranes in marine sediments is relatively rare. It has been suggested that their preservation results from anaerobic pore water and stabilization by coordination of heavy metals (Degens, 1971). The age at the base of this unit in the deep water of the Black Sea (about 2000 m) is about 7000 years B. P. The age tends to decrease higher up on the basin slope and a date of 6200 years B. P. has been measured on sediment collected from a depth of about 500 m. Decreasing age of the base of Unit 2 is consistent with Deuser's (1972) conclusion about the rising level of the H_2S–oxygen interface since 7300 years B. P.

In some cores, three distinct white-layered bands were noticed in this unit. The uppermost band, about 10 cm below the top of the unit, contained *Emiliania huxleyi* only. In the second band, about 20 cm from the top of the unit, the coccolithophore *Braarudosphaera bigelowi* is found (Fig. 7e–f) with resting cysts of *Peridinium trochoideum* (Figs. 7d and 8h), an autotrophic marine dinoflagellate. The lowest band, about 30 cm below the top of Unit 2, is composed entirely of aragonite (Fig. 7g–i). These bands can occasionally be correlated from one core to the other.

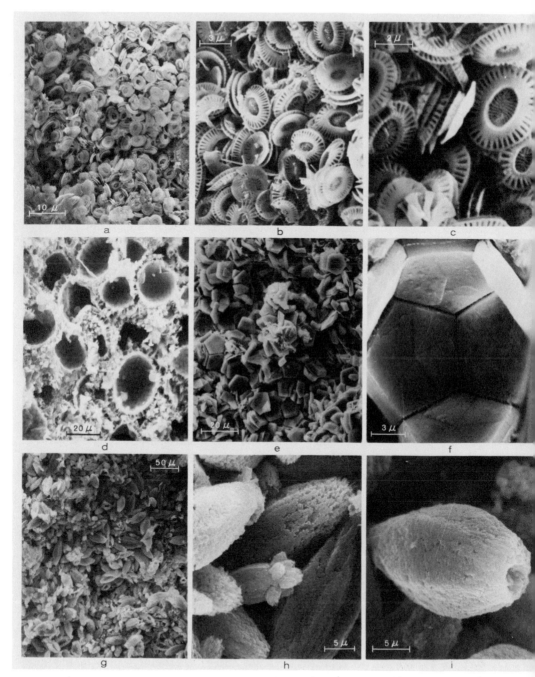

Fig. 7. Electron microscope photographs from Units 1 and 2. Pictures a–c, layers composed almost entirely of *Emiliania huxleyi;* d, calcitic cysts; e–f, layers composed of *Braarudosphaera bigelowi;* g–i, spongelike aragonite layer found in Unit 2 at depth of 60 cm. Photographs made by E.T. Degens. (From Ross and Degens, 1974.)

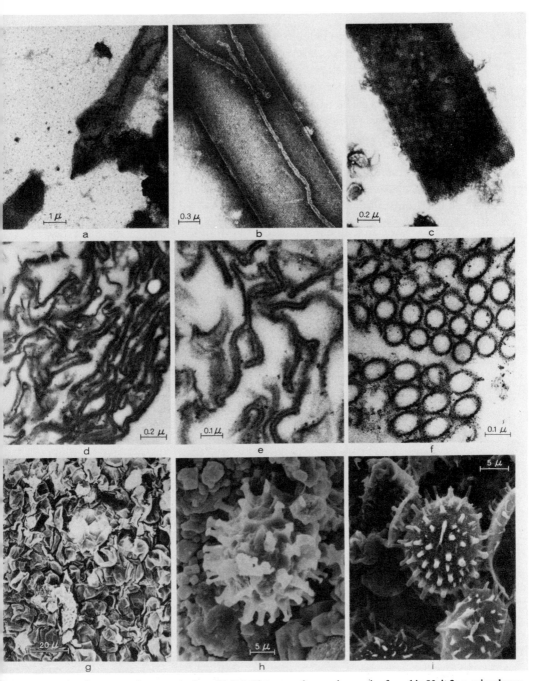

Fig. 8. Electron microscope photographs from Unit 2. Pictures a–f, organic remains found in Unit 2; g, microlayer composed of pollen; h, cyst of *Peridinium trochoideum;* i, cysts of *Gonyaulax polyhedra*. Photographs made by E.T. Degens. (From Ross and Degens, 1974.)

3. *Unit 3*

This unit has not been completely penetrated by any *Atlantis II* core. The sediment consists of an alternating sequence of light and dark lutite units, occasionally coarse-grained because of turbidites. Chemically, there is nothing distinctive about the unit; carbonate varies from 10 to 30% and organic carbon is usually 1% or less. The dark layers in the unit, according to Berner (1974), are due to the metastable iron sulfides, mackinawite and greigite, which oxidize when the cores are opened. This unit was deposited during brackish-water conditions.

None of the fauna of the upper two units were observed in Unit 3, although occasional fragments of older Eocene and Cretaceous coccoliths were found, which have apparently been reworked from land-based deposits.

Because oxygen is absent, the Recent bottom sediments of the Black Sea have not been disturbed by organisms and can thus be correlated over almost the entire Black Sea (Fig. 9). In some instances, individual layers, especially in Unit 1 on the order of 1 mm in thickness, can be correlated from one end of the basin to the other, a distance of over 1000 km.

The stratigraphic units of the *Atlantis II* work can be correlated with those found in the earlier deep-water work of Arkhangel'skiy and Strakhov (1938) and with the shallow-water units of Nevesskiy (1967). Correlations are shown in Table II.

General recognition throughout the Black Sea Basin of the 3000-year-old base of Unit 1 allows an estimate of sedimentation rates for the last 3000 years (Fig. 10). In this estimate, it is assumed that the age of the organic matter at the sediment–water interface is essentially zero (Ostlünd, 1974). Sedimentation rates in the northwestern part of the Black Sea (Fig. 10) are relatively low, probably due to the fact that most of the major rivers are now depositing sediment within their estuaries rather than on the shelf or in the deeper part of the basin. High sedimentation rates are found along the Turkish coast, apparently due to the high, yearly inflow of sediment from rivers and the relatively narrow shelf, which permits these sediments to be transported to the deeper portions of the basin, and because of the high frequency of turbidite deposition in the eastern part of the Black Sea. This point is fairly obvious when cores from these two areas are examined (Fig. 11).

Black Sea sediment characteristics allow establishment of a geologic history for the last 25,000 years, which can be correlated with known sea-level curves (Fig. 12). About 23,000 years ago, the oldest material our cores penetrated, the Black Sea was in the process of evolving from a marine basin to a more fresh-water environment. Sea level was probably lower than it is today and therefore isolated the Black Sea from the Mediterranean at the Bosporus. Based on the few cores that go back to that date, it appears that the Black Sea

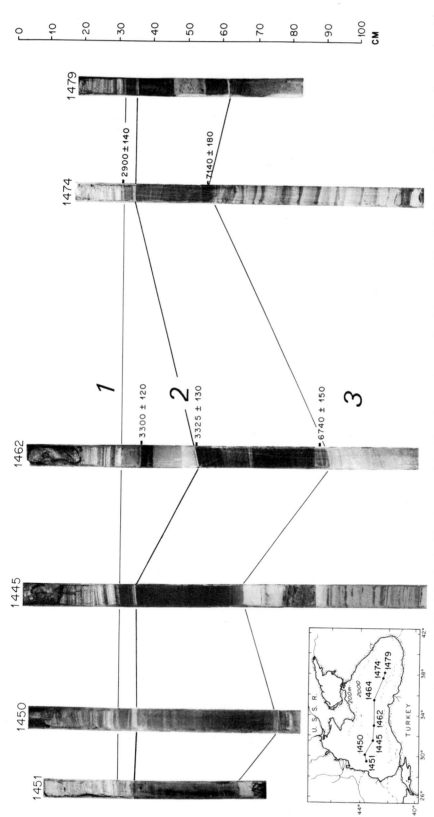

Fig. 9. Sediment-core profile across the Black Sea Basin. The three stratigraphic units can be correlated in essentially all depths greater than 200 m. Some carbon-14 dates are indicated. (Adapted from Ross *et al.*, 1970.)

TABLE II

Stratigraphic Correlation of Recent Black Sea Sediments

Arkhangel'skiy & Strakhov (1938)	Neveskiy (1967)	Ross, et al (1970)	
		Sediment units	Years B.P.
Recent deposit	Dzhemetinian	Unit 1: Coccolith ooze	2000
Old Black Sea beds	Kalamitian		3000
	Vityazevian	Unit 2: sapropel or organic-rich	5000
	Bugazian		6000
			7000
Neoeuxinian sediments	Neoeuxinian	Unit 3: banded lutite	10,000
	Karkinitian ? ? ? ? ? ? Tarkhankutian		11,000
			25,000

was then aerobic. The lake phase, initiated about 22,000 years ago, continued for about 12,000 to 13,000 years, during which Unit 3 was deposited. Beginning at about 9000 years B. P. and continuing to about 7000 years B. P., there was a gradual shift from a fresh-water to a marine environment, initiated by the

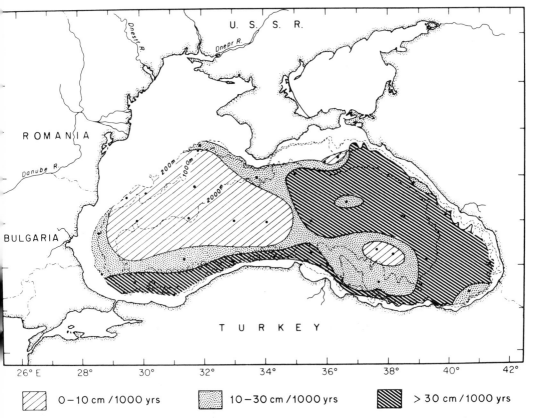

Fig. 10. Sedimentation rate in the Black Sea during the past 3000 years. (From Ross *et al.*, 1970.)

inflow of Mediterranean water. Concomitantly, the bottom water condition shifted from well-aerated to stagnant, and due to density stratification a reducing environment became established (Deuser, 1972, 1974). The reducing conditions favored preservation of organic material. It appears that about 7300 years B. P., the H_2S zone started to move upward from the bottom (Deuser, 1972). Sedimentary Unit 2 was deposited at this time. This environment continued until about 3000 years B. P. Sedimentation rates decreased within the last 3000 years or so because rising sea level caused estuarine sedimentation by the rivers. About 3000 years B. P., the environmental conditions found today became established in the Black Sea and deposition of Sedimentary Unit 1 started.

B. Sea of Azov

The sediments in the Sea of Azov are less well known than those of the Black Sea. Available studies are mainly descriptive, with emphasis on the sedimentary budget (Aleksandrov, 1964; Aleksina and Yedigaryan, 1971;

Relative Deposition Rate

high   low   high

10 cm

Fig. 11. Relative deposition rate in three representative sedimentary cores. Three kinds of patterns can be recognized in upper sections of cores: left core, high rate of sedimentation with occasional turbidite flow; middle core, low rate of sedimentation (about 10 cm/1000 years) showing alternate layers of white coccoliths and organic-rich bands; and right core, high rate of sedimentation at an evenly distributed rate. (From Ross *et al.*, 1970.)

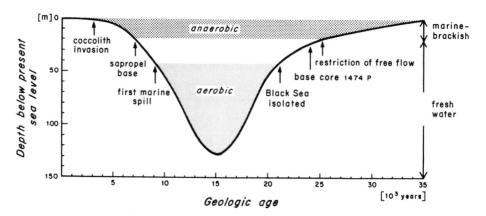

Fig. 12. Sea-level curve (after Milliman and Emery, 1968) showing major incidents in recent Black Sea history. (From Ross and Degens, 1974.)

TABLE III

Annual Supply of Sedimentary Material to Central Part of the Sea of Azov, and Ratio of Clay Minerals in Different Sources (adapted from Aleksina *et al.*, 1971)

Source	Sedimentary material (10^6 tons)	Percent less than 0.001 m	Amount less than 0.001 m (10^6 tons)	Percent clay minerals				
				Hydromica (illite)	Mixed layer	Montmorillonite	Kaolinite	Chlorite
River Don (suspension)	2.8	61.3	1.72	22	11	52	15	—
River Kuban (suspension)	4.9	25.0	1.22	48	6	22	12	12
Abrasion products from coastal loams	5.4	12.4	0.7	44	19	21	6	10
Abrasion products from bottom deposits	3.6	27.5	1.0	53	13	17	16	1
Biogenic carbonate	7.4	—	—	—	—	—	—	—
TOTAL	24.1		4.64	39	11	32	13	5

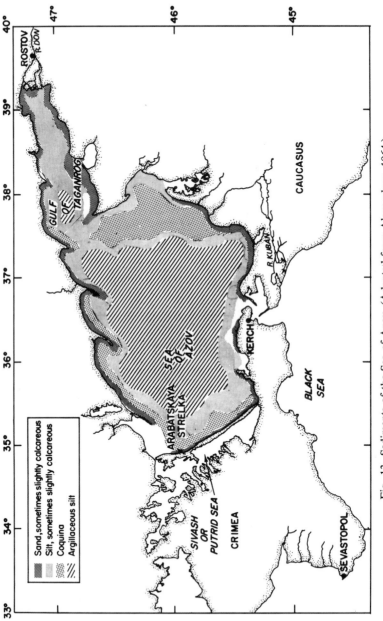

Fig. 13. Sediments of the Sea of Azov. (Adapted from Alexandrov, 1964.)

Yedigaryan *et al.*, 1970; Ratayev, 1964; Aleksina *et al.*, 1971). The Azov sedimentary basin is characterized by an extremely shallow depth (averages about 12 m), low salinity (about 11‰), and restricted water exchange with the Black Sea. Most of the sedimentary material entering the basin is the result of river inflow, coastal erosion, and biological activity (Table III). The main river is the Don, which enters the northern Taganrog Gulf.

Sands are the typical sediment along the coastal areas (Fig. 13); these grade seaward into a coquina type of deposit. The dominant sediment in the central portion of the sea is argillaceous silt, except in the Gulf of Taganrog where calcareous silt prevails. Montmorillonite and illite are the typical clay minerals (Ratayev, 1964). Estimates of sedimentation vary from 0.9 mm/yr (Fedosov, 1952) to 0.5 mm/year (Aksenov, 1956).

Little information was available to me concerning total sediment thickness and recent geological history of the Sea of Azov. However, it is clear that this sea was stranded above the Black Sea during Pleistocene low sea levels. At these times it was probably a fresh-water lake, and when the sea level was higher, exchange with the Black Sea produced a more saline environment.

V. STRUCTURE

A. Black Sea

The Black Sea has been the subject of numerous structural investigations, which have been summarized by Neprochnov *et al.* (1974). Deep seismic reflection data show that the crust beneath the central portion of the Black Sea is probably oceanic (Neprochnov *et al.*, 1974) and overlain by 18 to 24 km of sediments (Fig. 14). From seismic and magnetic data, Neprochnov *et al.*

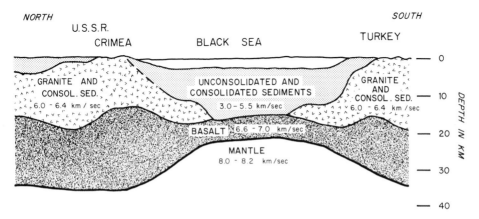

Fig. 14. Crustal north–south cross section across Black Sea. (Adapted from Subbotin *et al.*, 1968; Neprochnov, 1968; Neprochnov *et al.*, 1970.)

concluded that the principal anomaly-producing body in the central portion of the Black Sea is possibly a gabbroic layer having a velocity of 6.6 to 7.0 km/sec. They also suggest that the granitic layer with a velocity of 6.0 to 6.4 km/sec, encountered on the basin margins, is essentially nonmagnetic and may be composed of metamorphosed sedimentary units. This "granitic" layer reaches a thickness of up to 10 to 15 km (Neprochnov *et al.*, 1974) and may account, in part, for the tendency towards crustal thickening along the Black Sea margins (Neprochnov *et al.*, 1974). The same authors have shown that seismic velocities tend to change between the central Black Sea Basin and the marginal areas, and to change again toward the shallower shelf areas (Fig. 15).

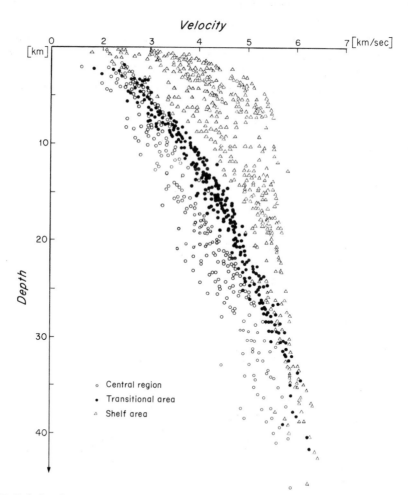

Fig. 15. Relation between average velocity and depth (depth calculated from sea-floor surface). Central region refers to "granite-free" area. Transitional area is the basin slope and basin apron area. (From Neprochnov *et al.*, 1974.)

Earthquake activity around the Black Sea is mainly shallow and restricted to the marginal seas. A distinct seismic zone runs parallel to the southern coast and to the northern part of the Anatolian Fault Zone. In this area focal depths are generally less than 30 km (Canitez and Toksöz, 1970). Another shallow earthquake belt extends southeastward from the eastern part of the Black Sea across the Caucasus Mountains and Caspian Sea into south-central Asia (Nowroozi, 1971; Tskhakaya, 1961).

A distinctive geophysical feature of the Black Sea is a large, positive gravity anomaly (Malovitskiy and Neprochnov, 1966; Subbotin *et al.*, 1968) that is especially common where the "granitic" material is absent or extremely thin (Mindeli *et al.*, 1965). Free-air anomalies tend to be negative, with occasional local highs in the eastern and western parts of the Black Sea (Ross *et al.*, 1974*a*). Recent syntheses of gravity observations (Ross *et al.*, 1974*a*; Bowin, C. O., 1974, personal communication) have shown that free-air anomalies can be divided into two main categories. The first occurs where the anomalies tend to follow variations in topography and sub-bottom structures as defined by continuous seismic profiles, and second, where they seem to bear no relationship to deeper structure or bathymetry (Fig. 16). The areas where gravity tends to correlate with structure are typical of the shelf and slope areas of the Black Sea, and variations in free-air anomalies tend to have short wave lengths (about 50 km). The second category of anomalies occurs in the deep-water portions of the Black Sea where the floor is relatively flat and free-air variations tend to have long wavelengths (several hundred kilometers). It is inferred that the source of the gravity anomalies lies at a shallow depth along the margins and at a greater depth in the central portion of the Black Sea. The rapid change from one pattern to another perhaps indicates that the basin of the Black Sea deepens rather abruptly near the basin slope. The residual magnetic field tends to parallel the surrounding Caucasus and Pontic Mountains, which suggests that some of these structures may extend into the Black Sea itself (Ross *et al.*, 1974*a*) (Fig. 17).

Heat-flow measurements in the Black Sea (Erickson and Simmons, 1974) have given relatively low values, averaging about $0.92 + 0.23$ μcal/cm^2 per second based on 16 observations. These authors conclude that a significant fraction of the geothermal flux is absorbed by the rapidly accumulating sediments and that their average value is only about 50% of the geophysically relevant heat flow. They likewise suggest that the lower thermal conductivity of the sediments has an essentially blanketing effect and may reduce flux by as much as 50%. They thus conclude that the average heat flow of the Black Sea, after correcting for these effects, is on the order of 2.2 μcal/cm^2 per second.

Numerous seismic profiles have been made recently in the Black Sea, both during the *Atlantis II* expedition and more recent Russian efforts (although

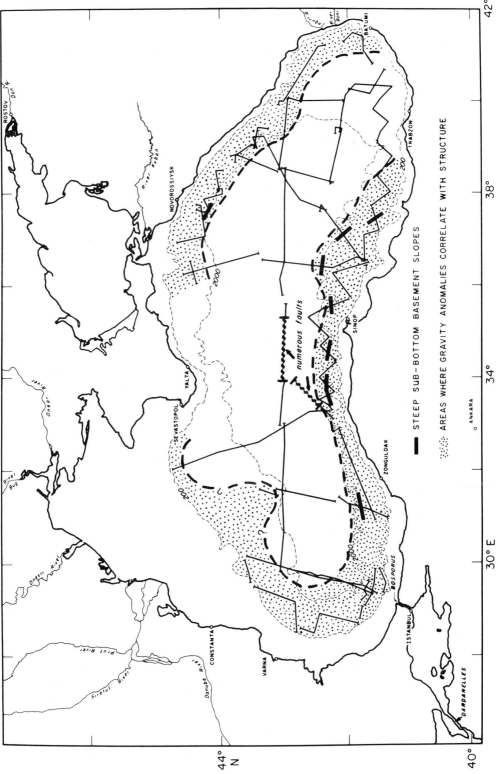

Fig. 16. Summary of gravity observations, location of steeply dipping basement, and location of small faults in deep basin. Based on results from *Atlantis II*

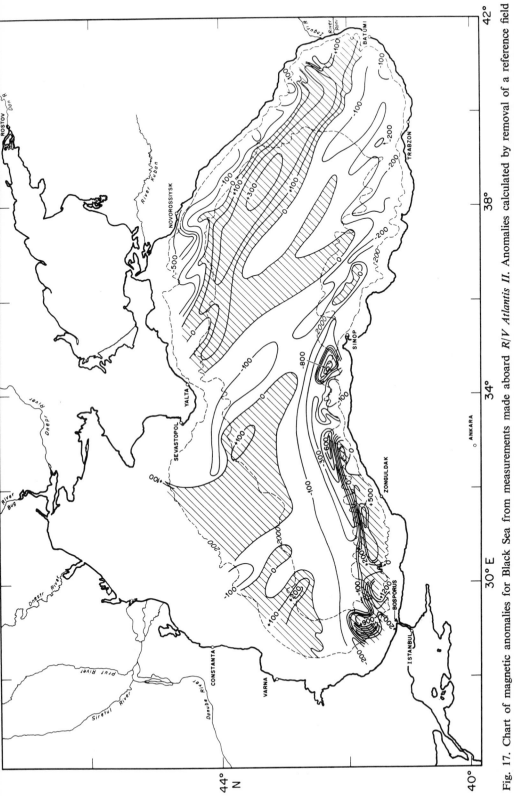

Fig. 17. Chart of magnetic anomalies for Black Sea from measurements made aboard *R/V Atlantis II*. Anomalies calculated by removal of a reference field (Cain *et al.*, 1968) from measurements of total intensity. Contour interval is 100 γ. Positive anomaly areas are shaded. (From Ross *et al.*, 1974*a*.)

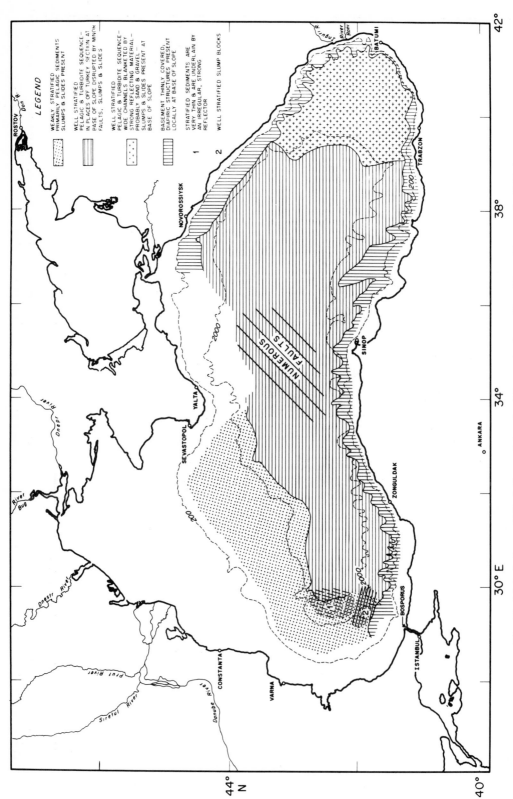

Fig. 18. Main features of Black Sea as determined from seismic profiles. (From Ross *et al.*, 1974*a*.)

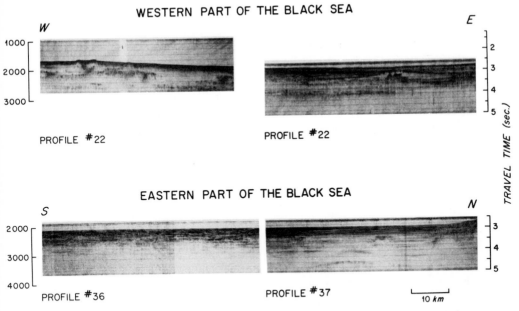

Fig. 19. Representative seismic profile taken from the western and eastern portions of the Black Sea. Note the transparent layers and irregularities in the profiles from the western portion of the Black Sea and the deep structure on Profile 37. For location of profiles, see Fig. 3.

these have not yet been published). The *Atlantis II* work shows (Ross *et al.*, 1974*a*) two distinct structural features in the Black Sea. The first is a small series of subsurface faults south of Yalta (Fig. 18). The other is a rather deeply buried anticline observed in the eastern Black Sea (Fig. 19, Profile 37). Seismic profiles off the Danube and western portions of the Black Sea tend to show many discontinuous layers, transparent zones, and indications of slumping, which are probably related to the depositional feature called the Danube Fan (Fig. 19). Seismic profiler reflectors from the eastern part of the Black Sea (Fig. 19) tend to have crisper reflectors, probably indicating a high incidence of turbidity currents in that area. On the lower part of Profile 37, you can see the deeply buried anticline mentioned above. A similar feature has been detected from some of the Russian seismic soundings, and it has been suggested that it is related to the Caucasus orogeny.

The age of the Black Sea is somewhat controversial with estimates ranging from the Precambrian (Milanovskiy, 1967) to the early Quaternary (Nalivkin, 1960). Brinkmann (1974) and others have argued more convincingly for a Middle or Upper Mesozoic age. A more complete elucidation of the recent history of the Black Sea area will be available in mid-1975 when the *Glomar Challenger* is scheduled to drill at least four holes in the area.

B. Sea of Azov

The Sea of Azov appears to have been developed in the Holocene. It is a platform basin that is tectonically heterogeneous (Panov, 1963). In its northern part, the coastal area is an eastern continuation of the Ukrainian Shield (Panov, 1965). There are occasional outcrops of crystalline basement, but generally it is covered by a thin (100–200 m) sedimentary mantle (Panov, 1965). The central portion is divided by fractures into a series of irregularly lowered blocks that are related to the Scythian Platform. Near the Kerch Peninsula the rocks are of Tertiary age, but elsewhere along the coast there are Quaternary marine deposits (Panov, 1965).

VI. ORIGIN OF THE BLACK SEA

The Black Sea is a geologically unusual feature, a marine basin situated between Alpine mountain systems. Its origin is still a matter of controversy. Although topographically the basin seems to be one continuous feature, the gravity data do show some differences. The center of the Black Sea is probably in isostatic equilibrium. The Bouguer anomaly highs in the western and eastern portions of the basin suggest that some of these areas are raised above an isostatic condition (Bowin, C. O., 1974, personal communication). If true, this suggests that additional deformation of the Black Sea is in progress.

The magnetic field of the Black Sea shows the extension of several northwest-trending patterns which parallel the adjacent mountain regions, suggesting that the structural patterns of the borderland probably continue offshore at least to the basin slope area and that this area has subsided to its present depth, as was suggested by Balavadze et al. (1966). More recent tectonic events can be noted on seismic profiles, which show numerous examples of slumping and sliding at portions of the basin slope and the upper basin apron, especially in the south and east.

A brief summary of ideas concerning the Black Sea was recently presented by Ross (1974). One of the important points to be considered is that the crustal structure is intermediate between that of continents and oceans. Although some evolutionary mechanisms have been suggested (for example, see Menard, 1967), the origin of the Black Sea and other marginal seas similar to it is not well understood. There is strong evidence suggesting that in the past the Black Sea was a topographic high from which Paleozoic sediments were derived (Brinkmann, 1974). The east–west alignment of the Caucasus and Pontic Mountains surrounding the Black Sea suggests that when these mountains were formed during the Mesozoic and Cenozoic, the Black Sea area was a zone of compression. Evidence has been presented concerning subsidence in the

Black Sea, and other studies along the Caucasus likewise indicate this (Bala-vadze *et al.*, 1966; Tsagareli, 1974). The occurrence of granitic-type material, at least as suggested by seismic velocity, may indicate prior erosion of a conti-nental crust followed by subsidence and sedimentation (Rezanov and Chamo, 1969). There are two possible explanations for the "granite-free" area in the central portion of the Black Sea. It is either a relict ocean crust, or it has been newly formed in place.

Any model for the Black Sea's origin must satisfy the presently observed structural features and be consistent with the origin of the surrounding moun-tain areas. Hypotheses that consider the changing of a continental crust into an intermediate or oceanic crust are of two basic types. These require either replacement of continental crust by upwelling mantle material (along cracks, faults, or extensions), or *in situ* conversion of low-density continental crust into a denser, thinner, and more oceanic crust. Neither of these possibilities is definitively favored by the available evidence. For example, the similarity of features around the Black Sea and their extension into the Black Sea as indi-cated by magnetic features would tend to argue against extension. Oceanization also presents several difficult mechanical and chemical problems, as does understanding the landward thickening sequence of 6.0–6.4 km/sec "granitic" or metamorphosed sedimentary material.

More recent hypotheses include that of Erickson and Simmons (1974) who

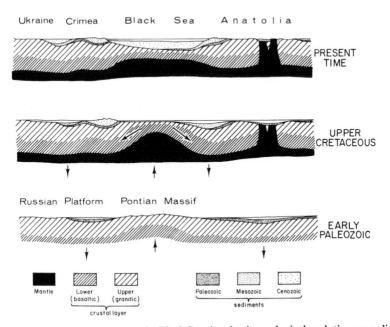

Fig. 20. Schematic sections across the Black Sea showing its geological evolution according to mechanisms proposed by Brinkmann (1974).

consider the Black Sea to be related to the initiation of convergence of the Eurasian and African plates. They suggest that subaerial erosion might have removed the upper part of the crust and that once subduction slowed or ceased, temperature decreases would cause the lithosphere to sink and eventually be covered by sediments. Brinkmann (1974) suggests a long period of uplift and erosion followed by crustal tension. He further speculates that increased tension would result in a suboceanic crust thickness, sinking, and formation of a "soft" mantle. Decreasing temperatures of the mantle and the crust would result in increased sinking of the basin accompanied by infilling (Fig. 20). At the present time, the available information does not permit a discrimination between the various hypotheses.

ACKNOWLEDGMENTS

The basic support for my studies in the Black Sea has come from the National Science Foundation (Grant GA-252-34), which sponsored the 1969 *Atlantis II* expedition to the Black Sea. Support for this paper comes from the Office of Naval Research (Grant NOOO14-74-CO262 NR 083-004). I would also like to acknowledge the help of the officers and crew of *R/V Atlantis II* and the numerous scientists and technicians who participated in this cruise and whose data I have used in this paper. E. Uchupi and Colin Summerhayes helped review this paper, and Miss Francy Forrestel typed the manuscript and helped with some of the research.

REFERENCES

Aksenov, A. A., 1956, On the drift of approach currents to the canal in the Sea of Azov, *Tr. Gos. Okeanogr. Inst.*, v. 31 (48), p. 58–71.

Aleksina, I. A., and Yedigaryan, Z. P., 1971, Analysis of the absolute mass of Holocene deposits in the Sea of Azov, *Dokl. Akad. Nauk SSSR*, v. 198 (3).

Aleksina, I. A., Korolev, Y. M., and Yedigaryan, Z. P., 1971, Clay minerals in Late Quaternary deposits of the Sea of Azov.

Alexandrov, A. M., 1964, Sediments of the Sea of Azov, *Okeanol. Akad. Nauk SSSR*, v. 4 (5), p. 856–865.

Andrusov, N. I., 1890, Predvaritel'nyi otchet ob uchastii v Chernomorskoi glubokomernoi ekspeditsii 1890 (Preliminary account on participation in the Black Sea deepwater expedition of 1890), *Izv. Russk. Geogr. Obva.*, v. 26, p. 380–409.

Andrusov, N. I., 1893, Einige Resultate der Tiefseeuntersuchungen im Schwarzen Meere, *Mitt. Geogr. Ges. Wien.*

Arkhangel'skiy, A. D., 1927, Ob osadkakh Chernogo morya i ikh znachenii v poznanii osadchnykh gornykh porod (On sediments of Black Sea and their significance in sedimentology), *Byull. Mosk. Obva. Ispyt. Prir. Otd. Geol.*, v. 5 (3–4), p. 199–289.

Arkhangel'skiy, A. D., 1928a, O novykh issledovaniyakh dna Chernogo morya (On new investigations in the Black Sea), *Geol. Vestn.*, no. 1–3.

Arkhangel'skiy, A. D., 1928*b*, Karta i razrezy osadkov dna Chernogo morya (Map and cross sections of Black Sea sediments, *Byull. Mosk. Obva. Ispyt. Prir. Otd. Geol.*, v. 6 (1), p. 77–108.

Arkhangel'skiy, A. D., 1930, Opolzanie osadkov na dne Chernogo morya i geologicheskoe z actremie etogo yavleniya (Slumping sediments on bottom of Black Sea and geological significance of this phenomenon), *Byull. Mosk. Obva. Ispyt. Prir. Otd. Geol.*, v. 8 (1–2), p. 32–79.

Arkhangel'skiy, A. D., and Strakhov, N. M., 1932, Geologicheskoye istoriya Chernogo morya (Geological history of the Black Sea), *Byull. Mosk. Obva. Ispyt. Prir. Otd. Geol.*, v. 10 (1), p. 3–104.

Arkhangel'skiy, A. D., and Strakhov, N. M., 1938, Geologicheskoye stroyeniye i istoriya razvitiya Chernogo morya (Geological structure and history of the evolution of the Black Sea), *Izv. Akad. Nauk SSSR*, Moscow–Leningrad.

Bacesco, M., and Dumitresco, H., 1958, Les Lagunes en formation aux embouchures du Danube et leur importance pour les poissons migrateurs, *Verh. Int. Ver. Limnol.*, v. 13, p. 699–709.

Balavadze, B. K., Tvaltradze, Y. K., Shengelaya, Y. S., Sikharulidze, D. I., and Kartrelishvili, K. M., 1966, Geophysical investigations of the earth's crust and upper mantle of the Caucasus, *Geotektonika*, v. 3, p. 30–40.

Barkovskaya, M. G., 1961*a*, Zakonomernosti raspredeleniya donnykh osadkov na shel'fe sovetskikh beregov Chernogo morya (Regularities in distribution of bottom sediments on shelf of Soviet shores of the Black Sea), *Tr. Inst. Okeanol. Akad. Nauk SSSR*, v. 53, p. 123–148.

Barkovskaya, M. G., 1961*b*, Zakonomernosti raspredeleniya terrigenogo materiala v priure-zovoy polose sovetskogo poberzh'ya Chernogo morya (Regularities in distribution of terrigenous material in littoral zone of Soviet shore of Black Sea), *Tr. Inst. Okeanol. Akad. Nauk SSSR*, v. 53, p. 64–94.

Berner, R. A., 1974, Iron sulfides in Pleistocene deep Black Sea sediments and their paleo-oceanographic significance, in: *The Black Sea—Geology, Chemistry, and Biology*, Degens, E. T., and Ross, D. A., eds., *Mem. Amer. Assoc. Petrol. Geol.*, no. 20, p. 524–531.

Brewer, P., 1971, Hydrographic and chemical data from the Black Sea, Woods Hole Ocean-ogr. Inst., Ref. no. 71–65.

Brewer, P. G., and Spencer, D. W., 1974, Distribution of some trace elements in Black Sea and their flux between dissolved and particulate phases, in: *The Black Sea—Geology, Chemistry, and Biology*, Degens, E. T., and Ross, D. A., eds., *Mem. Amer. Assoc. Petrol. Geol.*, no. 20, p. 137–143.

Brinkmann, R., 1974, Geologic relations between Black Sea and Anatolia, in: *The Black Sea—Geology, Chemistry, and Biology*, Degens, E. T., and Ross, D. A., eds., *Mem. Amer. Assoc. Petrol. Geol.*, no. 20, p. 63–76.

Buachidze, I. M., 1974, Black Sea shelf and littoral zone, in: *The Black Sea—Geology, Chemistry, and Biology*, Degens, E. T., and Ross, D. A., eds., *Mem. Amer. Assoc. Petrol. Geol.*, no. 20, p. 308–316.

Bukry, D., 1974, Coccoliths as paleosalinity indicators—evidence from Black Sea, in: *The Black Sea—Geology, Chemistry, and Biology*, Degens, E. T., and Ross, D. A., eds., *Mem. Amer. Assoc. Petrol. Geol.*, no. 20, p. 353–363.

Bukry, D., Kling, S. A., Horn, M. K., and Manheim, F. T., 1970, Geological significance of coccoliths in fine-grained postglacial carbonate bands of Black Sea sediments, *Nature*, v. 226, p. 156–158.

Cain, J. C., Hendricks, S., Daniels, W. E., and Jensen, D. C., 1968, Computation of the main geomagnetic field from spherical harmonic expressions, NASA Data Center, Data Users Note NSSDC 68–11, 46 p.

Canitez, N., and Toksöz, M. N., 1970, Source parameters of earthquakes and regional tectonics of the eastern Mediterranean (Abstract), *Trans. Amer. Geophys. Union*, v. 51, p. 420.

Caspers, H., 1957, Black Sea and Sea of Azov, *Mem. Geol. Soc. Amer.* no. 67, p. 803–890.

Degens, E. T., 1971, Sedimentological history of the Black Sea over the last 25,000 years, in: *Geology and History of Turkey*, Campbell, A. S., ed., Tripoli: Petrol. Expl. Soc. Libya, p. 407–429.

Degens, E. T., 1974, Cellular processes in Black Sea sediments, in: *Black Sea—Geology, Chemistry, and Biology*, Degens E. T., and Ross, D. A., eds., *Mem. Amer. Assoc. Petrol. Geol.*, no. 20, p. 296–307.

Degens, E. T., and Ross, D. A., eds., 1974, *The Black Sea—Geology, Chemistry, and Biology*, *Mem. Amer. Assoc. Petrol. Geol.*, no. 20, 600 p.

Degens, E. T., Watson, S. W., and Remsen, C. C., 1970, Fossil membranes and cell wall fragments from a 7000-year-old Black Sea sediment, *Science*, v. 168, p. 1207–1208.

Deuser, W. G., 1971, Organic-carbon budget of the Black Sea, *Deep-Sea Res.*, v. 18 (10), p. 995–1004.

Deuser, W. G., 1972, Late-Pleistocene and Holocene history of the Black Sea as indicated by stable-isotope studies, *J. Geophys. Res.*, v. 77 (6), p. 1071–1077.

Deuser, W. G., 1974, Evolution of anoxic conditions in Black Sea during the Holocene, in: *The Black Sea—Geology, Chemistry, and Biology*, Degens, E. T., and Ross, D. A., eds., *Mem. Amer. Assoc. Petrol. Geol.*, no. 20, p. 133–136.

Dobrzhanskaya, M. A., 1967, Vliyaniye dinamiki vodnykh mass na raspredeleniye gidrokhimicheskikh pokazatelei—na primere Chernogo morya (Influence of dynamics of water masses on distribution of hydrochemical indicators—in example of Black Sea), in: *Voprosy biookeanografii (Problems in Biooceanography)*, Vodyanitskiy, V. A., ed., Kiev: Izd. "Naukova Dumka," p. 31–36.

Erickson, A., and Simmons, G., 1974, Environmental and geophysical interpretation of heat-flow measurements in Black Sea, in: *The Black Sea—Geology, Chemistry, and Biology*, Degens, E. T., and Ross, D. A., eds., *Mem. Amer. Assoc. Petrol. Geol.*, no. 20, p. 50–62.

Erünal-Erentöz, L., 1956, La sédimentation actuelle dans la Mer Noire, *Inst. Etud. Rech. Min. Turquie (Ankara) Sér. B.*, v. 19.

Fedosov, M. V., 1952, The intensity of deposition in the Azov Sea, *Dokl. Akad. Nauk SSSR*, v. 84, p. 551–553.

Goncharov, V. P., 1958, Novyye dannyye o rel'yefe dna Chernogo morya (New data on topography of bottom of Black Sea), *Dokl. Akad. Nauk SSSR*, v. 121 (5), p. 830–833.

Goncharov, V. P., and Neprochnov, Y. P., 1967, Geomorphology of the bottom and tectonic problems in the Black Sea, in: *International Dictionary of Geophysics*, Runcorn, S. K., ed., Elmsford, New York: Pergamon Press, 2 v., 1728 p.

Goncharov, V. P., Neprochnova, A. F., and Neprochnov, Y. P., 1966, Geomorfologiya dna i glubinnoye stroyeniye Chernogo morya vpadiny (Geomorphology of the floor and deep structure of the Black Sea basin), in: *Glubinnoye stroyeniye Kavkaza (Deep Structure of the Caucasus)*, Moscow: Izd. Nauka.

Jipa, D. C., 1974, Graded bedding in recent Black Sea turbidites: a textural approach, in: *The Black Sea—Geology, Chemistry, and Biology*, Degens, E. T., and Ross, D. A., eds., *Mem. Amer. Assoc. Petrol. Geol.*, no. 20, p. 317–331.

Keller, G. H., 1974, Mass physical properties of some western Black Sea sediments, in: *The Black Sea—Geology, Chemistry, and Biology*, Degens, E. T., and Ross, D. A., eds., *Mem. Amer. Assoc. Petrol. Geol.*, no. 20, p. 332–337.

Kremling, K., 1974, Relation between chlorinity and conductometric salinity in Black Sea water, in: *The Black Sea—Geology, Chemistry, and Biology*, Degens, E. T., and Ross, D. A., eds., *Mem. Amer. Assoc. Petrol. Geol.*, no. 20, p. 151–154.

Krustalev, Y. O., and Scherbakov, P. A., 1968, On the balance of sedimentary material in the Sea of Azov, *Okeanol. Akad. Nauk SSSR*, v. 8 (3), p. 452–460 (In Russian; English abstract).

Laking, P. N., 1974, *The Black Sea—Its Geology, Chemistry, and Biology: A Bibliography*, Woods Hole, Massachusetts: Woods Hole Oceanogr. Inst., 368 p.

Malovitskiy, Y. P., and Neprochnov, Y. P., 1966, Sopostavleniye seysmicheskikh i gravimetricheskikh dannykh o stroyenii zemnoy kory Chernomorskoy vpadiny (Comparison of seismic and gravimetric data on crustal structure of Black Sea basin), in: *Stroyeniye Chernomorskoy vpadiny (Structure of the Black Sea Basin)*, Magnitskiy, V. A., *et al.*, eds., Moscow: Izk. "Nauka," p. 5–16.

Malovitskiy, Y. P., Uglov, B. D., and Osipov, G. V., 1969a, Geomagnitnoye pole Chernomorskoy vpadiny (Geomagnetic field of the Black Sea depression), *Akad. Nauk Ukr. SSR Geofiz. Geofiz. Sb.*, v. 32, p. 28–38.

Malovitskiy, Y. P., Neprochnov, Y. P., Garkelenko, I. A., Starskinov, E. A., Milashina, K. G., Komornaiam, Y., Ryunov, L. N., Klolopov, B. V., and Sedov, V. V., 1969b, Stroenie zemnoi kory v zapadnoi chasti Chernogo morya (Structure of the earth's crust in the western part of the Black Sea), *Dokl. Akad. Nauk SSSR*, v. 186 (4), p. 905–907.

McCoy, F. W., Jr., 1974, *Late Quaternary Sedimentation in the Eastern Mediterranean Sea*, Unpublished Ph. D. thesis, Harvard University, 132 p.

Manheim, F. T., and Chan, K. M., 1974, Interstitial waters of Black Sea sediments, in: *The Black Sea—Geology, Chemistry, and Biology*, Degens, E. T., and Ross, D. A., eds., *Mem. Amer. Assoc. Petrol. Geol.*, no. 20, p. 155–182.

Melikhov, V. P., Mirlin, E. G., Uglov, B. D., and Shreider, A. A., 1969, Otsenka raspredeleniya magnitovozumushchayuikh gel v kore globokovodnoi kotloviny Chernogo morya s pomoshch transformatsii v nizhnea poluprostranstvo (Estimate of the distribution of magnetic perturbing bodies in the crust of the deepwater basin of the Black Sea with the aid of transformations in the lower half-space), in: *Morskaya Geol. Geofiz.*, v. 2.

Menard, H. W., 1967, Transitional types of crust under small ocean basins, *J. Geophys. Res.*, v. 72 (12), p. 3061–3073.

Milanovskiy, Y. Y., 1967, Problema prioskhozhdeniya Chernomorskoy vpakiny i yeye mesto v strukture al'piyskogo poyasa, *Vestn. Mosk. Univ. Ser. Geol. 4*, v. 22 (1), p. 27–43; English translation (1967), Problem of origin of Black Sea Depression and its position in structure of the Alpine belt, *Int. Geol. Rev.*, v. 8 (1), p. 36–43.

Milliman, J. D., and Emery, K. O., 1968, Sea levels during the past 35,000 years, *Science*, v. 162 (3858), p. 1121.

Mindeli, P. S., Neprochnov, Y. P., and Pataraya, Y. I., 1965, Opredeleniye oblasti otsutstviya granitnogo sloya v Chernomorskoy vpadine po dannym GSZ i seysmologii, *Izv. Akad. Nauk SSSR Ser. Geol.*, no. 2, p. 7–15; English translation (1966), Granite-free area in Black Sea trough, from seismic data, *Int. Geol. Rev.*, v. 8 (1), p. 36–43.

Müller, G., and Blaschke, R., 1969, Zur Entstehung des Tiefsee-Kalkschlammes im Schwarzen Meer, *Naturwissenschaften*, v. 56, p. 561–562.

Müller, G., and Stoffers, P., 1974, Mineralogy and petrology of Black Sea basin sediments, in: *The Black Sea—Geology, Chemistry, and Biology*, Degens, E. T., and Ross, D. A., eds., *Mem. Amer. Assoc. Petrol. Geol.*, no. 20, p. 200–248.

Muratov, M. V., and Neprochnov, Y. P., 1967, Stroyeniye dna Chernomorskoy kotloviny i yeye proiskhozhdeniye (Structure of Black Sea basin and its origin), *Byull. Mosk. Obva. Ispyt. Prir. Otd. Geol.*, v. 42 (5), p. 40–59.

Nalivkin, D. V., 1960, *The Geology of the USSR* (English translation), Elmsford, New York: Pergamon Press, 170 p.

Neprochnov, Y. P., 1962, Resul'taty glubinnogo seysmicheskogo zondirovaniya na Cheromn more (On results of deep seismic sounding in the Black Sea), in: *Glubinnoye seysmicheskoye zondirovaniye zemnoy kory v SSSR* (*Deep Seismic Sounding of the Earth's Crust in the USSR*), Leningrad: Gostekhizdat, 271 p.

Neprochnov, Y. P., 1968, Structure of the earth's crust of epi-continental seas; Caspian, Black, and Mediterranean, in: *3rd Symposium on Continental Margins and Island Arcs, Zurich, 1967, Can. J. Earth Sci.*, v. 5 (4), pt. 2, p. 1037–1043.

Neprochnov, Y. P., Neprochnova, A. F., Zverev, S. M., Mironova, V. I., Bokuni, R. A., and Chekunov, A. V., 1964, Novye dannye o stroenii kory Chernomorskoi vpadiny k yugu ot Kryma (New data on the structure of the Black Sea crust south of the Crimea), *Dokl. Akad. Nauk SSSR*, v. 156 (3), p. 561–564.

Neprochnov, Y. P., Kosminskaya, I. P., and Malovitskiy, Y. P., 1970, Structure of the crust and upper mantle of the Black and Caspian Seas, *Tectonophysics*, v. 10, p. 517–538.

Neprochnov, Y. P., Neprochnova, A. F., and Mirlin, Y. G., 1974, Deep structure of Black Sea basin, in: *The Black Sea—Geology, Chemistry, and Biology*, Degens, E. T., and Ross, D. A., eds., *Mem. Amer. Assoc. Petrol. Geol.*, no. 20, p. 35–49.

Nevesskiy, E. N., 1967, *Protsessy osadkoobrazovaniya v pribrezhnoi zone morya* (*Processes of Sediment Formation in the Near-Shore Zone of the Sea*), Moscow: Izd. Nauka, 255 p.

Nowroozi, A. A., 1971, Seismo-tectonics of the Persian Plateau, eastern Turkey, Caucasus, and Hindu-Kush regions, *Bull. Seismol. Soc. Amer.*, v. 61, p. 317–341.

Ostlünd, G., 1974, Expedition "Odysseus 65": radiocarbon age of Black Sea deep water, in: *The Black Sea—Geology, Chemistry, and Biology*, Degens, E. T., and Ross, D. A., eds., *Mem. Amer. Assoc. Petrol. Geol.*, no. 20, p. 127–132.

Panin, N., 1974, Evolutia deltei dunarii in timpul Holocenului, *Stud. Geol. Cuatern. Ser. H.*, no. 5, p. 107–121.

Panin, N., and Panin, S., 1969, Sur la genèse des accumulations des minéraux lourds dans le delta du Danube, *Rev. Géogr. Phys. Géol. Dyn.* (2), v. 11, p. 511–522.

Panov, D. G., 1963, *Morfologiya dna Mirovogo okeana* (*Morphology of the Floor of the Seas and Oceans*), USSR Acad. Sci. Press.

Panov, D. G., 1965, The history of the Azov Sea development during the Holocene period *Okeanol. Akad. Nauk SSSR*, v. 5 (4), p. 673–683 (In Russian).

Panov, D. G., and Spichak, M. K., 1961, The sedimentation rate in the Sea of Azov, *Dokl. Akad. Nauk SSSR*, v. 137 (5), p. 1212–1213 (In Russian). Translation: *Consultants Bureau for Amer. Geol. Inst.*, v. 137 (1–7), p. 430–431.

Ratayev, M. A., 1964, Zakonomernosti i genezis glinistykh mineralov v sovremennykh i drevnikh morskikh basseinakh (*Distribution and Genesis of Clay Minerals in Modern and Ancient Marine Basins*), Moscow: Izv. Nauka, p. 67–73.

Raupach, F. von, 1952, Die rezente Sedimentation in Schwarzen Meer, im Kaspi und im Aral und ihre Gesetzmässigkeiten, *Geologie*, v. 1 (1–2), p. 78–132.

Rezanov, I. A., and Chamo, S. S., 1969, O prichinakh otsutstviya "granitnogo" sloya vo vpadinakh tipa Yuzhno-Kaspiyskoy i Chernomorskoy: *Izv. Akad. Nauk SSSR Ser. Geol.*, no. 3, p. 3–11; English translation (1969), Reasons for absence of a "granite" layer in basins of the south Caspian and Black Sea type, *Can. J. Earth Sci.*, v. 6 (4), pt. 1, p. 671–678.

Ross, D. A., 1974, The Black Sea, in: *The Geology of Continental Margins*, Burk, C. A., and Drake, C. L., eds., New York: Springer-Verlag, p. 669–682.

Ross, D. A., and Degens, E. T., 1974, Recent sediments of the Black Sea, in: *The Black Sea—*

Geology, Chemistry, and Biology, Degens, E. T., and Ross, D. A., eds., *Mem. Amer. Assoc. Petrol. Geol.*, no. 20, p. 183–199.

Ross, D. A., Degens, E. T., and MacIlvaine, J. C., 1970, Black Sea: recent sedimentary history, *Science*, v. 170 (3954), p. 163–165.

Ross, D. A., Uchupi, E., and Bowin, C. O., 1974*a*, Shallow structure of the Black Sea, in: *The Black Sea—Geology, Chemistry, and Biology*, Degens, E. T., and Ross, D. A., eds., *Mem. Amer. Assoc. Petrol. Geol.*, no. 20, p. 11–34.

Ross, D. A., Uchupi, E., Prada, K. E., and MacIlvaine, J. C., 1974*b*, Bathymetry and micro-topography of Black Sea, in: *The Black Sea—Geology, Chemistry, and Biology*, Degens, E. T., and Ross, D. A., eds., *Mem. Amer. Assoc. Petrol. Geol.*, no. 20, p. 1–10.

Shamov, G. I., 1951, Granulometricheskii sostav nanosov rek SSSR (Granulometric composition of alluvium of rivers of USSR), *Tr. Leningr. Gos. Gidrol. Inst.*, v. 18 (72).

Shimkus, K. M., and Trimonis, E. S., 1974, Modern sedimentation in the Black Sea, in: *The Black Sea—Geology, Chemistry, and Biology*, Degens, E. T., and Ross, D. A., eds., *Mem. Amer. Assoc. Petrol. Geol.*, no. 20, p. 249–278.

Strakhov, N. M., 1947, K poznaniyu zakonomernostei i mekhanizma morskoy sedimentatsii. 1. Chernom more (Toward understanding the mechanisms of marine sedimentation. 1. Black Sea), *Izv. Akad. Nauk SSSR Ser. Geol.*, v. 2, p. 49–90.

Strakhov, N. M., 1954, Osadkoobrazovaniye v Chernom more (Sediment formation in the Black Sea), in: *Obrazovaniye osadkov v sovremennykh vodoemakh* (*Formation of Sediments in Contemporary Basins*), Moscow: Akad. Nauk SSSR Izd., p. 81–136.

Strakhov, N. M., 1961, O znachenii serovodovodnogo zarazheniya naddonnoy vodoy basseyna dlya autigennogo mineraloobrazovaniya v ego osadkakh-na primere Chernogo morya (Significance of hydrogen sulfide contamination of water overlying the bottom in formation of authigenic materials in sediments in the Black Sea), in: *Sovremennye osadki morey i okeanov* (*Recent Sediments in Seas and Oceans*), Strakhov, N. M., ed., Moscow: Akad. Nauk SSSR Izd., p. 521–548.

Strakhov, N. M., 1967, *Principles of Lithogenesis*, 1, Edinburgh and London: Oliver and Boyd, 245 p.

Subbotin, S. I., Sollogub, V. B., Prosen, D., Dragasevic, T., Mituch, E., and Posgay, K., 1968, Junction of deep structures of the Carpatho-Balkan region with those of the Black and Adriatic seas, *Can. J. Earth Sci.*, v. 5, p. 1027–1035.

Trimonis, E. S., 1974, Some characteristics of carbonate sedimentation in Black Sea, in: *The Black Sea—Geology, Chemistry, and Biology*, Degens, E. T., and Ross, D. A., eds., *Mem. Amer. Assoc. Petrol. Geol.*, no. 20, p. 279–295.

Tsagareli, A. L., 1974, Geology of western Caucasus, in: *The Black Sea—Geology, Chemistry, and Biology*, Degens, E. T., and Ross, D. A., eds., *Mem. Amer. Assoc. Petrol. Geol.*, no. 20, p. 77–89.

Tshakaya, A. D., 1961, Seismicity of the Dzhavakhet (Alkhalkalaki) Highlands, in: *Earthquakes in the USSR*, Washington, D. C.: Department of Commerce Office of Technical Services, p. 200–313.

Yedigaryan, Z. P., Aleksina, I. A., and Glazunova, K. P., 1970, Stratigraphy of upper Quaternary bottom deposits of the Sea of Azov, *Byull. Kom. Izuch. Chetvertich. Perioda*, no. 37.

Zenkevich, L. A., 1963, *Biologiya morei SSSR*, Moscow: Akad. Nauk SSSR Izd., 793 p.; English translation (Zenkevitch), 1963, *Biology of the Seas of the USSR*, New York: Interscience Publishers, 955 p.

Zenkovich, V. P., 1958, *Berega Chernogo i Azovskogo morey* (*Shores of Black and Azov Seas*), Moscow: Geografgiz, 374 p.

INDEX